Local and Global Methods in Algebraic Geometry

CONTEMPORARY MATHEMATICS

712

Local and Global Methods in Algebraic Geometry

Conference in Honor of Lawrence Ein's 60th Birthday
Local and Global Methods in Algebraic Geometry
May 12–15, 2016
University of Illinois at Chicago, Chicago, IL

Nero Budur
Tommaso de Fernex
Roi Docampo
Kevin Tucker
Editors

AMS AMERICAN MATHEMATICAL SOCIETY

2010 *Mathematics Subject Classification.* Primary 13A35, 13C40, 14B05, 14D06, 14G05, 14J25, 14J60, 32S55, 53B21.

Library of Congress Cataloging-in-Publication Data

Names: Budur, Nero, 1977- editor. | De Fernex, Tommaso, 1970- editor. | Docampo, Roi, 1978- editor. | Tucker, Kevin (Mathematician), editor.

Title: Local and global methods in algebraic geometry : conference in honor of Lawrence Ein's 60th birthday, May 12-15, 2016, University of Illinois at Chicago, Chicago, Illinois / Nero Budur, Tommaso de Fernex, Roi Docampo, Kevin Tucker, editors.

Description: Providence, Rhode Island : American Mathematical Society, [2018] | Series: Contemporary mathematics ; volume 712 | Includes bibliographical references.

Identifiers: LCCN 2018003650 | ISBN 9781470434885 (alk. paper)

Subjects: LCSH: Ein, Lawrence. | Geometry, Algebraic–Congresses. | Festschriften. | AMS: Commutative algebra – General commutative ring theory – Characteristic p methods (Frobenius endomorphism) and reduction to characteristic p; tight closure. msc | Commutative algebra – Theory of modules and ideals – Linkage, complete intersections and determinantal ideals. msc | Algebraic geometry – Local theory – Singularities. msc | Algebraic geometry – Families, fibrations – Fibrations, degenerations. msc | Algebraic geometry – Arithmetic problems. Diophantine geometry – Rational points. msc | Algebraic geometry – Surfaces and higher-dimensional varieties – Special surfaces. msc | Algebraic geometry – Surfaces and higher-dimensional varieties – Vector bundles on surfaces and higher-dimensional varieties, and their moduli. msc | Several complex variables and analytic spaces – Singularities – Milnor fibration; relations with knot theory. msc | Differential geometry – Local differential geometry – Methods of Riemannian geometry. msc

Classification: LCC QA564 .L5985 2018 | DDC 516.3/5–dc23 LC record available at https://lccn.loc.gov/2018003650

Contemporary Mathematics ISSN: 0271-4132 (print); ISSN: 1098-3627 (online)

DOI: `http://dx.doi.org/10.1090/conm/712`

∞ The paper used in this book is acid-free and falls within the guidelines
established to ensure permanence and durability.
Visit the AMS home page at `http://www.ams.org/`

10 9 8 7 6 5 4 3 2 1 23 22 21 20 19 18

Dedicated to Lawrence Ein on the occasion of his 60th birthday

Contents

Preface

This volume results from the conference "Local and Global Methods in Algebraic Geometry" which was held at the University of Illinois at Chicago from May 12–15, 2016. The conference celebrated Lawrence Ein's 60th birthday, honoring his longstanding contribution to the field by featuring talks on a broad array of topics in algebraic geometry and nearby fields. There were a total of twelve talks by close collaborators and students of Lawrence Ein and other experts in the field. The conference attracted about 150 participants, both junior and senior mathematicians, from around the country and abroad, reflecting the wide impact Lawrence Ein's work has had in the community. The conference was made possible thanks to the support of the National Science Foundation via a Research and Training Grant and the assistance of the Department of Mathematics at the University of Illinois at Chicago.

This volume collects papers from speakers at the conference and other prominent members of the algebraic geometry community. The first article, written by a longtime collaborator of Lawrence Ein, gives an overview of his main contributions to the field. The other papers contribute to or report on recent progress on many topics, including birational geometry and moduli theory, geometry of surfaces, analytic and positive characteristic methods, singularity theory, hyper-Kähler geometry, rational points, and rational curves.

Nero Budur
Tommaso de Fernex
Roi Docampo
Kevin Tucker

Scientific program

Full details of the program are available at the conference homepage
`http://homepages.math.uic.edu/~ein60/`

Invited lectures

Nero Budur (KU Leuven): Cohomology jump loci

Tommaso de Fernex (University of Utah): Towards a link theoretic characterization of smoothness

Robin Hartshorne (University of California at Berkeley): Simple D-module components of local cohomology modules

Shihoko Ishii (University of Tokyo and Tokyo Woman's Christian University): Generic residual intersections and its applications

Ngaiming Mok (University of Hong Kong): Geometric substructures on uniruled projective manifolds

János Kollár (Princeton University): Celestial surfaces

Robert Lazarsfeld (Stony Brook University): Measures of irrationality for hypersurfaces of large degree

Mircea Mustaţă (University of Michigan): Hodge ideals and applications

Mihai Păun (University of Illinois at Chicago): Singular Hermitian metrics on vector bundles and extension of pluricanonical forms

Yum-Tong Siu (Harvard University): Revisiting analytic methods for algebraic geometric problems

Karen Smith (University of Michigan): Abhyankar valuations in prime characteristic

Claire Voisin (Pierre and Marie Curie University): Cubic fourfolds and O'Grady 10-dimensional examples

Contemporary Mathematics
Volume **712**, 2018
http://dx.doi.org/10.1090/conm/712/14339

Some remarks on the work of Lawrence Ein

Robert Lazarsfeld

Introduction. To set the stage for the conference, I gave an informal survey – seasoned with stories and reminiscenses – of Ein's mathematical work. The present pages constitute a write-up of these remarks. I'm not sure how well such a talk will translate into print, but I appreciate this opportunity to express my esteem for Lawrence's mathematics as well as my gratitude to him for our long collaboration.

Vector bundles. Lawrence arrived in Berkeley in 1977, where he joined an active group of students around Hartshorne and Ogus that included Ziv Ran, Mei-Chu Chang and Tim Sauer. At the time Hartshorne in particular was very interested in the geometry of vector bundles on projective space, and Lawrence's first papers were devoted to questions in this direction. In [**10**] – which became his thesis – Ein extended to characteristic $p > 0$ some of the classical results of Van de Ven and others about bundles on $\mathbf{P}^n(\mathbf{C})$. For those of us who think of Lawrence as mainly traveling in characteristic zero, it is interesting that this first paper involves some serious positive characteristic geometry. For example, Ein classifies uniform n-bundles on $\mathbf{P}^n(k)$, and finds that when char $k > 0$ some Frobenius pullbacks show up. In another paper [**11**], Ein, Hartshorne and Vogelaar prove analogues for rank three bundles of classical results of Barth on restruictions of rank 2 steble bundles to hyperplanes.

Starting in the mid 1980's, Lawrence shifted towards *applying* ideas involving vector bundles to study concrete geometric problems. This would become characteristic of some of his most interesting work during the 1980s. For example, in [**12**], Ein established a Noether-type theorem for the Picard groups of surfaces arising as the degeneracy locus associated to general sections of a suitably positive vector bundle. In the same paper he showed that the Le Potier vanishing theorem leads to a very quick proof of a result of Evans and Griffith [**28**] that if E is a vector bundle of rank e on a complex projective space $\mathbf{P}$ that satisfies the vanishing

$$H^i\big(\mathbf{P}, E(k)\big) \ = \ 0 \quad \text{for all } \ 0 < i < e \ , \ k \in \mathbf{Z},$$

then E is a direct sum of line bundles. Evans and Griffith had deduced this as a consequence of their deep algebraic results on syzygy modules, and it was very nice to have a quick geometric proof of the statement.

Research of the author partially supported by NSF grants DMS-1439285 and DMS-1701130.

Subvarieties of general hypersurfaces and varieties with small duals. Lawrence's most influential work during the period 1985–1990 are arguably his papers [**13**], [**14**] on varieties with small duals, and his results [**15**], [**16**] on subvarieties of very general complete intersections.

Let $X \subseteq \mathbf{P}^r$ be a smooth complex projective variety of dimension n. Recall that the *dual variety* of X is the set of hyperplanes tangent to X at some point:

$$X^* \ =_{\text{def}}\ \big\{ H \subseteq \mathbf{P}^r \mid H \text{ is tangent to } X \big\} \ \subseteq\ \mathbf{P}^{r^*}.$$

One expects X^* to be a hypersurface in $\mathbf{P}^{r*}$, but sometimes it has smaller dimension. Zak had established using the Fulton-Hansen connectedness theorem that in any event $\dim X^* \geq \dim X$. Zak's work brought renewed attention to the classical question of trying to say something about those smooth varieties whose duals are exceptionally small.

In his first paper [**13**], Lawrence shows that the only smooth varieties $X \subseteq \mathbf{P}^r$ with

$$\dim X \ =\ \dim X^* \ \leq\ \tfrac{2}{3}r$$

are four classically known examples.[1] In his second paper [**14**], he classifies all varieties of dimension < 6 whose duals are degenerate. Ein starts by fixing a general tangent hyperplane $H \subseteq \mathbf{P}^r$ to X. Then (as was classically understood) the contact locus

$$L \ =_{\text{def}}\ \big\{ x \in X \mid\ H \text{ is tangent to } X \text{ at } x\ \big\}$$

is a *linear* space of dimension $k = r - 1 - \dim X^*$. Lawrence's very nice idea is to study the normal bundle

$$N \ =\ N_{L/X}$$

to L in X. This is a bundle of rank $n - k$ on $L = \mathbf{P}^k$, and Ein proves that

$$(*) \qquad N \ \cong\ N^* \otimes \mathcal{O}_L(1),$$

which he then combines with input from the geometry of vector bundles on projective space. Note that (*) already implies Landman's theorem that if $k \geq 1$, then

$$k \ \equiv\ \dim X \quad (\text{mod } 2).$$

Some related ideas appear in Ein's paper [**17**] with Shepherd-Barron concerning special Cremona transformations.

In 1986, Herb Clemens [**3**] proved a lower bound on the geometric genus of a curve on a very general hypersurface in projective space. Ein's two very influential papers [**15**] and [**16**] gave a large generalization of this result.

THEOREM 0.1. *Let $X \subseteq \mathbf{P}^n$ be a very general complete intersection of multidegrees $m_1, \ldots, m_e$, and assume that*

$$\sum m_i \ \geq\ 2n + 1 - e$$

Then any subvariety of X is of general type.

---

[1]Hartshorne's famous conjecture on complete intersections predicts that if $\dim X > \frac{2}{3}r$ then X should be a complete intersection, and in particular will have a non-degenerate dual.

Very roughly speaking, Lawrence's idea was that the PGL action on projective space allows one to produce sections of a suitable twist of the normal or canonical bundle of such a subvariety. For example, he shows that if $Z \subset X$ is a smooth subvariety that deforms with X, and if $N = N_{Z/X}$ denotes the normal bundle of Z in X, then $N(1)$ is globally generated. By adjunction, it follows that X must be of general type in a suitable range of degrees. Voisin [**40**] subsequently gave a particularly clean formulation of the argument.

Linear series on higher dimensional varieties. I first got to know Lawrence during the 1981-82 academic year, when we were both at the IAS. We kept in close mathematical touch after that, but we only started actively collaborating in the late 1980's. The initial fruits of these efforts were the three papers [**2**], [**18**] and [**19**] (the first with Bertram) dealing with linear series on higher dimensional varieties. Ein and I had both been interested over the years in questions about linear series, and this was the time at which higher-dimensional geometry was really beginning to flower. So it seemed reasonable to see whether one could say something in higher dimensions about issues that had attracted attention for curves and surfaces, notably syzygies and Castelnuovo-Mumford regularity. The idea of [**2**] was to use vanishing theorems to study the regularity of varieties defined by equations of given degrees, while [**18**] proved analogues for all smooth varieties of results of Mark Green on syzygies of curves and Veronese varieties. This work was completed around 1990, and we were happy that it at least let us get our toes in the water of higher dimensions.

The paper [**19**], dealing with global generation of linear series on threefolds, gave us our first practice in using the cohomological techniques of Kawamata–Reid–Shokurov involving the Kawamata–Viehweg vanishing theorem for $\mathbf{Q}$-divisors. We came to this work through a fortuitous sequence of events, illustrating in an amusing way the role that chance sometimes plays in setting the direction of one's research. It's perhaps worth telling the story.

In the late 1980's, Fujita proposed a far-reaching generalization of the classical fact that a line bundle of degree $\geq 2g$ (or $\geq 2g+1$) on a curve of genus g is globally generated (or very ample):

CONJECTURE. *Let A be an ample line bundle on a smooth projective variety X of dimension n. Then:*

(*i*). *$K_X + (n+1)A$ is basepoint-free;*

(*ii*). *$K_X + (n+2)A$ is very ample.*

Fujita was also influenced by then recent work of Igor Reider [**37**], who had studied adjoint line bundles $K_X + L$ on surfaces, and whose results implied the $\dim X = 2$ case of the conjecture. The first breakthrough in Fujita's conjecture occured around 1990, when Demailly [**8**] applied L^2-methods to prove:

THEOREM 0.2 (Demailly). *In the situation of Fujita's conjecture,*

$$2K_X + 12n^n A \text{ is very ample.}$$

While the numerics were rather far from what one expected the result created a sensation, going as it did far beyond what algebraic methods gave at the time. Moreover, Demailly's work inaugurated a very fruitful interaction between analytic and algebraic geometry that continues to this day.

Y.-T. Siu:

A variation of Demailly's method of constructing holomorphic sections of line bundles

Demailly used singular solutions of the complex Monge-Ampere equation to study the following conjecture of Fujita: For any ample line bundle L over a compact complex manifold X of complex dimension n the line bundle $(n+2)L+K_X$ is very ample. He proved the very ampleness of $\beta_n L+2K_X$. In his method the coefficient 2 for K_X is needed, because in using the exterior product of positive currents he had to use a meromorphic vector field to differentiate the defining function of an effective Q-divisor approximating the current. In this talk we discussed the method of trying to remove the coefficient 2 by adding some ample divisor to move the effective Q-divisor.

FIGURE 1. Siu's 1991 Oberwolfach Abstract

In 1991, Lawrence and I attended a meeting at Oberwolfach at which Siu gave a report aimed at algebraic geometrers on Demailly's work and some extensions thereof. (Siu's abstracts reproduced in Figure 1.) During the course of his talk, Siu posed as a challenge the

EXERCISE. *Let L be an ample line bundle on a smooth projective n-fold X, and fix a point $x \in X$. Suppose that there exists $k \gg 0$ and a divisor $D \in |kL|$ such that*

$$\mathrm{mult}_x(D) \geq nk,$$

while $\mathrm{mult}_y(D) < k$ *for y in a punctured neighborhood of x. Then $K_X + L$ is free at x.*

As soon as Siu put the statement on the board, Hélène Esnault and Eckart Viehweg, who were in the audience, realized that one should be able to solve it using vanishng theorems for **Q**-divisors, along the lines of their paper [**27**] from some years back. By the end of the hour they had worked out the details, which they kindly explained to Lawrence and me over the following days. (At that time, the Kawamata–Viehweg vanishing theorem still seemed rather exotic, and its geometric content was not well understood.)

Lawrence and I noticed that a parameter count suggested that one might use Siu's exercise to reprove Reider's results on surfaces, and somewhat later – after we learned more of cohomological arguments of Kawamata–Reid–Shokurov – we got this working. But we assumed that the experts were aware of this approach, which discouraged us from trying to tackle higher dimensions. A couple of months later, however, I was chatting with János Kollár at a UCLA–Utah workshop, and when I mentioned to him in passing this well-known argument, János seemed rather surprised to hear about it.[2] At that point Lawrence and I realized that people actually hadn't tried to use Kawamata's "X-method" to give effective results, and after a couple of months of very intense work we were able to prove the global generation statement in dimension three. With the hindsight of subsequent developments such

[2]In fact he asked me to show him the proof, but it wasn't fresh on my mind and that part of the conversation didn't go very well...

as [**1**], [**34**], [**33**], the resulting paper [**19**] today looks hopelessly overcomplicated and dated, but we like to think it played a useful role by providing proof of concept for extracting reasonably sharp effective statements from the cohomological apparatus of the minimal model program.

Multiplier ideals. Work of Demailly, Siu and others (eg [**8**], [**1**], [**38**], [**39**]) gave dramatic evidence of the power of multiplier ideals and the vanishing theorems they satisfy. Starting in the mid 1990's, Lawrence and I made a systematic attempt to look for concrete applications of this machinery. A first one appeared in [**20**], where we combined generic vanishing theorems with some cousins of multiplier ideals to prove a conjecture about the singularities of theta divisors. Another application appears in [**21**], where we applied an algebro-geometric analogue of Skoda's theorem to establish variants of Kollár's results on the effective Nullstellensatz. But my favorite result from this period is the paper [**24**] with Karen Smith, which used multiplier ideals to prove a rather surprising comparison between the symbolic and ordinary powers of a radical ideal on a smooth variety.

One of the things that Ein teaches his collaborators is that it can be a useful exercise to try to use new methods to reprove old results. During Spring 2000, Lawrence visited Michigan while on sabbatical, and with Karen we decided to look for applications of the circle of ideas around the subadditivity theorem [**9**]. We came across an old result of Izumi [**32**] to the effect that ideals associated to a divisorial valuation are contained in growing powers of the ideal of its support, and saw that this followed immediately from subadditivity. But in fact, the argument seemed to prove more: namely, it showed that if $\mathfrak{q} \subseteq \mathcal{O}_X$ is the ideal sheaf of a reduced subvariety $Z \subseteq X$ of a smooth variety of dimension d, then

$$\mathfrak{q}^{(md)} \subseteq \mathfrak{q}^m,$$

where the symbolic power on the left denotes the germs of functions having multiplicity $\geq md$ at a general point of Z. At first it seemed unclear whether one could really expect such a statement to be true, but luckily the proof was only a couple of lines long and so there weren't many places for an error to hide. Soon thereafter, Hochster and Huneke [**31**] gave a different approach to this comparison via tight closure.[3]

Singularities. During the first decade of the 2000's, Lawrence's focus shifted to the geometry of the singularities that arise in birational geometry. Working with Mustaţă, de Fernex and others, he resolved a number of long-standing problems in the area. For example in [**26**] Ein, Mustaţă and Yasuda used the geometry of jets to prove the conjecture on inversion of adjunction on smooth varieties. At the end of this period, inspired by some ideas of Kollár, Ein, de Fernex and Mustaţă [**7**] proved the celebrated ACC conjecture on smooth varieties:

THEOREM 0.3. *The collection of rational numbers that can occur as log-canonical thresholds of divisors on a smooth n-fold does not contain any cluster points from the left.*

This was later extended to varieties with mild singularities by Hacon, McKernan and Xu [**30**].

---

[3]The ideas surrounding Izumi's theorem were further developed in [**25**].

Some other papers during these years established interesting new relations among classical invariants, and applied them to questions of birational rigidity. For example in [**5**] Ein, de Fernex and Mustaţă prove

THEOREM 0.4. *Let $\mathfrak{a} \subseteq \mathcal{O}_X$ be an ideal of finite colength on a smooth variety X of dimension n. Then*

$$e(\mathfrak{a}) \ \geq \ \left(\frac{n}{\operatorname{lct}(\mathfrak{a})} \right)^n .$$

Here $e(\mathfrak{a})$ denotes the classical Samuel multiplicity of $\mathfrak{a}$, and $\operatorname{lct}(\mathfrak{a})$ denotes the log-canonical threshold of a general element in $\mathfrak{a}$. The proof in [**5**] used a degeneration to reduce the question to the case of monomial ideals. Yuchen Liu [**35**] recently found a very different argument via the theory of K-stability. In [**6**], Ein, de Fernex and Mustaţă applied variants of their inequality to prove the birational rigidity of smooth hypersurfaces of degree n in $\mathbf{P}^n$ when $4 \leq n \leq 12$. De Fernex extended this result to all dimensions in [**4**].

Asymptotic syzygies. Over the last few years, Lawrence and I have returned to the circle of questions around the defining equations of projective varieties and the syzygies among them. Specifically, we have tried to understand what happens syzygetically as $d \to \infty$ when one takes a fixed projective variety X and embeds it by line bundles of the form

$$L_d \ = \ dA \ + \ P,$$

where A is ample and P is arbitrary. The paper [**22**] shows in effect that essentially all the Koszul cohomology groups that could be non-zero asymptotically become so. It was important technically to deal not just with the Koszul cohomology groups $K_{p,q}(X; L_d)$ of L_d, but to work more generally with the groups $K_{p,q}(X, B; L_d)$ in which one allows a fixed twisting line bundle. Combining this perspective with Voisin's Hilbert schematic approach to syzygies [**41**], [**42**] led to a remarkably simple proof in [**23**] of an old conjecture with Mark Green from [**29**] concerning the syzygies of curves of large degree:

THEOREM 0.5. *One can read off the gonality of a curve C from the grading of its resolution in any one embedding of sufficiently large degree.*

Rathmann [**36**] subsequently found a very nice argument giving an effective bound: it suffices in the Theorem to consider embeddings of degree $\geq 4g(C) + 1$.

Epilogue. This overview has omitted many of Ein's contributions, and we can be sure that there will be more to come in the future. Still, I hope these remarks have conveyed something of the breadth of Lawrence's work and vision. Through his research and his generosity in sharing an encyclopedic knowledge of the field, Ein has had an enormous and continuing impact on several generations of algebraic geometers. Working with Lawrence has certainly been the most important influence on my own mathematical career. It is a joy to have this occasion to wish him Happy Birthday!

References

[1] Urban Angehrn and Yum Tong Siu, *Effective freeness and point separation for adjoint bundles*, Invent. Math. **122** (1995), no. 2, 291–308. MR1358978

[2] Aaron Bertram, Lawrence Ein, and Robert Lazarsfeld, *Vanishing theorems, a theorem of Severi, and the equations defining projective varieties*, J. Amer. Math. Soc. **4** (1991), no. 3, 587–602. MR1092845

[3] Herbert Clemens, *Curves on generic hypersurfaces*, Ann. Sci. École Norm. Sup. (4) **19** (1986), no. 4, 629–636. MR875091

[4] Tommaso de Fernex, *Erratum to: Birationally rigid hypersurfaces [MR3049929]*, Invent. Math. **203** (2016), no. 2, 675–680. MR3455160

[5] Tommaso de Fernex, Lawrence Ein, and Mircea Mustaţă, *Multiplicities and log canonical threshold*, J. Algebraic Geom. **13** (2004), no. 3, 603–615. MR2047683

[6] Tommaso de Fernex, Lawrence Ein, and Mircea Mustaţă, *Bounds for log canonical thresholds with applications to birational rigidity*, Math. Res. Lett. **10** (2003), no. 2-3, 219–236. MR1981899

[7] Tommaso de Fernex, Lawrence Ein, and Mircea Mustaţă, *Shokurov's ACC conjecture for log canonical thresholds on smooth varieties*, Duke Math. J. **152** (2010), no. 1, 93–114. MR2643057

[8] Jean-Pierre Demailly, *A numerical criterion for very ample line bundles*, J. Differential Geom. **37** (1993), no. 2, 323–374. MR1205448

[9] Jean-Pierre Demailly, Lawrence Ein, and Robert Lazarsfeld, *A subadditivity property of multiplier ideals*, Michigan Math. J. **48** (2000), 137–156. Dedicated to William Fulton on the occasion of his 60th birthday. MR1786484

[10] Lawrence Ein, *Stable vector bundles on projective spaces in* char $p > 0$, Math. Ann. **254** (1980), no. 1, 53–72. MR597836

[11] Lawrence Ein, Robin Hartshorne, and Hans Vogelaar, *Restriction theorems for stable rank* 3 *vector bundles on* $\mathbf{P}^n$, Math. Ann. **259** (1982), no. 4, 541–569. MR660047

[12] Lawrence Ein, *An analogue of Max Noether's theorem*, Duke Math. J. **52** (1985), no. 3, 689–706. MR808098

[13] Lawrence Ein, *Varieties with small dual varieties. I*, Invent. Math. **86** (1986), no. 1, 63–74. MR853445

[14] Lawrence Ein, *Varieties with small dual varieties. II*, Duke Math. J. **52** (1985), no. 4, 895–907. MR816391

[15] Lawrence Ein, *Subvarieties of generic complete intersections*, Invent. Math. **94** (1988), no. 1, 163–169. MR958594

[16] Lawrence Ein, *Subvarieties of generic complete intersections. II*, Math. Ann. **289** (1991), no. 3, 465–471. MR1096182

[17] Lawrence Ein and Nicholas Shepherd-Barron, *Some special Cremona transformations*, Amer. J. Math. **111** (1989), no. 5, 783–800. MR1020829

[18] Lawrence Ein and Robert Lazarsfeld, *Syzygies and Koszul cohomology of smooth projective varieties of arbitrary dimension*, Invent. Math. **111** (1993), no. 1, 51–67. MR1193597

[19] Lawrence Ein and Robert Lazarsfeld, *Global generation of pluricanonical and adjoint linear series on smooth projective threefolds*, J. Amer. Math. Soc. **6** (1993), no. 4, 875–903. MR1207013

[20] Lawrence Ein and Robert Lazarsfeld, *Singularities of theta divisors and the birational geometry of irregular varieties*, J. Amer. Math. Soc. **10** (1997), no. 1, 243–258. MR1396893

[21] Lawrence Ein and Robert Lazarsfeld, *A geometric effective Nullstellensatz*, Invent. Math. **137** (1999), no. 2, 427–448. MR1705839

[22] Lawrence Ein and Robert Lazarsfeld, *Asymptotic syzygies of algebraic varieties*, Invent. Math. **190** (2012), no. 3, 603–646. MR2995182

[23] Lawrence Ein and Robert Lazarsfeld, *The gonality conjecture on syzygies of algebraic curves of large degree*, Publ. Math. Inst. Hautes Études Sci. **122** (2015), 301–313. MR3415069

[24] Lawrence Ein, Robert Lazarsfeld, and Karen E. Smith, *Uniform bounds and symbolic powers on smooth varieties*, Invent. Math. **144** (2001), no. 2, 241–252. MR1826369

[25] Lawrence Ein, Robert Lazarsfeld, and Karen E. Smith, *Uniform approximation of Abhyankar valuation ideals in smooth function fields*, Amer. J. Math. **125** (2003), no. 2, 409–440. MR1963690

[26] Lawrence Ein, Mircea Mustaţă, and Takehiko Yasuda, *Jet schemes, log discrepancies and inversion of adjunction*, Invent. Math. **153** (2003), no. 3, 519–535. MR2000468

[27] Hélène Esnault and Eckart Viehweg, *Sur une minoration du degré d'hypersurfaces s'annulant en certains points* (French), Math. Ann. **263** (1983), no. 1, 75–86. MR697332

[28] E. Graham Evans and Phillip Griffith, *The syzygy problem*, Ann. of Math. (2) **114** (1981), no. 2, 323–333. MR632842

[29] Mark Green and Robert Lazarsfeld, *On the projective normality of complete linear series on an algebraic curve*, Invent. Math. **83** (1986), no. 1, 73–90. MR813583

[30] Christopher D. Hacon, James McKernan, and Chenyang Xu, *ACC for log canonical thresholds*, Ann. of Math. (2) **180** (2014), no. 2, 523–571. MR3224718

[31] Melvin Hochster and Craig Huneke, *Comparison of symbolic and ordinary powers of ideals*, Invent. Math. **147** (2002), no. 2, 349–369. MR1881923

[32] Shuzo Izumi, *A measure of integrity for local analytic algebras*, Publ. Res. Inst. Math. Sci. **21** (1985), no. 4, 719–735. MR817161

[33] Yujiro Kawamata, *On Fujita's freeness conjecture for 3-folds and 4-folds*, Math. Ann. **308** (1997), no. 3, 491–505. MR1457742

[34] János Kollár, *Effective base point freeness*, Math. Ann. **296** (1993), no. 4, 595–605. MR1233485

[35] Yuchen Liu, The volume of singular Kähler-Einstein Fano manifolds, to appear, `arXiv:1605.01034`.

[36] Jürgen Rathmann, An effective bound for the gonality conjecture, to appear.

[37] Igor Reider, *Vector bundles of rank 2 and linear systems on algebraic surfaces*, Ann. of Math. (2) **127** (1988), no. 2, 309–316. MR932299

[38] Yum Tong Siu, *An effective Matsusaka big theorem* (English, with English and French summaries), Ann. Inst. Fourier (Grenoble) **43** (1993), no. 5, 1387–1405. MR1275204

[39] Yum-Tong Siu, *Invariance of plurigenera*, Invent. Math. **134** (1998), no. 3, 661–673. MR1660941

[40] Claire Voisin, *On a conjecture of Clemens on rational curves on hypersurfaces*, J. Differential Geom. **44** (1996), no. 1, 200–213. MR1420353

[41] Claire Voisin, *Green's generic syzygy conjecture for curves of even genus lying on a K3 surface*, J. Eur. Math. Soc. (JEMS) **4** (2002), no. 4, 363–404. MR1941089

[42] Claire Voisin, *Green's canonical syzygy conjecture for generic curves of odd genus*, Compos. Math. **141** (2005), no. 5, 1163–1190. MR2157134

Department of Mathematics, Stony Brook University, Stony Brook, New York 11794

Email address: `robert.lazarsfeld@stonybrook.edu`

Contemporary Mathematics
Volume **712**, 2018
http://dx.doi.org/10.1090/conm/712/14340

Contractible curves on a rational surface

Alberto Calabri and Ciro Ciliberto

Dedicated to Lawrence Ein on the occasion of his 60th birthday

ABSTRACT. In this paper we prove that if S is a smooth, irreducible, projective, rational, complex surface and D an effective, connected, reduced divisor on S, then the pair (S, D) is *contractible* (i.e., there is a birational map $\phi\colon S \dashrightarrow S'$ with S' smooth such that $\phi_*(D) = 0$) if $\operatorname{kod}(S, D) = -\infty$. More generally, we even prove that this contraction is possible without blowing up an assigned cluster of points on S. Using the theory of peeling, we are also able to give some information in the case D is not connected.

1. Introduction

Let (S, D) be a pair with S a smooth, irreducible, projective, complex surface and D an effective, reduced divisor on S. The pair (S, D) is said to be *contractible* if there is a birational map $\phi\colon S \dashrightarrow S'$ with S' smooth such that $\phi_*(D) = 0$, i.e., D is contracted to a finite set of points by ϕ. The *contractibility problem* consists in finding necessary and sufficient conditions for pairs (S, D) to be contractible.

The question of characterizing contractible pairs (S, D) is somehow trivial, unless S is a rational surface (see Proposition 2.21 below). If S is rational, the problem has its roots in the study of *Cremona geometry* of the complex projective plane $\mathbb{P}^2$ (see [**3**] for an historical account). Classical results (often with incomplete proofs), in the framework of the so called Italian school of algebraic geometry, go back to Castelnuovo–Enriques [**4**], Marletta [**12, 13**], Coolidge [**5**, p. 398].

The first result on the subject in modern times is due to Kumar and Murthy in 1982, cf. [**11**]. It can be stated as follows:

THEOREM 1.1 (Kumar–Murthy). *Let (S, D) be a pair with S rational and D smooth and irreducible. Then (S, D) is contractible if and only if the linear systems $|K_S + D|$ and $|2K_S + D|$ are both empty.*

Given the pair (S, D), for any non–negative integer m, the *m–log plurigenus* of (S, D) is

$$P_m(S, D) := h^0(S, \mathcal{O}_S(m(D + K_S))).$$

2010 *Mathematics Subject Classification.* Primary 14H50; Secondary 14E07, 14N20.

The authors are members of G.N.S.A.G.A. at I.N.d.A.M. and are partially supported by P.R.I.N. project "Geometry of Algebraic Varieties".

The second author received funding from the European Union's Horizon 2020 research and innovation programme under the Marie Sklodowska-Curie grant agreement No. 652782.

If $P_m(S, D) = 0$ for all $m \geq 1$, then one says that the *log Kodaira dimension* of the pair (S, D) is $\mathrm{kod}(S, D) = -\infty$. Otherwise

$$\mathrm{kod}(S, D) = \max\left\{\dim\left(\mathrm{Im}\left(\phi_{|m(D+K_S)|}\right)\right)\right\}$$

where $\phi_{|m(D+K_S)|}$ is the rational map determined by the linear system $|m(D+K_S)|$, whenever this is not empty. One sets $\mathrm{kod}(S) := \mathrm{kod}(S, 0)$, which is the *Kodaira dimension* of S.

If (S, D) is contractible, with S rational, and $\phi\colon S \dashrightarrow S'$ is a birational map which contracts D to a finite set of points, there is a commutative diagram

$$\begin{array}{ccc} & \bar{S} & \\ {}^{\alpha}\swarrow & & \searrow^{\beta} \\ S & \overset{\phi}{\dashrightarrow} & S' \end{array}$$

where α and β are birational morphisms and the proper transform $\bar{D}$ of D via α is contracted to a finite union of points by β. This implies that all irreducible components of D have geometric genus 0 and that $\mathrm{kod}(\bar{S}, \bar{D}) = -\infty$ (see Lemma 2.13).

If, in the above setting, D is smooth then one sees that $\mathrm{kod}(S, D) = \mathrm{kod}(\bar{S}, \bar{D}) = -\infty$ (see Lemma 2.10). So Theorem 1.1 implies that if S is rational and D is smooth and irreducible, then $\mathrm{kod}(S, D) = -\infty$ if and only if $P_2(S, D) = 0$, which can be seen as a log-analogue of Castelnuovo's rationality criterion for regular surfaces.

As for extensions of Kumar–Murthy's Theorem to reducible curves, the only known result so far was due to Iitaka [**7**, **8**], which can be stated as follows:

THEOREM 1.2 (Iitaka). *Let (S, D) be a pair with S rational and D with simple normal crossings and at most two irreducible components. Then $P_2(S, D) = 0$ if and only if* $\mathrm{kod}(S, D) = -\infty$ *and, if this happens, then (S, D) is contractible.*

Concerning reducible curves, the following theorem, though not immediately related to the contractibility problem, should also be recalled.

THEOREM 1.3 (Kojima–Takahashi, [**10**]). *Let (S, D) be a pair where S is a smooth rational surface and D a smooth, reduced curve on S with at most four irreducible components. Then,* $\mathrm{kod}(S, D) = -\infty$ *if and only if $P_6(S, D) = 0$.*

Furthermore, if $(\tilde{S}, \tilde{D})$ is an almost minimal model of (S, D) (see §4.2 below), and if $\tilde{D}$ is connected, then $\mathrm{kod}(S, D) = -\infty$ *if and only if $P_{12}(S, D) = 0$.*

A classical example of Pompilj's [**17**] (see also [**2**, Example 1]) shows that Kumar–Murthy's and Iitaka's theorems cannot be extended, as they stand, to curves with more than two components. In Pompilj's example one has a smooth curve D on a rational surface S with three irreducible components and $|K_S + D|$ and $|2K_S + D|$ both empty, but $|3K_S + D|$ non–empty. In this example each of the components of D is contractible, but D is not. This shows the difficulty in proving contractibility by induction on the number of irreducible components of the curve, as one may be tempted to do (see the historical account in [**3**]). The reason is that, after having contracted (if possible) some of the components of a reducible curve D, in order to make further contractions one may need to blow–up again points where previous components have been contracted, thus creating loops in the contraction process.

In [**2**] we posed the following:

PROBLEM 1. *Suppose* (S, D) *is a pair with* S *rational and* D *reduced. Then does* $\operatorname{kod}(S, D) = -\infty$ *imply that* (S, D) *is contractible?*

As a little evidence, in [**3**] we answered affirmatively to this question if (S, D) is the embedded resolution of $d \geqslant 12$ distinct lines in $\mathbb{P}^2$. However the problem is open in its full generality.

The present paper is devoted to give an affirmative answer to Problem 1 in some cases. After some preliminaries presented in §2, we prove in Theorem 3.1 that the answer to Problem 1 is affirmative if D is connected. This could be proved also as a consequence of the following deep:

THEOREM 1.4 (Miyanishi-Sugie, Fujita; cf. [**14**, Theorem 2.1.1]). *Let* D *be a reduced connected divisor on a rational surface* S *such that* $\operatorname{kod}(S, D) = -\infty$. *Then there exists a morphism* $h\colon S \setminus D \to J$, *with* J *a curve, such that any fibre of* h *is either isomorphic to* $\mathbb{A}^1$ *or to* $\mathbb{P}^1$.

However, we prove more. Indeed, in §2.3 we introduce the concept of a *marked triple* $(S, D, \mathfrak{K})$, where we add to the pair (S, D) a *cluster* $\mathfrak{K}$, i.e., a finite set of proper or infinitely near points on S (see §2.2 for a precise definition). We define the action of birational maps on pairs $(D, \mathfrak{K})$ (see again §2.2) and we introduce the concept of *contractible triples* $(S, D, \mathfrak{K})$ (see §2.6). Theorem 3.1 says that $(S, D, \mathfrak{K})$, with S rational and D connected, is contractible if $\operatorname{kod}(S, D) = -\infty$. The extension to marked triples is motivated by the need of keeping track of *previously contracted* components of D, as we mentioned above.

The proof of Theorem 3.1 uses standard techniques in surface theory. Mori's theory is however hidden in it, under the form of a lemma by Fujita's (see [**6**]). In Remark 3.2 we sketch the proof of (an extended version of) Iitaka's Theorem 1.2, which is not conceptually different from the original, but is definitely shorter.

The assumption of connectedness of D plays a central role in the proof of Theorem 3.1 and we have been unable to do without it. However we have been trying a different approach to the problem, which, though not exhaustive, gives some information even in the non–connected case. Indeed we prove a different contraction criterion, i.e., Theorem 5.1, in §5. This is based on Miyanishi–Tsunoda's *theory of peeling* (see [**14, 15**]), which we briefly recall in §4 for the reader's convenience. Theorem 5.1 basically says that if $(S, D, \mathfrak{K})$ is a marked triple, with (S, D) almost minimal, S rational and $\operatorname{kod}(S, D) = -\infty$, then (S, D) is contractible unless, perhaps, either there is a birational morphism $\phi\colon S \to \bar{S}$, contracting D to (singular) points, with $\bar{S}$ a normal logarithmic del Pezzo surface of rank 1 (this is called a *logarithmic del Pezzo surface of rank 1 with shrinkable boundary*, see the definition in §4.3), or D contains one, and only one, very specific connected component, called a *non–admissible fork* (see §4.1).

The classification of logarithmic del Pezzo surfaces of rank 1 is still an open problem in its generality[1]. Keel and McKernan gave in [**9**] a classification theorem (Theorem 23.2), which applies to all but a bounded family of rank one logarithmic del Pezzo surfaces. A case by case analysis (most likely to be quite hard) based on Keel–McKernan's results could possibly shed some more (though not decisive) light on the resolution of Problem 1.

[1]Note added in proof: there is a work in progress by DongSeon Hwang towards the complete classification of logarithmic del Pezzo surfaces of rank 1.

Acknowledgments. The authors warmly thank DongSeon Hwang for useful discussions and references.

Notation. We set, as usual, $\mathbb{F}_n := \mathbb{P}(\mathcal{O}_{\mathbb{P}^1} \oplus \mathcal{O}_{\mathbb{P}^1}(-n))$, with the structure morphism $f_n : \mathbb{F}_n \to \mathbb{P}^1$. We will denote by E an irreducible section of f_n with $E^2 = -n$ (which is unique if $n \neq 0$), and by F a fibre of f_n. One has $K_{\mathbb{F}_n} \sim -2E - (n+2)F$.

We use the symbol $\sim$ for linear equivalence. If $f\colon S \to S'$ is a birational morphism between smooth surfaces and P is a point of S' where f^{-1} is not defined, then one can consider the (-1)–*cycle* D on S (usually a non–reduced divisor) contracted to P by f: D is 1-connected, $K_S \cdot D = D^2 = -1$.

The concept of 1–connected effective divisors on a smooth surface is well known and we freely use it. For the rest, we use standard notation and concepts in algebraic geometry.

2. Preliminaries

Let S be a smooth, irreducible, projective, complex surface and D an effective divisor on S. The *support* $\mathrm{Supp}(D)$ of D is the reduced divisor sum of the irreducible components of D. The divisor D will usually be for us non–zero and reduced, i.e., $D = \mathrm{Supp}(D)$ in which case D is called a *curve*. We will often consider the case in which the curve D has *simple normal crossing singularities* (shortly, D is snc), i.e., each component of D is smooth and D has at most nodes. In this case we will say that the pair (S, D) is *log smooth*.

2.1. Infinitely near points. Let S and S' be smooth, irreducible, projective surfaces. Any birational morphism $\sigma\colon S' \to S$ is the composition of a certain number n of blow–ups $\sigma_i\colon S_i \to S_{i-1}$ at a point $P_i \in S_{i-1}$ with exceptional divisor E_i on S_i, for $i = 1, \ldots, n$

$$\sigma\colon S' = S_n \xrightarrow{\sigma_n} S_{n-1} \xrightarrow{\sigma_{n-1}} \cdots \xrightarrow{\sigma_2} S_1 \xrightarrow{\sigma_1} S_0 = S. \tag{2.1}$$

Let $P \in S$ be a point. One says that Q is an *infinitely near point to P of (vicinity) order n* on S, and we write $Q >^n P$ (or simply $Q > P$ if n is understood), if there exists a birational morphism $\sigma\colon S' \to S$ as in (2.1), such that $P_1 = P$, $\sigma_i(P_{i+1}) = P_i$, $i = 1, \ldots, n-1$, and $Q \in E_n$. Points of vicinity order 0 are the points of S, which are called *proper points*. We denote by $\mathfrak{P}(S)$ the set of infinitely near points on S and, abusing terminology, we refer to $\mathfrak{P}(S)$ as the set of *points on* S.

Given a curve C on S, one says that it *passes through the point* $Q > P$, with $P \in S$, $Q \in S'$ and $\sigma\colon S' \to S$ as above, if the proper transform C' of C on S' passes through Q. One also says that Q is infinitely near to P *along* C. The notion of infinitely near points which are base points for a linear system of curves on S is then clear.

2.2. Clusters. In this paper a *cluster* $\mathfrak{K}$ on S is a finite subset of $\mathfrak{P}(S)$ (note that in [**1**] a different definition of cluster is used). The *support* $\mathrm{Supp}(\mathfrak{K})$ of $\mathfrak{K}$ is the set of proper points P such that $Q > P$ for $Q \in \mathfrak{K}$. The points of $\mathrm{Supp}(\mathfrak{K})$ are not required to be in $\mathfrak{K}$. A cluster is *simple* if $\mathrm{Supp}(\mathfrak{K})$ consists of one point. Every cluster is a finite union of simple clusters. The concept of a curve passing through a cluster is clear. Clusters have a partial ordering $\mathfrak{K} \leqslant \mathfrak{K}'$ which is given by set inclusion. The *(vicinity) order* of a cluster is the maximum order of a point in the cluster.

Given a cluster $\mathfrak{K}$ on S, there is a surface $S_{\mathfrak{K}}$ and a birational morphism $\phi_{\mathfrak{K}}\colon S_{\mathfrak{K}} \to S$, where each point of the cluster has been blown–up and $S_{\mathfrak{K}}$ is minimal under this condition. We will denote by $E_{\mathfrak{K}}$ the sum of all proper transforms on $S_{\mathfrak{K}}$ of the exceptional divisors of blow–ups of points of $\mathfrak{K}$. The cluster $\mathfrak{K}$ on S is determined by the triple $(S_{\mathfrak{K}}, E_{\mathfrak{K}}, \phi_{\mathfrak{K}})$.

REMARK 2.1. Let Z be a zero–dimensional scheme on S. There is a smooth surface S', a birational morphism $\phi\colon S' \to S$ and a divisor E on S' such that $\phi_*(\mathcal{O}_{S'}(-E)) = \mathcal{I}_{Z,S}$. Then $\mathrm{Supp}(E)$ determines a cluster $\mathfrak{K}_Z$ on S, called the *supporting cluster* of Z. Note that E is reduced if and only if Z is *curvilinear*, i.e., Z is a subscheme of a smooth curve C on S. In this case Z is uniquely determined by $\mathfrak{K}_Z$.

REMARK 2.2. Given a triple (S', E', ϕ), with $\phi\colon S' \to S$ a birational morphism and E' an effective, reduced divisor on S' which is contracted to a union of points by ϕ, there is a unique cluster $\mathfrak{K}$ on S and a unique birational morphism $\varphi_{\mathfrak{K}} : S' \to S_{\mathfrak{K}}$ such that $\phi = \phi_{\mathfrak{K}} \circ \varphi_{\mathfrak{K}}$, $\varphi_{\mathfrak{K}}(E') = E_{\mathfrak{K}}$, and no component of E' is contracted to a point by $\varphi_{\mathfrak{K}}$. In particular $S_{\mathfrak{K}}$ is uniquely determined up to isomorphisms.

Let $f\colon S \dashrightarrow S'$ be a birational map between smooth, irreducible, projective surfaces. This induces a birational map $f_{\mathfrak{K}}\colon S_{\mathfrak{K}} \dashrightarrow S'$. We have a diagram

(2.2)

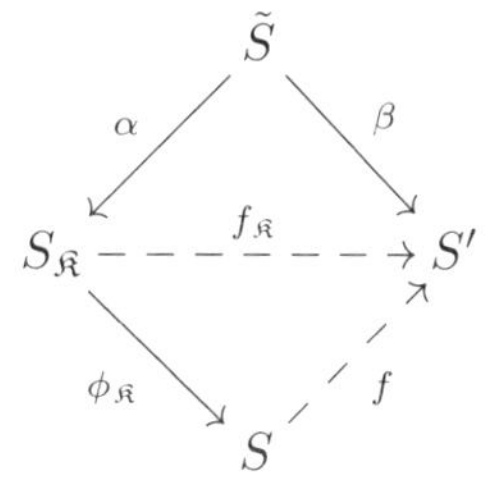

where α and β are sequences of blow–ups. Let $\tilde{E}_{\mathfrak{K}}$ be the proper transform of $E_{\mathfrak{K}}$ on $\tilde{S}$. We set

$$\mathrm{div}_f(\mathfrak{K}) = \beta_*(\tilde{E}_{\mathfrak{K}})$$

which is called the *divisorial part* of the *image* of $\mathfrak{K}$ via f. Let $\tilde{E}_{\mathfrak{K},0}$ be the maximal subdivisor of $\tilde{E}_{\mathfrak{K}}$ contracted to points by β. By Remark 2.2, this determines a cluster $\mathfrak{K}'$ on S'. We set

$$\mathrm{cl}_f(\mathfrak{K}) = \mathfrak{K}'$$

called the *image cluster* of $\mathfrak{K}$.

We will say that f *does not blow–up the cluster* $\mathfrak{K}$ if $\mathrm{div}_f(\mathfrak{K}) = 0$.

REMARK 2.3. Morphisms do not blow–up any cluster. Moreover, if f does not blow–up $\mathfrak{K}$ and $\mathfrak{K}' \leqslant \mathfrak{K}$, then f does not blow–up $\mathfrak{K}'$.

2.3. Markings. Let S be a smooth, irreducible, projective surface. A *marked pair* on S is a pair $(D, \mathfrak{K})$ with D an effective, reduced divisor on S and $\mathfrak{K}$ is a cluster on S, called the *marking* of the pair, whereas D is the *divisorial part* of the pair. The triple $(S, D, \mathfrak{K})$ will be called a *marked triple.* If the cluster $\mathfrak{K}$ is empty, we write (S, D) instead of $(S, D, \mathfrak{K})$.

Markings have a partial ordering

$$(D, \mathfrak{K}) \leqslant (D', \mathfrak{K}') \quad \overset{\text{def}}{\Longleftrightarrow} \quad D \leqslant D' \quad \text{and} \quad \mathfrak{K} \leqslant \mathfrak{K}'.$$

Let $f : S \dashrightarrow S'$ be a birational map and let $(D, \mathfrak{K})$ be a marking. We want to define $f_*(D, \mathfrak{K})$ which will be a marked pair $(D', \mathfrak{K}')$ on S', and we will then say that f maps $(S, D, \mathfrak{K})$ to $(S', f_*(D, \mathfrak{K}))$.

We have a diagram

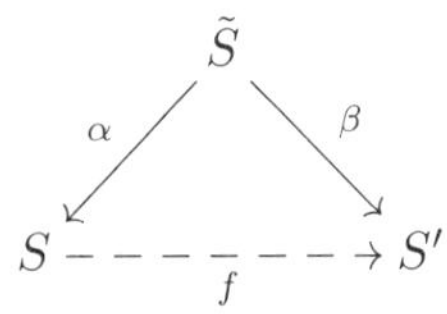

with α, β sequences of blow–ups. We let $\tilde{D}$ be the proper transform of D on $\tilde{S}$. We set

$$D' = \beta_*(\tilde{D}) + \mathrm{div}_f(\mathfrak{K}).$$

Let $\tilde{D}_0$ be the maximal subdivisor of $\tilde{D}$ contracted to points by β. By Remark 2.2, this determines a cluster $\mathfrak{D}$ of S', and we define

$$\mathfrak{K}' = \mathrm{cl}_f(\mathfrak{K}) \cup \mathfrak{D}.$$

2.4. Cremona transformations. Let S be a smooth, irreducible, projective surface. Consider a rational dominant map $f \colon S \dashrightarrow X$, with X an irreducible, projective variety. If X is non–degenerate in $\mathbb{P}^r$, there is a fixed components free linear system $\mathcal{L}$ of dimension r on S such that $\mathcal{L}$ is the pull–back via ϕ of the hyperplane linear system of $\mathbb{P}^r$ and f coincides with the map $\phi_{\mathcal{L}}$ determined by $\mathcal{L}$.

EXAMPLE 2.4. Let $f \colon S \dashrightarrow S'$ be a birational map between smooth, irreducible, projective surfaces and assume that $f = \phi_{\mathcal{L}}$. Let $\mathfrak{K}$ be a cluster on S. We want to give conditions under which f does not blow–up $\mathfrak{K}$. We assume $\mathfrak{K}$ simple, with support $P \in S$ (the non–simple case can be treated similarly). By Remark 2.3 we may assume that $P \in \mathfrak{K}$.

If f is a morphism, it does not blow–up $\mathfrak{K}$. Hence, if P is not a base point for $\mathcal{L}$, then f does not blow–up $\mathfrak{K}$.

Suppose that P is a base point for $\mathcal{L}$, that the general curve in $\mathcal{L}$ is smooth at P and that there is a smooth curve C through P such that the intersection multiplicity in P of C with the general curve in $\mathcal{L}$ is $m > 0$. In other words P, and its subsequent infinitely near points along C up to order m, are base points for $\mathcal{L}$. We claim that, if m is larger that the order of $\mathfrak{K}$, then f does not blow–up $\mathfrak{K}$.

Indeed, (2.2) specifies as follows

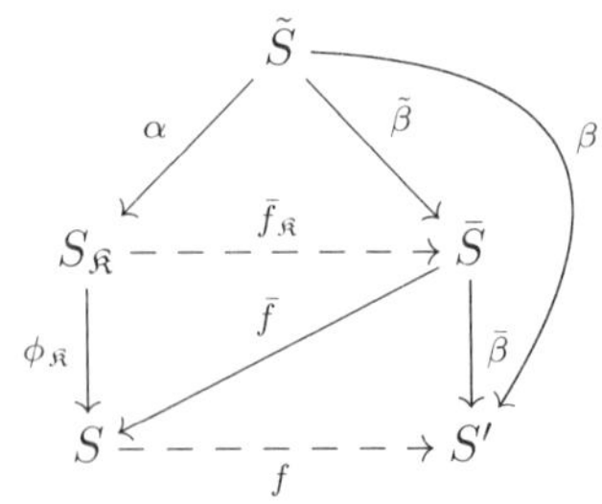

where:
(i) the lower right triangle is the resolution of the indeterminacies of f, hence it is

the composition of the blow–ups of P and of its subsequent infinitely near points along C up to order m, plus perhaps other blow–ups at points P' with $P' \not> P$. Hence the curve on $\bar{S}$ contracted by $\bar{f}$ to P is a chain E of rational curves $E_1 + \cdots + E_m$, where $E_i \cdot E_j = \delta_{i,j-1}$, $1 \leqslant i < j \leqslant m$, and $E_i^2 = -2$ if $i = 1, \ldots, m-1$ whereas $E_m^2 = -1$;
(ii) $f_{\mathfrak{K}} = f \circ \phi_{\mathfrak{K}}$ and $\bar{f}_{\mathfrak{K}}$ is determined by f;
(iii) since the order of $\mathfrak{K}$ is smaller than m, the curves in $\tilde{E}_{\mathfrak{K}}$ are either contracted by $\tilde{\beta}$ to points or are mapped to one of the curves $E_1, \ldots, E_{m-1}$;
(iv) $\bar{\beta}$ contracts $E_1 + \cdots + E_{m-1}$ to a point.

Our claim follows from (iii) and (iv).

EXAMPLE 2.5. We recall the *elementary transformations* of surfaces $\mathbb{F}_n$. Pick a point $P \in \mathbb{F}_n$ and consider the diagram

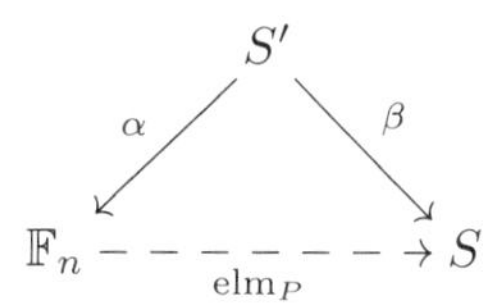

where α is the blow–up of $\mathbb{F}_n$ at P and β is the contraction of the proper transform of the fibre of $|F|$ passing through P on S'. Then

$$S = \begin{cases} \mathbb{F}_{n+1} & \text{if} \quad P \in E \\ \mathbb{F}_{n-1} & \text{if} \quad P \notin E. \end{cases}$$

The map elm_P is called the *elementary transformation based at* P.

More generally, one can make elementary transformations when we have a surface S and a base point free pencil $|F|$ of rational curves.

If S is rational, we have a birational map $\phi : S \dashrightarrow \mathbb{P}^2$, which is determined by a linear system $\mathcal{L}$ of dimension 2 whose general element is an irreducible curve of geometric genus 0 and two general curves of $\mathcal{L}$ intersect transversely at one point off the base points of $\mathcal{L}$. Any such a linear system is called a *homaloidal net.*

In particular, *Cremona transformations* of $\mathbb{P}^2$, i.e., birational maps $\phi : \mathbb{P}^2 \dashrightarrow \mathbb{P}^2$, are of the form $\phi_{\mathcal{L}}$, with $\mathcal{L}$ homaloidal nets of plane curves. The Cremona transformation $\phi_{\mathcal{L}}$ is said to be *based* at the base locus scheme of the homaloidal net $\mathcal{L}$.

EXAMPLE 2.6. Consider on $\mathbb{F}^n$ a complete linear $|E + dF|$ of sections of $\mathbb{F}_n \to \mathbb{P}^1$. If $d \geqslant n$, then $|E + dF|$ is base point free, of dimension $2d - n + 1$ and self–intersection $2d - n$, and the general curve in $|E + dF|$ is smooth, irreducible and rational.

Let us fix $C \in |E + dF|$ smooth, and an effective divisor D of degree $2d - n - 1$ on C. Let $\mathcal{L}$ be the linear system of curves in $|E + dF|$ which cut out on C a divisor containing D. Then $\mathcal{L}$ is a homaloidal net determining a birational map $\phi_{\mathcal{L}} : \mathbb{F}_n \dashrightarrow \mathbb{P}^2$ and $\mathcal{L}$ has the curvilinear base locus scheme D considered as a subscheme of C, which is determined by its supporting cluster $\mathfrak{K}_D$

(see Example 2.1). In the case $n = 1$, we have a diagram

$$\begin{array}{ccc} & \mathbb{F}_1 & \\ \pi \swarrow & & \searrow \phi_{\mathcal{L}} \\ \mathbb{P}^2 & \overset{}{\underset{\gamma}{\dashrightarrow}} & \mathbb{P}^2 \end{array}$$

where π is the blow–down of the curve E to a point $P \in \mathbb{P}^2$. The map γ is determined by a homaloidal net of plane curves of degree d with multiplicity $d-1$ at P and a further curvilinear base locus scheme, or cluster, of degree $2d-2$. A Cremona map of type γ is called a *De Jonquières transformation*. The case $d = 2$ is the case of *quadratic transformations*.

2.5. Curves on a surface. Consider a pair (S, D) with D a curve on S.

We will write $\ell(D)$ (or simply ℓ) to denote the number of irreducible components of D. We will denote by $\mathfrak{G}(D)$ the *vector weighted graph* of D, i.e., the graph:
(1) with ℓ vertices $\mathfrak{d}_1, \dots, \mathfrak{d}_\ell$ corresponding to the irreducible components $D_1, \dots, D_\ell$ of D, each with vector weight $(D_i^2, p_a(D_i))$, for $1 \leqslant i \leqslant \ell$;
(2) for each pair of indices (i, j) such that $1 \leqslant i < j \leqslant \ell$ and for each intersection point p of D_i and D_j, there is an edge $\mathfrak{e}_{i,j,p}$ joining $\mathfrak{d}_i$ and $\mathfrak{d}_j$, with weight given by the intersection multiplicity of D_i and D_j at p, so that the sum of the weights of edges joining d_i and d_j is $m_{ij} = D_i \cdot D_j$.

In the above setting, an edge of weight w has to be considered as the superposition of w simple edges, hence it contributes w to the valency of the vertices it joins and contributes to the homology of $\mathfrak{G}(D)$.

If the irreducible components of D are all smooth and rational, one may omit the second component of the vector weight of the vertices. We will sometimes denote by the same symbol $\mathfrak{G}(D)$ the unweighted graph.

A connected curve D is said to be a *tree* if $\mathfrak{G}(D)$ is a tree, i.e., all edges have weight 1 and $h^1(\mathfrak{G}(D), \mathbb{C}) = 0$ (then D is snc). A component of D corresponding to a vertex of valency 1 of $\mathfrak{G}(D)$ is called *terminal*.

The curve D is *connected* if and only if $\mathfrak{G}(D)$ is connected, in which case one has $p_a(D) \geqslant 0$.

Note that $\mathfrak{G}(D)$ does not identify D, even if all components of D are smooth. For example, if D consists of three lines in the plane, then $\mathfrak{G}(D)$ is the complete graph on three vertices, regardless to the fact that the three lines pass or do not pass through the same point.

LEMMA 2.7. *Let D be an effective, non–zero, reduced, connected divisor with h irreducible components on a smooth surface S. Then:*
(i) *if $p_a(D) = 0$ then $\mathfrak{G}(D)$ is a tree, all components of D are smooth and rational and D has $h-1$ nodes and no other singularity (in particular D is snc);*
(ii) *if $p_a(D) = 1$ then:*

(a) *either D has a component of arithmetic genus 1 (which can be either smooth, or rational nodal, or rational cuspidal), all other components are smooth rational and D has (further) $h-1$ nodes and no other singularity;*

(b) *or all components of D are smooth and rational and D has h nodes and no other singularity;*

(c) *or all components of D are smooth and rational, D has a tacnode and $h-2$ nodes and no other singularity;*

(d) *or all components of* D *are smooth and rational,* D *has an an ordinary triple point and* $h-3$ *nodes and no other singularity.*

In case (a) the graph $\mathfrak{G}(D)$ *is a tree, in all other cases it has a unique cycle.*

PROOF. The proof is standard, so we only sketch it in case (ii), giving for granted case (i), which can be proved similarly.

The assertion is clear if $h=1$: in this case (a) occurs. So we assume $h>1$ and proceed by induction on h. Since D is connected, there are singular points on D. Let P be one of them, and let m be its multiplicity. Let $\pi : S' \to S$ be the blow–up of P with exceptional divisor E, and take the proper transform D' of D. One has

$$p_a(D') = p_a(D) - \frac{m(m-1)}{2} = 1 - \frac{m(m-1)}{2}. \tag{2.3}$$

Assume first P is not a node. If D' is connected, then $p_a(D') \geqslant 0$, hence $m=2$, $p_a(D')=0$, and we can apply part (i) to D'. Since $m=2$, then $E\cdot D'=2$. Since P is not a node, then E intersects D' at only one point Q with intersection multiplicity 2. If Q is a smooth point of D', then D has a cusp, and we are in case (a). If Q is a node of D', then D has a tacnode, and we are in case (c). If D' is not connected, let k be the number of its connected components. One has $k \leqslant m$, hence $p_a(D') \geqslant -k+1 \geqslant -m+1$. By (2.3), one has $m \leqslant 3$. The case $m=2$ is not possible, because we assumed D' not connected and P not a node. Therefore $m=k=3$, P is an ordinary triple point and, by applying to the three connected components of D' part (i), we see we are in case (d).

Suppose now P is a node, hence $m=2$ and $p_a(D')=0$. If D' is connected then, applying part (i), we see we are in cases (a) or (b). If D' is not connected, then, since D is connected, D' consists of two connected components D'_1, D'_2. Then $D = D_1 + D_2$, with $D_i = \pi_*(D'_i)$, with $1 \leqslant i \leqslant 2$, and D_1 and D_2 intersect transversally at P. Since $1 = p_a(D) = p_a(D_1) + p_a(D_2)$, and $p_a(D_1), p_a(D_2)$ are both non–negative, we may assume that $p_a(D_1)=1, p_a(D_2)=0$. Then we conclude by applying part (i) for D_2 and induction for D_1. □

COROLLARY 2.8. *Let* (S,D) *be a pair with* S *a smooth, regular surface, and* D *an effective, reduced, non–zero divisor, such that* $|K_S+D|=\emptyset$ *(which is the case if* $\mathrm{kod}(S,D)=-\infty$*). Then* (S,D) *is log smooth.*

PROOF. The hypotheses yield that each connected component of D has $p_a(D)=0$. The assertion follows from Lemma 2.7. □

The following lemmata will be useful:

LEMMA 2.9. *Let* (S,D) *be a pair with* S *a smooth, rational surface,* $D = D_1+\cdots+D_h$ *an effective, reduced divisor, where* $D_1,\ldots,D_h$ *are the connected components of* D*. Assume that* $p_a(D_i)=0$ *for* $1 \leqslant i \leqslant h$*. Then*

(i) $D\cdot(D+K) = -2h$;

(ii) $h^0(S,\mathcal{O}_S(2K+D)) + h^0(S,\mathcal{O}_S(-K-D)) \geqslant K\cdot(K+D)-h+1 = (D+K)^2+h+1.$

PROOF. Part (i) is adjunction formula and part (ii) is Riemann–Roch theorem plus part (i). □

Let (S,D) and $(\bar{S},\bar{D})$ be pairs. If there is a birational morphism $\phi : \bar{S} \to S$ such that $\phi_*(\bar{D}) = D$, we write $(S,D) \leqslant (\bar{S},\bar{D})$.

LEMMA 2.10. *Let (S, D) and $(\bar{S}, \bar{D})$ be pairs with D and $\bar{D}$ effective and reduced. If $(S, D) \leqslant (\bar{S}, \bar{D})$, then $\mathrm{kod}(S, D) \geqslant \mathrm{kod}(\bar{S}, \bar{D})$, i.e., $\mathrm{kod}(S, D)$ is a decreasing function. In particular, if $\mathrm{kod}(S, D) = -\infty$ then also $\mathrm{kod}(\bar{S}, \bar{D}) = -\infty$.*

PROOF. It suffices to prove the assertion for the blow–up $\phi : \bar{S} \to S$ at a single point P, with exceptional divisor E. Then $\bar{D} = \phi^*(D) - aE$, with $a \geqslant -1$, hence $\bar{D} + K_{\bar{S}} \sim \phi^*(D + K_S) - (a-1)E$. So the assertion is clear if $a \geqslant 1$.

Assume $a = 0, -1$. Then $m(\bar{D} + K_{\bar{S}}) \sim \phi^*(m(D + K_S)) + m(1-a)E$ and we claim that in this case $P_m(S, D) = P_m(\bar{S}, \bar{D})$ for all positive integers m, hence $\mathrm{kod}(S, D) = \mathrm{kod}(\bar{S}, \bar{D})$. Indeed, if $m(\bar{D} + K_{\bar{S}})$ is not effective, then also $\phi^*(m(D + K_S))$ is not effective, hence $P_m(S, D) = P_m(\bar{S}, \bar{D}) = 0$. If $m(\bar{D} + K_{\bar{S}})$ is effective, then $m(\bar{D} + K_{\bar{S}}) \cdot E = m(a-1)$, hence $m(1-a)E$ is a fixed component of $|m(\bar{D} + K_{\bar{S}})|$, and again $P_m(S, D) = P_m(\bar{S}, \bar{D})$. □

REMARK 2.11. The proof of Lemma 2.10 shows that if $0 \leqslant a \leqslant 1$, then $P_m(S, D) = P_m(\bar{S}, \bar{D})$ for all positive integers m, hence $\mathrm{kod}(S, D) = \mathrm{kod}(\bar{S}, \bar{D})$.

In particular, in the setting of the proof of Lemma 2.10, if D has a double point and $\bar{D} = \phi^*(D) - E$, then $\bar{D}$ contains E, has the same number of connected components of D, and $P_m(S, D) = P_m(\bar{S}, \bar{D})$ for all positive integers m. In particular $\mathrm{kod}(S, D) = -\infty$ if and only if $\mathrm{kod}(\bar{S}, \bar{D}) = -\infty$.

Given (S, D) as in Lemma 2.10, we set

$$\overline{\mathrm{kod}}(S, D) = \min\{\mathrm{kod}(\bar{S}, \bar{D}) : (\bar{S}, \bar{D}) \geqslant (S, D)\}. \tag{2.4}$$

By Remark 2.11, the minimum in (2.4) is reached once we resolve the singularities of D.

2.6. Contractible triples. Let $(S, D, \mathfrak{K})$ be a marked triple. One says that $(S, D, \mathfrak{K})$ is *contractible* if there is a birational map $f : S \dashrightarrow S'$ such that $f_*(D, \mathfrak{K})$ has zero divisorial part. Similarly, one defines the concept of a *contractible pair* (S, D).

EXAMPLE 2.12. A triple $(\mathbb{P}^2, L, \mathfrak{K})$, where L is a line and $\mathfrak{K}$ is any cluster, is contractible, via a quadratic transformation based at two general points of D.

Similarly, a triple $(\mathbb{P}^2, D, \mathfrak{K})$, where D is a reduced conic and $\mathfrak{K}$ is any cluster, is contractible. If D is irreducible, we can reduce to the line case with a quadratic transformation based at three general points of D. If D is reducible, we also reduce to the line case, by applying a quadratic transformation based at two general points of one component of D and at another general point of the other component.

REMARK 2.13. Let (S, D) and (S', D') be pairs. Suppose there is a birational map $\phi : S \dashrightarrow S'$ such that ϕ [resp. ϕ^{-1}] does not contract any irreducible component of D [resp. of D'] and $\phi_*(D) = D'$ (hence $\phi_*^{-1}(D') = D$). Then (S, D) is contractible if and only if (S', D') is. Given this, there is no restriction, in the contractibility problem, to assume D to be snc or even smooth.

LEMMA 2.14. *If (S, D) is contractible, then $\overline{\mathrm{kod}}(S, D) = \mathrm{kod}(S)$. Moreover, all irreducible components of D have geometric genus 0.*

PROOF. If (S, D) is contractible, there is a commutative diagram

$$\begin{array}{ccc} & \bar{S} & \\ {}^{\alpha}\swarrow & & \searrow^{\beta} \\ S & \underset{\phi}{\dashrightarrow} & S' \end{array}$$

where α and β are birational morphisms. If $\bar{D}$ is the proper transform of D via α, then $\bar{D}$ is contracted to a union of points by β, thus, by Lemma 2.11, the assertion follows. □

The following is obvious.

LEMMA 2.15. *Let $(S, D, \mathfrak{K})$ and $(S', D', \mathfrak{K}')$ be triples, let $f : S \dashrightarrow S'$ be a birational map, such that $f_*(D, \mathfrak{K}) \leqslant (D', \mathfrak{K}')$ and assume that $(S', D', \mathfrak{K}')$ is contractible. Then $(S, D, \mathfrak{K})$ is contractible.*

PROPOSITION 2.16. *Let $(\mathbb{F}_n, D, \mathfrak{K})$ be a marked triple such that $D = \varepsilon E + D'$, with $0 \leqslant \varepsilon \leqslant 1$, and $D' \in |E + dF|$. Then $(\mathbb{F}_n, D, \mathfrak{K})$ is contractible.*

PROOF. By taking into account Examples 2.6 and 2.12 and by Lemma 2.15 the assertion is clear if $\varepsilon = 0$. So we focus on the case $\varepsilon = 1$.

We may write

$$D' = F_1 + \ldots + F_k + C \quad \text{with} \quad k \geqslant 0,$$

$F_1, \ldots, F_k \in |F|$ distinct and $C \in |E + (d-k)F|$ smooth and irreducible. We can make a series of elementary transformations based at general point of $F_1, \ldots, F_k$ and contract them. After having done this, the proper transform of E could have non–negative self intersection. However, we can make another series of elementary transformations either based at general points of the surface or at general points of the proper transform of E, so to reduce to the case $k = 0$ and $n = 1$, where $D = E + C$ with $E^2 = -1$ and $C \in |E + dF|$ smooth, irreducible.

Let $n_1Q_1 + \ldots + n_hQ_h$ be the degree $d - 1$ divisor cut out by E on C, with $Q_1, \ldots, Q_h$ distinct. Take non–negative integers $k_1, \ldots, k_h$, such that $m := k_1 + \ldots + k_h + 1 > d$, consider the linear system $|E + mF|$ of dimension $2m$ and its sublinear system $\mathcal{L}$ consisting of the curves:

- cutting out on C a divisor containing $(n_j + k_j)Q_j$, for all $j = 1, \ldots, h$;
- passing through $m - d$ further general points of E.

The total number of base points imposed to $\mathcal{L}$ is $2m - 2$ so that $\dim(\mathcal{L}) = 2$ and $\mathcal{L}$ is a homaloidal net, hence $\phi_{\mathcal{L}}$ birationally maps $\mathbb{F}_1$ to $\mathbb{P}^2$. It maps C to a line, maps E to a point P (which is not on C). By Example 2.4, we see that, by taking $k_1, \ldots, k_h$ sufficiently large, the map $\phi_{\mathcal{L}}$ does not blow–up the cluster $\mathfrak{K}$. So we are reduced to the case $(\mathbb{P}^2, L, \mathfrak{C})$, where L is a line and $\mathfrak{C}$ is a suitable cluster, which is contractible by Example 2.12. □

Recalling Example 2.6, as an immediate consequence we have:

COROLLARY 2.17. *Let $(\mathbb{P}^2, D, \mathfrak{K})$ where D is a reduced curve of degree d with a point of multiplicity at least $d - 1$. Then $(\mathbb{P}^2, D, \mathfrak{K})$ can be mapped via a De Jonquières transformation to a triple $(\mathbb{P}^2, L, \mathfrak{C})$, where L is a line, and so $(\mathbb{P}^2, D, \mathfrak{K})$ is contractible.*

We have also:

COROLLARY 2.18. *Let $(S, D, \mathfrak{K})$ be a marked triple, with S a minimal rational surface and* $\mathrm{kod}(S, D) = -\infty$. *Then $(S, D, \mathfrak{K})$ is contractible.*

PROOF. We assume D non–zero, otherwise there is nothing to prove.

If $S = \mathbb{P}^2$, then $\mathrm{kod}(S, D) = -\infty$ implies $D \sim kL$, with $1 \leqslant k \leqslant 2$ and the assertion follows by Example 2.12.

If $S = \mathbb{F}_n$, we have $D \sim aE + bF$, with $a, b \geqslant 0$ and $a + b > 0$.

If $n = 0$, then $\mathrm{kod}(S, D) = -\infty$ implies that either $0 \leqslant a \leqslant 1$ or $0 \leqslant b \leqslant 1$. We may assume that $0 \leqslant a \leqslant 1$, then the assertion follows from Proposition 2.16.

Assume next $n \geqslant 2$. If $a \leqslant 1$, the assertion follows again from Proposition 2.16. Suppose $a \geqslant 2$. If $b \geqslant n + 2$, then $K_S + D \sim (a - 2)E + (b - n - 2)F$ is effective, a contradiction. So $b \leqslant n + 1$. Suppose that $b < n(a - 1)$, then $D \cdot E = -an + b < -n$ which implies that $2E$ is contained in D, a contradiction, since D is reduced. Therefore we have $n + 1 \geqslant b \geqslant n(a - 1)$, hence we must have $a = 2$. But then $D \cdot E = -2n + b < 0$, so E splits off D and we may apply Proposition 2.16 to conclude. □

2.7. Small pairs. Let (S, D) be a pair as above. We will often write K to denote a canonical divisor K_S of S.

We say that the pair (S, D) is *small* if there is no (-1)–curve E on S such that $\epsilon := E \cdot D \leqslant 1$. Since D is reduced, one has $\epsilon \geqslant -1$, with equality if and only if E is a connected component of D, in which case we say that E is an *isolated component* of D.

Let E be a (-1)–curve offending smallness, let $\pi : S \to S'$ be the contraction of E to a point P' and let $\pi_*(D) = D'$.

If E is not contained in D, then D' is isomorphic to D and D' has multiplicity ϵ in P'. If E is contained in D, one has $E \cdot (D - E) = \epsilon + 1$, hence one has the following different possibilities:
(i) $\epsilon = -1$, i.e., E is an isolated component of D, then D' is isomorphic to $D - E$;
(ii) $\epsilon = 0$, then E intersects $D - E$ at a smooth point (i.e., E is a terminal component of D), hence D' is isomorphic to $D - E$ and P' is a smooth point of D';
(iii) $\epsilon = 1$ and E intersects $D - E$ at two distinct smooth points, hence D' acquires a node at P';
(iv) $\epsilon = 1$ and E intersects $D - E$ at a point P with intersection multiplicity 2, and $D - E$ has a *cusp of order* $k \geqslant 1$ at P (i.e., $D - E$ has, in a suitable neighborhood of P, local equation of the form $y^2 = x^{2k+1}$), in which case D' has at P' a cusp of order $k + 1$;
(v) $\epsilon = 1$ and E intersects $D - E$ at a double point P with intersection multiplicity 2, and $D - E$ has a *tacnode of order* $k \geqslant 1$ at P (i.e., $D - E$ has in a suitable neighborhood of P local equation of the form $y^2 = x^{2k}$), in which case D' has at P' a tacnode of order $k + 1$.

LEMMA 2.19. *In the above setting, one has:*
(i) *D' has the same number of connected components of D, unless E is an isolated component of D, in which case D' has one component less than D;*
(ii) $\pi^*(K_{S'} + D') = K_S + D - (1 - \epsilon)E$, *in particular $\kappa(S, D) = -\infty$ implies $\kappa(S', D') = -\infty$;*
(iii) $p_a(D) = p_a(D')$, *unless E is an isolated component of D, in which case* $p_a(D) = p_a(D') - 1$.

PROOF. Parts (i) and (ii) are obvious. As for part (iii), the assertion is clear if E is not contained in D, because then D and D' are isomorphic. If E is contained in D the assertion follows from part (ii) and adjunction formula. □

By iterating contractions of curves offending smallness, one arrives at a small pair (S_σ, D_σ), where $\pi : S \to S_\sigma$ is a birational morphism and $D_\sigma = \pi_*(D)$. The pair (S_σ, D_σ) is called a *small model* of (S, D).

REMARK 2.20. A small model of (S, D) is in general not unique, since it may depend on the (-1)–curves which one contracts first. For instance if $D = E_1 + E_2$, with E_1, E_2 two (-1)–curves such that $E_1 \cdot E_2 = 1$, then we may either contract E_1 or E_2, and the two resulting surfaces are obtained one from the other by an elementary transformation in a pencil of rational curves.

We will need to keep track of the components of D which is necessary to contract in order to come to a small model (S_σ, D_σ). This datum is encoded in the cluster $\mathfrak{K}_\pi$ determined by $\pi : S \to S_\sigma$ and by the curves contracted by π (recall Remark 2.2), or rather, in the marked triple $(S_\sigma, D_\sigma, \mathfrak{K}_\pi)$.

More generally, one can start with a marked triple $(S, D, \mathfrak{K})$. If (S, D) is small, also the triple $(S, D, \mathfrak{K})$ will be said to be *small*. In any case, let (S_σ, D_σ) be a small model of (S, D), with $\pi : S \to S_\sigma$ (observe that, since π is a morphism, it does not blow–up $\mathfrak{K}$). We define $\mathfrak{K}_\sigma$ to be the union of $\mathrm{cl}_\pi(\mathfrak{K})$ and of $\mathfrak{K}_\pi$. The small triple $(S_\sigma, D_\sigma, \mathfrak{K}_\sigma)$ will be said to be a *small model* of $(S, D, \mathfrak{K})$.

We finish by observing that the contractibility problem for a pair (S, D) is somehow trivial if S is not rational. Indeed, we have:

PROPOSITION 2.21. *Let (S, D) be a pair with S not rational and let (S_σ, D_σ) be a small model. Then (S, D) is contractible if and only if:*
(i) *either $D_\sigma = 0$ and S_σ is minimal, which is the case if* $\mathrm{kod}(S) \geqslant 0$,
(ii) *or, only if* $\mathrm{kod}(S) = -\infty$, *each irreducible component of D_σ is contained in a fibre of the Albanese morphism of S_σ.*

PROOF. If either (i) or (ii) hold, then clearly (S_σ, D_σ) is contractible (in case (ii) use elementary transformations), hence also (S, D) is. Assume next that (S, D) is contractible.

First suppose $\mathrm{kod}(S) \geqslant 0$. Let $\phi : S \dashrightarrow S'$ be a birational map such that $\phi_*(D) = 0$. Consider the commutative diagram

$$\begin{array}{ccc} S & \overset{\phi}{\dashrightarrow} & S' \\ {\scriptstyle p}\downarrow & & \downarrow{\scriptstyle p'} \\ S_{\min} & \underset{f}{\longrightarrow} & S_{\min} \end{array}$$

where p and p' are the birational morphisms to the unique minimal model $S_{\min}$ and f is an automorphism of $S_{\min}$. Since $(p' \circ \phi)_*(D) = 0$, then also $(f \circ p)_*(D) = 0$, hence $p_*(D) = 0$. This means that each connected component of D is contained in some (-1)–cycle on S and the assertion follows.

Assume now $\mathrm{kod}(S) = -\infty$. Since S is not rational, then the Albanese morphism factors through a morphism $a : S \to C$, with C a smooth curve of positive genus. Since all irreducible components of D have geometric genus 0 (see Lemma 2.14), the assertion follows. □

3. The contraction theorem

In this section we will prove the following:

THEOREM 3.1. *Let* $(S, D, \mathfrak{K})$ *be a triple with* S *rational,* D *connected and such that* $\operatorname{kod}(S, D) = -\infty$. *Then* $(S, D, \mathfrak{K})$ *is contractible.*

PROOF. By Lemma 2.19 we may and will assume that $(S, D, \mathfrak{K})$ is small and D is non–zero, otherwise there is nothing to prove. The hypotheses D connected and $\operatorname{kod}(S, D) = -\infty$ imply that $p_a(D) = 0$, so that Lemma 2.7, (i) applies.

By Corollary 2.18, we may assume that S is not minimal, so there is a (-1)–curve E on S, which, by the smallness of (S, D), is such that $D \cdot E \geqslant 2$.

CLAIM 1. *The divisor* $D + E$ *is* 1*-connected.*

PROOF OF THE CLAIM. The assertion is clear if D does not contain E. Write $D + E = A + B$ with A, B both effective, non–zero. There are two possibilities: (i) $A \geqslant 2E$ and $B \not\geqslant E$, so that $A = A' + 2E$ and $A' \not\geqslant E$; (ii) $A \geqslant E$ and $B \geqslant E$, so that $A = A' + E$, $B = B' + E$ and and $A' \not\geqslant E$, $B' \not\geqslant E$.

In case (i) we have $D = A' + B + E$. Then $A \cdot B = (A' + 2E) \cdot B = (A' + E) \cdot B + E \cdot B \geqslant 1$, because $(A' + E) \cdot B \geqslant 1$ and $E \cdot B \geqslant 0$.

In case (ii) we have $D = A' + B' + E$ hence $E \cdot (A' + B') \geqslant 1$ and we may assume $E \cdot B' \geqslant 1$. Then $A \cdot B = (A' + E) \cdot (B' + E) = A' \cdot (B' + E) + E \cdot (B' + E) = A' \cdot (B' + E) + E \cdot B' - 1 \geqslant A' \cdot (B' + E) \geqslant 1$. □

By Fujita's Lemma (see [**6**]), there is a non–negative integer m such that

$$|E + m(K + D)| \neq \emptyset, \text{ and } |E + (m + 1)(K + D)| = \emptyset. \tag{3.1}$$

CLAIM 2. *One has* $m > 0$.

PROOF OF THE CLAIM. One has $p_a(E + D) = p_a(E) + p_a(D) + D \cdot E - 1 \geqslant 1$ and $D + E$ is 1–connected by Claim 1. Hence $h^0(D + E, \omega_{D+E}) = p_a(D + E) \geqslant 1$. The adjunction exact sequence

$$0 \to \mathcal{O}_S(K) \to \mathcal{O}_S(K + D + E) \to \omega_{D+E} \to 0$$

yields $h^0(S, \mathcal{O}_S(K + D + E)) = h^0(D + E, \omega_{D+E}) > 0$, proving the assertion. □

We take an effective divisor $C \in |E + m(K + D)|$.

Case $C = 0$**.** Then $E \sim -m(K + D)$, hence $-1 = m^2(D + K)^2$, so that $m = 1$, i.e., $-K \sim D + E$, hence $D \cdot E = -E \cdot (K + E) = 2$.

Let $\pi : S \to S'$ be a series of blow–downs of (-1)–curves, the first one being E, with S' minimal. Let $D' = \pi_*(D)$, which is a connected reduced anticanonical divisor, singular at $\pi(E)$. Since π is a birational morphism, it does not blow–up $\mathfrak{K}$, hence $\pi_*(D, \mathfrak{K}) = (D', \mathfrak{K}')$.

Since S' is minimal, one has either $S' = \mathbb{P}^2$ or $S' = \mathbb{F}_n$, with $n \neq 1$. If $S' = \mathbb{P}^2$, then D' is a singular cubic and Corollary 2.17 implies that $(\mathbb{P}^2, D', \mathfrak{K}')$ is contractible. This proves the theorem in this case. Assume now $S' = \mathbb{F}_n$, with $n \neq 1$. If $n \neq 0$, we make a series of elementary transformations based at general points of a component of D' which is not in $|F|$, and reduce to the case $n = 0$. If $n = 0$, the linear system $\mathcal{L}$ of curves in $|E + F|$ with a base point general on a component of D', is a homaloidal system, the birational map $\phi_{\mathcal{L}} : \mathbb{F}_0 \dashrightarrow \mathbb{P}^2$ maps D' to a singular curve of degree 3, and we conclude as in the case $S' = \mathbb{P}^2$.

Case $C \neq 0$. By Lemma 2.9 and since $|2(K+D)| = \emptyset$, one has

$$C \cdot K = -1 + mK \cdot (D+K) \leqslant -1 + mh^0(S, \mathcal{O}_S(-K-D)).$$

If $C \cdot K \geqslant 0$, then $|-K-D| \neq \emptyset$, i.e., $-K-D \sim A$, with A effective and non–zero, because $D+K$ is not effective. On the other hand $C \sim E + m(K+D) \sim E - mA$, thus $m = 0$ against Claim 2.

So $C \cdot K < 0$, hence there is an irreducible component M of C such that $M \cdot K < 0$.

CLAIM 3. *All components of C are smooth and rational. Moreover $M^2 \geqslant 0$ and $M \cdot D \leqslant 1$.*

PROOF OF THE CLAIM. The first assertion follows from $|C+D+K| = \emptyset$.

In order to prove the rest, we first remark that it cannot be the case that $M \cdot D \geqslant 2$ and $M + D$ is 1–connected. Indeed, in this case one has

$$h^0(M+D, \omega_{M+D}) = p_a(M+D) = p_a(M) + p_a(D) + M \cdot D - 1 \geqslant 1. \tag{3.2}$$

Then the adjunction exact sequence

$$0 \to \mathcal{O}_S(K) \to \mathcal{O}_S(K+M+D) \to \omega_{M+D} \to 0 \tag{3.3}$$

yields

$$h^0(S, \mathcal{O}_S(K+D+C)) \geqslant h^0(S, \mathcal{O}_S(K+D+M)) = h^0(M+D, \omega_{M+D}) > 0, \tag{3.4}$$

a contradiction.

If $M^2 \geqslant 0$, then $\dim(|M|) \geqslant 1$, hence there is some irreducible curve $M' \sim M$ which is not contained in D. Then $M \cdot D \leqslant 1$ follows. Otherwise one has $M \cdot D \geqslant 2$ and $M + D$ is 1–connected, leading, as we saw, to a contradiction.

Next we argue by contradiction and assume $M^2 < 0$, hence $M^2 = -1$ because $K \cdot M < 0$.

If M is not contained in D, the same argument as above implies that $M \cdot D \leqslant 1$, against the smallness assumption. Hence we may assume that D contains M simply, because D is reduced. By smallness, we have $M \cdot D \geqslant 2$. Then we claim that $M+D$ is 1–connected, leading again to a contradiction.

To prove that $M+D$ is 1–connected, write $M+D = A+B$, with A, B effective and not zero. Note that $M+D$ contains M with multiplicity 2. If $2M$ is contained in A, then M is not contained in B, and we write $A = 2M + A'$, and $D = M + A' + B$. Since D is connected, we have $(M + A') \cdot B \geqslant 1$ and

$$A \cdot B = (2M + A') \cdot B = (M + A') \cdot B + M \cdot B \geqslant 1.$$

Otherwise, M is contained simply in both A and B, and we write $A = M + A'$, $B = M + B'$, $D = M + A' + B'$, and $A' \cdot B' \geqslant 0$, because A', B' have no common component. One has $2 \leqslant M \cdot D = M^2 + M \cdot (A' + B')$, hence $M \cdot (A' + B') \geqslant 3$. Then

$$A \cdot B = (M + A') \cdot (M + B') = M^2 + M \cdot (A' + B') + A' \cdot B' \geq 2.$$

□

We can now conclude the proof of the theorem. If $M^2 = 0$, then $|M|$ is a base point free pencil of rational curves which determines a morphism $\varphi_{|M|} : S \to \mathbb{P}^1$.

If $M \cdot D = 0$, then D, which is connected, is contained in a fibre. By making $\varphi_{|M|} : S \to \mathbb{P}^1$ *relatively minimal* (i.e., all fibres isomorphic to $\mathbb{P}^1$), we have a birational morphism $f : S \to \mathbb{F}_n$ for some $n \in \mathbb{N}$, which does not blow–up $\mathfrak{K}$.

Either f contracts D to a point, and we are done, or it maps D to a fibre F of $\mathbb{F}_n \to \mathbb{P}^1$, in which case we can still contract it by making an elementary transformation based at a general point of F.

If $M \cdot D = 1$, then $D = \sigma + \mathfrak{f}_1 + \ldots + \mathfrak{f}_h$, where σ is a section of $\varphi_{|M|} : S \to \mathbb{P}^1$ and $\mathfrak{f}_1, \ldots, \mathfrak{f}_h$ are disjoint connected components in different fibres of $\varphi_{|M|} : S \to \mathbb{P}^1$. By making $\varphi_{|M|} : S \to \mathbb{P}^1$ *relatively minimal*, we have a birational morphism $f : S \to \mathbb{F}_n$ for some $n \in \mathbb{N}$, which does not blow-up $\mathfrak{K}$, and $f_*(D, \mathfrak{K}) = (f_*(D), \mathfrak{K}')$, with $f_*(D)$ consisting of a section D' of $f_n : \mathbb{F}_n \to \mathbb{P}^1$ and of distinct fibres $F_1, \ldots, F_k$, with $k \leqslant h$. By applying Proposition 2.16, we see that $(\mathbb{F}_n, f_*(D), \mathfrak{K}')$ is contractible.

If $M^2 > 0$, then $|M|$ is base point free and the morphism $\varphi_{|M|}$ birationally maps S to a minimal rational surface without blowing-up $\mathfrak{K}$. Since $M \cdot D \leqslant 1$, then either $\varphi_{|M|}$ contracts D or it maps D to a line (plus perhaps points) and the contractibility of $(S, D, \mathfrak{K})$ follows again. □

REMARK 3.2. The proof of Theorem 3.1 can be easily adapted to prove also (a stronger version of) Iitaka's Theorem 1.2, to the effect that if $(S, D, \mathfrak{K})$ is a marked triple with S rational and D with snc, at most two irreducible components and $\mathrm{kod}(S, D) = -\infty$, then $(S, D, \mathfrak{K})$ is contractible. We briefly sketch the argument.

As in the proof of Theorem 3.1, we may assume that $(S, D, \mathfrak{K})$ is small. We may suppose D consists of two irreducible component D_1, D_2, with $D_1 \cdot D_2 = 0$, otherwise Theorem 3.1 applies. The curves D_1 and D_2 are smooth and rational. Again we may assume there is a (-1)-curve E such that $D \cdot E \geqslant 2$.

If $D_1 \cdot E = D_2 \cdot E = 1$, we consider the contraction $f : S \to S'$ of E. Then $D' := f_*(D)$ is connected and $\mathrm{kod}(S', D') = -\infty$ by Remark 2.11. So we apply Theorem 3.1 to (S', D') and finish.

Next we can assume that $E \cdot D_1 \geqslant 2$. Then $p_a(D_1 + E) > 0$, hence $h^0(S, \mathcal{O}_S(K + D + E)) \geqslant h^0(S, \mathcal{O}_S(K + D_1 + E)) > 0$. Consider m as in (3.1). Since $h^0(S, \mathcal{O}_S(K + D + E)) > 0$, Claim 2 holds.

Take again $C \in |E + m(K + D)|$, all irreducible components of which are smooth and rational. The discussion of the case $C = 0$ goes through as in the proof of Theorem 3.1. In the case $C \neq 0$, we still find an irreducible component M of C such that $M \cdot K < 0$. Since $0 = h^0(S, \mathcal{O}_S(K + D + C)) \geqslant h^0(S, \mathcal{O}_S(K + D + M))$, then $|K + D + M|$ is empty. This implies that one of the following occurs:
(i) M coincides with one of the curves D_1, D_2;
(ii) M is distinct from D_1 and D_2 and $M \cdot D \leqslant 1$;
(iii) M is distinct from D_1 and D_2 and $M \cdot D_1 = M \cdot D_2 = 1$.

In case (i) assume $M = D_1$. Then $D_1^2 \geqslant 0$, otherwise D_1 is a (-1)-curve, against smallness. By blowing-up points of D_1 we may assume $D_1^2 = 0$. Consider $\varphi_{|D_1|} : S \to \mathbb{P}^1$ so that D_1 is a fibre and D_2 is contained in another fibre. By making $\varphi_{|D_1|}$ relatively minimal and operating with suitable elementary transformations, we dispose of this case.

In case (ii), we have $M^2 \geqslant 0$ by smallness. As above we may assume that $M^2 = 0$. Consider $\varphi_{|M|} : S \to \mathbb{P}^1$, which either maps D_1 and D_2 to curves in fibres or to a curve in a fibre plus a unisecant curve to the fibres. By making $\varphi_{|M|}$ relatively minimal, making elementary transformations and finally applying Proposition 2.16 we finish in this case.

In case (iii), we may assume again $M^2 \geqslant 0$, otherwise we contract M and apply Theorem 3.1 as above. Again we may assume that $M^2 = 0$. Then $\varphi_{|M|} : S \to \mathbb{P}^1$

maps D_1 and D_2 to two unisecant curves to the fibres. By making $\varphi_{|M|}$ relatively minimal and operating with elementary transformations, we can obtain a birational map $f : S \dashrightarrow \mathbb{F}_1$ mapping D_1 to E and D_2 to another unisecant. Then, applying again Proposition 2.16 we finish in this case too.

REMARK 3.3. In [**3**, Proposition 4.9] we proved, by induction on the degree d and by using quadratic transformations, that the union of $d \geq 4$ distinct lines in $\mathbb{P}^2$, with a point P_0 of multiplicity $d-2$ and $2d-3$ nodes, is contractible. Theorem 3.1 allows us to give a faster and more conceptual proof.

Denote by $L_1, \ldots, L_{d-2}$ the lines through P_0 and let L_{d-1}, L_d be the other two lines. Set $P_{i,j} = L_i \cap L_j$ for $i \neq j$. Blow up P_0, $P_{1,d-1}$, $P_{2,d-1}$, $\ldots$, $P_{d-2,d-1}$. Denoting by $\tilde{L}_i$ the strict transform of L_i, $i = 1, \ldots, d$, on the blown-up surface S, it follows that $D = \tilde{L}_1 \cup \cdots \cup \tilde{L}_d$ is connected and $|m(D + K_S)|$ is the strict transform of the linear system of plane curves of degree $md - 3m$ with multiplicity $md - 3m$ at P_0 and multiplicy m at each one of the $d-2$ points $P_{1,d-1}$, $P_{2,d-1}$, $\ldots$, $P_{d-2,d-1}$, that is empty for each $m > 0$. This says that the hypotheses of Theorem 3.1 are satisfied. Hence, (S, D) is contractible.

According to Proposition 4.11 in [**3**], a union of $d \geq 9$ distinct lines in $\mathbb{P}^2$ with a point of multiplicity $d - 3$ and $3(d - 2)$ nodes is not contractible to a set of points by a plane Cremona transformation. Nonetheless one can show that such a configuration of lines is instead contractible to a set of points by a plane Cremona transformation when $d \leq 8$.

Clearly it suffices to show the assertion for $d = 8$. Denote by $L_1, \ldots, L_5$ the lines through the point P_0 of multiplicity $m_0 = 5$ and by L_6, L_7, L_8 the other three lines. Set $P_{i,j} = L_i \cap L_j$ for $i \neq j$. Blow up P_0 and

$$P_{1,7},\ P_{1,8},\ P_{2,7},\ P_{3,6},\ P_{4,6},\ P_{4,8},\ P_{5,6},\ P_{5,8},\ P_{6,7},\ P_{6,8} \text{ and } P_{7,8}. \tag{3.5}$$

Denoting by $\tilde{L}_i$ the strict transform of L_i, $i = 1, \ldots, 8$, it follows that $D = \tilde{L}_1 \cup \cdots \cup \tilde{L}_8$ is connected and $|m(D + K_S)|$, with m a positive integer, is the strict transform of the linear system $\mathcal{L}_m$ of plane curves of degree $5m$ with multiplicity at least $4m$ at P_0 and multiplicity at least m at the eleven points (3.5). One can show that the system $\mathcal{L}_m$ is empty for all positive integers m, proving that $\mathrm{kod}(S, D) = -\infty$ hence that (S, D) is contractible.

4. Basics of the theory of peeling

In this section we recall, for the reader's convenience, the basics of the theory of peeling, referring to [**14, 15**] for the proofs of the results we will mention and use.

4.1. A few definitions. Let (S, D) be a pair with D reduced. We denote by $M(D)$ the *intersection matrix* of D, i.e., the symmetric matrix of order $\ell = \ell(D)$ with entries $m_{ij} = D_i \cdot D_j$, for $1 \leqslant i, j \leqslant \ell$. To say that $M(D)$ is negative definite [resp. semidefinite], we write $M(D) < 0$ [resp $M(D) \leqslant 0$]. The matrix $M(D)$ depends on the ordering of the components of D, but its rank and its being negative definite (or semidefinite) do not.

For any curve $C \leqslant D$, one defines its *branching number* in D as $\beta_D(C) = C \cdot (D - C)$. If C is a component of D, this is the valence of the vertex corresponding to C in $\mathfrak{G}(C)$.

A tree D is a *chain* if each vertex of $\mathfrak{G}(D)$ has valency 1 or 2 (equivalently, D is a tree with only two terminal components). Given (S, D), a chain $C \leqslant D$ which is a connected component of D is called a *rod* of D. A chain $T \leqslant D$ is called a *twig* of D if $\beta_D(T) = 1$ and T meets $D - T$ in a point of a terminal component of T. A twig of D is *maximal* if it cannot be extended to a twig of D with more components.

One says that $C \leqslant D$ is a *fork* if C is a connected component of D, if all components of C are smooth and rational and if $\mathfrak{G}(C)$ is one of the graphs in [**15**, pp. 436–437], see also [**14**, Lemma 3.4.1].

A curve D is said to be *admissible* if all of its components C are smooth, rational with $C^2 \leqslant -2$. A smooth rational component C of D with $C^2 \geqslant -1$ is called *irrelevant*.

4.2. Peeling the bark and almost minimal models. From now on we will consider log smooth pairs (S, D) with S rational and all components of D rational.

Let $C \leqslant D$ be an admissible twig, rod or fork, with $\ell(C) = \ell$ and write $C = C_1 + \ldots + C_\ell$ as the sum of its irreducible components. Then one has $M(C) < 0$ and one can uniquely determine an effective $\mathbb{Q}$–divisor

$$\mathrm{Bk}(C) = \gamma_1 C_1 + \ldots + \gamma_\ell C_\ell,$$

called the *bark* of C, such that

$$(D - \mathrm{Bk}(C) + K_S) \cdot C_i = 0, \text{ for } 1 \leqslant i \leqslant \ell.$$

One has

$$0 < \gamma_i \leqslant 1, \text{ for } 1 \leqslant i \leqslant \ell \tag{4.1}$$

Property 1. The equality on the right in (4.1) holds for some $i \in \{1, \ldots, \ell\}$ if and only if $\gamma_i = 1$ for all $i \in \{1, \ldots, \ell\}$, in which case C is either a rod or a fork and $C_i^2 = -2$, for all $i \in \{1, \ldots, \ell\}$. In this case we say that C is a (-2)*–rod* or a (-2)*–fork*.

The process of subtracting $\mathrm{Bk}(C)$ from D is called the *peeling* of $\mathrm{Bk}(C)$ out of D.

Consider the sets $\{T_1, \ldots, T_t\}$, $\{R_1, \ldots, R_r\}$ and $\{F_1, \ldots F_f\}$ of maximal admissible twigs, of admissible rods and of admissible forks respectively. These curves are all pairwise disjoint, so we can peel their barks independently out of D, and we obtain

$$D = D^\sharp + \mathrm{Bk}(D),$$

where

$$\mathrm{Bk}(D) := \sum_{i=1}^{t} \mathrm{Bk}(T_i) + \sum_{i=1}^{r} \mathrm{Bk}(R_i) + \sum_{i=1}^{f} \mathrm{Bk}(F_i)$$

is called the *bark* of D.

Property 2. One has:

(i) $D^\sharp$ is an effective (perhaps 0) $\mathbb{Q}$–divisor and $\mathrm{Supp}(\mathrm{Bk}(D))$ contains no smooth rational curve C with $C^2 \geqslant -1$;

(ii) $M(\mathrm{Bk}(D)) < 0$;

(iii)

$$(K_S + D^\sharp) \cdot C = 0, \text{ for all components } C \text{ of } \mathrm{Supp}(\mathrm{Bk}(D)),$$

whereas

$$(K_S + D^\sharp) \cdot C \geqslant 0, \text{ for all components } C \text{ of } D - \mathrm{Supp}(\mathrm{Bk}(D)),$$

except for irrelevant components of non–admissible twigs, rods and forks;
(iv)

$$h^0(S, \mathcal{O}_S(n(D+K_S))) = h^0(S, \mathcal{O}_S([n(D^\sharp + K_S)])), \text{ for every integer } n \geqslant 0,$$

where [ ] denotes the *integral part.*

From

$$D + K_S = (D^\sharp + K_S) + \mathrm{Bk}(D),$$

and from the fact that $\mathrm{Bk}(D)$ is effective and that nef divisors are pseudo–effective, one has that

$$D^\sharp + K_S \text{ nef} \implies \mathrm{kod}(S, D) \geqslant 0. \tag{4.2}$$

The pair (S, D) is said to be *almost minimal* if, for every irreducible curve C on S, either $(D^\sharp + K) \cdot C \geqslant 0$ or $(D^\sharp + K) \cdot C < 0$ and $M(C + \mathrm{Bk}(D))$ is not negative definite.

PROPERTY 3. A curve C which offends almost minimality of (S, D) is a (-1)–curve which can be contracted without offending log smoothness.

Hence one has:

THEOREM 4.1. *Let (S, D) be log smooth, with S rational and all components of D rational. There is a log smooth, almost minimal pair $(\tilde{S}, \tilde{D})$, such that:*
(i) *there is a birational morphism $\mu : S \to \tilde{S}$ such that $\tilde{D} = \mu_*(D)$;*
(ii)

$$P_n(S, D) = P_n(\tilde{S}, \tilde{D}), \text{ for every integer } n \geqslant 0,$$

in particular

$$\mathrm{kod}(S, D) = \mathrm{kod}(\tilde{S}, \tilde{D}).$$

The pair $(\tilde{S}, \tilde{D})$ of Theorem 4.1 is called an *almost minimal model* of (S, D).

As at the end of §2.7, one can start with a marked triple $(S, D, \mathfrak{K})$ with (S, D) log smooth. If (S, D) is almost minimal, then $(S, D, \mathfrak{K})$ will be also said to be *almost minimal.* If $(\tilde{S}, \tilde{D})$ is an almost minimal model of (S, D), with $\mu : S \to \tilde{S}$, we define $\tilde{\mathfrak{K}}$ to be the union of $\mathrm{cl}_\mu(\mathfrak{K})$ and of $\mathfrak{K}_\mu$ (recall Remark 2.2 and §2.3). The almost minimal triple $(\tilde{S}, \tilde{D}, \tilde{\mathfrak{K}})$ will be said to be an *almost minimal model* of $(S, D, \mathfrak{K})$.

4.3. The bark contraction. By Property 2, (ii) and by [**14**, Lemma 5.2.3, Chapt. 1], there is a birational morphism $\phi : S \to \bar{S}$, such that:
(i) $\bar{S}$ is a normal projective surface, with rational singular points, ϕ induces an isomorphism between $S - \mathrm{Supp}(\mathrm{Bk}(D))$ and $\bar{S} - \mathrm{Sing}(\bar{S})$, and each singular point of $\bar{S}$ corresponds bijectively to a connected component of $\mathrm{Supp}(\mathrm{Bk}(D))$, which is contracted there;
(ii) there is a positive integer N such that for any Weil divisor Z of $\bar{S}$, NZ is a Cartier divisor. For any Weil divisor Z of $\bar{S}$, one defines the numerical equivalence class

$$\phi^*(Z) := \frac{\phi^*(NZ)}{N} \in \mathrm{NS}(S) \otimes_{\mathbb{Z}} \mathbb{Q}$$

which is independent of N, and the intersection of Weil divisors Z, Z' of $\bar{S}$ as

$$Z \cdot Z' = \frac{\phi^*(Z) \cdot \phi^*(Z')}{N^2} \in \mathbb{Z}\frac{1}{N^2}.$$

Thus, the *Néron–Severi* $\mathbb{Q}$–vector space $\mathrm{NS}(\bar{S})$, consisting of the numerical equivalence classes of Weil $\mathbb{Q}$–divisors on $\bar{S}$, is well defined, and its rank is, as usual, denoted by $\rho(\bar{S})$;
(iii) one has

$$K_{\bar{S}} = \phi_*(K_S), \quad D^\sharp + K_S = \phi^*(\bar{D} + K_{\bar{S}}), \quad \text{where } \bar{D} := \phi_*(D^\sharp). \tag{4.3}$$

The pair $(\bar{S}, \bar{D})$ is called the *bark-contraction* of (S, D).

Mori theory can be developed on $\bar{S}$. Let $\overline{\mathrm{NE}}(\bar{S}) \subset \mathrm{NS}(\bar{S}) \otimes_{\mathbb{Q}} \mathbb{R}$ be the *Mori cone*, i.e., the smallest convex cone (closed under multiplication by $\mathbb{R}_+$) containing the classes of all irreducible curves on $\bar{S}$. Let $\bar{L}$ be an ample Cartier divisor on $\bar{S}$. For any positive $\epsilon \in \mathbb{R}$, define

$$\overline{\mathrm{NE}}_\epsilon(\bar{D}, \bar{S}) := \{Z \in \overline{\mathrm{NE}}(\bar{S}) : (\bar{D} + K_{\bar{S}}) \cdot Z \geqslant -\epsilon(\bar{L} \cdot Z)\}.$$

One has:

THEOREM 4.2 (The Cone Theorem). *For any positive $\epsilon \in \mathbb{R}$, there are (not necessarily smooth) rational curves $\bar{\ell}_1, \dots, \bar{\ell}_r$ on $\bar{S}$, such that*

$$\overline{\mathrm{NE}}(\bar{S}) = \sum_{i=1}^{r} \mathbb{R}_+[\bar{\ell}_i] + \overline{\mathrm{NE}}_\epsilon(\bar{D}, \bar{S}) \tag{4.4}$$

and

$$0 > (\bar{D} + K_{\bar{S}}) \cdot \bar{\ell}_i \geqslant -3, \quad \textit{for } 1 \leqslant i \leqslant r.$$

Since $\overline{\mathrm{NE}}(\bar{S})$ is polyhedral in the half–space where $\bar{D} + K_{\bar{S}} + \epsilon\bar{L} < 0$, one can define the concept of *extremal ray* of $\overline{\mathrm{NE}}(\bar{S})$ as in the smooth case. Thus, if $r > 0$ in (4.4), we may assume that $\mathbb{R}_+[\bar{\ell}_i]$ are extremal rays for $1 \leqslant i \leqslant r$.

One has:

PROPOSITION 4.3. *In the above setting, let $\mathbb{R}_+[\bar{\ell}]$ be an extremal ray of $\overline{\mathrm{NE}}(\bar{S})$. Let ℓ be the proper transform of $\bar{\ell}$ on S. Then one of the following facts occurs:*
(i) $M(\ell + \mathrm{Bk}(D)) \leqslant 0$ but not $M(\ell + \mathrm{Bk}(D)) < 0$: then $\bar{\ell}^2 = 0$ and, for $n \gg 0$, the linear system $|nN\bar{\ell}|$ is base point free and composed with a pencil $|f|$ whose general member is isomorphic to $\mathbb{P}^1$. If $\mathfrak{f} = \phi^*(f)$, then $|\mathfrak{f}|$ is a base point free pencil of rational curves on S;
(ii) $\overline{\mathrm{NE}}(\bar{S}) \otimes \mathbb{Q}$ is generated by the class of $\bar{\ell}$, hence $\rho(\bar{S}) = 1$ and $-(\bar{D} + K_{\bar{S}})$ is ample. Then $-(D^\sharp + K_S)$ is nef and big and for an irreducible curve C of S one has $C \cdot (D^\sharp + K_S) = 0$ if and only if C is an irreducible component of $\mathrm{Bk}(D)$.

In case (ii) of Proposition 4.3, one says that (S, D) (or $(\bar{S}, \bar{D})$) is a *logarithmic del Pezzo surface of rank 1*. If $\bar{D} = 0$ (equivalently, if $D = \mathrm{Supp}(\mathrm{Bk}(D))$), we say that (S, D) has *shrinkable boundary* (in [**15**] is used a different terminology, which would be confusing here).

The classification of logarithmic del Pezzo surface of rank 1 is still an open problem in its generality. Some properties of those with shrinkable boundary are described in [**15**, §4]. Note, however, that [**15**, Conjecture 16] about them turned out to be false soon after, see [**16**].

Keel and McKernan gave in [**9**] a classification theorem (Theorem 23.2), which applies to all but a bounded family of rank one logarithmic del Pezzo surfaces.

5. The peeling approach to the contraction theorem

In this section we prove the following:

THEOREM 5.1. *Let $(S, D, \mathfrak{K})$ be a marked triple, with S rational and* $\mathrm{kod}(S, D) = -\infty$. *Then $(S, D, \mathfrak{K})$ is contractible unless, perhaps, $-(D^\sharp + K_S)$ is nef and big and one of the following occurs:*
(1) (S, D) is a logarithmic del Pezzo surface of rank 1 with shrinkable boundary;
(2) if $(\tilde{S}, \tilde{D})$ is an almost minimal model of (S, D), then a connected component of $\tilde{D}$ (and only one) is a non-admissible fork.

By Corollary 2.8, $\mathrm{kod}(S, D) = -\infty$ implies that (S, D) is log smooth and that, for any connected curve $C \leqslant D$, one has $p_a(C) = 0$. Since $\mu : S \to \tilde{S}$ is a morphism, which does not blow-up $\mathfrak{K}$, we can pretend that $(S, D, \mathfrak{K})$ is almost minimal (see the end of §4.2).

The proof of Theorem 5.1 will consist in a number of steps. The first one is to show that we are in position to apply Proposition 4.3.

LEMMA 5.2. *If (S, D) is almost minimal and* $\mathrm{kod}(S, D) = -\infty$, *then $\bar{D} + K_{\bar{S}}$ is not nef.*

PROOF. By (4.2), $D^\sharp + K_S$ is not nef. Hence there is an irreducible curve Z on S such that $(D^\sharp + K_S) \cdot Z < 0$. Set $\bar{Z} = \phi_*(Z)$ and note that $\bar{Z}$ is non-zero, because Z is not in $\mathrm{Supp}(\mathrm{Bk}(D))$ (see Property 2–(iii)). We claim that $(\bar{D} + K_{\bar{S}}) \cdot \bar{Z} < 0$. Indeed, by (4.3), we have

$$(\bar{D} + K_{\bar{S}}) \cdot \bar{Z} = \frac{1}{N}(D^\sharp + K_S) \cdot \phi^*(N\bar{Z}).$$

One has $\phi^*(N\bar{Z}) = NZ + E$, where E is contracted by ϕ, hence $\mathrm{Supp}(E) \leqslant \mathrm{Supp}(\mathrm{Bk}(D))$. By Property 2–(iii), one has $(D^\sharp + K_S) \cdot E = 0$, hence

$$(\bar{D} + K_{\bar{S}}) \cdot \bar{Z} = (D^\sharp + K_S) \cdot Z < 0.$$

□

By Lemma 5.2 and Proposition 4.3, there are extremal rays in $\mathrm{NS}(\bar{S})$. The proof of Theorem 5.1 consists in the discussion of the two cases corresponding to (i) and (ii) of Proposition 4.3.

5.1. The proof of Theorem 5.1 in case (i) of Proposition 4.3. As in (i) of Proposition 4.3, let $\mathfrak{f} = \phi^*(f)$. Since $|f|$ is base point free on $\bar{S}$, then $\mathfrak{f} \cdot \mathrm{Bk}(D) = 0$. From $f \cdot (\bar{D} + K_{\bar{S}}) < 0$ we have $\mathfrak{f} \cdot (D + K_S) = \mathfrak{f} \cdot (D^\sharp + K_S) < 0$. Since $\mathfrak{f} \cdot K_S = -2$ and $\mathfrak{f}$ is nef, we find $\mathfrak{f} \cdot D \leqslant 1$. The argument is now similar to the one at the end of the proof of Theorem 3.1.

Indeed, by making $\varphi_{|\mathfrak{f}|} : S \to \mathbb{P}^1$ *relatively minimal*, we have a birational morphism $f : S \to \mathbb{F}_n$ for some $n \in \mathbb{N}$, which does not blow-up $\mathfrak{K}$. Set $f_*(D, \mathfrak{K}) = (f_*(D), \mathfrak{K}')$. Since $\mathfrak{f} \cdot D \leqslant 1$, we have that:
(a) either $f_*(D) = 0$,
(b) or $f_*(D)$ consists of finitely many distinct curves of $|\mathfrak{f}|$,
(c) or $f_*(D)$ consists of finitely many (may be 0) curves of $|\mathfrak{f}|$, plus an irreducible curve Γ such that $\Gamma \cdot \mathfrak{f} = 1$.

In case (a), the proof of Theorem 5.1 is finished. In case (b) we can contract $f_*(D) = \mathfrak{f}_1 + \ldots + \mathfrak{f}_\ell$, with $\mathfrak{f}_i \in |\mathfrak{f}|$, for $1 \leqslant i \leqslant \ell$ with a series of elementary transformations based at general points of $\mathfrak{f}_1, \dots, \mathfrak{f}_\ell$ respectively. In case (c) we proceed as in the final part of the proof of Theorem 3.1.

5.2. The proof of Theorem 5.1 in case (ii) of Proposition 4.3. Here we are in the rank 1 logarithmic del Pezzo case. The shrinkable boundary case corresponds to (1) in Theorem 5.1. So we are left with the non–shrinkable boundary case.

From [**14**, Lemma 1.14.5, Lemma 3.14.6 and Theorem 3.15.1] and [**15**, Lemma 11 and Theorem 12], we have that $Y := D - \mathrm{Supp}(\mathrm{Bk}(D))$ is connected and consists of one or two irreducible components. Precisely:
(i) If Y is irreducible, then:
(i-a) either Y is an irrelevant component of a rod R, hence $m := Y^2 \geqslant -1$, and $R - Y$ consists of one or two admissible twigs;
(i-b) or Y is the central component of a non–admissible fork;
(ii) if Y has two irreducible components Y_1, Y_2, then $Y_1 \cdot Y_2 = 1$, $Y_1 + Y_2$ belongs to a rod R and $R - Y_1 - Y_2$ consists of one or two admissible twigs, accordingly $Y_i^2 \geqslant -1$ for $i = 1, 2$.

If (i-b) holds we are in case (2) of Theorem 5.1.

Suppose we are in case (i-a). Write $R = Y + T_1 + T_2$, where T_1, T_2 are two distinct admissible twigs, with $0 \leqslant \ell(T_1) \leqslant \ell(T_2)$. If $\ell(T_i) \neq 0$, we denote by Y_i the component of T_i which intersects Y, and by P_i the intersection point of Y_i with Y, for $i = 1, 2$.

Assume $m \geqslant 1$ and $\ell(T_i) > 0$ for $i = 1, 2$ (otherwise the proof is similar, but easier). By blowing up $m - 1$ general points of Y, we may assume that $m = 1$. Then the morphism $\phi_{|Y|} : S \to \mathbb{P}^2$ is birational, contracts $D - Y - Y_1 - Y_2$ to points and maps Y, Y_1, Y_2 to three lines C, C_1, C_2 respectively, and we are reduced to the case of $(\mathbb{P}^2, C + C_1 + C_2, \mathfrak{C})$. This is contractible by Corollary 2.17.

Let now $m = 0$, so that $|Y|$ is a base point free pencil of rational curves. By making a series of elementary transformations at P_1, we can increase Y_2^2 as much as we like and make it positive. Then, to finish he proof, we can argue as in the case $m \geqslant 1$.

Finally, assume $m = -1$. We note that, since (S, D) is almost minimal, $M(R)$ is not negative definite. Then we blow down Y and consecutively all (-1)–curves which appear in the image of R. Since $M(R)$ is not negative definite, after a finite number of steps we find a rod with a smooth rational component of non–negative self–intersection. Then we can finish as above.

Case (ii) can be treated in a similar way. Thus Theorem 5.1 is proved in this case too.

References

[1] Maria Alberich-Carramiñana, *Geometry of the plane Cremona maps*, Lecture Notes in Mathematics, vol. 1769, Springer-Verlag, Berlin, 2002. MR1874328

[2] Alberto Calabri and Ciro Ciliberto, *On Cremona contractibility*, Rend. Semin. Mat. Univ. Politec. Torino **71** (2013), no. 3-4, 389–400. MR3506393

[3] Alberto Calabri and Ciro Ciliberto, *On the Cremona contractibility of unions of lines in the plane*, Kyoto J. Math. **57** (2017), no. 1, 55–78. MR3621779

[4] Guido Castelnuovo and Federigo Enriques, *Sulle condizioni di razionalità dei piani doppi*, Rend. Circ. Mat. Palermo **14** (1900), 290–302.

[5] Julian Lowell Coolidge, *A treatise on algebraic plane curves*, Dover Publications, Inc., New York, 1959. MR0120551

[6] Takao Fujita, *On Zariski problem*, Proc. Japan Acad. Ser. A Math. Sci. **55** (1979), no. 3, 106–110. MR531454

[7] Shigeru Iitaka, *Characterization of two lines on a projective plane*, Algebraic geometry (Tokyo/Kyoto, 1982), Lecture Notes in Math., vol. 1016, Springer, Berlin, 1983, pp. 432–448. MR726437

[8] Shigeru Iitaka, *Classification of reducible plane curves*, Tokyo J. Math. **11** (1988), no. 2, 363–379. MR976572

[9] Seán Keel and James McKernan, *Rational curves on quasi-projective surfaces*, Mem. Amer. Math. Soc. **140** (1999), no. 669, viii+153. MR1610249

[10] Hideo Kojima and Takeshi Takahashi, *Reducible curves on rational surfaces*, Tokyo J. Math. **29** (2006), no. 2, 301–317. MR2284974

[11] N. Mohan Kumar and M. Pavaman Murthy, *Curves with negative self-intersection on rational surfaces*, J. Math. Kyoto Univ. **22** (1982/83), no. 4, 767–777. MR685529

[12] Giuseppe Marletta, *Sulla identità cremoniana di due curve piane*, Rend. Circolo Mat. Palermo, **24** (1907), no. 1, 223–242.

[13] Giuseppe Marletta, *Sui sistemi aggiunti dei varii indici alle curve piane*, Rend. del R. Ist. Lombardo di Scienze e Lettere, (2) **43** (1910), 781–804.

[14] Masayoshi Miyanishi, *Open algebraic surfaces*, CRM Monograph Series, vol. 12, American Mathematical Society, Providence, RI, 2001. MR1800276

[15] Masayoshi Miyanishi and Shuichiro Tsunoda, *Open algebraic surfaces with Kodaira dimension* $-\infty$, Algebraic geometry, Bowdoin, 1985 (Brunswick, Maine, 1985), Proc. Sympos. Pure Math., vol. 46, Amer. Math. Soc., Providence, RI, 1987, pp. 435–450. MR927966

[16] Masayoshi Miyanishi and De-Qi Zhang, *Gorenstein log del Pezzo surfaces of rank one*, J. Algebra **118** (1988), no. 1, 63–84. MR961326

[17] Giuseppe Pompilj, *Sulle trasformazioni cremoniane del piano che posseggono una curva di punti uniti*, Rend. del Sem. Mat. dell'Univ. di Roma (4) **2** (1937), 3–43.

Dipartimento di Matematica e Informatica, Università degli Studi di Ferrara, Via Machiavelli 30, 44121 Ferrara, Italy

Email address: `alberto.calabri@unife.it`

Dipartimento di Matematica, Università degli Studi di Roma "Tor Vergata", Via della Ricerca Scientifica, 00133 Roma, Italy

Email address: `cilibert@mat.uniroma2.it`

Contemporary Mathematics
Volume **712**, 2018
http://dx.doi.org/10.1090/conm/712/14341

On the canonical map of some surfaces isogenous to a product

Fabrizio Catanese

Dedicated to Lawrence Ein on the occasion of his 60th birthday.

Abstract. We construct several families (indeed, connected components of the moduli space) of surfaces S of general type with $p_g = 5, 6$ whose canonical map has image Σ of very high degree, $d = 48$ for $p_g = 5$, $d = 56$ for $p_g = 6$. And a connected component of the moduli space consisting of surfaces S with $K_S^2 = 40, p_g = 4, q = 0$ whose canonical map has always degree ≥ 2, and, for the general surface, of degree 2 onto a canonical surface Y with $K_Y^2 = 12, p_g = 4, q = 0$.

The surfaces we consider are SIP 's, i.e. surfaces S isogenous to a product of curves $(C_1 \times C_2)/G$; in our examples the group G is elementary abelian, $G \cong (\mathbb{Z}/m)^k$. We also establish some basic results concerning the canonical maps of any surface isogenous to a product, basing on elementary representation theory.

Our methods and results are a first step towards answering the question of existence of SIP 's S with $p_g = 6, q = 0$ whose canonical map embeds S as a surface of degree 56 in $\mathbb{P}^5$.

Contents

2010 *Mathematics Subject Classification.* Primary 14J29, 14J10, 14M07.

Key words and phrases. Canonical maps, canonical surfaces, Galois coverings, Surfaces isogenous to a product.

The present work took place in the framework of the ERC Advanced grant n. 340258, 'TADMICAMT'.

1. Introduction and state of the art on canonical surfaces of high degree in $\mathbb{P}^3, \mathbb{P}^4, \mathbb{P}^5$

Let S be a complex minimal projective surface of general type, and denote by $p_g(S) := dimH^0(\mathcal{O}_S(K_S))$, and by Φ the rational map to $\mathbb{P}^{p_g-1}$ associating to a point $x \in S$ the point in $\mathbb{P}(H^0(\mathcal{O}_S(K_S)))^\vee$ obtained by evaluation of sections of $H^0(\mathcal{O}_S(K_S))$ in x.

If the image of Φ is a surface Σ (this is the case if the canonical system $|K_S|$ has no base points) one has a well known inequality:

$$K_S^2 \geq deg(\Phi)deg(\Sigma),$$

equality holding iff $|K_S|$ has no base points.

DEFINITION 1.1. We define the **canonical degree** of S as $d := deg(\Sigma)$ if Σ is a surface, $d = 0$ otherwise, and we define the **canonical volume** of S to be K_S^2. Hence $K_S^2 \geq d$, equality holding if and only if $|K_S|$ has no base points and Φ is birational onto its image (i.e., $deg(\Phi) = 1$).

S is said to be a **(simple) canonical surface** if Φ is birational onto its image Σ. This obviously implies that $p_g(S) \geq 4$.

QUESTION 1.2. What is the maximal canonical degree d for a fixed value of p_g?

What is the maximal canonical volume K_S^2?

In particular, for $p_g = 4, 5, 6$?

Recall the Castelnuovo inequality, holding if Φ is birational (onto its image Σ):

$$K_S^2 \geq 3p_g(S) - 7,$$

and the Bogomolov-Miyaoka-Yau inequality

$$K_S^2 \leq 9\chi(S) = 9 + 9p_g(S) - 9q(S),$$

where $q := dimH^1(\mathcal{O}_S)$ (if $q = 0$ S is said to be **regular**).

By virtue of these inequalities, under the assumptions of question 1.2 one must have:

$$d \leq K_S^2 \leq 9(1 + p_g).$$

In particular, the upper bound for the volume is $K_S^2 \leq 45, 54, 63$ for $p_g = 4, 5, 6$.

In order to have high volume it is better to have $q(S) = 0$, even if also the canonical maps of irregular surfaces are a very interesting subject of investigation, see [**C-S02**]. For instance Cesarano in his Ph.D. work considers surfaces S which are a polarization of type $(1, 2, 2)$ on an Abelian threefold, which are canonical surfaces in $\mathbb{P}^5$ with $K_S^2 = 24$, and he seems to have shown that for a general such surface the canonical map is an embedding.

QUESTION 1.3. What is the maximal canonical degree d of a surface for which Φ is an embedding?

In particular, for $p_g = 6$?

(Note that in case of an embedding $d = K^2$).

Indeed, question 1.3 is easily answered for $p_g = 4$, here only the case $d = 5$ is possible, while, for $p_g = 5$, it must be $d = 8, 9$, and in all the three cases Σ is a complete intersection, of respective types $(5), (4, 2), (3, 3)$, see [**Cat97**], propositions 6.1 and 6.2, corollary 6.3., and [**Cat16**].

Concerning the case $p_g = 6$, in the article [**Cat97**] methods of homological algebra were used to construct embedded canonical surfaces with low degree $11 \leq K_S^2 \leq 17$, and also to attempt a classification of them. Recently, M. and G. Kapustka constructed in [**Kap-Kap15**] such canonical surfaces of degree $K_S^2 = 18$, using the method of bilinkage. In [**Cat16**] we constructed a family of embedded canonical surfaces with $d = 24, q = 0$.

Concerning question 1.2, in the case $p_g = 4$, against Enriques' expectation $K_S^2 \leq 24$, in [**BC08**] we constructed a surface with $p_g(S) = 4, q(S) = 0, K_S^2 = 45$ (and with K_S ample being a ball quotient): but here the canonical system has base curves, S is canonical and we have $d = 19$. Surfaces with $p_g(S) = 6, q(S) = 2, K_S^2 = 45 > d$ (again ball quotients) were constructed in [**BC08**], but the canonical degree d was not calculated.

For $p_g = 4$, the current record for the canonical degree is $d = 28$ [**Cat99**], and all possible degrees in the interval $[5, 28]$ occur: Ciliberto got the interval $[5, 10]$ in [**Cil81**], Burniat the interval $[11, 16]$ in [**Burn66**], while the interval $[11, 28]$ was gotten in [**Cat99**]. Liedtke [**Lied03**] got $K_S^2 = 31$, but $d = 12$.

In this paper we construct surfaces with very high canonical degrees, $p_g = 5, d = 48$, and $p_g = 6, d = 56$.

Indeed, the maximum degree a priori possible $d = 9(p_g + 1)$ can only occur for ball quotients with $q = 0$, but these are difficult to construct, and it is even more difficult to describe their canonical map.

We obtain $d = 8(p_g + 1)$ using surfaces isogenous to a product with $q = 0$, for these the volume is $K^2 = 8(p_g + 1)$, and the only difficulty is to describe their canonical maps.

Recall that surfaces isogenous to a product are quotients $(C_1 \times C_2)/G$ for the free action of a finite group G. Their classification is a hard enterprise (but it was achieved for $p_g = 0$ in [**BCG08**], see also [**Frap13**]).

In our paper we consider easier cases where the group G is Abelian (but we do not attempt a complete classification of the case G Abelian and $p_g = 4, 5, 6$). We lay moreover the foundations for the description of the canonical maps of surfaces isogenous to a product.

For this purpose a first step that we achieve is to decide when does a subrepresentation $V \subset \mathbb{C}[G]$ (i.e., a $\mathbb{C}[G]$-submodule) yield a projective embedding of the finite group G. Another open question is:

QUESTION 1.4. Does there exist a surface S isogenous to a product and with $p_g(S) = 6$ such that the canonical map $\Phi : S \to \mathbb{P}^5$ is an embedding?

Our main results in this article are the following ones, here written without full details (these will be given in the later sections).

Theorem 4.3. *I) There exists a connected component of the moduli space of minimal surfaces of general type with $p_g = 6, q = 0, K_S^2 = 56$, consisting of surfaces isogenous to a product with $G = (\mathbb{Z}/2)^3$, such that the canonical map for each surface S is a non injective birational morphism onto a canonical image Σ of degree 56 in $\mathbb{P}^5$.*

II) There exists a connected component of the moduli space of the minimal surfaces of general type with $p_g = 5, q = 0, K_S^2 = 48$, consisting of surfaces isogenous to a product with $G = (\mathbb{Z}/2)^3$, such that the canonical map of S is a non injective *birational morphism onto a canonical image* Σ *of degree* 48 *in* $\mathbb{P}^4$, *for general choice of the branch points for* $C_2 \to \mathbb{P}^1$.

There are other families of surfaces, beyond those mentioned in theorem 4.3, that achieve the same d. The families are here indicated, but the details are not worked out for them.

The next theorem provides new features of an example which in the past had not been considered possible (see [**Bab34**], [**Bea79**], [**Cat81**]): a canonical map of degree greater or equal to 2 onto a canonical surface. Previous examples (sometimes the same) appeared in [**G-Z77**], [**Bea79**], [**Cat81**], [**Par91**], [**Tan92**], [**CMLR15**], [**Bea13**]. In the previous examples the canonical map Φ was ramified only on a finite set, and Φ was birational for a general deformation of the surfaces.

Theorem 5.1. *There is a connected component of the moduli space of minimal surfaces of general type with* $p_g = 4, q = 0, K_S^2 = 40$, *consisting of surfaces isogenous to a product with* $G = (\mathbb{Z}/2)^3$, *such that for each of them there is a nontrivial involution* $\psi := (e, 1) \in G \times G$ *which acts trivially on the canonical system, and with quotient a nodal surface* Z *with* $K_Z^2 = 12$, *and with* K_Z *ample.*

The canonical map of S *factors through the canonical map of* Z, *which is a morphism, and is birational onto its image for general choice of the six branch points of* $C_1 \to C_1/G = \mathbb{P}^1$.

The fixed locus of ψ *consists of* 14 *isolated points, and two curves* $F_1' + F_2'$ *which form the base locus of the canonical system .*

The last result we mention in the introduction is meant to give a flavour of the methods developed in section 2:

Proposition 2.3. *A subrepresentation* $V \subset \mathbb{C}[G]$ *yields a map of* G *into* $\mathbb{P}(V^\vee)$ *iff* $V \neq 0$; *it embeds* G *only if the following properties i) and ii) hold.*

i) Writing $V = \sum_{\chi \in Irr} V_\chi, V_\chi = W_\chi \otimes U_\chi$, *the representation*

$$\rho := \otimes_{U_\chi \neq 0} \rho_\chi$$

is faithful.

ii) Let $\chi_1, \dots, \chi_m$ *be the irreducible representations for which* $U_\chi \neq 0$, *and notice that* ρ_j *sends the centre* $Z(G)$ *of* G *to the set of scalar multiples of the identity, via a character* ψ_j *of* $Z(G)$. *Then the characters* $\psi_1, \dots, \psi_j$ *are affine generators of the group of characters* $Z(G)^*$ *(i.e.,* $\frac{\psi_j}{\psi_1}$ *generate the group* $Z(G)^*$*).*

If moreover V *is full (each space* $U_\chi \neq 0$ *is* $= W_\chi^\vee$*), the two conditions i) and ii) are also sufficient.*

If V *is not full, we obtain a projective embedding of* G *if* V *is projectively general (i.e., each space* $U_\chi \neq 0$ *is projectively general).*

2. Representations yielding an affine, respectively a projective embedding, of a finite group G

In this section G is a finite group, and V is a finite dimensional complex vector space which is a representation of G (indeed, one could replace $\mathbb{C}$ by an algebraically closed field K of characteristic prime to the order of G).

Assume that we have a linear map of representations

$$\epsilon : V \to \mathbb{C}[G] = \{\sum_{g \in G} a_g g | a_g \in \mathbb{C}\},$$

where we view the group algebra $\mathbb{C}[G]$ as the space of complex valued functions on G, and g denotes for brevity the characteristic function of an element $g \in G$. Equivalently, V is a left $\mathbb{C}[G]$-module and ϵ is a homomorphism of $\mathbb{C}[G]$-modules.

2.1. Embedding G. The first question is: when does V separate points of G? I.e., we ask when is the map $\epsilon^* : G \to V^\vee$, obtained composing the inclusion $G \subset \mathbb{C}[G]^\vee$ with the dual map $\epsilon^\vee$, injective?

The second question is: when does $\mathbb{P}(\epsilon^*)$ yield a projective embedding of G inside $\mathbb{P}(V^\vee)$?

Replacing V by its image inside $\mathbb{C}[G]$, we may assume in this section, without loss of generality, that $V \subset \mathbb{C}[G]$.

A useful observation is that if $V \subset V' \subset \mathbb{C}[G]$ and V separates points of G, then so does a fortiori V'; and similarly for the projective embedding since $V'^\vee \to V^\vee$ is surjective.

We denote by $\mathcal{A}$ the group algebra $\mathbb{C}[G]$; its centre $Z_\mathcal{A} := Z(\mathbb{C}[G])$ is the space of class functions, the complex vector space with basis $\{v_\mathcal{C}\}$, indexed by the set of conjugacy classes $\mathcal{C}$ of G, and where

$$v_\mathcal{C} := \sum_{g \in \mathcal{C}} g.$$

Another more useful basis is indexed by the irreducible complex representations W_χ of G, and their characters χ (the irreducible characters form an orthonormal basis for the space of class functions). We shall, by the way, denote by $\rho_\chi : G \to GL(W_\chi)$ the corresponding representation homomorphism, and observe that there is a canonical isomorphism (dual map) $GL(W_\chi) \cong GL(W_\chi^\vee)$.

One associates as usual, to an irreducible character χ, an idempotent in the group algebra:

$$e(\chi) := \frac{\chi(1)}{|G|} \sum_{g \in G} \chi(g^{-1}) g.$$

The elements $e(\chi)$ are the principal idempotents of the group algebra $\mathbb{C}[G]$, and, from the fact that their sum is equal to 1, follows that the group algebra is semisimple, indeed

$$\mathcal{A} = \mathbb{C}[G] = \sum_{\chi \in Irr} \mathbb{C}[G] e(\chi) =: \sum_{\chi \in Irr} \mathcal{A}_\chi.$$

By the same token, each representation V splits as a direct sum of irreducible representations,

$$V = \sum_{\chi \in Irr} V_\chi, V_\chi = W_\chi \otimes (\mathbb{C}^{n(\chi)}),$$

where $\mathbb{C}^{n(\chi)}$ is a trivial representation of G (the orthogonality of characters imply that $e(\chi)$ acts as the identity on V_χ and acts as zero on $V_{\chi'}$ for $\chi' \neq \chi$).

The lemma of Schur says that $Hom_\mathcal{A}(W_\chi, W_{\chi'}) = 0$ for $\chi' \neq \chi$, and $\cong \mathbb{C}$ for $\chi' = \chi$.

If V is a left $\mathbb{C}[G]$-module, $V^\vee := Hom_\mathbb{C}(V, \mathbb{C})$ is a right $\mathbb{C}[G]$-module for the (duality preserving) operation such that

$$v(fg) := (g^{-1}v)f, \ v \in V, \ f \in V^\vee \Rightarrow (gv)(fg) = vf.$$

Since there exists on V a G-invariant positive definite Hermitian product, the matrices of elements of G in a unitary basis are unitary, so that, viewing $V^\vee$ as a left

$\mathbb{C}[G]$-module for the operation $f \mapsto fg^{-1}$, the character of $V^\vee$ is the conjugate of the one of V.

Moreover,

$$\mathbb{C}[G] = \sum_{\chi \in Irr} \mathcal{A}\chi, \ \mathcal{A}\chi \cong End(W_\chi) = W_\chi \otimes W_\chi^\vee.$$

And, in the last isomorphism, $e(\chi)$ corresponds to the identity tensor. In fact, $\mathcal{A} = \mathbb{C}[G]$ is also a right $\mathbb{C}[G]$-module, and $\mathcal{A}_\chi = e(\chi)\mathcal{A} = \mathcal{A}e(\chi)$, so $\mathcal{A}_\chi$ is naturally of the form $W_\chi \otimes W'$ and of the form $W'' \otimes W_\chi^\vee$. Since the dimension of $\mathcal{A}_\chi$ equals the square of the dimension of W_χ, then necessarily $\mathcal{A}_\chi = W_\chi \otimes W_\chi^\vee$.

REMARK 2.1. Replacing $\mathbb{C}$ by any other field K of characteristic zero, one sees that the semisimple decomposition of $K[G] =: \mathcal{A}_K$ is given by the direct sum over $\mathcal{A}_K e_K(\chi)$, where

$$e_K(\chi) := \sum_{\sigma \in Gal(K(\chi), K)} e(\chi^\sigma)$$

and where $K(\chi)$ is the extension of K generated by the values of the character χ.

Again, $\mathcal{A}_K e_K(\chi)$ is a matrix algebra over a division algebra D_χ.

DEFINITION 2.1. 1) We shall say that a left $\mathbb{C}[G]$- submodule $V \subset \mathbb{C}[G]$ is a **full** subrepresentation if it is a direct sum $\mathcal{A}\chi_1 \oplus \cdots \oplus \mathcal{A}\chi_m$.

This holds automatically if G is Abelian, since in this case each irreducible representation is 1-dimensional.

2) In general, for $V \subset \mathbb{C}[G]$, $V_\chi = W_\chi \otimes U_\chi$, for U_χ a vector subspace of $W_\chi^\vee$.

2.1) We shall say that U_χ is **linearly general** if, setting $G' := \rho_\chi(G)$, U_χ is not contained in any of the proper subspaces

$$Ker(\gamma - \gamma'), \ for \ \gamma \neq \gamma' \in G''$$

(then it is not contained in their union).

We shall say that V is **linearly general** if every $U_\chi \neq 0$ is linearly general.

2.2) We shall say that U_χ is **projectively general** if, setting

$$G'' := \mathbb{P}(\rho_\chi(G)) \subset \mathbb{P}GL(W_\chi) = \mathbb{P}GL(W_\chi^\vee),$$

$\mathbb{P}(U_\chi)$ is not contained in any of the proper subspaces

$$Fix(\gamma), \quad for \ \gamma \in G'' \setminus \{Id\}.$$

Equivalently, U_χ is not contained in a proper eigenspace for the action of an element $g \in G$ on $W_\chi^\vee$.

We shall say that V is **projectively general** if every $U_\chi \neq 0$ is projectively general.

It is convenient to define $E(\chi) := \sum_{g \in G} \chi(g^{-1})g$, and rewrite $\mathcal{A}\chi$ as a linear span

$$\mathcal{A}\chi = \langle \sum_{g \in G} \chi(g^{-1})\gamma g \rangle_{\gamma \in G} = \langle \sum_{g \in G} \chi(\gamma g^{-1}) g \rangle_{\gamma \in G}.$$

The subspace $\mathcal{A}\chi$ maps each element $h \in G$ to the vector

$$(\chi(\gamma h^{-1}))_{\gamma \in G} \in (\mathcal{A}\chi)^\vee \subset \mathcal{A}^\vee.$$

Given elements h^{-1}, k^{-1}, they map to the same vector if and only if

$$\chi(\gamma h) = \chi(\gamma k), \ \forall \gamma \in G \Leftrightarrow \chi(\gamma') = \chi(\gamma' h^{-1} k), \ \forall \gamma' \in G.$$

Set $h' := h^{-1}k$: then our condition amounts to

$$\chi(\gamma h') = \chi(\gamma), \ \forall \gamma \in G.$$

We specialize our formula to the case where $\gamma = 1$:

$$\chi(h') = \chi(1) = r := dim_{\mathbb{C}} W_\chi.$$

LEMMA 2.2. *If we have a group representation ρ_χ of a finite group G, with character χ, the condition $\chi(h') = \chi(1) =: r$ holds if and only $\rho_\chi(h')$ is the identity.*

PROOF. The eigenvalues of $\rho_\chi(h')$ are r numbers which are roots of unity, hence their real part is at most 1, and is equal to 1 if and only if the eigenvalue is 1. Since the sum of the real parts equals r, their real part must equal 1, therefore all the eigenvalues are equal to 1.

Since h' has finite order, we conclude that $\rho_\chi(h')$ must be equal to the identity. □

We summarize our previous discussion.

PROPOSITION 2.2. a) A subrepresentation $V \subset \mathbb{C}[G]$ separates points of G only if, writing $V = \sum_{\chi \in Irr} V_\chi, V_\chi = W_\chi \otimes U_\chi$,

(i) the representation

$$\rho := \otimes_{U_\chi \neq 0} \rho_\chi$$

is faithful, i.e.

$$\cap_{n(\chi)\neq 0}(Ker \rho_\chi) = \{1\}.$$

b) If V is full, or V is linearly general, then condition (i) is necessary and sufficient.

c) In general, $V \subset \mathbb{C}[G]$ separates points of G if and only if, setting $\epsilon_\chi^* : G \to Hom(U_\chi, W_\chi^\vee) = V_\chi^\vee \subset V^\vee$

$$G_{\chi,V} := \{(g,g') \in G \times G | \epsilon_\chi^*(g) = \epsilon_\chi^*(g')\}$$

$$\cap_\chi G_{\chi,V} = \{(g,g)\}.$$

PROOF. Since every such representation is contained in a unique minimal full one, in order to prove a) it suffices to prove that (i) is necessary and sufficient in the case where V is full.

Indeed, we have just shown that two points h, k are separated by $\mathcal{A}\chi$ if and only if they lie in different cosets of $Ker\rho_\chi$.

This shows that condition (1) is sufficient, while it is obviously necessary: if $h \in \cap_{n(\chi)\neq 0}(Ker\rho_\chi)$ then h and 1 are not separated.

The proof of b) in the case where V is linearly general shall follow from the forthcoming lemma 2.4.

Finally, c) is a straightforward reformulation of the requirement that ϵ^* separates elements of G, since $\epsilon^*(g) = \epsilon^*(g') \Leftrightarrow \epsilon_\chi^*(g) = \epsilon_\chi^*(g') \ \forall \chi$. □

For non full representations, condition (1) is not sufficient, as shown by the following example, which explains in concrete terms the notion of a general subrepresentation $V \subset \mathbb{C}[G]$.

EXAMPLE 2.3. Consider the dihedral group $D_n \subset O(2, \mathbb{R})$, the group of linear transformations of $\mathbb{C}$ of the form

$$x^j : z \mapsto \zeta^j z, \quad x^j y : z \mapsto \zeta^j \bar{z},$$

where ζ is a primitive n-th root of unity. We denote by ρ the standard representation $D_n \subset O(2,\mathbb{C})$, where, by a suitable change of basis

$$x(e_1) = \zeta e_1, x(e_2) = \zeta^{n-1} e_2, \quad y(e_1) = e_2, y(e_2) = e_1.$$

The group algebra is $\mathcal{A} = \mathbb{C}[x]/(x^n - 1) \oplus [\mathbb{C}[x]/(x^n-1)]y$, where $y^2 = 1, yx = x^{n-1}y$. The character χ of ρ has the property that

$$\chi(x^j) = \zeta^j + \zeta^{-j}, \chi(x^j y) = 0,$$

hence

$$E(\chi) = \sum_j (\zeta^j + \zeta^{-j}) x^j.$$

We see easily that $\mathcal{A}\chi$ has a basis

$$e_1 := \sum_j \zeta^{-j} x^j, \; e_2 := \sum_j \zeta^{-j} y x^j = y(e_1),$$

$$e_2' := \sum_j \zeta^j x^j, \; e_1' := \sum_j \zeta^j y x^j = y e_2'.$$

Moreover $x(e_1) = \zeta e_1, \; x(e_1') = \zeta e_1', \; x(e_2) = \zeta^{-1} e_1, \; x(e_2') = \zeta^{-1} e_2'$.

Hence, for each choice of $(a,b) \in \mathbb{C}^2 \setminus \{(0,0)\}$ we get a copy of the standard representation inside $\mathcal{A}\chi$, namely, the span

$$V =< ae_1 + be_1', ae_2 + be_2' > .$$

Does V separate points of D_n? Calculation shows that

$$x^j \mapsto (a\zeta^{-j}, b\zeta^j), \; yx^j \mapsto (b\zeta^j, a\zeta^{-j}).$$

The only possibility for non separation is that x^j, yx^h have the same image, and this holds precisely when

$$a = b\zeta^{h+j},$$

i.e., exactly when (a,b) is an eigenvector for some reflexion yx^t, with eigenvalue $+1$.

LEMMA 2.4. *Let χ be an irreducible character, and let $V_\chi = W_\chi \otimes U_\chi$, for U_χ a vector subspace of $W_\chi^\vee$. Denote by G' be the factor group $G/(Ker(\rho_\chi)$. G' is separated by V_χ if U_χ is a linearly general subspace.*

PROOF. G' is separated by $\mathcal{A}_\chi = W_\chi \otimes W_\chi^\vee$, and we have a composition of maps

$$G \to \mathbb{C}[G]^\vee \to End(W_\chi^\vee) \to Hom(U_\chi, W_\chi^\vee).$$

Hence G' is separated by V_χ if and only if there is no pair of distinct elements $\gamma, \gamma' \in \rho_\chi(G) =: G' \subset End(W_\chi^\vee)$ such that their restriction to U_χ coincide, equivalently $U_\chi \subset Ker(\gamma - \gamma')$. □

PROPOSITION 2.3. A subrepresentation $V \subset \mathbb{C}[G]$ yields a map of G into $\mathbb{P}(V^\vee)$ iff $V \neq 0$; it embeds G only if the following properties i) and ii) hold.

i) Writing $V = \sum_{\chi \in Irr} V_\chi, V_\chi = W_\chi \otimes U_\chi$, the representation

$$\rho := \otimes_{U_\chi \neq 0} \rho_\chi$$

is faithful.

ii) Let $\chi_1, \dots, \chi_m$ be the irreducible representations for which $U_\chi \neq 0$, and notice that ρ_j sends the centre $Z(G)$ of G to the set of scalar multiples of the identity, via a character ψ_j of $Z(G)$. Then the characters $\psi_1, \dots, \psi_j$ are affine generators of the group $Z(G)^*$ of characters of $Z(G)$ (i.e., $\frac{\psi_j}{\psi_1}$ generate the group $Z(G)^*$).

If moreover V is full, the two conditions i) and ii) are also sufficient.

If V is not full, we obtain a projective embedding of G if V is projectively general (i.e., if each space $U_\chi \neq 0$ is projectively general).

iii) In general, $V \subset \mathbb{C}[G]$ provides a projective embedding of G if and only if, setting $\epsilon_\chi^* : G \to Hom(U_\chi, W_\chi^\vee) = V_\chi^\vee \subset V^\vee$

$$G_{\mathbb{P}(V)} := \{g \in G | \mathbb{P}(\oplus_{U_\chi \neq 0} \epsilon_\chi^*)(g) = Id \text{ on } \mathbb{P}(\oplus_{U_\chi \neq 0} U_\chi)\}$$

consists only of the neutral element $1 \in G$.

PROOF. First of all, if $V \neq 0$, there exists $V_\chi \neq 0, V_\chi = W_\chi \otimes U_\chi$, and since every element of G yields an element in $GL(W_\chi)$, $g(U_\chi^\vee) \neq 0$.

We observe once more that if $V \subset V'$ and V yields a projective embedding of G, then a fortiori also V' does. Therefore we first deal with the case where V is full, and establish the necessity of conditions i) and ii) in all cases.

Assume that V is full. Now, h, k map to two proportional vectors inside the dual space $(V)^\vee$ if and only if 1 and $h' := h^{-1}k$ map to proportional vectors.

So, wlog, we may assume that $k = 1$, and our condition of non injectivity is equivalent to the existence of $\lambda \in \mathbb{C}^*$ such that:

$$\chi_j(\gamma h) = \lambda \chi_j(\gamma), \ \forall \gamma \in G, \ \forall j = 1, \dots, m.$$

Again here, setting $\gamma = 1$, we see that

$$\chi_j(h) = \lambda \chi_j(1) = \lambda r_j, \ r_j := dim(W_{\chi_j}),$$

while setting $\gamma = h$, we get

$$\chi_j(h^2) = \lambda \chi_j(h) = \lambda^2 \chi_j(1) \Rightarrow \chi_j(h^n) = \lambda^n \chi_j(1), \ \forall n \in \mathbb{N}.$$

If d denotes the order of h, then $\lambda^d = 1$, so that λ is a root of unity. At this point the equation $\chi_j(h) = \lambda r_j$ implies that $|\chi_j(h)| = r_j$, hence all the eigenvalues of $\rho_j(h)$ are equal to λ and $\rho_j(h)$ is a multiple of the identity, $\rho_j(h) = \lambda Id$.

Hence $\rho_j(h)$ is in the centre of $\rho_j(G)$. Conversely, by Schur's lemma, the centre of $\rho_j(G)$ acts via scalar multiples of the identity (the scalar being given via a character ψ_j of $Z(\rho_j(G))$.

The same argument in the general case shows iii).

Let us now prove the sufficiency of conditions i), ii).

Since ρ is a faithful representation, if h is not in the centre of G, there is a j such that $\rho_j(h)$ is not in the centre of $\rho_j(G)$. Hence $1, h$ are separated.

If instead h is in the centre, $\rho_j(h)$ is in the centre of $\rho_j(G)$ for all j, hence $\rho_j(h) = \psi_j(h) Id$. By condition ii), if $h \neq 1$, there exist j, j' such that $\psi_j(h) \neq \psi_{j'}(h)$, so that $1, h$ are separated.

Conversely, if ρ is not faithful, there exists an element $h \neq 1$ such that $\chi_j(h) = \chi_j(1)$, $\forall j$, and $1, h$ are not separated. In instead ii) is not fulfilled, there is an element $h \in Z(G)$ such that $\chi_j(h) = \lambda$, $\forall j$, hence again $1, h$ are sent to proportional vectors and we do not have projective separation.

The case where V is not full but projectively general.

Let $h \neq k$ be elements of G. Since $U_\chi \neq 0$ is projectively general, h, k are separated unless their image in $\mathbb{P}GL(W_\chi)$ coincide. If this is the case for all such χ, then $h^{-1}k$ lies in the centre of G.

We proceed now as in the full case, using property ii). □

EXAMPLE 2.5. Anyone of the full representations of D_n associated to an irreducible 2-dimensional representation separates points of D_n.

The remaining nontrivial irreducible representations are just the elements of D_n^*: the determinant $\psi_1 : D_n \to \{\pm 1\}$ is the only one for n odd. For $n = 2k$ there are also ψ_2 and $\psi_1 \otimes \psi_2$, where ψ_2 is trivial on reflections, and takes values -1 on x and on each generator of the subgroup of rotations.

Therefore, for $n \geq 3$, a full representation V separates points if and only if some irreducible 2-dimensional representation occurs in V.

The centre of D_n is trivial for n odd, while for n even it consists of the transformation $z \mapsto -z$, which we shall call -1.

Hence for n odd V yields a projective embedding of D_n if some irreducible 2-dimensional representation occurs in V. If n is even, since in every irreducible 2-dimensional representation -1 maps to $-Id$, we need moreover a 1-dimensional character ψ such that $\psi(-1) = 1$.

If $n = 4r$, any such ψ does the job, if instead $n = 2k$ with k odd, only ψ_1 and the trivial character do the job.

For the reader's convenience, we rewrite the above results in special cases.

COROLLARY 2.4. Assume that $V \subset \mathbb{C}[G]$ is a (left) $\mathbb{C}[G]$-submodule, $V = V_{\chi_1} \oplus \cdots \oplus V_{\chi_m}$. Then

I) If G is Abelian, V separates points of G if and only if the characters $\chi_1, \ldots, \chi_m$ generate the group $G^* := Hom(G, \mathbb{C}^*)$ of characters of G.

II) If G is Abelian, V yields and embedding of G inside $\mathbb{P}(V^\vee)$ if and only if the characters $\chi_1, \ldots, \chi_m$ affinely generate the group G^* of characters of G.

III) If G is simple and $V \neq 0$ is full, then V separates points of G and yields an embedding of G inside $\mathbb{P}(V^\vee)$. Similarly, if $V \neq 0$ is linearly general, respectively projectively general.

PROOF. In the first two cases we can apply the previous proposition 2.3 since V is full, $Z(G) = G$, and $\rho := \chi_1 \times \cdots \times \chi_m$ embeds G.

In case III), $Z(G) = 1$ and every nontrivial homomorphism of G is injective, so i) and ii) hold. □

2.2. Embedding $(G \times G)/G$**.** In general, if we have a homogeneous space G/H, which we view as the family of left cosets gH of H, and a representation $V \subset \mathbb{C}[G]$, the map $\epsilon^* : G \to V^\vee$ does not factor through the quotient map $G \to G/H$. However, if we consider the subspace V^H of H-invariants, composing ϵ with the projection $V^\vee \to (V^H)^\vee$ we obtain a well defined map. It is in general not so easy to see when do we get an embedding of G/H, or a projective embedding of G/H, except of course in the case where H is a normal subgroup, and V^H can be split as a direct sum of summands V_χ^H, where we only take irreducible characters χ such that $H \subset Ker(\rho_\chi)$. In this way we can apply the result of the previous subsection, viewing V^H as a representation of the quotient group $G' = G/H$.

We shall treat now the following geometrically interesting case.

Assume that we let $G \subset (G \times G)$ be the diagonal subgroup $\{(g, g) | g \in G\}$, graph of the identity map of G.

We consider the homogeneous space $G^0 := (G \times G)/G$, the space of left cosets of the diagonal subgroup, and we are given two representations $V_1 \subset \mathbb{C}[G], V_2 \subset \mathbb{C}[G]$.

As in the previous subsection $(G \times G)$ maps into $V_1^\vee \otimes V_2^\vee$ while G^0 maps into the subspace of invariants $(V_1^\vee \otimes V_2^\vee)^G = ((V_1 \otimes V_2)^G)^\vee$.

We ask once more whether V_1, V_2 produce in this way an affine, respectively a projective embedding of G^0.

Set for typographical convenience $V_1 =: V(1), V_2 =: V(2)$.

We have first of all by Schur's lemma a decomposition

$$(V(1) \otimes V(2))^G = \oplus_\chi (V(1)_\chi \otimes V(2)_{\chi^\vee})^G \subset \oplus'_\chi (\mathcal{A}_\chi \otimes \mathcal{A}_{\chi^\vee})^G$$

where the symbol $\oplus'_\chi$ denotes summing over the (irreducible) characters χ such that $V(1)_\chi \otimes V(2)_{\chi^\vee} \neq 0$.

From the abstract point of view, we have the decomposition

$$(\mathcal{A}_\chi \otimes \mathcal{A}_{\chi^\vee})^G = [(W_\chi \otimes W_\chi^\vee) \otimes (W_\chi^\vee . \otimes W_\chi)]^G = W_\chi^\vee \otimes W_\chi,$$

where we have performed contraction on the first and third terms in view of Schur's lemma.

We derive that $(V(1)_\chi \otimes V(2)_{\chi^\vee})^G \subset (\mathcal{A}_\chi \otimes \mathcal{A}_{\chi^\vee})^G$ is then isomorphic to

$$(**)\ U(1)_\chi \otimes U(2)_{\chi^\vee} \subset W_\chi^\vee \otimes W_\chi.$$

There remains the problem of identifying the right hand side as a space of functions on $G^0 = (G \times G)/G$. To be then concrete, let us write as in the previous subsection

$$(\mathcal{A}_\chi \otimes \mathcal{A}_{\chi^\vee}) =< (\sum_{g_1 \in G} \chi(\gamma_1 g_1^{-1}) g_1) \otimes (\sum_{g_2 \in G} \chi^\vee(\gamma_2 g_2^{-1}) g_2) >_{\gamma_1, \gamma_2 \in G}$$

and observe that the G-invariants are just obtained applying the averaging operator $\sum_\gamma \gamma \cdot$, so that

$$(\mathcal{A}_\chi \otimes \mathcal{A}_{\chi^\vee})^G =< \sum_{\gamma \in G} (\sum_{g_1 \in G} \chi(\gamma_1 g_1^{-1}) \gamma g_1) \otimes (\sum_{g_2 \in G} \chi^\vee(\gamma_2 g_2^{-1}) \gamma g_2) >_{\gamma_1, \gamma_2 \in G} =$$

$$=< (\sum_{g_1, g_2 \in G} [\sum_{\gamma \in G} \chi(\gamma_1 g_1^{-1} \gamma) \chi^\vee(\gamma_2 g_2^{-1} \gamma)] g_1 \otimes g_2) >_{\gamma_1, \gamma_2 \in G}.$$

In the same way we can explicitly calculate $(V(1)_\chi \otimes V(2)_{\chi^\vee})^G$.

We observe now that for each element of G^0 there is a unique representative of the form $(h, 1)$, and we evaluate the basis tensors on $(h, 1)$, obtaining by restriction the following elements in the group algebra

$$\sum_{h \in G} F_{\chi, \gamma_1, \gamma_2}(h) h := \sum_{h \in G} [\sum_{\gamma \in G} \chi(\gamma_1 h^{-1} \gamma) \chi^\vee(\gamma_2 \gamma)] h, \ \forall \gamma_1, \gamma_2 \in G.$$

We have

$$F_{\chi, \gamma_1, \gamma_2}(h) = \sum_{\gamma \in G} \chi(\gamma_1 h^{-1} \gamma) \chi^\vee(\gamma \gamma_2) = \sum_{\gamma \in G} \chi(\gamma_1 h^{-1} \gamma \gamma_2^{-1}) \chi^\vee(\gamma) =$$

$$= \sum_{\gamma \in G} \chi(\gamma_2^{-1} \gamma_1 h^{-1} \gamma) \chi^\vee(\gamma).$$

Hence the above functions depend only upon $g := \gamma_2^{-1} \gamma_1$ and we get

$$\sum_{h \in G} F_{\chi, g}(h) h := \sum_{h \in G} [\sum_{\gamma \in G} \chi(g h^{-1} \gamma) \chi^\vee(\gamma)] h, \ \forall g \in G.$$

Given two elements h, k the two vectors $F_{\chi,g}(h)$ and $F_{\chi,g}(k)$ are equal (resp. linearly dependent) if and only if, defining

$$L_x\chi(\gamma) := \chi(x\gamma)$$

the following Hermitian scalar products are equal

$$< L_{gh^{-1}}\chi, \chi >=< L_{gk^{-1}}\chi, \chi > \ \forall g \in G,$$

respectively they are proportional

$$\exists \lambda \ s.t. < L_{gh^{-1}}\chi, \chi >= \lambda < L_{gk^{-1}}\chi, \chi > \ \forall g \in G.$$

This holds true for $h = k$, in general we set $x := kh^{-1}$, and rewrite the previous conditions as

$$< L_{gx}\chi, \chi >= \lambda < L_g\chi, \chi > \ \forall g \in G,$$

$$\exists \lambda \ s.t. < L_{gx}\chi, \chi >= \lambda < L_g\chi, \chi > \ \forall g \in G.$$

The second condition, taking $g = 1$ implies

$$< L_x\chi, \chi >= \lambda, < L_{x^2}\chi, \chi >= \lambda^2, \dots \Rightarrow \exists m \ s.t. \ \lambda^m = 1.$$

We can now use the following lemma in order to infer that if h, k are not separated, then $\rho_\chi(h) = \rho_\chi(k)$, respectively to infer that if h, k are not projectively separated, then $\rho_\chi(h^{-1}k)$ lies in the centre of $G' := \rho_\chi(G)$.

LEMMA 2.6. *Let G be a finite group, ρ an irreducible representation, χ the character of ρ. Then*

$$< L_x\chi, \chi >= 1 \Leftrightarrow x \in ker(\rho).$$

If instead $< L_x\chi, \chi >= \lambda$, $|\lambda| = 1$, then $\rho(x)$ is in the centre of $G' := \rho(G)$.

PROOF. Without loss of generality we can assume that ρ is faithful, and we show then that in the first case $x = 1$.

By the Cauchy-Schwarz inequality, since $|\chi| = |L_x\chi| = 1$ (translation invariance of the counting measure), we obtain $| < L_x\chi, \chi > | = 1$ iff the two functions are proportional, $L_x\chi = \lambda\chi$. Now $< L_x\chi, \chi >= 1$ implies $\lambda = 1$, so that $\chi(x) = 1$, hence, by lemma 2.2 x acts as the identity, as we wanted to show.

Assume now that $< L_x\chi, \chi >= \lambda$, $|\lambda| = 1$: then the same argument shows that $L_x\chi = \lambda\chi$, hence $\rho(x)$ is a scalar multiple of the identity and lies in the centre of G'. □

We obtain therefore the simplest type of result, which is sufficient to treat the case where the group G is Abelian. We shall pursue elsewhere the analysis of the general case.

PROPOSITION 2.5. Assume that V_1, V_2 contain respectively full subrepresentations $V[1], V[2]$ such that, letting $\chi_1, \dots, \chi_m$ be the irreducible representations such that $V[1] = \sum_{j=1,\dots,m} \mathcal{A}_{\chi_j}$, the following hold:

i) the representation

$$\rho := \otimes_{j=1,\dots,m} \rho_{\chi_j}$$

of G is faithful.

ii) Let ψ_j be the character of the centre $Z(G)$ of G such that $\rho_j := \rho_{\chi_j}$ sends $Z(G)$ to the set of scalar multiples of the identity, via the character ψ_j. Then the characters $\psi_1, \dots, \psi_j$ are affine generators of $Z(G)$ ($\frac{\psi_j}{\psi_1}$ generate the group $Z(G)$).

iii) $V[2] = \sum_{j=1,\dots,m} \mathcal{A}_{\chi_j^\vee}$.

Then $V_1 \otimes V_2$ yields a projective embedding of $G^0 = (G \times G)/G$.

REMARK 2.6. A useful variant is the one where H is a normal subgroup of G, we let $G' := G/H$, and we consider the orbit space $\bar{G}' := (G' \times G)/G$, which is in an obvious bijection with $G' \times \{1\}$.

We obtain a projective embedding of the orbit space $\bar{G}'$ each time we are given representations V_1, V_2 of G' satisfying the three properties stated in proposition 2.5.

Just observe that a representation of G' is in a natural way a representation of G!

3. Canonical maps of Galois coverings and of surfaces isogenous to a product

3.1. Galois coverings. The first situation in which we can apply the results of the previous section is the case where we have a finite flat Galois morphism $f : X \to Y$ between projective varieties. For simplicity, we make the assumption that both X, Y are smooth, and we denote by G the Galois group, so that $Y = X/G$.

For every locally free sheaf ω of rank r on X such that the action of G on X extends to ω, the direct image sheaf $\mathcal{F} := f_*(\omega)$ is a locally free $\mathcal{O}_Y$-module and a $\mathcal{O}_Y[G]$-module and accordingly splits as

$$\mathcal{F} = \oplus_{\chi \in Irr(G)} \mathcal{F}_\chi.$$

If $y \in Y$ is a point such that $f^{-1}(y)$ is in bijection with G, then evaluation at the point y (algebraically, this is the operation of tensoring with the $\mathcal{O}_Y$-module $\mathcal{O}_Y/\mathcal{M}_y$, $\mathcal{M}_y$ being the maximal ideal of the point y)) yields the surjective homomorphism

$$\mathcal{F} = \oplus_{\chi \in Irr(G)} \mathcal{F}_\chi \to \mathbb{C}[G]^r = \oplus_{\chi \in Irr(G)} (\mathcal{A}_\chi)^r, \ r = rk(\omega).$$

In the present article we are particularly focused on the case of the canonical sheaf of X, $\omega = \omega_X = \mathcal{O}_X(K_X)$ (here $r = 1$).

The G-representation we shall consider from now on is

$$V := H^0(X, \mathcal{O}_X(K_X)) = H^0(Y, f_*\mathcal{O}_X(K_X)) =$$

$$= \oplus_{\chi \in Irr(G)} H^0(Y, \mathcal{F}_\chi) = \oplus_{\chi \in Irr(G)} V_\chi.$$

We want to transform basic questions as: is the canonical map of X an embedding, is it birational, into necessary or sufficient conditions on the vector bundles $\mathcal{F}_\chi$, keeping an eye to the examples we shall discuss.

PROPOSITION 3.1. Let $f : X \to Y$ be a finite flat Galois morphism between projective varieties, where X is smooth. Let G be the Galois group, and assume that the stabilizers of points of x are normal subgroups H (this holds if G is Abelian, or if f is unramified).

The canonical system $|K_X|$ is base-point free iff for each point $y \in Y$ the representation V has non trivial image via the composition $V \to \mathcal{F} \otimes \mathbb{C}_y \to \mathbb{C}[G/H]$ (the last homomorphism being associated to the inclusion of the reduced fibre $f^{-1}(y)_{red} \subset f^{-1}(y)$).

Moreover, in this case the canonical morphism Φ of X is injective if:

(1) for each point $y \in Y$ the representation V embeds the reduced fibre $f^{-1}(y) \cong G/H$, in particular if the associated representation $V \to \mathbb{C}[G/H]$ fulfills properties (i) and (ii) and is projectively general;

(2) $H^0(Y, \mathcal{F})$ separates pairs of points of Y.

The canonical map Φ of X is birational if (1) holds for some point $y \in Y$, and (2) holds for general points $y, y' \in Y$ (we do not need here the condition that H be a normal subgroup).

The canonical map Φ of X is an embedding if:

(1') for each point $y \in Y$ the representation V embeds the non reduced fibre $f^{-1}(y)$: this holds in particular if $\mathcal{F}$ is generated by global sections;

(2') $H^0(Y, \mathcal{F}) \to \mathcal{F} \otimes \mathcal{O}_\eta$ is surjective for each length two subscheme η of Y.

PROOF. The first assertion follows from the first assertion of proposition 2.3.

Injectivity of Φ means that pairs of distinct points x, x' have different image.

In the case where $f(x) = f(x')$ (1) guarantees that this is the case, and for the second assertion we simply use proposition 2.3 (recall that a full subrepresentation is projectively general!).

In the case where $y := f(x)$ and $y' := f(x')$ are distinct points, we use (2) implying that there is a section $s \in H^0(Y, \mathcal{F})$ with $s(y) = 0, s(y') \neq 0$. To s corresponds a section $\sigma \in H^0(X, \mathcal{O}_X(K_X))$ with $\sigma(x) = 0, \sigma(x') \neq 0$.

Birationality of Φ means that two general points x, x' are separated, and it suffices to show this for $x, x' \in f^{-1}(\mathcal{U})$, for $\mathcal{U}$ a Zariski open set of Y. If (1) holds for some point y, being an open condition, it holds on an a Zariski open set $\mathcal{U} \subset Y$, and this settles the case where $f(x) = f(x')$, whereas the case where $y := f(x)$ and $y' := f(x')$ are distinct points is handled as before.

To show embedding, one must moreover show that, for each length two subscheme $\xi \subset X$, $H^0(X, \omega_X) \to \omega_X \otimes \mathcal{O}_\xi$ is surjective, and it suffices to treat the case where ξ is supported at one point x, once we have stablished the injectivity of Φ. The schematic image of ξ is either a length two subscheme $\eta \subset Y$, in which case we use (2'); or ξ is contained in a fibre, and we use (1'). □

EXAMPLE 3.1. Let X, Y be curves, and assume that x is a ramification point of f, so that, in local coordinates, $t = f(z) = z^m$. We see that in this case the stabilizer H of x is cyclic of order m, and that the stalk of $\mathcal{F}$ at $y = f(x)$ contains $1, z, \dots, z^{m-1}$ but indeed $1, z$ suffice to embed a tangent vector to X at x.

This example shows that the condition that $\mathcal{F}$ is globally generated is too strong.

COROLLARY 3.2. Same notation as in proposition 3.1 and assume that G is Abelian and $Y = Q := \mathbb{P}^1 \times \mathbb{P}^1$.

Then the canonical map is birational if, writing $\mathcal{F}_\chi = \mathcal{O}_Q(n_\chi, r_\chi)$,

a) $H^0(\mathcal{F}_\chi) \neq 0$ ($\Leftrightarrow n_\chi, r_\chi \geq 0$) holds for a set of characters χ which affinely generate G^*;

b) there $\exists \chi, \chi'$ such that $n_\chi \geq 1, r_{\chi'} \geq 1$.

PROOF. a) implies that condition (1) of proposition 3.1 holds.

If b) holds for $\chi = \chi'$ (there is a χ with $n_\chi \geq 1, r_\chi \geq 1$) then $H^0(\mathcal{F}_\chi)$ separates two general points y, y'.

Otherwise, by b), there are χ, χ' such that $n_\chi \geq 1, r_\chi = 0, n_{\chi'} = 0, r_{\chi'} \geq 1$, whence $H^0(\mathcal{F}_\chi)$ separates general points y, y' with different first projection to $\mathbb{P}^1$, and $H^0(\mathcal{F}_{\chi'})$ separates general points y, y' with different second projection to $\mathbb{P}^1$. □

3.2. Canonical maps of SIP 's = surfaces isogenous to a product. In this section and in the sequel we shall consider surfaces isogenous to a product of unmixed type (see [**Cat00**] for general properties of SIP ' s and proofs for many statements which we shall make).

We have two curves C_1, C_2 of respective genera $g_1, g_2 \geq 2$ and a faithful action of a group G on both curves, we denote by $C_i' = C_i/G$ and by $p_i : C_i \to C_i'$ the quotient morphism.

Clearly $G \times G$ acts on the surface $C_1 \times C_2$, and we consider $G \subset G \times G$ the diagonal subgroup. The basic assumption is that G acts freely on the product $C_1 \times C_2$ (this boils down to the fact that, denoting by Σ_j the set of stabilizers for the action of G on C_j, $\Sigma_1 \cap \Sigma_2$ consists only of $1 \in G$).

The quotient $S := (C_1 \times C_2)/G$ is then smooth and we have a sequence of morphisms

$$\pi : C_1 \times C_2 \to S,\ p : S \to C_1' \times C_2' = (C_1 \times C_2)/(G \times G) =: Y$$

which are uniquely determined by the requirement that G acts faithfully on both curves.

We denote by $V_j := H^0(C_j, \mathcal{O}_{C_j}(K_{C_j}))$, so that

$$H^0(S, \mathcal{O}_S(K_S)) = (V_1 \otimes V_2)^G = \oplus_{\chi \in Irr(G)} (V_{1,\chi} \otimes V_{2,\chi^\vee})^G.$$

Observe that for $y = (y_1, y_2) \in Y$ the inverse image $p^{-1}(y)$ is the quotient of the $(G \times G)$-orbit of a point $x = (x_1, x_2) \in C_1 \times C_2$ by the action of the diagonally embedded G.

Since we assume that this action is free, either x is a point where $p \circ \pi$ is unramified, and then $p^{-1}(y) \cong (G \times G)/G$, or x is a point where $p \circ \pi$ is not étale, equivalently one of the projections p_j is ramified, and then if for instance $j = 1$, $p^{-1}(y) \cong (G/H \times G)/G$, where H is the subgroup of G stabilizing x_1.

Assumption N. We shall assume, as in proposition 3.1, that H is a normal subgroup for each point with a nontrivial stabilizer.

The projection $p : S \to Y$ is Galois if and only if G is Abelian, since $G \subset G \times G$ is a normal subgroup iff G is Abelian. We can define as usual

$$\mathcal{F} := p_* \mathcal{O}_S(K_S), \epsilon_y : (V_1 \otimes V_2)^G = H^0(S, \mathcal{O}_S(K_S)) = H^0(Y, \mathcal{F}) \to \mathcal{F} \otimes \mathbb{C}_y.$$

In this way we are reduced to the situation considered in proposition 2.5 and remark 2.6.

Indeed, evaluation on the orbit of (x_1, x_2) yields maps

$$(V_1 \otimes V_2) \to V'[1] \otimes V'[2] \subset \mathbb{C}[G/H_1] \otimes \mathbb{C}[G/H_2],$$

(here one of the subgroups H_j is trivial, by the assumption that the action of G is free) inducing

$$(V_1 \otimes V_2)^G \to (V[1] \otimes V[2])^G \subset (\mathbb{C}[G/H_1] \otimes \mathbb{C}[G/H_2])^G,$$

where $V[1], V[2]$ are determined by property iii) of proposition 2.5.

Define moreover H to be trivial when both H_1, H_2 are trivial, otherwise H is the only nontrivial normal subgroup H_j. Then $V[1], V[2]$ are representations of $G' := G/H$. Observe finally that $\mathbb{C}[G/H] = (\mathbb{C}[G/H_1] \otimes \mathbb{C}[G/H_2])^G$ is just the space of functions on the reduced fibre $f^{-1}(y)_{red} \subset f^{-1}(y))$.

PROPOSITION 3.3. Let S be a surface isogenous to a product of unmixed type and let $p : S \to Y = C_1' \times C_2'$ be the natural finite flat morphism. Keep assumption N.

Then the canonical system $|K_S|$ is base-point free iff, for each point $y \in Y$, $V := (V_1 \otimes V_2)^G$ has non trivial image $(V[1] \otimes V[2])^G$ via the composition $V \to \mathcal{F} \otimes \mathbb{C}_y \to \mathbb{C}[G/H]$, (the last homomorphism being associated to the inclusion of the reduced fibre $f^{-1}(y)_{red} \subset f^{-1}(y)$).

Moreover, in this case the canonical morphism Φ of S is injective if:

(1) for each point $y \in Y$, V embeds the reduced fibre $f^{-1}(y)_{red} \cong G/H$, and this holds in particular if $V[1], V[2]$ are full and (i), (ii) of prop. 2.5 are fulfilled.

(2) $H^0(Y, \mathcal{F})$ separates pairs of points of Y.

The canonical map Φ of X is birational if (1) holds for some point $y \in Y$, and (2) holds for general points $y, y' \in Y$.

The canonical map Φ of X is an embedding if:

(1') for each point $y \in Y$ the representation V embeds the non reduced fibre $f^{-1}(y)$: this holds in particular if $\mathcal{F}$ is generated by global sections;

(2') $H^0(Y, \mathcal{F}) \to \mathcal{F} \otimes \mathcal{O}_\eta$ is surjective for each length two subscheme η of Y.

PROOF. The proof is the same as the one of proposition 3.1, using proposition 2.5 and remark 2.6. □

REMARK 3.4. Since π is unramified, in order to verify that the canonical map is a local embedding, it suffices to verify that it gives a local embedding on $C_1 \times C_2$.

In particular, for (1'), it suffices to show that $V[1] \times V[2]$ embeds every tangent vector which is in the kernel of the projection $p \circ \pi$.

4. Surfaces with record winning high canonical degree

4.1. Regular SIP 's with $p_g = 4, 5, 6$. If we have a SIP, then

$$K_S^2 = 8(1 - q(S) + p_g(S)) = 8\frac{1}{|G|}(g_1 - 1)(g_2 - 1)$$

and, in order to make K_S^2 as high as possible, we require that $q(S) = g_1' + g_2' = 0$.

This means that $C_i' = \mathbb{P}^1, i = 1, 2$, $Y = \mathbb{P}^1 \times \mathbb{P}^1$, and then

$$(1 + p_g(S))|G| = (g_1 - 1)(g_2 - 1).$$

Recall that each covering $p_j : C_j \to \mathbb{P}^1$ is determined by the branch points set $\mathcal{B}_j := \{P_1, \dots, P_{r_j}\}$ and, once a geometric basis of $\pi_1(\mathbb{P}^1 \setminus \mathcal{B}_j)$ is chosen, by a spherical system of generators for the group G. A spherical system of generators for p_1 is the choice of elements $\gamma_1, \dots, \gamma_{r_1} \in G$ such that $\prod_1^{r_1} \gamma_j = 1$. Similarly for p_2 we get r_2 points and $\gamma_1', \dots, \gamma_{r_2}' \in G$ such that $\prod_1^{r_2} \gamma_j' = 1$.

One denotes by m_j the order of γ_j, and by m_j' the order of γ_j'.

Then Hurwitz' formula yields

$$g_1 - 1 = \frac{1}{2}|G|[-2 + \sum_1^{r_1}(1 - \frac{1}{m_j})],$$

$$g_2 - 1 = \frac{1}{2}|G|\{-2 + \sum_1^{r_2}(1 - \frac{1}{m_j'})\},$$

hence

$$(*)\ 4(1 + p_g(S)) = |G|[-2 + \sum_1^{r_1}(1 - \frac{1}{m_j})]\{-2 + \sum_1^{r_2}(1 - \frac{1}{m_j'})\}.$$

These formulae, even if we restrict to the values $p_g = 4, 5, 6$ lead to a nontrivial determination problem.

Our first search is for uniform Abelian groups $G = (\mathbb{Z}/m)^k$.

The m_j, m'_i 's are divisors of the exponent m of the group,

$$m = a_j m_j, \ m = b_i m'_i.$$

Formula $(*)$ rewrites as an equality among integers:

$$(**) \ 4(1 + p_g(S)) = m^{k-2}[-2m + \sum_1^{r_1}(m - a_j)]\{-2m + \sum_1^{r_2}(m - b_j)\}.$$

And also as

$$(***) \ 4(1 + p_g(S)) = m^k[-2 + \sum_1^{r_1}(1 - \frac{1}{m_j})]\{-2 + \sum_1^{r_2}(1 - \frac{1}{m'_j})\} =: m^k A_1 A_2.$$

Claim 1: if $k \geq 3$ then $m \leq 3$.

Since we have a spherical system of generators, $r_j \geq k + 1$. If $r_1, r_2 \geq 5$, in formula (***) we have $A_j \geq 1/2$, hence

$$m^k \leq 16(1 + p_g),$$

and for $k \geq 3$ this implies $m \leq 4$. Actually, $k \leq 6$, and for $k = 4$ $m \leq 3$, for $k = 5, 6$, $m = 2$.

If $r_1 = 4$, then $k = 3$ and all the a_j's are equal to 1 (three elements are a basis, so also their sum has order m) hence the term in square brackets is $[2m - 4]$; hence if $r_1 = 4, r_2 \geq 5$ we obtain

$$m^2(m - 2) \leq 4(1 + p_g) \Rightarrow m \leq 3.$$

While, for $r_1 = r_2 = 4$ we get $m(m - 2)^2 = (1 + p_g)$ and this case is not possible for divisibility reasons.

Assume now $m = 4, k = 3$: then at least four of the a_j's equal 1, and the terms [..], {..} are at least 6, so again this case is not possible, and our claim is established.

□

Case i) $k \geq 3$, $m = 2, 3$.

Since $2, 3$ are prime numbers, $a_j = b_i = 1, \forall i, j$, and then formula $(**)$ rewrites as

$$(****) \ 4(1 + p_g(S)) = m^{k-2}[-2m + r_1(m - 1)]\{-2m + r_2(m - 1)\}.$$

Since $k \geq 3$, then:

$$p_g = 4 \Rightarrow m = 2, k = 3, 4$$

$$p_g = 5 \Rightarrow m = 3, k = 3, \ \textit{or} \ m = 2, k = 3, 4, 5$$

$$p_g = 6 \Rightarrow m = 2, k = 3, 4.$$

Recall moreover that $r_j \geq k + 1$, because G is not $(k - 1)$-generated. Observe moreover that if $m = 3$ we must have $r_1 = 4$, $r_2 = 5$, since $2r_j - 6 \geq 2$.

Consider now the case $m = 2$.

If $m = 2$, $k = 2$ there are at most 3 non trivial elements in G hence we cannot get a free action. Else, we have

$$(***) \ 4(1 + p_g(S)) = 2^{k-2}[-4 + r_1]\{-4 + r_2\},$$

leading to many possibilities:

$$p_g = 4, m = 2, k = 3, r_1 = 6, r_2 = 9, \ \textit{or} \ r_1 = 5, r_2 = 14,$$

$$p_g = 4, m = 2, k = 4, r_1 = 5, r_2 = 9,$$

$$p_g = 5, m = 2, k = 3, r_1 = 5, r_2 = 16, \ or \ r_1 = 6, r_2 = 10, \ or \ r_1 = 7, r_2 = 8,$$

$$p_g = 5, m = 2, k = 4, r_1 = 5, r_2 = 10, \ or \ r_1 = 6, r_2 = 7,$$

$$p_g = 5, m = 2, k = 5, r_1 = 5, r_2 = 7,$$

$$p_g = 6, m = 2, k = 3, r_1 = 6, r_2 = 11, \ or \ r_1 = 5, r_2 = 18,$$

$$p_g = 6, m = 2, k = 4, r_1 = 5, r_2 = 11.$$

If we want the canonical map of S to be birational, then at least $k+1$ distinct characters must appear in V, therefore $p_g \geq k+1$, and we can exclude the second and the fifth case.

Case ii) $k = 2$, $m \geq 3$ a prime number implies $m = 3$ or $m = 5, p_g = 4, 6$.

Here $4(1+p_g(S)) = [(r_1 - 2)(m-1) - 2]\{(r_2 - 2)(m-1) - 2\}$, hence $m \leq 8$. Since m is odd, we can rewrite as

$$1 + p_g(S) = [(r_1 - 2)\frac{m-1}{2} - 1]\{(r_2 - 2)\frac{m-1}{2} - 1\}.$$

For $m = 7$ the terms in brackets are 2 or 5 or 8, or more, hence the equation is not solvable by divisibility conditions. For $m = 5$ the terms in brackets are $1, 3, 5, 7 \ldots$, so the only solutions are $p_g = 4, r_1 = 3, r_2 = 5$ and $p_g = 6, r_1 = 3, r_2 = 6$. For $m = 3$ there are several solutions $(r_1 - 3)(r_2 - 3) = 1 + p_g$: either $r_1 = 4$ and $r_2 = 4 + p_g$, or $p_g = 5$ and $r_1 = 5, r_2 = 6$.

4.2. Examples with $m = 2$. We assume in the following for the sake of simplicity that $m = 2$. Recall moreover that the condition that G acts freely boils down to the condition that the set Σ_1 of nontrivial elements in the union of the cyclic subgroups generated by the elements γ_j does not intersect the set Σ_2 of nontrivial elements in the union of the cyclic subgroups generated by the elements γ'_j.

REMARK 4.1. If $k = 3$, $m = 2$, then $G \setminus \{0\}$ has precisely 7 elements, and Σ_j has at least 3 elements, hence $|\Sigma_j| = 3, 4$ and the cardinalities $|\Sigma_j|$ cannot both equal 4.

If $|\Sigma_1| = 3$, then the monodromy elements γ_j are just three basis vectors counted with multiplicity, and these multiplicities must be even, hence r_1 must be even ≥ 6. Similarly if $|\Sigma_2| = 3$.

Hence, if $|\Sigma_1| = 3$, $|\Sigma_2| = 4$, the first spherical generating system Σ_1 is of the form, for a suitable basis, (we use here the classical notation γ^r to denote that the vector γ occurs r times)

$$e_1^{a_1}, \ e_2^{a_2}, e_3^{a_3}, \ a_j \equiv 0 \ (mod \ 2),$$

while Σ_2 is of the form (here $e := e_1 + e_2 + e_3$):

$$e^b, \ (e_1 + e_2)^{b_3}, \ (e_1 + e_3)^{b_2}, \ (e_2 + e_3)^{b_1}, b \equiv 0 \ (mod \ 2), b_1 \equiv b_2 \equiv b_3 \ \ (mod \ 2).$$

For instance, for $r_2 = 5$, the only possibility is e^2, $(e_1 + e_2)$, $(e_1 + e_3)$, $(e_2 + e_3)$.

If r_2 is even, then all the b_j's are even, and $r_2 \geq 8$.

REMARK 4.2. Observe that if $r_j = 5$, and $k = 3$, then $g_j = 3$ and there are at most 3 characters occurring in V_1. Hence the canonical map of S cannot be birational since condition (ii) of proposition 2.5 is certainly violated. We can exclude this case.

Moreover, condition (2) in proposition 3.3 suggests that the easier cases are those for which either one character contributes to a subspace of dimension $4 = dimH^0(\mathcal{O}_Q(1,1))$ (here we use the notation already introduced $Q := \mathbb{P}^1 \times \mathbb{P}^1$) or two contribute to a subspace of dimension $2 = dimH^0(\mathcal{O}_Q(1,0)) = dimH^0(\mathcal{O}_Q(0,1))$.

We are left with the possibly easy case:

$$p_g = 6, m = 2, k = 3, r_1 = 6, r_2 = 11,$$

and with the cases:

$$p_g = 4, m = 2, k = 3, r_1 = 6, r_2 = 9,$$

$$p_g = 5, m = 2, k = 3, r_1 = 6, r_2 = 10, \ \ or\ r_1 = 7, r_2 = 8,$$

$$p_g = 5, m = 2, k = 4, r_1 = 6, r_2 = 7.$$

For the sake of uniformity of treatment we make the assumption that $r_1 = 6, k = 3$ (the first three cases).

LEMMA 4.1. *To a character* $\chi \in ((\mathbb{Z}/2)^3)^\vee$, *we associate the dimension* $n_1(\chi) = dim(V_{1,\chi})$.

In the case $r_1 = 6$, $|\Sigma_1| = 3$, *letting* $e_i^\vee, i = 1, 2, 3$, *be the dual basis, and letting* $e^\vee := e_1^\vee + e_2^\vee + e_3^\vee$, *we have:*

$$n_1(e_i^\vee) = 0,\ n_1(e_i^\vee + e_j^\vee) = 1,\ n_1(e^\vee) = 2 \quad \forall i \neq j,\ i, j \in \{1, 2, 3\}.$$

If more generally $|\Sigma_1| = 3$, *and the generating system is of the form* $e_1^{2A_1}$, $e_2^{2A_2}$, $e_3^{2A_3}$, $(A_j \geq 1)$, *then* $\forall i \neq j,\ i, j \in \{1, 2, 3\}$,

$$n_1(e_i^\vee) = A_i - 1,\ n_1(e_i^\vee + e_j^\vee) = A_i + A_j - 1,\ n_1(e^\vee) = A_1 + A_2 + A_3 - 1.$$

PROOF. $n_1(\chi) = dim(V_{1,\chi})$ is the genus of the double cover of $\mathbb{P}^1$ obtained as $C_1/Ker(\chi)$.

In the first cases we have a covering branched in two points, in the second ones the covering is branched on four points, in the third on six points. The corresponding genera are then $0, 1, 2$.

The proof of the second assertion is similar. □

With the same method we determine V_2 using the normal form developed in remark 4.1.

LEMMA 4.2. *In the case* $|\Sigma_2| = 4$, Σ_2 *is of the form*

$$e^b,\ (e_1 + e_2)^{b_3},\ (e_1 + e_3)^{b_2},\ (e_2 + e_3)^{b_1}, b > 0, b \equiv 0\ (mod\ 2), b_1 \equiv b_2 \equiv b_3\ \ (mod\ 2).$$

Then if $b = 2B \geq 2, b_j = 1 + 2B_j$, *then*

$$n_2(e_i^\vee) = B + B_j + B_k,\ n_2(e_i^\vee + e_j^\vee) = B_i + B_j,\ n_2(e^\vee) = B - 1,\ \{i, j, k\} = \{1, 2, 3\}.$$

If instead $b = 2\beta \geq 2, b_j = 2\beta_j \geq 2$,

$$n_2(e_i^\vee) = \beta + \beta_j + \beta_k - 1,\ n_2(e_i^\vee + e_j^\vee) = \beta_i + \beta_j - 1,\ n_2(e^\vee) = \beta - 1, \{i, j, k\} = \{1, 2, 3\}.$$

Corollary 4.3. In the case where $|\Sigma_1| = 3$, $|\Sigma_2| = 4$, letting

$$V = \sum_{\chi} V_{1,\chi} \otimes V_{2,\chi}, \ N(\chi) := dim(V_{1,\chi} \otimes V_{2,\chi})$$

we have, in the case $b = 2B \geq 2, b_j = 1 + 2B_j$,

$$N(e_i^\vee) = (A_i - 1)(B + B_j + B_k),$$
$$N(e_i^\vee + e_j^\vee) = (A_i + A_j - 1)(B_i + B_j), \{i, j, k\} = \{1, 2, 3\}.$$
$$N(e^\vee) = (A_1 + A_2 + A_3 - 1)(B - 1).$$

Whereas, in the case $b = 2\beta \geq 2, b_j = 2\beta_j \geq 2$, we get

$$N(e_i^\vee) = (A_i - 1)(\beta + \beta_j + \beta_k - 1), \ N(e_i^\vee + e_j^\vee) = (A_i + A_j - 1)(\beta_i + \beta_j - 1),$$
$$N(e^\vee) = (A_1 + A_2 + A_3 - 1)(\beta - 1), \ \ \{i, j, k\} = \{1, 2, 3\}.$$

Proposition 4.4. In the case

$$p_g = 6, m = 2, k = 3, r_1 = 6, r_2 = 11$$

if the canonical map of S is birational, then $|\Sigma_2| = 4$, $|\Sigma_1| = 3$, $A_i = 1, \forall i = 1, 2, 3$, $b = 4, b_1 = 3, b_2 = 3, b_3 = 1$, and

$$N(e^\vee) = 2, \ N(e_1^\vee + e_2^\vee) = 2, \ N(e_1^\vee + e_3^\vee) = 1, \ N(e_2^\vee + e_3^\vee) = 1.$$

In the case

$$p_g = 4, m = 2, k = 3, r_1 = 6, r_2 = 9,$$

the canonical map of S cannot be birational.

In the case

$$p_g = 5, m = 2, k = 3, r_1 = 6, r_2 = 10,$$

if the canonical map of S is birational, then we must have $|\Sigma_2| = 4$, $|\Sigma_1| = 3$, $A_i = 1, \forall i = 1, 2, 3$, $b = 4, b_1 = b_2 = b_3 = 2$, and in this case

$$N(e^\vee) = 2, \ N(e_i^\vee + e_j^\vee) = 1, \ \forall i \neq j.$$

In the case

$$p_g = 5, m = 2, k = 3, \ r_1 = 7, r_2 = 8,$$

the canonical map of S cannot be birational.

Proof. If $p_g = 6, m = 2, k = 3, r_1 = 6, r_2 = 11$ since r_2 is odd, $|\Sigma_2| = 4$, hence $|\Sigma_1| = 3$, $A_i = 1, \forall i = 1, 2, 3$, and we get $b = 2B \geq 2, b_j = 1 + 2B_j$, $B + B_1 + B_2 + B_3 = 4$. Since we need four characters to appear, we must have $B = 2$, $B_1 = B_2 = 1$, $B_3 = 0$. Then $N(e^\vee) = 2$, $N(e_1^\vee + e_2^\vee) = 2$, $N(e_1^\vee + e_3^\vee) = 1$, $N(e_2^\vee + e_3^\vee) = 1$.

In the case $p_g = 4, m = 2, k = 3, r_1 = 6, r_2 = 9$, again r_2 is odd, $|\Sigma_2| = 4$, hence $|\Sigma_1| = 3$, $A_i = 1, \forall i = 1, 2, 3$, and we get $b = 2B \geq 2, b_j = 1 + 2B_j$, $B + B_1 + B_2 + B_3 = 3$: here at most three characters can appear, so this case should be excluded.

If $p_g = 5, m = 2, k = 3, r_1 = 6, r_2 = 10$, again if $|\Sigma_1| = 3$, $|\Sigma_2| = 4$, $A_i = 1, \forall i = 1, 2, 3$, then $\beta + \sum_j \beta_j = 5$, hence there occur four different characters only if $\beta = 2, \beta_j = 1 \ \forall j$, and then $N(e^\vee) = 2$, $N(e_i^\vee + e_j^\vee) = 1$, $\forall i \neq j$.

If instead $|\Sigma_1| = 3$, $|\Sigma_2| = 3$, then we use the same formulae as before, except that now one among $\beta_1, \beta_2, \beta_3$ is zero ($\beta \geq 1$ since the other three vectors are linearly dependent) and only three characters can occur. In fact, if $\beta_3 = 0$, we should have $\beta = \beta_1 = \beta_2 = 2$, a contradiction.

If instead $|\Sigma_1| = 4$, $|\Sigma_2| = 3$, the fact that $r_1 = 6, r_2 = 10$ contradicts remark 4.1.

In the case $p_g = 5, m = 2, k = 3,\ r_1 = 7, r_2 = 8$, let us replace the order and treat $r_1 = 8, r_2 = 7$. Since r_2 is odd, $|\Sigma_2| = 4$, hence $|\Sigma_1| = 3$, $A_1 = 2, A_2 = 1, A_3 = 1$. We have $b = 2B \geq 2, b_j = 1 + 2B_j$, $B + B_1 + B_2 + B_3 = 2$. If $B = 2$, then $N(e^\vee) = 3(B-1) = 3$: since $p_g = 5$, at most three characters can occur. If instead $B = 1$, then only one $B_j = 1$, and there are two cases: $B_1 = 1$, or $B_2 = 1$. In both cases we see that only three characters occur (with dimensions $2, 2, 1$). □

THEOREM 4.3. *In the two possible cases of proposition* 4.4

I) $p_g = 6, G = (\mathbb{Z}/2)^3, r_1 = 6, r_2 = 11$, *with generating systems* e_1^2, e_2^2, e_3^2, *respectively* $e^4, (e_2+e_3)^3, (e_1+e_3)^3, (e_1+e_2)$, *we have*

$$N(e^\vee) = 2\cdot 1,\ N(e_1^\vee + e_2^\vee) = 1\cdot 2,\ N(e_1^\vee + e_3^\vee) = 1\cdot 1,\ N(e_2^\vee + e_3^\vee) = 1\cdot 1,$$

and the canonical map of S *is a non injective birational morphism onto a canonical image* Σ *of degree* 56 *in* $\mathbb{P}^5$.

II) $p_g = 5, G = (\mathbb{Z}/2)^3, r_1 = 6, r_2 = 10$, *with generating systems* e_1^2, e_2^2, e_3^2, *respectively* $e^4, (e_2+e_3)^2, (e_1+e_3)^2, (e_1+e_2)^2$, *we have*

$$N(e^\vee) = 2\cdot 1,\ N(e_i^\vee + e_j^\vee) = 1\cdot 1,\ \forall i \neq j,$$

and the canonical map of S *is a non injective birational morphism onto a canonical image* Σ *of degree* 48 *in* $\mathbb{P}^4$, *for general choice of the branch points for* $C_2 \to \mathbb{P}^1$.

PROOF. Let for convenience $u_{ij} = 0$ be the reduced divisor on C_1 corresponding to the inverse image of the two points with local monodromy e_k, for $\{1,2,3\} = \{i,j,k\}$.

Similarly define $\sigma_{ij} = 0$ as the reduced divisor on C_2 corresponding to the points with local monodromy $e_i + e_j$, respectively $\sigma = 0$ as the reduced divisor on C_2 corresponding to the points with local monodromy $e = e_1 + e_2 + e_3$. Define $v_{ij} := \sigma\sigma_{ij}$ and $v_{123} := \prod_{i\neq j}\sigma_{ij}$.

Let us first treat case I) : $p_g = 6$. Here we can write the canonical system as

$$V = z_{123}H^0(\mathcal{O}_Q(1,0)) \oplus z_{12}H^0(\mathcal{O}_Q(0,1)) \oplus \mathbb{C}z_{13} \oplus \mathbb{C}z_{23} =$$

$$V = v_{123}H^0(\mathcal{O}_Q(1,0)) \oplus u_{12}v_{12}H^0(\mathcal{O}_Q(0,1)) \oplus \mathbb{C}u_{13}v_{13} \oplus \mathbb{C}u_{23}v_{23}.$$

First step: The canonical system has no base points.

Indeed, the base locus is defined by the equations:

$$v_{123} = u_{12}v_{12} = u_{13}v_{13} = u_{23}v_{23} = 0,$$

equivalently

$$\prod_{i\neq j}\sigma_{ij} = u_{12}\sigma\sigma_{12} = u_{13}\sigma\sigma_{13} = u_{23}\sigma\sigma_{23} = 0,$$

It is crucial to observe that the divisors of the σ_{ij}'s and of σ are pairwise disjoint, and the same occurs for the divisors of the u_{ij}'s.

Hence the locus $\sigma = 0$ is away from the base locus, and if $\sigma_{ij} = 0$, then we use that $u_{ik} = 0, u_{jk} = 0$ defines the empty set. This proves our first claim.

Moreover, we see immediately that, since the four above characters are affine generators, condition (1) of proposition 3.3 is fulfilled for a general point of Q; likewise we see that (2) holds (see corollary 3.2 for a similar argument).

To see that the canonical map is not injective, consider the curve $D = div(\sigma)$, consisting of the inverse image of 4 points on the second copy of $\mathbb{P}^1$. Here the canonical map sends each component of D to $\mathbb{P}^1$, hence the canonical map has degree at least 4 (indeed 16 since $C_1/e \to C_1' = \mathbb{P}^1$ has degree 4).

Let us now treat the more difficult case II) : $p_g = 5$. In this case a basis of V is provided by

$$V = z_{123}H^0(\mathcal{O}_Q(1,0) \oplus \mathbb{C}z_{12} \oplus \mathbb{C}z_{13} \oplus \mathbb{C}z_{23} =$$

$$V = v_{123}H^0(\mathcal{O}_Q(1,0) \oplus \mathbb{C}u_{12}v_{12} \oplus \mathbb{C}u_{13}v_{13} \oplus \mathbb{C}u_{23}v_{23}.$$

Here u_{ij}, $v_{ij} = \sigma\sigma_{ij}$ are exactly as before. Hence the same argument shows that the canonical system has no base points.

Likewise, condition (1) of proposition 3.3 is again fulfilled for a general point of Q; to see that (2) holds we observe that two general points y, y' are separated if they have a different first coordinate $y_1 \neq y_1'$.

Assuming that $y_1 = y_1'$, then the fibre over a general such point is just C_2, and the restriction of the canonical map Φ to C_2 is projectively equivalent to the map $F : C_2 \to \mathbb{P}^3$ given by

$$(v_{123}, v_{12}, v_{13}, v_{23}) = (\sigma_{12}\sigma_{13}\sigma_{23}, \sigma\sigma_{12}, \sigma\sigma_{13}, \sigma\sigma_{23}).$$

The map F is equivariant with respect to the action of G on C_2 and the action on $\mathbb{P}^3$ given by the indicated characters of G.

Therefore, if we let $s : \mathbb{P}^3 \to \mathbb{P}^3$ be the map which squares the coordinates, $s \circ F$ induces a map of $F' : \mathbb{P}^1 = C_2/G \to \mathbb{P}^3$, given by

$$(\sigma_{12}^2\sigma_{13}^2\sigma_{23}^2, \sigma^2\sigma_{12}^2, \sigma^2\sigma_{13}^2, \sigma^2\sigma_{23}^2) = (P_{12}P_{13}P_{23}, PP_{12}, PP_{13}, PP_{23}),$$

where P, P_{ij} are polynomials on $\mathbb{P}^1$ (of respective degrees $4, 2$) vanishing on the branch points which are the respective images of $\sigma = 0, \sigma_{ij} = 0$.

A projection of F' is the rational map

$$F'' = (P_{12}, P_{13}, P_{23}),$$

given by three polynomials of degree 2. Either F'' is an embedding of $\mathbb{P}^1$ as a conic, and our claim is established, or there is a linear relation among P_{12}, P_{13}, P_{23}. This is however impossible if the three quadratic polynomials P_{12}, P_{13}, P_{23} are taken to be general. □

REMARK 4.5. There remain the cases (with $m = 2$)

$$p_g = 5, m = 2, k = 4, r_1 = 6, r_2 = 7,$$

$$p_g = 6, m = 2, k = 4, r_1 = 5, r_2 = 11.$$

In the latter case (this was the first example found) the canonical map is again a non injective birational morphism for the following choice of generating systems: e_1, e_2, e_3, e_4, e and $(e_1+e_2), (e_1+e_3)^2, (e_1+e_4), (e_2+e_3), (e_2+e_4), (e_3+e_4), (e_1+e_2+e_3), (e_1+e_3+e_4), (e_1+e_2+e_4)^2$. There occur here 5 characters, with dimensions $2, 1, 1, 1, 1$, so that calculations here are similar to case II) of theorem 4.3, but slightly more complicated.

5. New double canonical surfaces

Let us now consider the case $G = (\mathbb{Z}/2)^3$, $p_g = 4, q = 0$, $r_1 = 6, r_2 = 9$. Therefore we can use the standard argument yielding generating systems e_1^2, e_2^2, e_3^2, respectively $e^{2B}, (e_2+e_3)^{1+2B_1}, (e_1+e_3)^{1+2B_2}, (e_1+e_2)^{1+2B_3}$. Here, $B+B_1+B_2+B_3 = 3$, hence the only choice in order that the image is not a rational surface is easily seen to be $B = B_1 = B_2 = 1$.

The only characters which occur are the three characters $e_i^\vee + e_j^\vee$, and we get

$$N(e_1^\vee + e_2^\vee) = 1 \cdot 2,\ N(e_1^\vee + e_3^\vee) = 1 \cdot 1,\ N(e_2^\vee + e_3^\vee) = 1 \cdot 1.$$

It is now immediate to observe that the involution $(e, 1) \in G \times G$ $(e = e_1 + e_2 + e_3)$ acts on the canonical system as multiplication by $+1$.

Let Z be the quotient surface $S/(e,1)$, so that the canonical map Φ of S factors through Z. Z is a $(\mathbb{Z}/2)^2$ covering of Q with branch loci of bidegrees $(2,3), (2,3), (2,1)$ made of unions of vertical and horizontal lines, so Z has $14 = 6+6+2$ nodes, which are the images of the 14 isolated fixed points for the involution $(e,1)$ on S. This involution has moreover a fixed divisor, consisting of the reduced inverse image $F_1' + F_2'$ of the two points in the second copy of $\mathbb{P}^1$ where the local monodromy is e.

Since the involution acts as the identity on $H^0(K_S)$, it follows that the isolated fixed points occur in the base locus of the canonical system with even multiplicity, while the fixed curves occur in the base locus with odd multiplicities k_1, k_2.

Hence if $q : S \to Z$ is the quotient morphism, we claim that $K_S = q^*(K_Z) + F_1 + F_2$. In fact $40 = K_S^2 = q^*(K_Z + \frac{k_1}{2}F_1 + \frac{k_2}{2}F_2)^2$, with $k_1, k_2 \geq 1$, hence, since $K_Z^2 = 12$, $F_j K_Z = 4$ (the two fibrations on Z have respective genera $g_1 = 4, g_2 = 3$, $20 = 12 + 4(k_1 + k_2) \Rightarrow k_1 = k_2 = 1$.

As in [**Cat99**] we see that

$$H^0(Z, \mathcal{O}_Z(K_Z)) = H^0(Q, \mathcal{O}_Q(0,1))u_{12}\sigma_{12} \oplus \mathbb{C}u_{13}\sigma_{13} \oplus \mathbb{C}u_{23}\sigma_{23},$$

here again $z_{ij} = u_{ij}\sigma\sigma_{ij}$, and the canonical system of Z has no base points.

Moreover, the canonical map of Z is birational if the three quadratic forms u_{ij} are linearly independent (see proposition 4, pages 104 and 105 of [**Cat99**]).

We summarize our discussion in the following

THEOREM 5.1. *Surfaces isogenous to a product with* $G = (\mathbb{Z}/2)^3$, $p_g = 4, q = 0$, $r_1 = 6, r_2 = 9$, *and generating systems* e_1^2, e_2^2, e_3^2, *respectively* $e^2, (e_2+e_3)^3, (e_1+e_3)^3, (e_1+e_2)^1$ *yield a connected component of the moduli space of the minimal surfaces of general type with* $p_g = 4, q = 0, K_S^2 = 40$.

For each of them there is a nontrivial involution $\psi := (e,1)$ *which acts trivially on the canonical system, and with quotient a 14-nodal surface* Z *with* $K_Z^2 = 12$, *with* K_Z *ample.*

Thus the canonical map of S *factors through the canonical map of* Z, *which is a morphism, and is birational onto its image for general choice of the six branch points of* $C_1 \to C_1/G = \mathbb{P}^1$.

The fixed locus of ψ *consists of* 14 *isolated points, and two curves which form the reduced inverse image* $F_1' + F_2'$ *of the two points in the second copy of* $\mathbb{P}^1$ *where the local monodromy is* e.

The canonical system of S *has base locus equal to the divisor* $F_1' + F_2'$.

PROOF. In [**Cat00**] it was proven that SIP's with given ramification data form a connected component of the moduli space.

The base points of $|K_S|$ come from the fact that

$$H^0(Z, \mathcal{O}_Z(K_Z)) = H^0(Q, \mathcal{O}_Q(0,1))z_{12} \oplus \mathbb{C}z_{13} \oplus \mathbb{C}z_{23},$$

and $z_{ij} = u_{ij}\sigma\sigma_{ij}$, so that we get $\sigma = 0$ as base locus, and the rest was already shown. □

Ackowledgments. The author gratefully acknowledges the support of the ERC Advanced grant n. 340258, 'TADMICAMT', and of KIAS Seoul, where he has been Research scholar in April 2016.

Thanks to Jong Hae Keum and Yongnam Lee for organizing a Workshop 'Algebraic Surfaces and Moduli' at KAIST Daejeon, March 2016, where a first example was presented.

References

[Bab34] D. W. Babbage, *Multiple canonical surfaces.* **Math. Proc. Cambridge Philos. Soc. 30** (1934), 297–308.

[BC08] Ingrid C. Bauer and Fabrizio Catanese, *A volume maximizing canonical surface in 3-space*, Comment. Math. Helv. **83** (2008), no. 2, 387–406. MR2390050

[BCG08] I. C. Bauer, F. Catanese, and F. Grunewald, *The classification of surfaces with* $p_g = q = 0$ *isogenous to a product of curves*, Pure Appl. Math. Q. **4** (2008), no. 2, Special Issue: In honor of Fedor Bogomolov., 547–586. MR2400886

[Bea79] Arnaud Beauville, *L'application canonique pour les surfaces de type général* (French), Invent. Math. **55** (1979), no. 2, 121–140. MR553705

[Bea13] Arnaud Beauville, *Surfaces algébriques complexes* (French), Société Mathématique de France, Paris, 1978. Avec une sommaire en anglais; Astérisque, No. 54. MR0485887

[Burn66] Pol Burniat, *Sur les surfaces de genre* $P_{12} > 1$ (French), Ann. Mat. Pura Appl. (4) **71** (1966), 1–24. MR0206810

[Cat81] F. Catanese, *Babbage's conjecture, contact of surfaces, symmetric determinantal varieties and applications*, Invent. Math. **63** (1981), no. 3, 433–465. MR620679

[Cat97] Fabrizio Catanese, *Homological algebra and algebraic surfaces*, Algebraic geometry—Santa Cruz 1995, Proc. Sympos. Pure Math., vol. 62, Amer. Math. Soc., Providence, RI, 1997, pp. 3–56. MR1492517

[Cat99] Fabrizio Catanese, *Singular bidouble covers and the construction of interesting algebraic surfaces*, Algebraic geometry: Hirzebruch 70 (Warsaw, 1998), Contemp. Math., vol. 241, Amer. Math. Soc., Providence, RI, 1999, pp. 97–120. MR1718139

[Cat00] Fabrizio Catanese, *Fibred surfaces, varieties isogenous to a product and related moduli spaces*, Amer. J. Math. **122** (2000), no. 1, 1–44. MR1737256

[Cat16] Fabrizio Catanese, *Canonical surfaces of higher degree,* dedicated to the 60th birthday of Philippe Ellia. Published online on the **Rend. Circ.Mat. Palermo**, http://link.springer.com/article/10.1007/s12215-016-0274-6.

[C-S02] Fabrizio Catanese and Frank-Olaf Schreyer, *Canonical projections of irregular algebraic surfaces*, Algebraic geometry, de Gruyter, Berlin, 2002, pp. 79–116. MR1954059

[Cil81] Ciro Ciliberto, *Canonical surfaces with* $p_g = p_a = 4$ *and* $K^2 = 5, \cdots, 10$, Duke Math. J. **48** (1981), no. 1, 121–157. MR610180

[CMLR15] Ciro Ciliberto, Margarida Mendes Lopes, and Xavier Roulleau, *On Schoen surfaces*, Comment. Math. Helv. **90** (2015), no. 1, 59–74. MR3317333

[Frap13] Davide Frapporti, *Mixed quasi-étale surfaces, new surfaces of general type with* $p_g = 0$ *and their fundamental group*, Collect. Math. **64** (2013), no. 3, 293–311. MR3084399

[Kap-Kap15] Grzegorz Kapustka and Michał Kapustka, *Bilinkage in codimension 3 and canonical surfaces of degree 18 in* $\mathbb{P}^5$, Ann. Sc. Norm. Super. Pisa Cl. Sci. (5) **16** (2016), no. 3, 767–787. MR3618076

[Lied03] Christian Liedtke, *Singular abelian covers of algebraic surfaces*, Manuscripta Math. **112** (2003), no. 3, 375–390. MR2067044

[Par91] Rita Pardini, *Canonical images of surfaces*, J. Reine Angew. Math. **417** (1991), 215–219. MR1103913

[Tan92] Sheng Li Tan, *Surfaces whose canonical maps are of odd degrees*, Math. Ann. **292** (1992), no. 1, 13–29. MR1141782

[G-Z77] G. van der Geer and D. Zagier, *The Hilbert modular group for the field* $\mathbf{Q}(\sqrt{13})$, Invent. Math. **42** (1977), 93–133. MR0485704

Lehrstuhl Mathematik VIII, Mathematisches Institut der Universität Bayreuth, NW II, Universitätsstr. 30, 95447 Bayreuth

Email address: fabrizio.catanese@uni-bayreuth.de

Contemporary Mathematics
Volume **712**, 2018
http://dx.doi.org/10.1090/conm/712/14342

Degeneration of differentials and moduli of nodal curves on $K3$ surfaces

C. Ciliberto, F. Flamini, C. Galati, and A. L. Knutsen

It is our pleasure to dedicate this paper to Lawrence Ein on the occasion of his sixtieth birthday.

ABSTRACT. We consider, under suitable assumptions, the following situation: $\mathcal{B}$ is a component of the moduli space of polarized surfaces and $\mathcal{V}_{m,\delta}$ is the universal Severi variety over $\mathcal{B}$ parametrizing pairs (S, C), with $(S, H) \in \mathcal{B}$ and $C \in |mH|$ irreducible with exactly δ nodes as singularities. The moduli map $\mathcal{V} \to \mathcal{M}_g$ of an irreducible component $\mathcal{V}$ of $\mathcal{V}_{m,\delta}$ is generically of maximal rank if and only if certain cohomology vanishings hold. Assuming there are suitable semistable degenerations of the surfaces in $\mathcal{B}$, we provide sufficient conditions for the existence of an irreducible component $\mathcal{V}$ where these vanishings are verified. As a test, we apply this to $K3$ surfaces and give a new proof of a result recently independently proved by Kemeny and by the present authors.

1. Introduction

Let (S, H) be a smooth, projective, polarized, complex surface, with H an ample line bundle such that the linear system $|H|$ contains smooth, irreducible curves. We set

$$p := p_a(H) = \frac{1}{2}(H^2 + K_S \cdot H) + 1,$$

the arithmetic genus of any curve in $|H|$. For any integer $m \geqslant 1$, we set

$$\ell(m) := \dim(|mH|) \text{ and } p(m) := p_a(mH) = \frac{m(m-1)}{2}H^2 + m(p-1).$$

One has

$$\ell(m) = \chi(\mathcal{O}_S) + m^2H^2 - p(m), \text{ for } m \gg 0.$$

For any integer $\delta \in \{0, \ldots, \ell(m)\}$, consider the locally closed, functorially defined subscheme of $|mH|$

$$V_{m,\delta}(S, H) \text{ (simply } V_{m,\delta}(S) \text{ or } V_{m,\delta} \text{ when } H \text{ or } (S, H) \text{ are understood),}$$

2010 *Mathematics Subject Classification.* Primary 14H10, 14B05, 14D06; Secondary 14B07, 14J28, 14N05.

Key words and phrases. Severi varieties, moduli map, nodal curves, $K3$ surfaces.

Acknowledgments. The first three authors have been supported by the GNSAGA of Indam and by the PRIN project "Geometry of projective varieties", funded by the Italian MIUR. The authors warmly thank the referee for the careful reading, the very positive report as well as for making several suggestions that improved the readability of the paper.

which is the parameter space for the universal family of irreducible curves in $|mH|$ having only δ nodes as singularities; this is called the (m, δ)–*Severi variety* of (S, H).

We will assume that there exists a Deligne–Mumford *moduli stack* $\mathcal{B}$ parametrizing isomorphism classes of polarized surfaces (S, H) as above. Since we will basically deal only with local properties, we can get rid of the stack structure. Indeed, up to replacing $\mathcal{B}$ with an étale finite type representable cover, we may pretend that $\mathcal{B}$ is a *fine moduli scheme.* Although not necessary, we will assume that $\mathcal{B}$ is irreducible (otherwise one may replace $\mathcal{B}$ with one of its components).

Then we may consider the scheme $\mathcal{V}_{m,\delta}$, called the (m, δ)–*universal Severi variety* over $\mathcal{B}$, which is endowed with a morphism

$$\phi_{m,\delta} : \mathcal{V}_{m,\delta} \to \mathcal{B},$$

whose fiber over $(S, H) \in \mathcal{B}$ is $V_{m,\delta}(S, H)$. A point in $\mathcal{V}_{m,\delta}$ can be identified with a pair (S, C), with $(S, H) \in \mathcal{B}$ and $C \in V_{m,\delta}(S, H)$.

We make the following:

ASSUMPTION 1.1. (i) $\mathcal{B}$ is smooth;
(ii) for all $(S, H) \in \mathcal{B}$, the surface S is *regular*, i.e., $h^1(S, \mathcal{O}_S) = 0$, and $h^0(S, \mathcal{T}_S) = 0$, i.e., S has no positive dimensional automorphism group;
(iii) for any $m \geqslant 1$ and $\delta \in \{0, \ldots, \ell(m)\}$ and for all $(S, H) \in \mathcal{B}$, $\ell(m)$ is constant and the Severi variety $V_{m,\delta}(S, H)$ is smooth, of pure (and *expected*) dimension $\ell(m) - \delta$ (hence $\mathcal{V}_{m,\delta}$ is smooth, of pure dimension $\dim(\mathcal{B}) + \ell(m) - \delta$, and $\phi_{m,\delta}$ is smooth and surjective).

REMARK 1.2. In Assumption 1.1(i) we could have asked $\mathcal{B}$ to be *generically* smooth, and in (iii) we could have asked that for the *general* $(S, H) \in \mathcal{B}$, the Severi variety $V_{m,\delta}(S, H)$ is smooth, of pure dimension $\ell(m) - \delta$. But under these weaker assumptions, (i) and (iii) hold on a Zariski dense open subset. Since we will be interested only in what happens at the general point of $\mathcal{B}$, we may replace $\mathcal{B}$ with this open subset. The hypotheses in (ii) are technical and not strictly necessary for our purposes, but they make things easier for us.

Conditions (i)–(iii) hold in some important cases, e.g., for polarized $K3$ surface of genus p (in which case the moduli stack is usually denoted by $\mathcal{K}_p$, is of dimension 19, and $\ell(m) = p(m)$ for any $m \geqslant 1$, cf., e.g., [**5**, **12**, **15**, **16**]). Moduli spaces exist also for polarized Enriques surfaces (cf. [**14**]) and degenerations of (polarized) Enriques surfaces are also studied (cf. e.g. [**8**, **18**, **20**, **22**]).

Another relevant class is the one of minimal, regular surfaces of general type (S, H) whose moduli space has at least one (generically) smooth component $\mathcal{B}$ with points (S, H) verifying, for some m and δ, the conditions in [**3**, **10**, **11**] ensuring smoothness and expected dimension of any component of $V_{m,\delta}(S, H)$. Particular cases are, for some m and δ, surfaces in $\mathbb{P}^3$ of degree $d \geqslant 5$ (cf. [**2**]) and complete intersections of general type in $\mathbb{P}^N$.

Consider now the *moduli map*

$$\psi_{m,\delta} : \mathcal{V}_{m,\delta} \to \mathcal{M}_g, \text{ where } g = p(m) - \delta,$$

and where $\mathcal{M}_g$ denotes the moduli space of smooth, genus–g curves: $\psi_{m,\delta}$ sends a curve to the isomorphism class of its normalization. In this set–up, one is interested in the following general problem: find conditions on m and δ ensuring the existence of a component $\mathcal{V}$ of $\mathcal{V}_{m,\delta}$ such that $\psi_{m,\delta}|_{\mathcal{V}}$ is either *generically finite* onto its image or *dominant* onto $\mathcal{M}_g$. By taking into account Assumption 1.1, in principle, one may expect

dominance if	$\dim(\mathcal{B}) + \ell(m) - \delta \geqslant \dim(\mathcal{M}_g)$,
generic finiteness onto its image if	$\dim(\mathcal{B}) + \ell(m) - \delta \leqslant \dim(\mathcal{M}_g)$.

The typical example is the case of polarized $K3$ surfaces studied by various authors (cf., e.g., [**5, 12, 15–17, 19**]). In particular, [**5**] and Kemeny in [**17**] independently show that, as expected, $\psi_{m,\delta}$ is generically finite on some component for all $g = p(m) - \delta \geqslant 11$ with only a few finite possible exceptions (m, g) with $m \leqslant 4$, for fixed p; moreover [**5**] shows that $\psi_{m,\delta}$ is dominant, as expected, for $g \leqslant 11$ with only a few finite possible exceptions (m, g) with $m \leqslant 4$, for fixed p. The precise result for $m = 1$ is the following:

THEOREM 1.3. *Let $\mathcal{V}_{m,\delta}$ be the universal Severi variety over the moduli space $\mathcal{K}_p$ of polarized $K3$ surfaces (S, H) of genus p. For $m = 1$ and $g = p - \delta$ one has:*
(a) [**5, 17**] *if $g \geqslant 15$ there is a component $\mathcal{V}$ of $\mathcal{V}_{1,\delta}$ such that ${\psi_{1,\delta}}_{|\mathcal{V}}$ is generically finite onto its image;*
(b) [**5**] *if $g \leqslant 7$ there is a component $\mathcal{V}$ of $\mathcal{V}_{1,\delta}$ such that ${\psi_{1,\delta}}_{|\mathcal{V}}$ is dominant onto $\mathcal{M}_g$.*

In case (a) Kemeny's result is stronger in the sense that he may weaken the assumptions on g for infinitely many p's.

The proofs in [**5, 17**], although different, both rely on studying the fibers of the moduli map on curves on *special* $K3$ surfaces. Kemeny's proof is inspired by ideas of [**19**], and uses appropriate curves on $K3$ surfaces with high rank Picard group. The approach in [**5**] is by specialization to a reducible $K3$ surface in a partial compactification of $\mathcal{K}_p$ and therefore uses an extension of the moduli map to an appropriate partial compactification of the Severi variety containing reducible curves, with target space $\overline{\mathcal{M}}_g$.

In the present paper we want to present a different approach to the aforementioned general problem. This approach relies on two different techniques. Firstly, it is based on the analysis of first order deformations of pairs $(S, C) \in \mathcal{V}_{m,\delta}$ as in [**12**, § 4]. The strategy in [**12**], which requires Assumption 1.1, was originally introduced for polarized $K3$ surfaces and for $m = 1$, but can be easily adapted to $m \geqslant 1$ and to the case where the canonical bundle is not necessarily trivial (cf. also [**15**, Thm. 1.1(ii)]). The upshot is the following. In the above setting, take $(S, C) \in \mathcal{V}_{m,\delta}$ and set $Z := \mathrm{Sing}(C)$. Arguing as in [**12**, §§ 4-5], the differential $\mathrm{d}\psi_{(S,C)}$ of $\psi := \psi_{m,\delta}$ at (S, C) can be identified with a suitable cohomology map (the $H^1(\tau)$ in [**12**, (4.21)]). In particular, if $\mu_Z : \tilde{S} \to S$ is the blowing-up of S at Z, $\tilde{C}$ is the strict transform of C and $\mathcal{T}_S$ and $\mathcal{T}_{\tilde{C}}$ are the tangent bundles of S and $\tilde{C}$ respectively, then

$$\mathrm{coker}(\mathrm{d}\psi_{(S,C)}) \simeq H^2(\mu_Z^*(\mathcal{T}_S)(-\tilde{C}))$$

and

$$\ker(\mathrm{d}\psi_{(S,C)}) \simeq H^1(\mu_Z^*(\mathcal{T}_S)(-\tilde{C}))/H^0(\mathcal{T}_{\tilde{C}}).$$

Moreover, by the Serre duality theorem and Leray isomorphism, as in [**12**, Proof of Thm (5.1)]), one has that

$$H^2(\mu_Z^*(\mathcal{T}_S)(-\tilde{C})) \simeq H^0(\Omega_S(mH + K_S) \otimes \mathcal{J}_{Z/S})$$

and

$$H^1(\mu_Z^*(\mathcal{T}_S)(-\tilde{C})) \simeq H^1(\Omega_S(mH+K_S)\otimes \mathcal{J}_{Z/S}).$$

To summarize, if (S,C) belongs to a component $\mathcal{V}$ of $\mathcal{V}_{m,\delta}$, the differential of the moduli map $\psi_{m,\delta}$ at (S,C) is

$$\text{(1.1)}\qquad \begin{array}{lll} \textit{surjective} & \textit{if and only if} & h^0(\Omega_S(mH+K_S)\otimes \mathcal{J}_{Z/S})=0,\\ \textit{injective} & \textit{if} & h^1(\Omega_S(mH+K_S)\otimes \mathcal{J}_{Z/S})=0. \end{array}$$

Finally, the vanishing of $H^1(\Omega_S(mH+K_S)\otimes \mathcal{J}_{Z/S})$ is equivalent to the injectivity of the differential of $\psi_{m,\delta}$ at (S,C) if $g \geq 2$.

Unfortunately, the vanishings (1.1) are, in general, not so easy to be proved, even if one assumes $\delta = 0$ (cf. [**1**]). This is where the second tool of our approach enters the scene (see §2). In order to prove the above vanishings, we propose to use degenerations. We assume in fact that the surfaces in $\mathcal{B}$ and nodal curves on them possess good semistable degenerations with limiting surfaces that are reducible in two components (the more general case of reducibility in more components could be treated, but, for simplicity, we do not dwell on this here). Then we look at the *limits* of the relevant cohomology spaces. The latter are driven by the so–called *abstract log complex* (see [**13**, § 3]). Using this we arrive at sufficient conditions for the vanishings in (1.1) to hold, expressed in terms of cohomological properties of suitable sheaves of forms on the two components of the limit surface (cf. § 2.2). These properties are hopefully easier to prove than the vanishings in (1.1), since these components are simpler than S. The results are summarized in Proposition 2.7, which is the main result of this note.

In the rest of the paper (i.e. §§3 and 4) we test our approach in the (known) case of $K3$ surfaces for $m = 1$, giving a new proof of Theorem 1.3. We do not claim that this is easier than the proofs in [**5**, **17**], but it works quite nicely and gives good hopes to fruitfully apply the same method in other unexplored cases, like the ones mentioned at the end of Remark 1.2. We also mention that the approach of this paper can be applied to the $m > 1$ case (using a slightly different degeneration, namely the one in [**4**]), but we leave this out as the bounds we obtain depend linearly on m and are thus considerably weaker than the bounds in [**5**, **17**]. Furthermore, although the present approach gives the same result as [**5**] for $m = 1$, the analysis of this case in [**5**] is finer (as it studies the family of degenerate $K3$s on which the curves in the fibers of the moduli map live) and is needed in the proof of the $m > 1$ case. Thus, the approach in this paper cannot replace the proof of the $m = 1$ case in [**5**].

Terminology and conventions. We work over $\mathbb{C}$. For X any Gorenstein, projective variety, we denote by $\mathcal{O}_X$ and $\omega_X \simeq \mathcal{O}_X(K_X)$ the structural and the canonical line bundle, respectively, where K_X is a canonical divisor. We denote by $\mathcal{T}_X$ and Ω_X the tangent sheaf and the sheaf of 1-forms on X, respectively. For $Y \subset X$ any closed subscheme, $\mathcal{J}_{Y/X}$ (or simply $\mathcal{J}_Y$ if X is intended) will denote its ideal sheaf and $\mathcal{N}_{Y/X}$ its normal sheaf. We use $\sim$ to denote linear equivalence of divisors. We often abuse notation and identify divisors with the corresponding line bundles, using the additive and the multiplicative notation interchangeably. Finally, we use the convention that if $\mathcal{F}$ is a sheaf on a scheme X and $Y \subset X$ is a subscheme, then $H^0(\mathcal{F})_{|Y}$ is the image of the restriction map $H^0(\mathcal{F}) \to H^0(\mathcal{F}\otimes \mathcal{O}_Y)$.

2. Semistable degenerations and the abstract log complex

In this section we will provide a tool for proving the vanishings of the cohomology groups occurring in (1.1) by degeneration of the surface and semicontinuity. The main results are summarized in Proposition 2.7 below.

2.1. Semistable degenerations. We recall some basic facts concerning semistable degenerations of compact complex surfaces and the associated abstract log complex (see [**13**, § 3]). This complex allows to define flat limits of the sheaves occurring in (1.1).

DEFINITION 2.1. (i) Let R and $\mathfrak{R}$ be connected, complex analytic varieties. Let $\Delta = \{t \in \mathbb{C} | \mid t \mid < 1\}$. A proper, flat morphism $\alpha : \mathfrak{R} \to \Delta$ is said to be a deformation of R if R is the scheme theoretical fibre of α over 0. Accordingly, R is said to be a flat limit of R_t, the scheme theoretical fibre of α over $t \neq 0$.
(ii) If there is a line bundle $\mathcal{H}$ on $\mathfrak{R}$, set $H := \mathcal{H}_{|R}$ and $H_t := \mathcal{H}_{|R_t}$ for $t \neq 0$. Then the pair (R, H) is said to be a limit of (R_t, H_t) for $t \neq 0$.
(iii) The deformation (or, equivalently, the degeneration) is *semistable* if $\mathfrak{R}$ is smooth (so that we may assume that R_t is smooth for $t \neq 0$) and R has at most normal crossing singularities.
(iv) Assume that there are δ disjoint sections $s_1, \ldots, s_\delta$ of α and that their images $\mathcal{Z}_1, \ldots, \mathcal{Z}_\delta$ are smooth curves in $\mathfrak{R}$, disjoint from $\mathrm{Sing}(R)$. Set $\mathcal{Z} := \bigcup_{i=1}^{\delta} \mathcal{Z}_i$. Then we say that $Z := \mathcal{Z}_{|R}$ is a limit of $Z_t := \mathcal{Z}_{|R_t}$ for $t \neq 0$.
(v) If conditions (i)-(iv) are satisfied, we will say that (R, H, Z) is a semistable degeneration of (R_t, H_t, Z_t), for $t \neq 0$, or that (R_t, H_t, Z_t), for $t \neq 0$, admits the semistable degeneration (R, H, Z).

Let $\mathfrak{R} \xrightarrow{\alpha} \Delta$ be a semistable degeneration of surfaces as in Definition 2.1. Assume that all components of R are smooth and R has no triple point. Then $\mathrm{Sing}(R)$ consists of the transversal intersection points of pairs of components of R.

Consider the sheaf $\Omega_{\mathfrak{R}/\Delta}(\log R)$ on $\mathfrak{R}$ defined by the exact sequence (cf. [**23**] and [**7**, § 3.3])

$$0 \longrightarrow \alpha^*(\Omega_\Delta(0)) \xrightarrow{\iota} \Omega_{\mathfrak{R}}(\log R) \longrightarrow \Omega_{\mathfrak{R}/\Delta}(\log R) \longrightarrow 0. \tag{2.1}$$

The map ι in (2.1) has rank one at every point, whence $\Omega_{\mathfrak{R}/\Delta}(\log R)$ is locally free of rank 2, as recalled in the following remark.

REMARK 2.2. Away from R, one has the isomorphisms $\Omega_{\mathfrak{R}}(\log R) \simeq \Omega_{\mathfrak{R}}$ and $\Omega_{\mathfrak{R}/\Delta}(\log R) \simeq \Omega_{\mathfrak{R}/\Delta}$. In particular

$$\Omega_{\mathfrak{R}/\Delta}(\log R) \otimes \mathcal{O}_{R_t} \simeq \Omega_{R_t}, \text{ for any } t \neq 0.$$

Let $P \in R - \mathrm{Sing}(R)$, and let $x,\ y,\ z$ be local coordinates on $\mathfrak{R}$ around P. Let t be the coordinate on Δ and assume that α is locally defined by $t = x$ around p. Then $\Omega_{\mathfrak{R}}(\log R)$ is locally free generated by $\frac{dx}{x}$, dy, dz, the map ι is defined by

$$\frac{dt}{t} \longrightarrow \frac{dx}{x}$$

hence $\Omega_{\mathfrak{R}/\Delta}(\log R)$ is locally free generated by dy, dz (cf. [**9**, Prop. 2.2.c]).
Let now $P \in \mathrm{Sing}(R)$. From our assumptions, we may assume that α is locally defined by $t = xy$. Then $\Omega_{\mathfrak{R}}(\log R)$ is locally free generated by $\frac{dx}{x}$, $\frac{dy}{y}$, dz, the

map ι is defined by

$$\frac{dt}{t} \longrightarrow \frac{dx}{x} + \frac{dy}{y}$$

and so $\Omega_{\mathfrak{R}/\Delta}(\log R)$ is locally free generated by $\frac{dx}{x} = -\frac{dy}{y}$, dz.

Following [**13**, § 3], we set

$$\Lambda^1_R := \Omega_{\mathfrak{R}/\Delta}(\log R)|_R,$$

which is locally free on R.

LEMMA 2.3. *Let (R, H, Z) be a semistable degeneration of (R_t, H_t, Z_t) for $t \neq 0$, as in Definition 2.1. Assume furthermore that all components of R are smooth and R has no triple points. Then for all $m \in \mathbb{Z}$, $i \in \mathbb{N}$ and $t \neq 0$, one has*

$$h^i(\Lambda^1_R \otimes \mathcal{O}_R(mH + K_R) \otimes \mathcal{J}_{Z/R}) \geqslant h^i(\Omega_{R_t} \otimes \mathcal{O}_{R_t}(mH_t + K_{R_t}) \otimes \mathcal{J}_{Z_t/R_t}).$$

PROOF. The statement follows by semicontinuity as $\Omega_{\mathfrak{R}/\Delta}(\log R)$ is flat over Δ and the ideal sheaf $\mathcal{J}_{\mathcal{Z}/\mathfrak{R}}$ is flat over Δ by [**21**, Prop. 4.2.1(ii)]. □

2.2. Degenerations of differentials. From now on we assume that $R = R_1 \cup R_2$, with R_1, R_2 smooth, irreducible, projective surfaces, with transversal intersection along a smooth, irreducible curve $E := R_1 \cap R_2$. Then

$$K_{R|_{R_i}} = (K_{\mathfrak{R}} + R)_{|_{R_i}} = (K_{\mathfrak{R}} + R_1 + R_2)_{|_{R_i}} = K_{R_i} + E, \; i = 1, 2. \tag{2.2}$$

In this situation there are exact sequences involving $\Lambda^1_R \otimes \mathcal{O}_R(mH + K_R) \otimes \mathcal{J}_{Z/R}$, which allow us to compute its cohomology by conducting computations on R_1 and R_2.

Consider the exact sequences

$$0 \longrightarrow \Omega_{R_i} \longrightarrow \Omega_{R_i}(\log E) \xrightarrow{\rho_i} \mathcal{O}_E \longrightarrow 0 , \quad i = 1, 2, \tag{2.3}$$

and

$$0 \longrightarrow \Omega_{R_i}(\log E) \otimes \mathcal{O}_{R_i}(-E) \longrightarrow \Omega_{R_i} \xrightarrow{r_i} \omega_E \longrightarrow 0, \quad i = 1, 2, \tag{2.4}$$

where ρ_i is the *residue map* and r_i is the *trace map* of differential forms, cf. [**9**, § 2].

For the reader's convenience, we recall how ρ_i and r_i are defined locally around a point of E. We may assume that locally, in an open subset of $\mathbb{C}^4$ with coordinates x, y, z, t, the equation of $\mathfrak{R}$ is $xy - t = 0$. We let R_1 be given by $x = t = 0$ and R_2 by $y = t = 0$, so that E is given by $x = y = t = 0$. In this chart

$$\mathcal{O}_{R_1} \simeq \mathbb{C}[[y, z]], \; \Omega_{R_1} \simeq \mathbb{C}[[y, z]]\, dy \oplus \mathbb{C}[[y, z]]\, dz,$$

$$\Omega_{R_1}(\log E) \simeq \mathbb{C}[[y, z]]\, \frac{dy}{y} \oplus \mathbb{C}[[y, z]]\, dz, \; \mathcal{O}_E \simeq \mathbb{C}[[z]] \text{ and } \omega_E \simeq \mathbb{C}[[z]]\, dz,$$

$$\rho_1\Big(f(y, z)\frac{dy}{y} + g(y, z)\, dz\Big) := f(0, z), \qquad r_1\Big(f(y, z)\, dy + g(y, z)\, dz\Big) := g(0, z)\, dz$$

and similarly for R_2.

Let $a : R_1 \sqcup R_2 \to R$ be the desingularization of R; consider the exact sequence

$$0 \longrightarrow \mathcal{A} \longrightarrow a_*\Big(\Omega_{R_1}(\log E) \oplus \Omega_{R_2}(\log E)\Big) \xrightarrow{\rho} \mathcal{O}_E \longrightarrow 0, \tag{2.5}$$

defining $\mathcal{A}$, where $\rho := \rho_1 + \rho_2$. As in [**13**, Pf. of Lemma 3.1, p.94-95], Λ^1_R fits in the exact sequence

$$0 \longrightarrow \Lambda^1_R \longrightarrow \mathcal{A} \xrightarrow{\ r\ } \omega_E \longrightarrow 0, \tag{2.6}$$

where r is locally defined as follows: if

$$(\omega_1, \omega_2) = \Big(f_1(y,z)\frac{dy}{y} + g_1(y,z)dz,\ f_2(x,z)\frac{dx}{x} + g_2(x,z)dz\Big)$$

is a local section of $\mathcal{A}$, then

$$r(\omega_1, \omega_2) := \big(g_1(0,z) - g_2(0,z)\big)dz.$$

REMARK 2.4. Notice that

$$a_*(\Omega_{R_1} \oplus \Omega_{R_2}) \hookrightarrow \mathcal{A} \text{ and } r|_{a_*(\Omega_{R_1} \oplus \Omega_{R_2})} = r_1 - r_2, \tag{2.7}$$

where r_1 and r_2 are the maps in (2.4). One may verify that the map r_i does not extend to a map on $\mathcal{A}$.

Having in mind (1.1) and Lemma 2.3, we are looking for conditions ensuring the vanishing of $h^i(\Lambda^1_R \otimes \mathcal{O}_R(mH + K_R) \otimes \mathcal{J}_{Z/R})$, for $i = 0, 1$. To this end, consider the maps

$$\begin{array}{rcl} \mathfrak{F} & := & \rho \otimes \mathcal{O}_R(mH + K_R) \otimes \mathcal{J}_{Z/R}, \\ q & := & r \otimes \mathcal{O}_R(mH + K_R) \otimes \mathcal{J}_{Z/R}, \\ \mathfrak{F}_i & := & \rho_i \otimes \mathcal{O}_{R_i}(mH + K_{R_i} + E) \otimes \mathcal{J}_{Z_i} \text{ and} \\ q_i & := & r_i \otimes \mathcal{O}_{R_i}(mH + K_{R_i} + E) \otimes \mathcal{J}_{Z_i}. \end{array}$$

Then by (2.3), (2.4) and the fact that Z_i avoids E, for $i = 1, 2$ we have

(2.8)

$$\begin{array}{ccccccc} & & & 0 & & & \\ & & & \downarrow & & & \\ 0 \to \Omega_{R_i}(\log E)(mH + K_{R_i}) \otimes \mathcal{J}_{Z_i} & \longrightarrow & & \Omega_{R_i}(mH + K_{R_i} + E) \otimes \mathcal{J}_{Z_i} & \xrightarrow{q_i} & \omega_E(mH + K_{R_i} + E) \to 0 \\ & & & \downarrow & & & \\ & & & \Omega_{R_i}(\log E)(mH + K_{R_i} + E) \otimes \mathcal{J}_{Z_i} & & & \\ & & & \downarrow{\scriptstyle \mathfrak{F}_i} & & & \\ & & & \mathcal{O}_E(mH + K_{R_i} + E) & & & \\ & & & \downarrow & & & \\ & & & 0 & & & \end{array}$$

Moreover, by (2.5)–(2.7), observing $\mathfrak{F} = \mathfrak{F}_1 + \mathfrak{F}_2$ we have
(2.9)

$$\begin{array}{ccccccc}
0 \longrightarrow \mathcal{A}\otimes\mathcal{O}_R(mH+K_R)\otimes\mathcal{J}_{Z/R} & \longrightarrow & a_*\Big(\oplus_{i=1}^2 \Omega_{R_i}(\log E)(mH+K_{R_i}+E)\otimes\mathcal{J}_{Z_i}\Big) & \stackrel{\mathfrak{F}}{\longrightarrow} & \mathcal{O}_E(mH+K_R) \longrightarrow 0 \\
 & & \uparrow & & \\
 & & a_*\Big(\oplus_{i=1}^2 \Omega_{R_i}(mH+K_{R_i}+E)\otimes\mathcal{J}_{Z_i}\Big) & \stackrel{q_1-q_2}{\searrow} & \\
 & & \downarrow & & \\
0 \longrightarrow \Lambda_R^1(mH+K_R)\otimes\mathcal{J}_{Z/R} & \longrightarrow & \mathcal{A}\otimes\mathcal{O}_R(mH+K_R)\otimes\mathcal{J}_{Z/R} & \stackrel{q}{\longrightarrow} & \omega_E(mH+K_R) \longrightarrow 0.
\end{array}$$

The next two lemmas provide sufficient conditions for the vanishings of the two first cohomology groups of $\Lambda_R^1(mH+K_R)\otimes\mathcal{J}_{Z/R}$.

LEMMA 2.5. *Assume that*

$$\operatorname{im}(H^0(\mathfrak{F}_1)) \cap \operatorname{im}(H^0(\mathfrak{F}_2)) = \{0\}. \tag{2.10}$$

Then $h^0(\Lambda_R^1(mH+K_R)\otimes\mathcal{J}_{Z/R}) = 0$ *if and only if*

$$\operatorname{im}(H^0(q_1)) \cap \operatorname{im}(H^0(q_2)) = \{0\} \tag{2.11}$$

and

$$h^0(\Omega_{R_i}(\log E)(mH+K_{R_i})\otimes\mathcal{J}_{Z_i}) = 0 \ \textit{ for } i = 1,2. \tag{2.12}$$

PROOF. By (2.9), we have that $h^0(\Lambda_R^1(mH+K_R)\otimes\mathcal{J}_{Z/R}) = 0$ if and only if $H^0(q)$ is injective. Moreover, as $H^0(\mathfrak{F}) = H^0(\mathfrak{F}_1) + H^0(\mathfrak{F}_2)$, we get from (2.10), (2.9) and the vertical sequence in (2.8) that

$$H^0(\mathcal{A}\otimes\mathcal{O}_R(mH+K_R)\otimes\mathcal{J}_{Z/R}) = \ker(H^0(\mathfrak{F})) = \bigoplus_{i=1}^2 \ker(H^0(\mathfrak{F}_i)) =$$

$$= \bigoplus_{i=1}^2 H^0(\Omega_{R_i}(mH+K_{R_i}+E)\otimes\mathcal{J}_{Z_i}).$$

Hence, by (2.9) again, we have $H^0(q) = H^0(q_1) + H^0(q_2)$. The statement follows by the isomorphism $\ker(H^0(q)) \simeq \Big(\oplus_i \ker(H^0(q_i)\Big) \bigoplus \Big(\operatorname{im}(H^0(q_1)) \cap \operatorname{im}(H^0(q_2))\Big)$ and the horizontal sequence in (2.8). □

LEMMA 2.6. *Assume that*

$$\sum_{i=1}^2 H^0(q_i) \ \textit{ is surjective,} \tag{2.13}$$

$$\sum_{i=1}^2 H^0(\mathfrak{F}_i) \ \textit{ is surjective} \tag{2.14}$$

and

$$h^1(\Omega_{R_i}(\log E)(mH+K_{R_i}+E)\otimes\mathcal{J}_{Z_i}) = 0, \ \textit{ for } i = 1,2. \tag{2.15}$$

Then $h^1(\Lambda_R^1(mH+K_R)\otimes\mathcal{J}_{Z/R}) = 0$.

PROOF. This follows from (2.9). □

We summarize our main results as follows:

PROPOSITION 2.7. *Let S' be a smooth, projective surface, H' a line bundle on S' and $Z' \subset S'$ a reduced zero-dimensional scheme. Assume that (S', H', Z') admits a semistable degeneration (R, H, Z) with $R = R_1 \cup R_2$, R_1, R_2 smooth, with transversal intersection $E = R_1 \cap R_2$, and $Z = Z_1 \cup Z_2$, with $Z_i \subset R_i - E$, $i = 1, 2$.*

If (2.10), (2.11) and (2.12) hold, one has $h^0(\Omega_{S'}(mH' + K_{S'}) \otimes \mathcal{J}_{Z'/S'}) = 0$.

If (2.13), (2.14) and (2.15) hold, one has $h^1(\Omega_{S'}(mH' + K_{S'}) \otimes \mathcal{J}_{Z'/S'}) = 0$.

PROOF. This follows from the two preceding lemmas and Lemma 2.3. □

3. The $K3$ case

In the rest of the note, we will show how to apply Proposition 2.7 to semistable degenerations of smooth, primitively polarized $K3$ surfaces, thus giving, via (1.1), a new proof of Theorem 1.3.

3.1. A semistable degeneration of K3 surfaces. This degeneration is well known (see [**6**]) and we recall it to fix notation.

Let $p = 2n + \varepsilon \geqslant 3$ be an integer, with $n \geqslant 1$ and $\varepsilon \in \{0, 1\}$, and let $E' \subset \mathbb{P}^p$ be a smooth, elliptic normal curve of degree $p + 1$. Consider two general line bundles $L_1, L_2 \in \mathrm{Pic}^2(E')$ with $L_1 \neq L_2$. In particular there is no relation between L_1, L_2 and $\mathcal{O}_{E'}(1)$ in $\mathrm{Pic}(E')$.

We denote by R'_1 and R'_2 the rational normal scrolls of degree $p - 1$ in $\mathbb{P}^p$ described by the secant lines of E' generated by the divisors in $|L_1|$ and $|L_2|$, respectively. We have

$$R'_i \simeq \mathbb{F}_{1-\varepsilon} = \begin{cases} \mathbb{P}^1 \times \mathbb{P}^1 & \text{if } p = 2n + 1, \\ \mathbb{F}_1 & \text{if } p = 2n \end{cases}$$

($\mathbb{F}_{1-\varepsilon}$ is called the *type* of the scrolls R'_1 and R'_2) and R'_1 and R'_2 transversely intersect along E', which is anticanonical on both (cf. [**6**]).

Denoting by $\mathfrak{p}_i : R'_i \to \mathbb{P}^1$ the structural morphism and by σ_i and F_i a section with minimal self-intersection and a fiber of $\mathfrak{p}_i$, respectively, we have $\sigma_i^2 = \varepsilon - 1$, $\sigma_i \cdot F_i = 1$, $F_i^2 = 0$ and $\mathrm{Pic}(R'_i) \simeq \mathbb{Z}[\sigma_i] \oplus \mathbb{Z}[F_i]$. One has

$$\mathcal{O}_{R'_i}(1) \simeq \mathcal{O}_{R'_i}(\sigma_i + nF_i) \text{ and } K_{R'_i} \sim -2\sigma_i - (3 - \varepsilon)F_i \sim -E', \tag{3.1}$$

and $\Omega_{R'_i}$ fits in the exact sequence

(3.2)
$$0 \longrightarrow \mathfrak{p}_i^*(\omega_{\mathbb{P}^1}) \simeq \mathcal{O}_{R'_i}(-2F_i) \longrightarrow \Omega_{R'_i} \longrightarrow \Omega_{R'_i/\mathbb{P}^1} \simeq \mathcal{O}_{R'_i}(-E' + 2F_i) \longrightarrow 0,$$

which splits if $\varepsilon = 1$.

Set $R' := R'_1 \cup R'_2$. The *first cotangent sheaf* $T^1_{R'}$ (cf. [**13**, § 1]) is the degree 16 line bundle on E'

$$T^1_{R'} \simeq \mathcal{N}_{E'/R'_1} \otimes \mathcal{N}_{E'/R'_2} \simeq \mathcal{O}_{E'}(4) \otimes (L_1 \otimes L_2)^{\otimes(3-p)}, \tag{3.3}$$

the last isomorphism coming from (3.1).

The Hilbert point of R' sits in the smooth locus of the component $\mathcal{X}_p$ of the Hilbert scheme whose very general point represents a smooth $K3$ surface of degree $2p - 2$ in $\mathbb{P}^p$ having Picard group generated by the hyperplane section (cf. [**6**, Thms. 1, 2]).

The fact that $T^1_{R'}$ is non-trivial on E' implies that R' does not admit any semistable deformation (cf. [**13**, Prop. 1.11]). Indeed, the total space of a general flat deformation of R' in $\mathbb{P}^p$ is singular along 16 points on E' that are the zeros of a global section of $T^1_{R'}$ (cf. [**13**, § 2]). More precisely if $\mathfrak{R}' \xrightarrow{\alpha'} \Delta$ is a (general) embedded deformation of R' in $\mathbb{P}^p$ corresponding to a (general) section $\tau \in H^0(R', \mathcal{N}_{R'/\mathbb{P}^p})$, then the total space $\mathfrak{R}'$ has double points at the 16 (distinct) points of the divisor $W := \{s_\tau = 0\} \in |T^1_{R'}|$, where s_τ is the image of τ by the standard map $H^0(\mathcal{N}_{R'/\mathbb{P}^p}) \to H^0(T^1_{R'})$, which is surjective by [**6**, Cor. 1]. By blowing up $\mathfrak{R}'$ along these singular points and contracting every exceptional divisor on one of the two irreducible components of the strict transform of R', one obtains a small resolution of singularities $\Pi : \mathfrak{R} \to \mathfrak{R}'$ and a semistable degeneration $\mathfrak{R} \xrightarrow{\alpha} \Delta$ of $K3$ surfaces, with central fiber $R = R_1 \cup R_2$, where $R_i = \Pi^{-1}(R'_i)$, $i = 1, 2$. Then $E := R_1 \cap R_2 = \mathrm{Sing}(R)$ is such that $E' = \Pi(E) \simeq E$ and $T^1_R \simeq \mathcal{O}_E$. The curve E' [resp. E] is anticanonical on R'_i [resp. on R_i], for $i = 1, 2$, hence both R' and R have trivial dualizing sheaf (see [**13**, Rem. 2.11]). On $\mathfrak{R}'$ there is a line bundle $\mathcal{H}'$ restricting to the hyperplane bundle on each fiber. We set $\mathcal{H} = \Pi^*(\mathcal{H}')$.

The map $\pi_i := \Pi_{|R_i} : R_i \to R'_i$ is the contraction of k_i disjoint (-1)-curves $\mathfrak{e}_{i,1}, \ldots, \mathfrak{e}_{i,k_i}$, such that $\mathfrak{e}_{i,j} \cdot E = 1$, to distinct points $x_{i,1}, \ldots, x_{i,k_i}$ on E'. We set $W_i = x_{i,1} + \cdots + x_{i,k_i}$ and $\mathfrak{e}_i := \sum_{j=1}^{k_i} \mathfrak{e}_{i,j}$, for $i = 1, 2$. Then $W = W_1 + W_2 \in |T^1_{R'}|$ is general, hence reduced.

If $\mathfrak{R}' \xrightarrow{\alpha'} \Delta$ is general in the above sense, we will accordingly say that $\mathfrak{R} \xrightarrow{\alpha} \Delta$ is *general* and (R_t, H_t), for $t \neq 0$, can be thought of as the general point of $\mathcal{K}_p$ (cf. Remark 1.2).

3.2. Technical lemmas. We will now develop tools to verify the conditions (2.10)-(2.12) and (2.13)-(2.15) in Proposition 2.7.

Consider the relative cotangent sequence of the map $\pi_i : R_i \to R'_i$, together with the dual of the exact sequence defining its normal sheaf $\mathcal{N}_{\pi_i}$ (c.f. e.g. [**21**, Ex. 3.4.13(iv)]); we have $\Omega_{R_i/R'_i} \simeq \mathfrak{Ext}^1(\mathcal{N}_{\pi_i}, \mathcal{O}_{R_i}) \simeq \mathfrak{Ext}^1(\mathcal{O}_{\mathfrak{e}_i}(-\mathfrak{e}_i), \mathcal{O}_{R_i})$. One easily verifies that $\mathfrak{Ext}^1(\mathcal{O}_{\mathfrak{e}_i}(-\mathfrak{e}_i), \mathcal{O}_{R_i}) \simeq \oplus_{j=1}^{k_i} \omega_{\mathfrak{e}_{i,j}}$, whence

$$0 \longrightarrow \pi_i^*(\Omega_{R'_i}) \longrightarrow \Omega_{R_i} \longrightarrow \Omega_{R_i/R'_i} \simeq \oplus_{j=1}^{k_i} \omega_{\mathfrak{e}_{i,j}} \longrightarrow 0. \tag{3.4}$$

On each (-1)-curve $\mathfrak{e}_{i,j}$ on R_i, with $i = 1, 2$ and $j = 1, \ldots, k_i$, we can consider the two points $Y_{i,j}$ and $Y'_{i,j}$ respectively cut out on $\mathfrak{e}_{i,j}$ by the strict transform on R_i of the ruling of R'_i through $x_{i,j}$ and by E. Note that $Y_{i,j} = Y'_{i,j}$ if and only if $x_{i,j}$ is a ramification point on E' of the linear series $|L_i|$. This will not be the case if $W \in |T^1_{R'}|$ is general. We will consider the 0–dimensional scheme $Y_i = \sum_{j=1}^{k_j} Y_{i,j}$ on R_i, for $i = 1, 2$. Since $\pi_i(Y_{i,j}) = \pi_i(Y'_{i,j}) = x_{i,j}$, we have $W_i = \pi_i(Y_i)$, for $i = 1, 2$. Then:

LEMMA 3.1. *Let $W \in |T^1_{R'}|$ be general. We have an exact sequence*

$$0 \longrightarrow \mathcal{O}_{R_i}(-2\pi_i^*(F_i)) \longrightarrow \Omega_{R_i} \longrightarrow \mathcal{O}_{R_i}(-E + 2\pi_i^*(F_i)) \otimes \mathcal{J}_{Y_i} \longrightarrow 0. \tag{3.5}$$

PROOF. The injective map in (3.5) is the composition

$$\pi_i^*(\mathcal{O}_{R'_i}(-2F_i)) \hookrightarrow \pi_i^*(\Omega_{R'_i}) \hookrightarrow \Omega_{R_i}$$

obtained from (3.2) and (3.4). To study its cokernel, we simplify notation and set $S = R_i, S' = R'_i, \pi = \pi_i$, $F = F_i$, $L = L_i$, for $i = 1, 2$. By the local nature of the question, we may and will assume that $\pi : S \to S'$ is the blow–up at only one point $w \in E'$, which, by the generality assumption, is not any of the four ramification points of the pencil $|L|$. Denote by $\mathfrak{e}$ the π–exceptional divisor.

Choose a chart $U \subset S'$ centered at w with coordinates (x, z), such that the map $S' \to \mathbb{P}^1$ is given on U by $(x, z) \mapsto x$. Then $\Omega_{S'|_U}$ is generated by dx, dz, the sheaf $\mathcal{O}_{S'}(-2F)_{|U}$ is generated by dx, and dz is the local generator for the quotient line bundle.

Consider $\widetilde{U} \subset U \times \mathbb{P}^1$ the blow-up of U at $w = (0, 0)$. If $[\epsilon, \eta]$ are homogeneous coordinates on $\mathbb{P}^1$, an equation for $\widetilde{U}$ in $U \times \mathbb{P}^1$ is $x\eta = \epsilon z$. As usual, we have two charts defined by $\eta \neq 0$ and $\epsilon \neq 0$. In the latter chart, where the coordinates are (x, η), the inclusion in (3.5) reads $1 \mapsto dx$, hence the quotient is locally free generated by $d\eta$. In the former, with coordinates (z, ϵ) and $x = z\epsilon$, the inclusion is $1 \mapsto \epsilon dz + z d\epsilon$, i.e., the map looks like $\mathcal{O} \xrightarrow{(\epsilon,z)} \mathcal{O} \oplus \mathcal{O}$, whose quotient is the ideal of the point $Y = \{z = \epsilon = 0\}$, which is the intersection of $\mathfrak{e}$ with the strict transform on S of the ruling of S' passing through w, the total transform having equation $x = 0$, i.e., $z\epsilon = 0$.

We have therefore proved that the cokernel of the inclusion in (3.5) is of the form $\mathcal{L} \otimes \mathcal{J}_{Y_i}$ for some line bundle $\mathcal{L}$ on R_i. Taking first Chern classes in the sequence, we get that $\mathcal{L} = \mathcal{O}_{R_i}(-E + 2\pi_i^*(F_i))$. □

For $i = 1, 2$, we denote by $X_i \in |2L_i|$ the ramification divisor of $|L_i|$ on E', or, by abuse of notation, on E.

LEMMA 3.2. *Same assumptions as in Lemma* 3.1. *Then the composed map*

$$\mathcal{O}_{R_i}(-2\pi_i^*(F_i)) \hookrightarrow \Omega_{R_i} \xrightarrow{r_i} \omega_E,$$

given by (2.4) *and* (3.5), *is non–zero, for* $i = 1, 2$. *Its image is the line bundle* $\omega_E(-X_i)$.

PROOF. We use the same simplified notation as in the proof of Lemma 3.1. Accordingly we will write X for X_i.

Take a point P in X on E. By the generality assumption on W, S and S' are isomorphic around P. So, if we work locally, we may do it on S'. Choose a chart U on S' centered at P, with coordinates (x, z), such that the structural map $S' \to \mathbb{P}^1$ is given on U by $(x, z) \mapsto x$ and E' has equation $x = z^2$. So we may take z as the coordinate on E'.

On U the sheaf injection in (3.5) is the same as the one in (3.2) which, by the proof of Lemma 3.1, is given by $1 \mapsto dx$. Composing this injection with the trace map on E' gives $1 \mapsto dx = 2zdz$, which shows that the map is non–zero and its image is a differential form on E vanishing at P.

The proof is accomplished by making similar local computation at points $P \in E$ that are not on X. This can be left to the reader. □

By Lemmas 3.1 and 3.2, we have the following commutative diagram

(3.6)

$$\begin{array}{ccccccccc}
 & & 0 & & 0 & & 0 & & \\
 & & \downarrow & & \downarrow & & \downarrow & & \\
0 & \to & \mathcal{J}_{Z_i}(mH - 2\pi_i^*(F_i) - E) & \to & \Omega_{R_i}(\log E)(mH - E) \otimes \mathcal{J}_{Z_i} & \to & \mathcal{J}_{X_i \cup Y_i \cup Z_i}(mH - E + 2\pi_i^*(F_i)) & \to & 0 \\
 & & \downarrow & & \downarrow & & \downarrow & & \\
0 & \longrightarrow & \mathcal{J}_{Z_i}(mH - 2\pi_i^*(F_i)) & \longrightarrow & \Omega_{R_i}(mH) \otimes \mathcal{J}_{Z_i} & \longrightarrow & \mathcal{J}_{Y_i \cup Z_i}(mH - E + 2\pi_i^*(F_i)) & \longrightarrow & 0 \\
 & & \downarrow & & \downarrow{\scriptstyle q_i} & & \downarrow & & \\
0 & \longrightarrow & \omega_E(mH)(-X_i) & \longrightarrow & \omega_E(mH) & \longrightarrow & \mathcal{O}_{X_i} & \longrightarrow & 0 \\
 & & \downarrow & & \downarrow & & \downarrow & & \\
 & & 0 & & 0 & & 0 & &
\end{array}$$

where the second vertical and horizontal exact sequences are respectively (2.4), (3.5) tensored by $\mathcal{O}_{R_i}(mH) \otimes \mathcal{J}_{Z_i}$, and where we used the isomorphism $\mathcal{O}_E \simeq \omega_E$.

Next we claim that we have the following commutative diagram with exact rows and columns

(3.7)

$$\begin{array}{ccccccccc}
 & & & & 0 & & 0 & & \\
 & & & & \downarrow & & \downarrow & & \\
0 & \to & \mathcal{O}_{R_i}(mH - 2\pi_i^*(F_i)) \otimes \mathcal{J}_{Z_i} & \longrightarrow & \Omega_{R_i}(mH) \otimes \mathcal{J}_{Z_i} & \longrightarrow & \mathcal{O}_{R_i}(mH - E + 2\pi_i^*(F_i)) \otimes \mathcal{J}_{Y_i \cup Z_i} & \to & 0 \\
 & & \| & & \downarrow & & \downarrow & & \\
0 & \to & \mathcal{O}_{R_i}(mH - 2\pi_i^*(F_i)) \otimes \mathcal{J}_{Z_i} & \to & \Omega_{R_i}(\log E)(mH) \otimes \mathcal{J}_{Z_i} & \to & \mathcal{O}_{R_i}(mH + 2\pi_i^*(F_i)) \otimes \mathcal{J}_{X_i \cup Y_i \cup Z_i} & \to & 0 \\
 & & & & \downarrow{\scriptstyle \mathfrak{F}_i} & & \downarrow & & \\
 & & & & \mathcal{O}_E(mH) & = & \mathcal{O}_E(mH) & & \\
 & & & & \downarrow & & \downarrow & & \\
 & & & & 0 & & 0 & &
\end{array}$$

The existence of the first and second horizontal exact rows follows from diagram (3.6), whereas the central vertical exact column follows from (2.3). The inclusion in the central vertical column restricts to the identity on $\mathcal{O}_{R_i}(mH - 2\pi_i^*(F_i))$. The rest follows from the snake lemma.

We want to describe $\operatorname{im}(H^0(\mathfrak{F}_i))$ and $\operatorname{im}(H^0(q_i))$ in the case $m = 1$. To this end we first define the following subspaces of $H^0(\mathcal{O}_E(H)) \simeq H^0(\omega_E(H))$. Recalling the conventions and notation as at the end of the introduction, from the right-most vertical sequence in (3.7), we set, for $i = 1, 2$,

$$V_i := H^0(\mathcal{O}_{R_i}(H + 2\pi_i^*(F_i)) \otimes \mathcal{J}_{X_i \cup Y_i \cup Z_i})_{|E} \subseteq H^0(\mathcal{O}_E(H)). \tag{3.8}$$

The inclusion works as follows: take a (non–zero) section $s \in V_i$ (which vanishes along a curve C containing X_i), restrict it to E, then divide by fixed local equations of the points in X_i (i.e., remove X_i from the divisor cut out by C on E).

Similarly, from the left-most vertical sequence in (3.6), we define, for $i = 1, 2$,

$$U_i := H^0(\mathcal{O}_{R_i}(H - 2\pi_i^*(F_i)) \otimes \mathcal{J}_{Z_i})_{|E} \subseteq H^0(\omega_E(H)(-X_i)) \subset H^0(\omega_E(H)). \tag{3.9}$$

At divisor level, the inclusion is given by taking a divisor in $|\mathcal{O}_{R_i}(H - 2\pi_i^* F_i) \otimes \mathcal{J}_{Z_i}|$, restricting it to E, and then adding the points X_i.

LEMMA 3.3. *If $m = 1$, then*

$$\mathrm{im}(H^0(q_i)) = U_i \simeq H^0(\mathcal{O}_{R_i}(H - 2\pi_i^*(F_i)) \otimes \mathcal{J}_{Z_i}) \tag{3.10}$$

and

$$\mathrm{im}(H^0(\mathfrak{F}_i)) \subseteq V_i \simeq H^0(\mathcal{O}_{R_i}(H + 2\pi_i^*(F_i)) \otimes \mathcal{J}_{X_i \cup Y_i \cup Z_i}), \tag{3.11}$$
with equality if $h^1(\mathcal{O}_{R_i}(H - 2\pi_i^*(F_i)) \otimes \mathcal{J}_{Z_i}) = 0.$

PROOF. This follows from (3.6) and (3.7) with $m = 1$, and the fact that $h^0(\mathcal{O}_{R_i}(H - E + a\pi_i^*(F_i))) = 0$ for any integer a. □

3.3. A new proof of Theorem 1.3. We consider $V_\delta(R)$ the locally closed subscheme of the linear system $|H|$ on R parametrizing the universal family of curves $C \in |H|$ having only nodes as singularities, exactly δ of them (called the *marked nodes*) off the singular locus E of R, and such that the partial normalization of C at the δ marked nodes is connected, i.e., the marked nodes are *non-disconnecting* nodes (cf. [**4**, §1.1]). Under a semistable deformation $\alpha : \mathfrak{R} \to \Delta$ of R as in Definition 2.1, it is possible to deform such a curve C to a δ-nodal curve on the fibres R_t of $\alpha : \mathfrak{R} \to \Delta$, for $t \neq 0$, preserving its marked nodes and smoothing the remaining $g + 1$ nodes of C located at its intersection with E (see [**4**, Lemma 1.4]).

Usually we will assume the deformation $\alpha : \mathfrak{R} \to \Delta$ to be general. Then, given $C \in V_\delta(R)$, we may find a pair (C', S') general in some component $\mathcal{V} \subseteq \mathcal{V}_{1,\delta}$, where (S', H') is general in $\mathcal{K}_p$ and $C' \in V_{1,\delta}(S', H')$, such that C is a flat limit of C'. Let Z [resp. Z'] be the scheme of the δ marked nodes of C [resp. of C']. Then (possibly after shrinking Δ further) the triple (R, H, Z) is a semistable degeneration of (S', H', Z') and we may apply Proposition 2.7 to show the desired vanishings in (1.1) needed to prove Theorem 1.3. For this we need the following result, whose proof we postpone until the next section.

PROPOSITION 3.4. *There exist $W \in |T^1_{R'}|$ and $C \in V_\delta(R)$, with Z its scheme of δ marked nodes, such that:*
(i) *if $g := p - \delta \geqslant 15$, then the maps $\sum_{i=1}^2 H^0(q_i)$ and $\sum_{i=1}^2 H^0(\mathfrak{F}_i)$ are surjective and*

$$h^1(\mathcal{O}_{R_i}(H + 2\pi_i^*(F_i)) \otimes \mathcal{J}_{X_i \cup Y_i \cup Z_i}) = h^1(\mathcal{O}_{R_i}(H - 2\pi_i^*(F_i)) \otimes \mathcal{J}_{Z_i}) = 0, \text{ for } i = 1, 2;$$

(ii) *if $g := p - \delta \leqslant 7$, then*

$$\mathrm{im}(H^0(\mathfrak{F}_1)) \cap \mathrm{im}(H^0(\mathfrak{F}_2)) = \mathrm{im}(H^0(q_1)) \cap \mathrm{im}(H^0(q_2)) = \{0\}.$$

PROOF OF THEOREM 1.3. We will prove the desired vanishings in (1.1) using Proposition 2.7.

If $g \geqslant 15$, conditions (2.13) and (2.14) in Proposition 2.7 are satisfied by Proposition 3.4(i), whereas condition (2.15) is satisfied by the middle horizontal sequence in (3.7) and the vanishings of h^1 in Proposition 3.4(i).

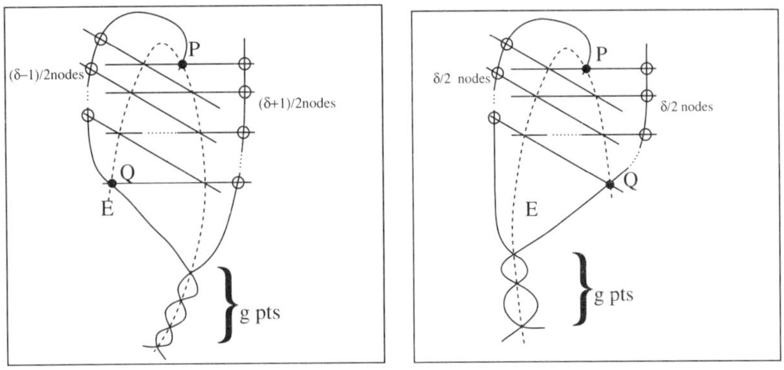

FIGURE 1. Members of $W_\delta(R)$ when δ is odd (left) and even (right)

If $g \leqslant 7$, conditions (2.10) and (2.11) in Proposition 2.7 are satisfied by Proposition 3.4(ii), whereas condition (2.12) is satisfied by the upper horizontal sequence in (3.6) and the fact that $h^0(\mathcal{O}_{R_i}(H - E \pm 2\pi_i^*(F_i))) = 0$, for $i = 1, 2$. □

4. Proof of Proposition 3.4

With a slight abuse of terminology, we will call *lines* the curves on R_i in the pencil $|\pi^*(F_i)|$, for $i = 1, 2$. The following component of $V_\delta(R)$ has been introduced in [**5**].

DEFINITION 4.1. For any $0 \leqslant \delta \leqslant p-1$, we define $W_\delta(R)$ to be the set of curves C in $V_\delta(R)$ such that:
(i) C does not contain any of the exceptional curves $\mathfrak{e}_{i,j}$ of the contractions $\pi_i : R_i \to R_i'$, $i = 1, 2$;
(ii) C has exactly $\delta_1 := \lfloor \frac{\delta}{2} \rfloor$ nodes on $R_1 - E$ and $\delta_2 := \lceil \frac{\delta}{2} \rceil$ nodes on $R_2 - E$, hence it splits off δ_i lines on R_i, for $i = 1, 2$;
(iii) the union of these $\delta = \delta_1 + \delta_2$ lines is connected.
For any curve C in $W_\delta(R)$, we denote by $\mathfrak{C}$ the connected union of δ lines as in (iii), called the *line chain of length* δ of C, and by γ_i the irreducible component of the residual curve to $\mathfrak{C}$ on R_i, for $i = 1, 2$.

It has been proved in [**5**, Prop. 4.2] that $W_\delta(R)$ is a smooth open subset of a component of $V_\delta(R)$, with the nodes described in (ii) as the marked nodes of any of its members. We write $W_\delta(R')$ for the set of images in R' of the curves in $W_\delta(R)$. Without further notice, we will denote the image of $C \in W_\delta(R)$ and its line chain $\mathfrak{C}$ and components γ_i by C', $\mathfrak{C}'$ and γ_i', respectively.

Members of $W_\delta(R)$ are shown in Figure 1 below.

The points P and Q in the picture (the *starting point* and the *end point* of $\mathfrak{C}$) satisfy the following relation on E:

$$P + Q \sim \frac{\delta+1}{2} L_2 - \frac{\delta-1}{2} L_1, \text{ when } \delta \text{ is odd;} \tag{4.1}$$

$$P - Q \sim \frac{\delta}{2}(L_2 - L_1), \text{ when } \delta \text{ is even.} \tag{4.2}$$

We denote by $\mathfrak{c}$ the intersection of $\mathfrak{C}$ with E, considered as a reduced divisor on E. This consists of $\delta + 1$ points, i.e., $P + Q$ plus the $\delta - 1$ double points of $\mathfrak{C}$

which are all located on E. One has:

$$\mathfrak{c} \sim P+Q+\frac{\delta-1}{2}L_1 \text{ when } \delta \text{ is odd;} \tag{4.3}$$

$$\mathfrak{c} \sim P+\frac{\delta}{2}L_1 \sim Q+\frac{\delta}{2}L_2 \text{ when } \delta \text{ is even.} \tag{4.4}$$

Recalling (3.8), we have:

LEMMA 4.2. *There exist* $W \in |T^1_{R'}|$ *and* $C \in V_\delta(R)$ *such that* $\dim(V_1 \cap V_2) = \max\{0, g-15\}$.

PROOF. One has $Y_i, Z_i \notin E$ whereas $X_i \in E$, cf. Lemma 3.2. By abuse of notation, we will identify Z_i and X_i with their images via the maps $\pi_i : R_i \to R'_i$. As indicated before Lemma 3.1, one has $W_i = \pi_i(Y_i)$, for $i = 1, 2$.

By the Leray spectral sequence and (3.8), we have

$$V_i \simeq H^0(\mathcal{O}_{R'_i}(H'+2F_i) \otimes \mathcal{J}_{X_i \cup W_i \cup Z_i})_{|E'} \subseteq H^0(\mathcal{O}_{E'}(H')),$$

with the inclusion explained right after (3.8): at divisor level, take a divisor in $|\mathcal{O}_{R'_i}(H'+2F_i) \otimes \mathcal{J}_{X_i \cup W_i \cup Z_i}|$, restrict it to E' and then remove the points X_i, $i = 1, 2$.

Case 1: $p \leqslant 6$. Since $\deg(W) = 16$, there is $h \in \{1, 2\}$ such that $k_h = \deg(W_h) \geqslant 8$. Moreover, $\dim(V_h) \leqslant \dim H^0(\mathcal{O}_{R'_h}(H'+2F_h) \otimes \mathcal{J}_{X_h \cup W_h})_{|E'}$. From

$$0 \to \mathcal{O}_{R'_h}(H'+2F_h-E') \to \mathcal{O}_{R'_h}(H'+2F_h) \otimes \mathcal{J}_{X_h \cup W_h} \to \mathcal{O}_{E'}(H'+2F_h)(-X_h-W_h) \to 0$$

and $\deg(\mathcal{O}_{E'}(H'+2F_h)(-X_h-W_h)) = p+1-k_h \leqslant -1$, we see that $V_h = \{0\}$ and the assertion follows.

Case 2: $p \geqslant 7$. If we project $R'_i \subset \mathbb{P}^p$ from X_i (which consists of four points on different fibers of $|F_i|$), we obtain a smooth, rational normal scroll $R''_i \subset \mathbb{P}^{p-4}$ of the same type as R'_i. Indeed, since $p \geqslant 7$, in this projection E' is mapped isomorphically to an elliptic normal curve E'' in $\mathbb{P}^{p-4}$. Hence the scrolls are mapped to the rational normal scrolls swept out by the lines joining points in divisors of the two linear series $|L_1|$ and $|L_2|$ on E'', which are still smooth of the same type.

The linear system $|\mathcal{O}_{R''_i}(1)|$ on $R''_i \subset \mathbb{P}^{p-4}$ is given (with obvious notation) by $|\sigma''_i + (n-2)F''_i|$, which corresponds on R'_i to $|\mathcal{O}_{R'_i}(H') \otimes \mathcal{J}_{X_i}|$. Likewise, the linear system $|\mathcal{O}_{R'_i}(H'+2F_i) \otimes \mathcal{J}_{X_i \cup W_i \cup Z_i}|$ on R'_i corresponds to $|\mathcal{O}_{R''_i}(\sigma''_i + nF''_i) \otimes \mathcal{J}_{W_i \cup Z_i}|$ on R''_i (by abuse of notation we denote by the same symbols the images of W_i and Z_i under the projection from X_i).

Hence, the space V_i corresponds on R''_i to $H^0(\mathcal{O}_{R''_i}(\sigma''_i + nF''_i) \otimes \mathcal{J}_{W_i \cup Z_i})_{|E''}$, where $E'' \simeq E'$ is the image of E' via the projection from the points X_i. This shows that we can directly work on R'_i and make the identification

$$V_i = H^0(\mathcal{O}_{R'_i}(H') \otimes \mathcal{J}_{W_i \cup Z_i})_{|E'} \subseteq H^0(\mathcal{O}_{E'}(H')). \tag{4.5}$$

Since the surface R'_i is linearly normal, we have

$$H^0(\mathcal{O}_{R'_i}(H') \otimes \mathcal{J}_{W_i \cup Z_i})_{|E'} = H^0(\mathcal{O}_{\mathbb{P}^p}(1) \otimes \mathcal{J}_{(W_i \cup Z_i)/\mathbb{P}^p})_{|E'},$$

hence

$$\begin{aligned} V_1 \cap V_2 &= H^0(\mathcal{O}_{\mathbb{P}^p}(1) \otimes \mathcal{J}_{(W \cup Z)/\mathbb{P}^p})_{|E'} \\ &\simeq H^0(\mathcal{O}_{R'}(1) \otimes \mathcal{J}_{W \cup Z})_{|E'} \simeq H^0(\mathcal{O}_{R'}(1) \otimes \mathcal{J}_{W \cup Z}). \end{aligned}$$

Subcase 2A: $g \leqslant 14$. Since 15 of the 16 points in W are general on E', and since $\dim(|\mathcal{O}_{R'}(1) \otimes \mathcal{J}_Z|) = g$ for any $C \in W_\delta(R')$ with Z its scheme of δ marked not disconnecting nodes (cf. [4, Rem. 1.1]), one has $|\mathcal{O}_{R'}(1) \otimes \mathcal{J}_{W \cup Z}| = \emptyset$, i.e., $V_1 \cap V_2 = \{0\}$, as desired.

Subcase 2B: $p = g = 15$. The argument is similar to the one in the previous case. Indeed, the fact that $\mathcal{O}_{E'}(W) \simeq T^1_{R'} \not\simeq \mathcal{O}_{E'}(H')$ when $p = 15$ (cf. (3.3)) implies that $|\mathcal{O}_{R'}(H') \otimes \mathcal{J}_W|$ is empty, hence again $V_1 \cap V_2 = \{0\}$.

Subcase 2C: $g \geqslant 15$ and $p \geqslant 16$. Since $\deg(\mathcal{O}_{E'}(H' - W)) = p - 15 \geqslant 1$, we have $\dim(|\mathcal{O}_{R'}(H') \otimes \mathcal{J}_W|) = p - 16$.

If $\delta \leqslant 5$ (whence $3\delta \leqslant p$), then we can choose a curve $C \in V_\delta(R)$ whose δ nodes map to general points on one of the two components of R'. It follows that $\dim(|\mathcal{O}_{R'}(1) \otimes \mathcal{J}_{W \cup Z}|) = \max\{-1, p - 16 - \delta\}$, as desired.

Hence we may assume $\delta \geqslant 6$, which yields $p = g + \delta \geqslant 21$ and we will construct a curve $C \in W_\delta(R)$ verifying the assertion.

Let $P \in E'$ be a general point, and consider the line chain $\mathfrak{C}'$ starting at P. Set $\Sigma := |\mathcal{O}_{\mathbb{P}^p}(1) \otimes \mathcal{I}_{\mathfrak{C}'/\mathbb{P}^p}|$. The intersection of $\mathfrak{C}'$ with E' consists of a divisor $\mathfrak{c}$ of degree $\delta + 1$, i.e., $P + Q$ plus the $\delta - 1$ double points of the line chain $\mathfrak{C}'$. One has $\dim(\Sigma) = g - 1$, and the hyperplanes in Σ cut out E' in $\mathfrak{c}$ plus a divisor D of degree g. Hence the linear system Σ cuts out on E', off $\mathfrak{c}$, the complete, base point free linear system $|D|$ of degree g.

Assume $g \geqslant 17$. Take a general hyperplane $H' \in \Sigma$, let $C' = \mathfrak{C}' + \gamma_1' + \gamma_2'$ be the curve in $W_\delta(R')$ cut out by H' on R' (see Definition 4.1 and Figure 1), and let D be divisor cut out by H' on E' off $\mathfrak{c}$. Take D' any effective divisor of degree 15 contained in D. Since $g \geqslant 17$, D' is a general effective divisor of degree 15 on E'.

Let P' be the unique point on E' such that $W := P' + D' \in |T^1_{R'}|$. We claim that

$$P' \notin C', \text{ i.e. } P' \notin D + \mathfrak{c}. \tag{4.6}$$

Indeed if $P' \in \mathfrak{c}$, by an analogue of (4.1) or (4.2) there would be an integer $k \geqslant 1$ such that either

$$P + P' \sim kL_2 - (k-1)L_1, \text{ or} \tag{4.7}$$

$$P - P' \sim k(L_2 - L_1). \tag{4.8}$$

Since $T^1_{R'} \simeq \mathcal{O}_{E'}(D' + P')$, then (3.3) and (4.7) combined yield

$$\mathcal{O}_{E'}(P) \simeq \mathcal{O}_{E'}(D') \otimes \mathcal{O}_{E'}(-4) \otimes L_1^{(p-k-2)} \otimes L_2^{(p+k-3)}.$$

This uniquely determines P once D', L_1, L_2 and $\mathcal{O}_{E'}(1)$ have been generically chosen (as they can), which is a contradiction, since P is also general on E'. Similarly by combining (3.3) and (4.8). This proves that $P' \notin \mathfrak{c}$.

To prove that $P' \notin D$ we note that, by the generality of D', the divisor $P' + D'$ is general in $|T^1_{R'}|$, hence it is reduced, thus $P' \notin D'$. Moreover, since $|D - D'|$ is base point free, because $\deg(D - D') \geqslant 2$, we may also assume that $P' \notin D - D'$. This ends the proof of (4.6).

Next we project R_i' from D'. By the generality of D', we obtain a smooth rational normal scroll $R_i'' \subset \mathbb{P}^{p-15}$ for $i = 1, 2$ (smoothness follows from $p \geqslant 21$ and the fact that the projection is an internal projection of the original scroll from the linear span of D'), and R_1'' and R_2'' intersect transversally along a smooth elliptic curve $E'' \simeq E'$ (the projection of E'). Under this projection, the image C'' of C' is isomorphic to the curve obtained by normalizing C' at D'. Thus C'' still has δ

marked nodes on the smooth locus of $R'' = R_1'' \cup R_2''$, whose set we denote by Z as the nodes of C'. Since $g \geqslant 17$, the curves γ_1'' and γ_2'', images of γ_1' and γ_2', intersect in at least two points on E''. Hence, the normalization of C'' at the δ marked nodes is connected. By [**4**, Rem. 1.1] we have $\dim(|\mathcal{O}_{R''}(1) \otimes \mathcal{J}_Z|) = (p-15) - \delta = g - 15$. Since the hyperplane sections of R'' are in one-to-one correspondence with the hyperplane sections of R' passing through D', this yields

$$\dim(|\mathcal{O}_{R'}(1) \otimes \mathcal{J}_{D' \cup Z}|) = g - 15.$$

To accomplish the proof we have to exclude that

$$|\mathcal{O}_{R'}(1) \otimes \mathcal{J}_{W \cup Z}| = |\mathcal{O}_{R'}(1) \otimes \mathcal{J}_{D' \cup Z}|,$$

which would mean that any hyperplane passing through Z and D' contains also P'. If this were the case, this would in particular happen for the curve C', against (4.6). This ends the proof in the case $g \geqslant 17$.

For $g = 16$ the proof runs exactly as above. There is only one minor change in the proof of (4.6). The proof that $P' \notin \mathfrak{c} + D'$ works with no change. If $P' \in D - D'$, then $P' = D - D'$, hence $D \in |T_{R'}^1|$. Since $\mathfrak{c} + D \sim \mathcal{O}_{E'}(1)$, then (3.3) yields

$$\mathfrak{c} \sim \mathcal{O}_{E'}(-3) \otimes L_1^{\otimes(p-3)} \otimes L_2^{\otimes(p-3)}.$$

By (4.3) and (4.4) we have

$$\begin{aligned} P + Q &\sim \mathcal{O}_{E'}(-3) \otimes L_1^{\otimes(p-3-\frac{\delta-1}{2})} \otimes L_2^{\otimes(p-3)} && \text{when } \delta \text{ is odd;} \\ P &\sim \mathcal{O}_{E'}(-3) \otimes L_1^{\otimes(p-3-\frac{\delta}{2})} \otimes L_2^{\otimes(p-3)} && \text{when } \delta \text{ is even.} \end{aligned}$$

The latter relation contradicts the generality of P. The former gives, together with (4.3), the relation

$$\mathcal{O}_{E'}(3) \sim L_1^{\otimes(p-\delta-2)} \otimes L_2^{\otimes(p-3-\frac{\delta+1}{2})},$$

contradicting the general choices of L_1 and L_2.

Let us finally consider the case $g = 15$. The basic idea of the proof is the same, so we will be brief. We let D' be any effective divisor of degree 14 contained in D, so that D' is a general divisor of degree 14 on E'. Let P' be the unique point of E' such that $W := D' + P + P' \in T_{R'}^1$. We claim that (4.6) still holds. The proof is similar to the ones in the previous cases and can be left to the reader. Then we project R' from D'. The projection C'' of C' is connected and has δ marked nodes on the smooth locus of the projection R'' of R', whose set we denote by Z as the nodes of C'. Then $\dim(|\mathcal{O}_{R''}(1) \otimes \mathcal{J}_Z|) = (p - 14) - \delta = 1$, hence $\dim(|\mathcal{O}_{R'}(1) \otimes \mathcal{J}_{D' \cup Z}|) = 1$. We claim that

$$|\mathcal{O}_{R'}(1) \otimes \mathcal{J}_{D' \cup \{P\} \cup Z}| = \{C'\}.$$

Indeed, a curve in $|\mathcal{O}_{R'}(1) \otimes \mathcal{J}_{D' \cup \{P\} \cup Z}|$ clearly contains the line cycle $\mathfrak{C}'$, hence it cuts on E' a divisor of degree $p+1$ which contains $\mathfrak{c} + D'$, whose degree is p. Hence, this curve is uniquely determined. Since $C' \in |\mathcal{O}_{R'}(1) \otimes \mathcal{J}_{D' \cup \{P\} \cup Z}|$, the assertion follows. Finally, by (4.6), we see that $|\mathcal{O}_{R'}(1) \otimes \mathcal{J}_{W \cup Z}| = \emptyset$, proving the assertion in this case. □

COROLLARY 4.3. *If $g \geqslant 15$, then, for $i = 1, 2$, we have:*

$$\dim(V_i) = h^0(\mathcal{O}_{R_i}(H + 2\pi_i^*(F_i)) \otimes \mathcal{J}_{X_i \cup W_i \cup Z_i}) = p - \delta_i - k_i + 1$$

and

$$h^1(\mathcal{O}_{R_i}(H + 2\pi_i^*(F_i)) \otimes \mathcal{J}_{X_i \cup W_i \cup Z_i}) = 0.$$

PROOF. By (3.11) and the Leray spectral sequence, we have

(4.9) $\dim(V_i) = h^0(\mathcal{O}_{R'_i}(H'+2F_i)\otimes\mathcal{J}_{X_i\cup W_i\cup Z_i}) \geqslant p+1-\delta_i-k_i \geqslant 0$, for $i=1,2$,

because $p+1-\delta_i-k_i \geqslant p-\delta-16+1 = g-15 \geqslant 0$. As $\dim(V_1\cap V_2) = g-15 \geqslant 0$ and $V_1+V_2 \subseteq H^0(\mathcal{O}_{E'}(1))$, the latter of dimension $p+1$, by the Grassmann formula equality must hold in (4.9). The statement about h^1 then follows. □

Recalling (3.9), we have:

LEMMA 4.4. *There exist $W \in |T^1_{R'}|$ and $C \in V_\delta(R)$, with Z its scheme of δ marked nodes, such that* $\dim(U_1\cap U_2) = \max\{0, g-7\}$.

PROOF. It is similar to the one of Lemma 4.2. We have

$$U_i = H^0(\mathcal{O}_{R'_i}(H'-2F_i)\otimes\mathcal{J}_{Z_i})_{|E'} \subset H^0(\omega_{E'}(H')) = H^0(\mathcal{O}_{E'}(H'));$$

as explained right after (3.9), at the level of divisors the inclusion is given by taking a divisor in $|\mathcal{O}_{R'_i}(H'-2F_i)\otimes\mathcal{J}_{Z_i}|$, restricting it to E', then adding the ramification divisor X_i, for $i=1,2$.

As in the proof of Lemma 4.2, where we showed that we could make the identification (4.5), we can reduce to making the identification

$$U_i = H^0(\mathcal{O}_{R'_i}(H')\otimes\mathcal{J}_{X_i\cup Z_i})_{|E'} \subset H^0(\mathcal{O}_{E'}(H'))$$

(recall that $X_i \in |2F_i|$). Since E' and R'_i are linearly normal, we thus have

$$\begin{aligned} U_1\cap U_2 &= H^0(\mathcal{O}_{\mathbb{P}^p}(1)\otimes\mathcal{J}_{(X\cup Z)/\mathbb{P}^p})_{|E'} \\ &= H^0(\mathcal{O}_{R'}(1)\otimes\mathcal{J}_{X\cup Z})_{|E'} \simeq H^0(\mathcal{O}_{R'}(1)\otimes\mathcal{J}_{X\cup Z}). \end{aligned}$$

where $X := X_1+X_2$ as a divisor on E'. We have the exact sequence

$$0 \longrightarrow \mathcal{O}_{R'}(1)\otimes\mathcal{J}_{E'} \longrightarrow \mathcal{O}_{R'}(1)\otimes\mathcal{J}_X \longrightarrow \mathcal{O}_{E'}(H'-X) \longrightarrow 0.$$

Since the curve E' is non–degenerate, one has $h^0(\mathcal{O}_{R'}(1)\otimes\mathcal{J}_{E'}) = 0$. Moreover, $\deg(\mathcal{O}_{E'}(H'-X)) = p-7$ and, when $p=7$, one has $\mathcal{O}_{E'}(2X) \simeq L_1^{\otimes 4}\otimes L_2^{\otimes 4} \not\simeq \mathcal{O}_{E'}(2)$. So $h^0(\mathcal{O}_{E'}(H'-X)) = 0$ for $p \leqslant 7$, hence in this case $h^0(\mathcal{O}_{R'}(1)\otimes\mathcal{J}_X) = 0$ and we are done. We may therefore assume that $p \geqslant 8$, in which case we have $h^0(\mathcal{O}_{R'}(1)\otimes\mathcal{J}_X) = h^0(\mathcal{O}_{E'}(H'-X)) = p-7$.

If $\delta \leqslant 1$, we may find a curve $C \in V_\delta(R)$ whose marked node (if any) maps to a general point on one of the two components of R'. Then it follows that $\dim(|\mathcal{O}_{R'}(1)\otimes\mathcal{J}_{X\cup Z}|) = \max\{0, p-8-\delta\}$, and we are done.

We may henceforth assume that $\delta \geqslant 2$ and we will construct a curve $C \in W_\delta(R)$ satisfying the assertion. Assume L_1 and L_2 are fixed and general on E' as an abstract elliptic curve. Then also X is fixed. We will accomplish the proof by finding a suitable embedding of E' as an elliptic normal curve of degree $p+1$ in $\mathbb{P}^p$, thus defining R'_1, R'_2 and $R' := R'_1 \cup R'_2$ via $|L_1|$ and $|L_2|$, and then finding $C' \in W_\delta(R')$ with its scheme Z of marked nodes such that $U_1\cap U_2$ satisfies the desired condition.

Pick a point $P \in E'$. We assume that P is general if $g \geqslant 8$ whereas we take $P \in X_1$ if $g \leqslant 7$. Consider the line chain $\mathfrak{C}'$ of length δ with starting point at P and the divisor $\mathfrak{c} = P+P_1+\ldots+P_\delta$ cut out by $\mathfrak{C}'$ on E', so that the lines of $\mathfrak{C}'$ are spanned by $P+P_1$, P_1+P_2, etc. We claim that

(4.10) if $g \geqslant 8$, the divisor $\mathfrak{c}$ is reduced and $\mathfrak{c}\cap X = \emptyset$;

(4.11) if $g \leqslant 7$, the divisor $\mathfrak{c}$ is reduced and $\mathfrak{c}\cap X = P$.

We prove only (4.10), since the proof of (4.11) is similar and can be left to the reader. For (4.10), by generality, P is not in X. Moreover, for each $k \in \{1, \dots, \delta\}$, one has

$$P + P_k \sim \frac{k+1}{2}L_2 - \frac{k-1}{2}L_1 \text{ if } k \text{ is odd,} \tag{4.12}$$

$$P - P_k \sim \frac{k}{2}(L_2 - L_1) \text{ if } k \text{ is even,} \tag{4.13}$$

(see (4.1) and (4.2)). Hence, if $P_k = P_h$ for $1 \leqslant k < h \leqslant \delta$, then there is a non–trivial relation between L_1, L_2 and P, a contradiction. Similarly, if $2P_k \sim L_i$ for $i = 1$ or $i = 2$, then (4.12) and (4.13) yield a non–trivial relation between L_1, L_2 and P, a contradiction again. The same if $P = P_k$ for some $k \in \{1, \dots, \delta\}$. Hence (4.10) is proved.

Fix now an effective subdivisor $\hat{X}$ of X of degree $\min\{7, g-1\}$, not containing P if $g \leqslant 7$, and pick a general effective divisor D on E' containing P with

$$\deg(D) = g + 1 - \deg(\hat{X}) \geqslant 2.$$

The divisor $F := \hat{X} + D + P_1 + \cdots + P_\delta$ has degree $p+1$ and $|\mathcal{O}_{E'}(F)|$ gives an embedding of E' as an elliptic normal curve of degree $p+1$ in $\mathbb{P}^p$. The hyperplane cutting out the divisor F on E' cuts out on R' a curve $C' \in W_\delta(R')$ with line chain $\mathfrak{C}'$. With notation as in Definition 4.1, $\gamma_1' \cap \gamma_2' = \hat{X} + D - P$.

When we project R_i' from $\hat{X}$ we obtain a smooth rational normal scroll R_i'' in $\mathbb{P}^r$ with $r = \max\{p - 7, \delta + 1\} \geqslant 3$ (here we use $\delta \geqslant 2$ and the fact that the projection is an internal projection of a scroll). As above, R_1'' and R_2'' intersect transversally along a smooth elliptic curve $E'' \simeq E'$. Under this projection, the image C'' of C' is isomorphic to the curve obtained by normalizing C' at $\hat{X}$. It still has δ nodes on the smooth locus of R'', which we denote by Z as the nodes of C'. The images of γ_1' and γ_2' under the projection intersect in at least $\deg(D - P) \geqslant 1$ points on E''. Hence, the normalization of C'' at the nodes lying on the smooth locus of $R'' = R_1'' \cup R_2''$ is connected, i.e., the marked nodes are not disconnecting. By [**4**, Rem. 1.1], we have

$$\dim(|\mathcal{O}_{R''}(1) \otimes \mathcal{J}_Z|) = r - \delta = \max\{g - 7, 1\}. \tag{4.14}$$

Since the hyperplane sections of R'' are in one-to-one correspondence with the hyperplane sections of R' passing through $\hat{X}$, (4.14) yields

$$\dim(|\mathcal{O}_{R'}(1) \otimes \mathcal{J}_{\hat{X} \cup Z}|) = \max\{g - 7, 1\}. \tag{4.15}$$

If $g \geqslant 8$, we are done unless

$$\dim(|\mathcal{O}_{R'}(1) \otimes \mathcal{J}_{X \cup Z}|) = \dim(|\mathcal{O}_{R'}(1) \otimes \mathcal{J}_{\hat{X} \cup Z}|) = g - 7,$$

which means that if a hyperplane in $\mathbb{P}^p$ contains $\hat{X} \cup Z$, it also contains X. In this case we would have that $X \subset C'$, because $C' \in |\mathcal{O}_{R'}(1) \otimes \mathcal{J}_{\hat{X} \cup Z}|$, so that

$$X < F = \hat{X} + D + P_1 + \cdots + P_\delta = \hat{X} + D + (\mathfrak{c} - P).$$

This contradicts either (4.10) or the generality of D.

If $g \leqslant 7$, we are done unless $X - \hat{X}$, which has degree $9 - g \geqslant 2$ and contains P, imposes $c \leqslant 1$ condition to the 1–dimensional system $|\mathcal{O}_{R'}(1) \otimes \mathcal{J}_{\hat{X} \cup Z}|$. If $c = 0$, then, as above, we would have $X < F$ and we find a contradiction.

Suppose $c = 1$. Then P is not in the base locus B of $|\mathcal{O}_{R'}(1) \otimes \mathcal{J}_{\hat{X} \cup Z}|$. Otherwise $\mathfrak{C}'$ sits in B, and (as at the end of the proof of Lemma 4.2) $|\mathcal{O}_{R'}(1) \otimes \mathcal{J}_{\hat{X} \cup Z}| = \{C'\}$, contradicting (4.15). Thus $c = 1$ yields

$$|\mathcal{O}_{R'}(1) \otimes \mathcal{J}_{\hat{X} \cup \{P\} \cup Z}| = |\mathcal{O}_{R'}(1) \otimes \mathcal{J}_{X \cup Z}| = \{C'\},$$

implying again $X < F$, which leads to a contradiction as above. □

COROLLARY 4.5. *If $g \geqslant 7$, then, for $i = 1, 2$, we have:*

$$\dim(U_i) = h^0(\mathcal{O}_{R_i}(H - 2\pi_i^*(F_i)) \otimes \mathcal{J}_{Z_i}) = p - 3 - \delta_i$$

and

$$h^1(\mathcal{O}_{R_i}(H - 2\pi_i^*(F_i)) \otimes \mathcal{J}_{Z_i}) = 0.$$

PROOF. By (3.10) and Leray, $\dim(U_i) = h^0(\mathcal{O}_{R'_i}(H' - 2F_i) \otimes \mathcal{J}_{Z_i}) \geqslant p - 3 - \delta_i$. As $\dim(U_1 \cap U_2) = g - 7 \geqslant 0$, by the Grassmann formula each U_i must have the expected dimension. The statement about h^1 then follows. □

PROOF OF PROPOSITION 3.4. (i) If $g \geqslant 15$, then Corollary 4.5 yields that $h^1(\mathcal{O}_{R_i}(H - 2\pi_i^*(F_i)) \otimes \mathcal{J}_{Z_i}) = 0$. Hence, by (3.11), we have $H^0(\mathfrak{F}_i) = V_i$ and, by Corollary 4.3, we have $h^1(\mathcal{O}_{R_i}(H + 2\pi_i^*(F_i)) \otimes \mathcal{J}_{X_i \cup W_i \cup Z_i}) = 0$. Therefore, the part of the statement Proposition 3.4(i) concerning the vanishings of the h^1s is proved. The rest follows by Lemma 4.2, Corollary 4.3 and the Grassmann formula.

(ii) If $g \leqslant 7$, the statement follows by Lemmas 4.4, 3.3 and 4.2. □

References

[1] Arnaud Beauville, *Fano threefolds and K3 surfaces*, The Fano Conference, Univ. Torino, Turin, 2004, pp. 175–184. MR2112574

[2] L. Chiantini and C. Ciliberto, *On the Severi varieties on surfaces in* $\mathbf{P}^3$, J. Algebraic Geom. **8** (1999), no. 1, 67–83. MR1658208

[3] L. Chiantini and E. Sernesi, *Nodal curves on surfaces of general type*, Math. Ann. **307** (1997), no. 1, 41–56. MR1427674

[4] Ciro Ciliberto and Andreas Leopold Knutsen, *On k-gonal loci in Severi varieties on general K3 surfaces and rational curves on hyperkähler manifolds*, J. Math. Pures Appl. (9) **101** (2014), no. 4, 473–494. MR3179751

[5] Ciro Ciliberto, Flaminio Flamini, Concettina Galati, and Andreas Leopold Knutsen, *Moduli of nodal curves on K3 surfaces*, Adv. Math. **309** (2017), 624–654. MR3607287

[6] Ciro Ciliberto, Angelo Lopez, and Rick Miranda, *Projective degenerations of K3 surfaces, Gaussian maps, and Fano threefolds*, Invent. Math. **114** (1993), no. 3, 641–667. MR1244915

[7] Pierre Deligne, *Équations différentielles à points singuliers réguliers* (French), Lecture Notes in Mathematics, Vol. 163, Springer-Verlag, Berlin-New York, 1970. MR0417174

[8] Federigo Enriques, *Le Superficie Algebriche* (Italian), Nicola Zanichelli, Bologna, 1949. MR0031770

[9] Hélène Esnault and Eckart Viehweg, *Lectures on vanishing theorems*, DMV Seminar, vol. 20, Birkhäuser Verlag, Basel, 1992. MR1193913

[10] F. Flamini, *Some results of regularity for Severi varieties of projective surfaces*, Comm. Algebra **29** (2001), no. 6, 2297–2311. MR1845111

[11] F. Flamini, *Moduli of nodal curves on smooth surfaces of general type*, J. Algebraic Geom. **11** (2002), no. 4, 725–760. MR1910265

[12] Flaminio Flamini, Andreas Leopold Knutsen, Gianluca Pacienza, and Edoardo Sernesi, *Nodal curves with general moduli on K3 surfaces*, Comm. Algebra **36** (2008), no. 11, 3955–3971. MR2460395

[13] Robert Friedman, *Global smoothings of varieties with normal crossings*, Ann. of Math. (2) **118** (1983), no. 1, 75–114. MR707162

[14] V. Gritsenko and K. Hulek, *Moduli of polarized Enriques surfaces*, K3 surfaces and their moduli, Progr. Math., vol. 315, Birkhäuser/Springer, [Cham], 2016, pp. 55–72. MR3524164

[15] Mihai Halic, *Modular properties of nodal curves on K3 surfaces*, Math. Z. **270** (2012), no. 3-4, 871–887. MR2892928
[16] Mihai Halic, *Erratum to: Modular properties of nodal curves on K3 surfaces [MR2892928]*, Math. Z. **280** (2015), no. 3-4, 1203–1211. MR3369374
[17] Michael Kemeny, *The moduli of singular curves on K3 surfaces* (English, with English and French summaries), J. Math. Pures Appl. (9) **104** (2015), no. 5, 882–920. MR3405539
[18] Vik. S. Kulikov, *Degenerations of K3 surfaces and Enriques surfaces* (Russian), Izv. Akad. Nauk SSSR Ser. Mat. **41** (1977), no. 5, 1008–1042, 1199. MR0506296
[19] Shigefumi Mori and Shigeru Mukai, *The uniruledness of the moduli space of curves of genus* 11, Algebraic geometry (Tokyo/Kyoto, 1982), Lecture Notes in Math., vol. 1016, Springer, Berlin, 1983, pp. 334–353. MR726433
[20] David R. Morrison, *Semistable degenerations of Enriques' and hyperelliptic surfaces*, Duke Math. J. **48** (1981), no. 1, 197–249. MR610184
[21] Edoardo Sernesi, *Deformations of algebraic schemes*, Grundlehren der Mathematischen Wissenschaften [Fundamental Principles of Mathematical Sciences], vol. 334, Springer-Verlag, Berlin, 2006. MR2247603
[22] Jayant Shah, *Projective degenerations of Enriques' surfaces*, Math. Ann. **256** (1981), no. 4, 475–495. MR628229
[23] Joseph Steenbrink, *Limits of Hodge structures*, Invent. Math. **31** (1975/76), no. 3, 229–257. MR0429885

Dipartimento di Matematica, Università di Roma Tor Vergata, Via della Ricerca Scientifica, 00173 Roma, Italy
Email address: cilibert@mat.uniroma2.it

Dipartimento di Matematica, Università di Roma Tor Vergata, Via della Ricerca Scientifica, 00173 Roma, Italy
Email address: flamini@mat.uniroma2.it

Dipartimento di Matematica e Informatica, Università della Calabria, via P. Bucci, cubo 31B, 87036 Arcavacata di Rende (CS), Italy
Email address: galati@mat.unical.it

Department of Mathematics, University of Bergen, Postboks 7800, 5020 Bergen, Norway
Email address: andreas.knutsen@math.uib.no

Contemporary Mathematics
Volume **712**, 2018
http://dx.doi.org/10.1090/conm/712/14343

Weak Brill-Noether for rational surfaces

Izzet Coskun and Jack Huizenga

To Lawrence Ein, with gratitude and admiration, on the occasion of his sixtieth birthday

ABSTRACT. A moduli space of sheaves satisfies weak Brill-Noether if the general sheaf in the moduli space has no cohomology. Göttsche and Hirschowitz prove that on $\mathbb{P}^2$ every moduli space of Gieseker semistable sheaves of rank at least two and Euler characteristic zero satisfies weak Brill-Noether. In this paper, we give sufficient conditions for weak Brill-Noether to hold on rational surfaces. We completely characterize Chern characters on Hirzebruch surfaces for which weak Brill-Noether holds. We also prove that on a del Pezzo surface of degree at least 4 weak Brill-Noether holds if the first Chern class is nef.

CONTENTS

1. Introduction

Let X be a smooth, complex projective surface and let H be an ample divisor on X. Let $\mathbf{v} \in K_{\text{num}}(X)$ be the (numerical) Chern character of a Gieseker semistable sheaf on X such that the Euler characteristic satisfies $\chi(\mathbf{v}) = 0$. Let $M_{X,H}(\mathbf{v})$ denote the moduli space of Gieseker semistable sheaves on X with Chern character $\mathbf{v}$. We will call $\mathbf{v}$ *stable* if there is a semistable sheaf of character $\mathbf{v}$.

DEFINITION 1.1. The moduli space $M_{X,H}(\mathbf{v})$ satisfies *weak Brill-Noether* if there exists a sheaf $E \in M_{X,H}(\mathbf{v})$ such that $H^i(X,E) = 0$ for all i.

2010 *Mathematics Subject Classification.* Primary 14J60, 14J26; Secondary 14D20, 14F05.

Key words and phrases. Moduli spaces of sheaves, Brill-Noether theory, rational surfaces, Hirzebruch and del Pezzo surfaces.

During the preparation of this article the first author was partially supported by the NSF CAREER grant DMS-0950951535 and NSF grant DMS-1500031, and the second author was partially supported by a National Science Foundation Mathematical Sciences Postdoctoral Research Fellowship DMS-1204066 and an NSA Young Investigator Grant H98230-16-1-0306.

By semicontinuity, if E is any sheaf with no cohomology, then the cohomology also vanishes for the general sheaf in any component of $M_{X,H}(\mathbf{v})$ that contains E. Since the cohomology vanishes identically, for weak Brill-Noether to hold, we must have $\chi(\mathbf{v}) = 0$.

The weak Brill-Noether property is the key ingredient in constructing effective theta divisors on the moduli spaces $M_{X,H}(\mathbf{v})$ and plays a central role in describing effective cones of moduli spaces and strange duality. Assume that $M_{X,H}(\mathbf{v})$ is irreducible. The locus

$$\Theta = \{E \in M_{X,H}(\mathbf{v}) | h^1(X,E) \neq 0\}$$

is called the *theta locus* and is an effective divisor when weak Brill-Noether holds for $M_{X,H}(\mathbf{v})$. In this paper, we study when Chern characters $\mathbf{v}$ satisfy weak Brill-Noether on rational surfaces X. We completely classify Chern characters $\mathbf{v}$ satisfying weak Brill-Noether on Hirzebruch surfaces. For general rational surfaces, we give sufficient conditions on Chern characters that guarantee that weak Brill-Noether holds. In particular, we show that if X is a del Pezzo surface of degree at least 4 and $\mathrm{ch}_1(\mathbf{v})$ is nef, then weak Brill-Noether holds. We now summarize our results in greater detail and give some examples.

1.1. Weak Brill-Noether for rank one sheaves. When the sheaves have rank one, the story is particularly simple.

PROPOSITION 1.2. *Let L be a line bundle on a smooth projective surface X such that $n = \chi(L) \geq 0$. Let $Z \subset X$ be a general zero-dimensional scheme of length n, and set $\mathbf{v} = \mathrm{ch}(L \otimes I_Z)$, so that $\chi(\mathbf{v}) = 0$. Then $H^i(X, L \otimes I_Z) = 0$ for all i if and only if $H^1(X,L) = H^2(X,L) = 0$. In particular, weak Brill-Noether holds for $M_{X,H}(\mathbf{v})$ if and only if there exists a line bundle M with $\mathrm{ch}_1(M) = \mathrm{ch}_1(\mathbf{v})$ and $H^1(X,M) = H^2(X,M) = 0$.*

PROOF. Any torsion-free rank one sheaf on a surface is isomorphic to $L \otimes I_Z$ for a line bundle L and an ideal sheaf I_Z of a zero-dimensional scheme. First, suppose $H^1(X,L) = H^2(X,L) = 0$ and $h^0(X,L) = n$. Let $X^{[n]}$ denote the Hilbert scheme of n points on X and let $Z \in X^{[n]}$ be a general subscheme of length n. Consider the restriction sequence

$$0 \to L \otimes I_Z \to L \to L|_Z \to 0.$$

Since Z is general, the map $H^0(X,L) \to H^0(X, L|_Z)$ is an isomorphism, and we find $L \otimes I_Z$ has no cohomology.

Conversely, suppose $H^1(X,L)$ or $H^2(X,L)$ is nonzero. If $h^0(X,L) > n$, then $h^0(X, L \otimes I_Z) > 0$ for every $Z \in X^{[n]}$. Thus we may assume $h^0(X,L) \leq n$. Then since at least one of $H^1(X,L)$ or $H^2(X,L)$ is nonzero and $\chi(L) = n$, we find $H^2(X,L) \neq 0$. But $H^2(X, L \otimes I_Z) \cong H^2(X,L)$ for any $Z \in X^{[n]}$. In particular, weak Brill-Noether holds for $M_{X,H}(\mathbf{v})$ if and only if there exists a line bundle M with $\mathrm{ch}_1(M) = \mathrm{ch}_1(\mathbf{v})$ and $H^1(X,M) = H^2(X,M) = 0$. □

EXAMPLE 1.3. On $\mathbb{P}^2$, weak Brill-Noether fails for moduli spaces of rank one sheaves when the slope is -3 or less. This failure is intimately tied to the fact that the general rank one sheaf is not locally free. In contrast, Göttsche and Hirschowitz [**GHi94**] prove that if $\mathbf{v}$ has $\mathrm{rk}(\mathbf{v}) \geq 2$ and $\chi(\mathbf{v})$ is arbitrary, then the general sheaf $E \in M_{\mathbb{P}^2, \mathcal{O}(1)}(\mathbf{v})$ has at most one nonzero cohomology group. The signs of the Euler characteristic and the slope determine which cohomology group is nonzero.

In particular, if $\chi(\mathbf{v}) = 0$, then for the general stable sheaf, all cohomology groups vanish and weak Brill-Noether holds. The purpose of this paper is to generalize this theorem to other rational surfaces.

Our first main result classifies Chern characters on Hirzebruch surfaces that satisfy weak Brill-Noether. Let H be any ample class on the Hirzebruch surface $\mathbb{F}_e = \mathbb{P}(\mathcal{O}_{\mathbb{P}^1} \oplus \mathcal{O}_{\mathbb{P}^1}(e))$, $e \geq 0$. Let $\mathbf{v}$ be a stable Chern character of rank at least 2 on $\mathbb{F}_e$ such that $\chi(\mathbf{v}) = 0$. Let

$$\nu(\mathbf{v}) = \frac{c_1(\mathbf{v})}{r(\mathbf{v})} = \frac{k}{r}E + \frac{l}{r}F$$

denote the total slope of $\mathbf{v}$, where E is the section of self-intersection $-e$ and F is a fiber. By Walter's Theorem [**Wal98**, Theorem 1], the moduli spaces $M_{\mathbb{F}_e,H}(\mathbf{v})$ are irreducible and the general sheaf is locally free. By Serre duality, we may assume that

$$\frac{k}{r} \geq -1 \quad \text{and if} \quad \frac{k}{r} = -1, \text{ then } \frac{l}{r} \geq -1 - \frac{e}{2}.$$

THEOREM 1.4. *Let $\mathbf{v}$ be a stable Chern character that satisfies these inequalities. Then $M_{\mathbb{F}_e,H}(\mathbf{v})$ satisfies weak Brill-Noether if and only if*

$$\frac{l - ke}{r} = \nu(\mathbf{v}) \cdot E \geq -1.$$

As Theorem 1.4 demonstrates, the existence of an effective curve C on X such that

$$\mathbf{w} = \mathrm{ch}(\mathcal{O}_X(C)), \quad \chi(\mathbf{w}, \mathbf{v}) > 0, \quad \text{and} \quad \nu(\mathbf{v}) \cdot H > (K_X + C) \cdot H$$

provides an obstruction to weak Brill-Noether for $M_{X,H}(\mathbf{v})$. By Serre duality and stability,

$$\mathrm{Ext}^2(\mathcal{O}_X(C), \mathcal{E}) \cong \mathrm{Hom}(\mathcal{E}, \mathcal{O}_X(K_X + C))^* = 0.$$

Since $\chi(\mathcal{O}_X(C), \mathcal{E}) > 0$, we have $\mathrm{Hom}(\mathcal{O}_X(C), \mathcal{E}) \neq 0$. Composing with the natural map

$$\mathcal{O}_X \longrightarrow \mathcal{O}_X(C),$$

we see that $\mathrm{Hom}(\mathcal{O}_X, \mathcal{E}) = H^0(X, \mathcal{E}) \neq 0$ for every $\mathcal{E} \in M_{X,H}(\mathbf{v})$. We remark that if $M_{X,H}(\mathbf{v})$ is nonempty, C must also satisfy $\nu(\mathbf{v}) \cdot H \geq C \cdot H$.

In order to prove weak Brill-Noether theorems, we need to ensure that these obstructions vanish. For a line bundle D, a torsion-free sheaf $\mathcal{E}$ is called *D-prioritary* if $\mathrm{Ext}^2(\mathcal{E}, \mathcal{E} \otimes D^{-1}) = 0$. Let $\mathcal{P}_{X,D}(\mathbf{v})$ denote the stack of D-prioritary sheaves on X. For blowups of $\mathbb{P}^2$, our sharpest result is the following.

THEOREM 1.5. *Let X be a blowup of $\mathbb{P}^2$ at k distinct points $p_1, \ldots, p_k$. Let L be the pullback of the hyperplane class of $\mathbb{P}^2$ and let E_i be the exceptional divisor over p_i. Let $\mathbf{v} \in K(X)$ with $r = r(\mathbf{v}) > 0$, and write*

$$\nu(\mathbf{v}) := \frac{c_1(\mathbf{v})}{r(\mathbf{v})} = \delta L - \alpha_1 E_1 - \cdots - \alpha_k E_k,$$

so that the coefficients $\delta, \alpha_i \in \mathbb{Q}$. Assume that $\delta \geq 0$ and $\alpha_i \geq 0$ for all i. Suppose that the line bundle

$$\lfloor \delta \rfloor L - \lceil \alpha_1 \rceil E_1 - \cdots - \lceil \alpha_k \rceil E_k$$

has no higher cohomology. If $\chi(\mathbf{v}) = 0$, then the stack $\mathcal{P}_{X,L-E_1}(\mathbf{v})$ is nonempty and the general $\mathcal{E} \in \mathcal{P}_{X,L-E_1}(\mathbf{v})$ has no cohomology.

In particular, when H is an ample divisor on X that satisfies $H \cdot (K_X + L - E_1) < 0$ and $\mathbf{v}$ is an H-stable Chern character satisfying the assumptions of Theorem 1.5, then $M_{X,H}(\mathbf{v})$ satisfies weak Brill-Noether. On del Pezzo surfaces of large degree we obtain sharper results.

THEOREM 1.6. *Let X be a del Pezzo surface of degree at least 4. Let $\mathbf{v} \in K(X)$ with $\chi(\mathbf{v}) = 0$, and suppose $c_1(\mathbf{v})$ is nef. Then the stack $\mathcal{P}_{X,L-E_1}(\mathbf{v})$ is nonempty and a general $\mathcal{E} \in \mathcal{P}_{X,L-E_1}(\mathbf{v})$ has no cohomology.*

The following conjecture may be thought of as a higher rank analogue of the celebrated Segre-Harbourne-Gimigliano-Hirschowitz conjecture [**Seg60, Harb84, Gim87, Hir89**].

CONJECTURE 1.7. *Assume X is a general blowup of $\mathbb{P}^2$ and $c_1(\mathbf{v})$ is nef. Let $F = L - E_1$. If H is an ample class such that $H \cdot (K_X + F) < 0$, then $M_{X,H}(\mathbf{v})$ satisfies weak Brill-Noether.*

We use two techniques to prove weak Brill-Noether Theorems. First, we give a resolution of the general sheaf on $M_{X,H}(\mathbf{v})$ in terms of a strong exceptional collection satisfying certain cohomology vanishing properties. This method allows us to prove Theorem 1.4 and show weak Brill-Noether on arbitrary rational surfaces provided $\mathbf{v}$ satisfies certain inequalities. The advantage of this method is that it gives a convenient resolution of the general sheaf in $M_{X,H}(\mathbf{v})$. As a consequence, it shows that the moduli space is unirational. The disadvantage is that this method is only applicable when the surface X admits a strong exceptional collection of the desired form.

Second, we construct an explicit prioritary sheaf with vanishing cohomology as a sum of line bundles. Walter [**Wal98**, Proposition 2] proves that on a birationally ruled surface the stack parameterizing sheaves prioritary with respect to the fiber class is smooth and irreducible. Assuming that the stable sheaves in $M_{X,H}(\mathbf{v})$ are prioritary, to prove weak Brill-Noether, it suffices to exhibit one prioritary sheaf with vanishing cohomology. Constructing prioritary sheaves is much easier than constructing stable sheaves. In particular, under suitable assumptions, one may construct prioritary sheaves as sums of line bundles. The problem then reduces to the combinatorial problem of finding a prioritary combination of line bundles with no higher cohomology that has the same rank and first Chern class as $\mathbf{v}$. We solve this problem explicitly for del Pezzo surfaces of degree at least 4. Both of these techniques are applicable much more generally. However, to minimize the combinatorial complexity, we make additional assumptions on the Chern character $\mathbf{v}$ when convenient.

1.2. The organization of the paper. In §2, we collect basic facts about moduli spaces of sheaves and the cohomology of line bundles on rational surfaces. In §3, we introduce our first method for proving weak Brill-Noether and characterize Chern characters on Hirzebruch surfaces that satisfy weak Brill-Noether. In §4 and §5, we introduce our second method and show that Chern characters with nef ch_1 satisfy weak Brill-Noether on del Pezzo surfaces of degree at least 4.

1.3. Acknowledgments. We would like to express our gratitude to Lawrence Ein whose unfailing support has been invaluable in our careers. We would also like to thank Daniel Levine and the referee for their suggestions and corrections.

2. Preliminaries

In this section, we recall standard facts concerning Hirzebruch and del Pezzo surfaces and cohomology of line bundles on rational surfaces. We refer the reader to [**Bea83**], [**Cos06a**], [**Cos06b**] or [**Hart77**] for more detailed expositions.

2.1. Hirzebruch surfaces. Let $e \geq 0$ be a nonnegative integer. Let $\mathbb{F}_e$ denote the Hirzebruch surface $\mathbb{P}(\mathcal{O}_{\mathbb{P}^1} \oplus \mathcal{O}_{\mathbb{P}^1}(e))$. When $e \geq 1$, let E be the class of the unique section of self intersection $E^2 = -e$ and let F denote the class of a fiber of the projection to $\mathbb{P}^1$. The surface $\mathbb{F}_0$ is isomorphic to $\mathbb{P}^1 \times \mathbb{P}^1$. In that case, let E and F denote the classes of the two rulings. Then

$$\mathrm{Pic}(\mathbb{F}_e) \cong \mathbb{Z}E \oplus \mathbb{Z}F \quad \text{with} \quad E^2 = -e, \quad E \cdot F = 1, \quad F^2 = 0.$$

By adjunction,

$$K_{\mathbb{F}_e} = -2E - (e+2)F.$$

Consequently, the Riemann-Roch Theorem implies that

$$\chi(\mathcal{O}_{\mathbb{F}_e}(aE + bF)) = (a+1)(b+1) - e\frac{a(a+1)}{2}.$$

The effective cone of $\mathbb{F}_e$ is generated by E and F, consequently

$$H^0(\mathbb{F}_e, \mathcal{O}_{\mathbb{F}_e}(aE + bF)) \neq 0 \quad \text{if and only if} \quad a, b \geq 0.$$

By Serre duality,

$$H^2(\mathbb{F}_e, \mathcal{O}_{\mathbb{F}_e}(aE + bF)) \neq 0 \quad \text{if and only if} \quad a \leq -2 \text{ and } b \leq -2 - e.$$

Hence, to compute the cohomology of all line bundles, it suffices to assume that $a \geq -1$. In this case, since H^2 vanishes and we have computed the Euler characteristic, specifying h^0 determines the dimension of all cohomology groups. The following theorem (see [**Cos06a**], [**Hart77**, §V.2]) summarizes the answer.

THEOREM 2.1. *Let $\mathcal{O}_{\mathbb{F}_e}(aE + bF)$ be a line bundle on the Hirzebruch surface $\mathbb{F}_e$ with $a \geq -1$. Then:*

(1) *$h^i(\mathbb{F}_e, \mathcal{O}_{\mathbb{F}_e}(-E + bF)) = 0$ for $0 \leq i \leq 2$ and all b.*
(2) *$h^0(\mathbb{F}_e, \mathcal{O}_{\mathbb{F}_e}(bF)) = b + 1$ if $b \geq -1$ and 0 otherwise. In particular, $h^i(\mathbb{F}_e, \mathcal{O}_{\mathbb{F}_e}(-F)) = 0$ for $0 \leq i \leq 2$.*
(3) *We may assume that $a \geq 1$ and $b \geq 0$. If $b < ae$, then*

$$h^0(\mathbb{F}_e, \mathcal{O}_{\mathbb{F}_e}(aE + bF)) = h^0(\mathbb{F}_e, \mathcal{O}_{\mathbb{F}_e}((a-1)E + bF)).$$

If $b \geq ae$, then

$$h^0(\mathbb{F}_e, \mathcal{O}_{\mathbb{F}_e}(aE + bF)) = \chi(\mathcal{O}_{\mathbb{F}_e}(aE + bF)) = (a+1)(b+1) - e\frac{a(a+1)}{2}.$$

PROOF. We have already observed that $H^i(\mathcal{O}_{\mathbb{F}_e}(-E + bF)) = 0$ for $i = 0, 2$. The case $i = 1$ follows from the fact that the Euler characteristic vanishes. This proves (1).

Next, since F is a pullback from the base, we have $H^i(\mathbb{F}_e, \mathcal{O}_{\mathbb{F}_e}(bF)) \cong H^i(\mathbb{P}^1, \mathcal{O}_{\mathbb{P}^1}(b))$. Part (2) of the theorem follows.

We can therefore assume that $a \geq 1$. If $b < 0$, then $h^i(\mathbb{F}_e, \mathcal{O}_{\mathbb{F}_e}(aE + bF)) = 0$ for $i = 0, 2$ and h^1 is determined by the Euler characteristic. Hence, we may assume that $a \geq 1$ and $b \geq 0$. If $E \cdot (aE + bF) = b - ae < 0$, then E is in the base locus of the linear system and the map given by multiplication by a section s_E of $\mathcal{O}_{\mathbb{F}_e}(E)$

$$H^0(\mathbb{F}_e, \mathcal{O}_{\mathbb{F}_e}((a-1)E + bF)) \xrightarrow{s_E} H^0(\mathbb{F}_e, \mathcal{O}_{\mathbb{F}_e}(aE + bF))$$

induces an isomorphism. Repeating this process inductively, we reduce to the case when $ae \leq b$. Consider the exact sequence

$$0 \longrightarrow \mathcal{O}_{\mathbb{F}_e}((a-1)E + bF) \longrightarrow \mathcal{O}_{\mathbb{F}_e}(aE + bF) \longrightarrow \mathcal{O}_{\mathbb{P}^1}(b - ae) \longrightarrow 0.$$

If $b - ae \geq -1$, we have a surjection

$$H^1(\mathbb{F}_e, \mathcal{O}_{\mathbb{F}_e}((a-1)E + bF)) \to H^1(\mathbb{F}_e, \mathcal{O}_{\mathbb{F}_e}(aE + bF)) \to 0.$$

By inductively reducing a to 0, we conclude that $h^1(\mathbb{F}_e, \mathcal{O}_{\mathbb{F}_e}(aE + bF)) = 0$. Consequently, $h^0(\mathbb{F}_e, \mathcal{O}_{\mathbb{F}_e}(aE + bF)) = \chi(\mathcal{O}_{\mathbb{F}_e}(aE + bF))$. □

2.2. Blowups of $\mathbb{P}^2$. We next record several basic facts concerning the cohomology of line bundles on blowups of $\mathbb{P}^2$. Let X be the blowup of $\mathbb{P}^2$ at k distinct points $p_1, \dots, p_k$. Let L denote the pullback of the hyperplane class on $\mathbb{P}^2$ and let E_i denote the exceptional divisor lying over p_i. Then

$$\operatorname{Pic}(X) \cong \mathbb{Z}L \oplus \bigoplus_{i=1}^{k} \mathbb{Z}E_i \text{ with } L^2 = 1,\ L \cdot E_i = 0,\ E_i \cdot E_j = -\delta_{i,j},$$

where $\delta_{i,j}$ is the Krönecker delta function. Let $D = \delta L - \sum_{i=1}^{k} \alpha_i E_i$ be an integral class on X. Since $K_X = -3L + \sum_{i=1}^{k} E_i$, by Riemann-Roch

$$\chi(\mathcal{O}_X(D)) = \frac{(\delta+2)(\delta+1)}{2} - \sum_{i=1}^{k} \frac{\alpha_i(\alpha_i+1)}{2}.$$

If D is effective, then $\delta \geq 0$. Otherwise, a general line with class L would be a moving curve with $L \cdot D < 0$. In particular, by Serre duality, $H^2(X, \mathcal{O}_X(D)) = 0$ if $\delta \geq -2$.

EXAMPLE 2.2 (Del Pezzo surfaces). Del Pezzo surfaces are smooth complex surfaces X with ample anti-canonical bundle $-K_X$. They consist of $\mathbb{P}^1 \times \mathbb{P}^1$ and the blowup of $\mathbb{P}^2$ in fewer than 9 points in general position. Since $\mathbb{P}^1 \times \mathbb{P}^1$ is also a Hirzebruch surface, we will concentrate on the surfaces D_n, the blowup of $\mathbb{P}^2$ at $9 - n$ general points. The effective cone of curves on D_n is spanned by the (-1)-curves and the nef cone is the dual cone consisting of classes that intersect (-1)-curves nonnegatively. The classes of (-1)-curves $C = aL - \sum_{i=1}^{9-n} b_i E_i$ on D_n can be obtained by solving the equations

$$C^2 = a^2 - \sum_{i=1}^{9-n} b_i^2 = -1, \quad -K_{D_n} \cdot C = 3a - \sum_{i=1}^{9-n} b_i = 1.$$

For our purposes, it suffices to know that on D_n for $n \geq 3$, the (-1)-curves are E_i for $1 \leq i \leq 9 - n$, $L - E_i - E_j$ for $i \neq j$ and $2L - E_a - E_b - E_c - E_d - E_e$, where a, b, c, d, e are 5 distinct indices (whenever the number of points is large enough for these classes to exist) (see [**Cos06b**], [**Hart77**, §V.4]).

We will need the following cohomology computations.

LEMMA 2.3. *Let $I \subset \{1, \dots, k\}$ be a possibly empty index set. Then:*

(1) *We have $H^i(X, \mathcal{O}_X(D)) = 0$ for all i if D is one the following*

$$-2H + \sum_{i \in I} E_i, \quad -H + \sum_{i \in I} E_i, \quad -E_j + \sum_{i \in I, i \neq j} E_i.$$

(2) *Assume $H^i(X, \mathcal{O}_X(D)) = 0$ for $i > 0$. If $D \cdot E_j \geq 0$ (respectively, $D \cdot L \geq -2$), then $H^i(X, \mathcal{O}_X(D + E_j)) = 0$ (respectively, $H^i(X, \mathcal{O}_X(D + L)) = 0$) for $i > 0$.*

PROOF. If an effective class is represented by a smooth rational curve C on a smooth rational surface X, then we claim that $\mathcal{O}_X(-C)$ has no cohomology. Since $H^i(X, \mathcal{O}_X) = H^i(C, \mathcal{O}_C) = 0$ for $i \geq 1$, the natural sequence

$$0 \to \mathcal{O}_X(-C) \to \mathcal{O}_X \to \mathcal{O}_C \to 0$$

implies that $H^i(X, \mathcal{O}_X(-C)) = 0$ for all i. Since E_i, H and $2H$ can be represented by the exceptional curve, a line and a conic, respectively, the proposition is true when $I = \emptyset$. If D is a class such that $H^i(X, \mathcal{O}_X(D)) = 0$ for all i and $E_j \cdot D = 0$, then $H^i(X, \mathcal{O}_X(D + E_j)) = 0$ for all i. To see this, consider the exact sequence

$$0 \to \mathcal{O}_X(D) \to \mathcal{O}_X(D + E_j) \to \mathcal{O}_{\mathbb{P}^1}(-1) \to 0.$$

Since $\mathcal{O}_X(D)$ and $\mathcal{O}_{\mathbb{P}^1}(-1)$ have no cohomology, $\mathcal{O}_X(D + E_j)$ has no cohomology. Similar sequences imply the last statement. □

2.3. Blowups of Hirzebruch surfaces. Since the blowup of $\mathbb{F}_0$ at one point is isomorphic to the blowup of $\mathbb{P}^2$ at 2 points and $\mathbb{F}_1$ is isomorphic to the blowup of $\mathbb{P}^2$ at one point, we may assume that $e \geq 2$. Let X be the blowup of $\mathbb{F}_e$ along k distinct points $p_1, \dots, p_k$ which are not contained in the exceptional curve E. Then the Picard group of X is the free abelian group generated by $E, F, E_1, \dots, E_k$, where E and F are the pullbacks of the two generators from $\mathbb{F}_e$ and $E_1, \dots, E_k$ are the exceptional divisors lying over $p_1, \dots, p_k$. The same argument as in Lemma 2.3 proves the following.

LEMMA 2.4. *Let $I \subset \{1, \dots, k\}$ be a possibly empty index set. Then $H^i(X, \mathcal{O}_X(D)) = 0$ for all i if D is one the following*

$$-E + mF + \sum_{i \in I} E_i \quad (m \in \mathbb{Z}), \qquad -F + \sum_{i \in I} E_i, \qquad -E_j + \sum_{i \in I, i \neq j} E_i.$$

Moreover, assume $H^i(X, \mathcal{O}_X(D)) = 0$ for $i > 0$ and C is a rational curve with $C \cdot D \geq -C^2 - 1$. Then $H^i(X, \mathcal{O}_X(D + C)) = 0$ for $i > 0$.

2.4. Moduli spaces of vector bundles. Next, we recall some basic facts concerning moduli spaces of Gieseker semistable sheaves and prioritary sheaves. We refer the reader to [**CH15**], [**Hui17**], [**HuL10**] and [**LeP97**] for details.

Let (X, H) be a polarized, smooth projective surface. All the sheaves we consider will be pure-dimensional and coherent. If $\mathcal{E}$ is a pure d-dimensional, coherent sheaf, then the Hilbert polynomial has the form

$$P_{\mathcal{E}}(m) = \chi(\mathcal{E}(mH)) = a_d \frac{m^d}{d!} + \text{l.o.t.}$$

The reduced Hilbert polynomial of $\mathcal{E}$ is defined by $p_{\mathcal{E}} = P_{\mathcal{E}}/a_d$. A sheaf $\mathcal{E}$ is *Gieseker semistable* if for every proper subsheaf $\mathcal{F} \subsetneq \mathcal{E}$, we have $p_{\mathcal{F}} \leq p_{\mathcal{E}}$, where polynomials are compared for sufficiently large m. The sheaf is called *Gieseker stable* if for every proper subsheaf the inequality is strict. By theorems of Gieseker, Maruyama and Simpson, there exist projective moduli spaces $M_{X,H}(\mathbf{v})$ parameterizing S-equivalence classes of Gieseker semistable sheaves on X with Chern character $\mathbf{v}$ (see [**HuL10**] or [**LeP97**]).

It is often hard to verify the stability of a sheaf. The following notion provides a more flexible alternative.

DEFINITION 2.5. Let F be a line bundle on X. A torsion-free coherent sheaf $\mathcal{E}$ is *F-prioritary* if $\mathrm{Ext}^2(\mathcal{E}, \mathcal{E} \otimes F^{-1}) = 0$.

We denote the stack of F-prioritary sheaves on X with Chern character $\mathbf{v}$ by $\mathcal{P}_{X,F}(\mathbf{v})$. The stack $\mathcal{P}_{X,F}(\mathbf{v})$ is an open substack of the stack of coherent sheaves. In this paper, we will consider F-prioritary sheaves on (blowups of) $\mathbb{F}_e$ and blowups of $\mathbb{P}^2$, where F is the fiber class on $\mathbb{F}_e$ and the class $L - E_1$ on a blowup of $\mathbb{P}^2$. The class F endows these surfaces with the structure of a birationally ruled surface. The following theorem of Walter will be crucial to our arguments.

THEOREM 2.6 ([**Wal98**, Proposition 2]). *Let X be a birationally ruled surface, let F be the fiber class on X and let $\mathbf{v}$ be a fixed Chern character of rank at least 2. Then the stack $\mathcal{P}_{X,F}(\mathbf{v})$ of F-prioritary sheaves is smooth and irreducible.*

In particular, if H is an ample divisor on a birationally ruled surface X such that $H \cdot (K_X + F) < 0$ and $\mathbf{v}$ is a stable Chern character of rank at least 2, then the moduli space $M_{X,H}(\mathbf{v})$ is irreducible and normal [**Wal98**, Theorem 1]. The inequality $H \cdot (K_X + F) < 0$ guarantees that Gieseker semistable sheaves are F-prioritary. When Walter's Theorem applies, we can construct a prioritary sheaf with no cohomology to deduce that general Gieseker semistable sheaves with the same invariants have no cohomology. The advantage is that prioritary sheaves are much easier to construct than semistable sheaves.

In our computations, we will use the following consequence of Riemann-Roch repeatedly.

LEMMA 2.7. *Let $\mathcal{E}$ be a sheaf of rank r on a surface X such that $\chi(\mathcal{E}) = 0$ and let M be a line bundle. Then*

$$\chi(\mathcal{E} \otimes M) = \mathrm{ch}_1(E) \cdot \mathrm{ch}_1(M) + r(\chi(M) - \chi(\mathcal{O}_X)).$$

PROOF. By the Hirzebruch-Riemann-Roch Theorem

$$\chi(\mathcal{E} \otimes M) = \int_X \mathrm{ch}(\mathcal{E} \otimes M)\,\mathrm{td}(X) = \int_X \mathrm{ch}(\mathcal{E})\,\mathrm{ch}(M)\,\mathrm{td}(X).$$

The formula follows immediately by expanding this expression. □

3. Strong exceptional collections and resolutions

In this section, we introduce our first method for proving weak Brill-Noether theorems. This method provides a resolution of the general sheaf of the moduli space in terms of a strong exceptional collection that satisfies certain cohomological properties. The method gives a unirational parameterization of (a component of) the moduli space. The disadvantage is that it is only applicable when a suitable strong exceptional collection exists. We begin by recalling some standard terminology.

DEFINITION 3.1. A sheaf A is *exceptional* if $\mathrm{Hom}(A,A)=\mathbb{C}$ and $\mathrm{Ext}^i(A,A)=0$ for $i\neq 0$. An ordered collection $(A_1,\ldots,A_m)$ of exceptional sheaves on a projective variety X is *an exceptional collection* if

$$\mathrm{Ext}^i(A_t,A_s)=0 \text{ for } 1\le s<t\le m \text{ and all } i.$$

The exceptional collection is *strong* if in addition

$$\mathrm{Ext}^i(A_s,A_t)=0 \text{ for } 1\le s<t\le m \text{ and all } i>0.$$

EXAMPLE 3.2. On the Hirzebruch surface $\mathbb{F}_e$, the collection of line bundles

$$\mathcal{O}_{\mathbb{F}_e}(-E-(e+1)F),\quad \mathcal{O}_{\mathbb{F}_e}(-E-eF),\quad \mathcal{O}_{\mathbb{F}_e}(-F),\quad \mathcal{O}_{\mathbb{F}_e}$$

is a strong exceptional collection. If $1\le s<t\le 4$,

$$\mathrm{Ext}^i(A_t,A_s)\cong H^i(\mathbb{F}_e,-F) \text{ or } H^i(\mathbb{F}_e,-E+bF) \text{ for some } b.$$

Since these cohomology groups vanish by Theorem 2.1, we conclude that the collection is exceptional. Similarly,

$$\mathrm{Ext}^i(A_s,A_t)\cong H^i(\mathbb{F}_e,F) \text{ or } H^i(\mathbb{F}_e,E+bF) \text{ for some } b\ge e-1.$$

Since these cohomology groups vanish for $i>0$ by Theorem 2.1, we conclude that the collection is a strong exceptional collection.

EXAMPLE 3.3. Let $\Gamma=\{p_1,\ldots,p_k\}$ be a set of k distinct points on $\mathbb{P}^2$ and let X be the blowup of $\mathbb{P}^2$ along Γ. Let E_i denote the exceptional divisor lying over p_i and let L be the pullback of the hyperplane class from $\mathbb{P}^2$. Then

$$\mathcal{O}_X(-2L),\quad \mathcal{O}_X(-L),\quad \mathcal{O}_X(-E_1),\quad \mathcal{O}_X(-E_2),\quad \ldots\quad \mathcal{O}_X(-E_k),\quad \mathcal{O}_X$$

is a strong exceptional collection on X. This can be checked as follows (see also Bondal's Theorem [**Bon89**], [**KO95**]). Let $I\subset\{1,\ldots,k\}$ be an index set. By Lemma 2.3, $H^j(X,\mathcal{O}_X(D))=0$ for all j and D of the form

$$-2L+\sum_{i\in I}E_i,\quad -L+\sum_{i\in I}E_i,\quad -E_i,\quad \text{or}\quad E_l-E_i, l\neq i.$$

Since for $1\le s<t\le m$, each $\mathrm{Ext}^i(A_t,A_s)$ is isomorphic to one of these cohomology groups, we conclude that the collection is exceptional. Similarly, $H^j(X,\mathcal{O}_X(D))=0$ for $j>0$ and D of the form

$$2L,\quad 2L-E_i,\quad L,\quad L-E_i,\quad E_l-E_i \quad\text{or}\quad E_l.$$

Since the groups $\mathrm{Ext}^i(A_s,A_t)$ for $1\le s<t\le m$ are isomorphic to one of these groups, we conclude that the collection is a strong exceptional collection.

NOTATION 1. Throughout this section, let X be a smooth projective surface and let $(A_1,\ldots,A_m,\mathcal{O}_X)$ be a strong exceptional collection on X. Suppose that a sheaf $\mathcal{E}$ has a resolution of the form

$$0\longrightarrow\bigoplus_{i=1}^{j}A_i^{\oplus a_i}\xrightarrow{\phi}\bigoplus_{i=j+1}^{m}A_i^{\oplus a_i}\longrightarrow\mathcal{E}\longrightarrow 0. \tag{1}$$

Notice that $\mathcal{O}_X$ is the last member of the strong exceptional collection and does not occur in the resolution of $\mathcal{E}$.

LEMMA 3.4. *Let $\mathcal{E}$ be a sheaf with a resolution given by Sequence (1). Then $H^i(X,\mathcal{E})=0$ for all i. The exponents a_i are determined by the following relations:*

$$a_s = -\chi(\mathcal{E}, A_s) - \sum_{i=1}^{s-1} a_i \hom(A_i, A_s) \quad \text{for} \quad 1 \leq s \leq j \quad \text{and}$$

$$a_t = \chi(A_t, \mathcal{E}) - \sum_{i=t+1}^{m} a_i \hom(A_t, A_i) \quad \text{for} \quad j+1 \leq t \leq m.$$

PROOF. Since $(A_1, \ldots, A_m, \mathcal{O}_X)$ is a strong exceptional collection, $\mathrm{Ext}^i(\mathcal{O}_X, A_s) = 0$ for all $1 \leq s \leq m$ and all i. Applying $\mathrm{Ext}(\mathcal{O}_X, -)$ to the Sequence (1), we conclude that

$$\mathrm{Ext}^i(\mathcal{O}_X, \mathcal{E}) = H^i(X, \mathcal{E}) = 0.$$

To compute the exponents a_s with $1 \leq s \leq j$, we apply $\mathrm{Ext}(-, A_s)$ to the same sequence. Since $\mathrm{Ext}^k(A_i, A_s) = 0$ for all k if $i > s$ and $\mathrm{Ext}^k(A_i, A_s) = 0$ for $k > 0$, we obtain the relation

$$\chi(\mathcal{E}, A_s) + \sum_{i=1}^{s} a_i \hom(A_i, A_s) = 0.$$

The desired formula follows from the fact that $\hom(A_s, A_s) = 1$. Similarly, to compute the exponents a_t with $j < t < m$, we apply $\mathrm{Ext}(A_t, -)$ to the sequence. We deduce that

$$\chi(A_t, \mathcal{E}) = a_t + \sum_{i=t+1}^{m} a_i \hom(A_t, A_i).$$

This concludes the proof of the lemma. □

LEMMA 3.5. *Let $\mathcal{E}$ be a locally free sheaf with a resolution given by Sequence (1) and let F be a line bundle on X. Assume that*

(1) $\mathrm{Ext}^1(A_i, A_s \otimes F^{-1}) = 0$ *for* $1 \leq i \leq j$ *and* $j < s \leq m$, *and*
(2) $\mathrm{Ext}^2(A_i, A_s \otimes F^{-1}) = 0$ *for* $j < i, s \leq m$.

Then $\mathcal{E}$ is F-prioritary.

PROOF. We need to check that $\mathrm{Ext}^2(\mathcal{E}, \mathcal{E} \otimes F^{-1}) = 0$. By applying $\mathrm{Ext}(\mathcal{E}, -)$ to the sequence

$$0 \longrightarrow \bigoplus_{i=1}^{j} A_i^{\oplus a_i} \otimes F^{-1} \longrightarrow \bigoplus_{i=j+1}^{m} A_i^{\oplus a_i} \otimes F^{-1} \longrightarrow \mathcal{E} \otimes F^{-1} \longrightarrow 0,$$

it suffices to check $\mathrm{Ext}^2(\mathcal{E}, A_s \otimes F^{-1}) = 0$ for $s > j$. We now apply $\mathrm{Ext}(-, A_s \otimes F^{-1})$ to Sequence (1) to obtain

$$\bigoplus_{1 \leq i \leq j} \mathrm{Ext}^1(A_i, A_s \otimes F^{-1})^{\oplus a_i} \longrightarrow \mathrm{Ext}^2(\mathcal{E}, A_s \otimes F^{-1}) \longrightarrow \bigoplus_{j < i \leq m} \mathrm{Ext}^2(A_i, A_s \otimes F^{-1})^{\oplus a_i}.$$

Since the terms on the right and left are zero by assumption, we conclude that $\mathcal{E}$ is prioritary with respect to F. □

PROPOSITION 3.6. *Let $\mathcal{E}$ be a locally free sheaf with a resolution given by Sequence (1). Assume that $\mathcal{E}$ is F-prioritary and has Chern character* $\mathbf{v}$. *Set*

$$U = \bigoplus_{i=1}^{j} A_i^{\oplus a_i} \quad \text{and} \quad V = \bigoplus_{i=j+1}^{m} A_i^{\oplus a_i}.$$

Then the open set $S \subset \mathrm{Hom}(U,V)$ parameterizing locally free F-prioritary sheaves is a complete family of F-prioritary sheaves. If the stack $\mathcal{P}_{X,F}(\mathbf{v})$ is irreducible, then the general sheaf in $\mathcal{P}_{X,F}(\mathbf{v})$ has no cohomology.

PROOF. To show that the family is complete, we need to show that the Kodaira-Spencer map is surjective. The Kodaira-Spencer map

$$\kappa : T_\phi S = \mathrm{Hom}(U,V) \to \mathrm{Ext}^1(\mathcal{E},\mathcal{E})$$

factors as the composition of the two maps [**HuL10**], [**LeP97**]

$$\mathrm{Hom}(U,V) \xrightarrow{\mu} \mathrm{Hom}(U,\mathcal{E}) \xrightarrow{\nu} \mathrm{Ext}^1(\mathcal{E},\mathcal{E}),$$

where μ and ν are the morphisms that appear in the natural exact sequences obtained by applying $\mathrm{Ext}(U,-)$ and $\mathrm{Ext}(-,\mathcal{E})$, respectively:

$$\mathrm{Hom}(U,V) \xrightarrow{\mu} \mathrm{Hom}(U,E) \longrightarrow \mathrm{Ext}^1(U,U)$$
$$\mathrm{Hom}(U,\mathcal{E}) \xrightarrow{\nu} \mathrm{Ext}^1(\mathcal{E},\mathcal{E}) \longrightarrow \mathrm{Ext}^1(V,\mathcal{E}).$$

Since $(A_1,\dots,A_m,\mathcal{O}_X)$ is a strong exceptional collection, we have that

$$\mathrm{Ext}^1(U,U)=0, \quad \mathrm{Ext}^1(V,V)=0 \quad \text{and} \quad \mathrm{Ext}^2(V,U)=0$$

We conclude that μ is surjective. Applying $\mathrm{Ext}(V,-)$ to Sequence (1), we also conclude that

$$\mathrm{Ext}^1(V,\mathcal{E})=0$$

and the map ν is surjective. Therefore, the Kodaira-Spencer map is surjective. Since being F-prioritary is an open condition, we conclude that we have a complete family of F-prioritary sheaves. If $\mathcal{P}_{X,F}(\mathbf{v})$ is irreducible, then prioritary sheaves having a resolution of the form (1) give a Zariski-dense open subset of $\mathcal{P}_{X,F}(\mathbf{v})$. By Lemma 3.4, these sheaves have no cohomology. This concludes the proof of the proposition. □

In view of Proposition 3.6, it is useful to know when $\mathcal{E}$ given by a resolution of the form (1) is locally free. We recall a standard Bertini-type theorem for the reader's convenience (see [**Hui16**, Proposition 2.6]).

LEMMA 3.7. *Assume that $\mathcal{H}om(A_s,A_t)$ is globally generated for $1 \leq s \leq j$ and $j < t \leq m$ and that $\mathrm{rk}(\mathcal{E}) \geq 2$. Then a general sheaf $\mathcal{E}$ given by a resolution of the form (1) is locally free.*

3.1. Weak Brill-Noether for Hirzebruch surfaces. As an application of our discussion, we determine when weak Brill-Noether holds for Hirzebruch surfaces. Let $\mathbf{v}$ be a Chern character of rank $r(\mathbf{v}) \geq 2$ and Euler characteristic $\chi(\mathbf{v})=0$ on a Hirzebruch surface $\mathbb{F}_e$. Let H be an ample divisor on $\mathbb{F}_e$. Since $\mathbb{F}_e$ and H will be fixed in this subsection, we denote $M_{\mathbb{F}_e,H}(\mathbf{v})$ by $M(\mathbf{v})$. By Walter's Theorem 2.6, if the moduli space $M(\mathbf{v})$ is nonempty, then it is irreducible and the general sheaf is locally free. In particular, Serre duality gives a birational map between $M(\mathbf{v})$ and $M(\mathbf{v}^D)$, where $\mathbf{v}^D$ is the Serre dual Chern character. Weak Brill-Noether for $M(\mathbf{v})$ and $M(\mathbf{v}^D)$ are equivalent problems.

We write

$$\nu(\mathbf{v}) = \frac{c_1(\mathbf{v})}{r(\mathbf{v})} = \frac{k}{r}E + \frac{l}{r}F$$

for the "total slope" of $\mathbf{v}$, so $\mu_H(\mathbf{v}) = \nu(\mathbf{v}) \cdot H$. Since

$$K_{\mathbb{F}_e} = -2E - (2+e)F,$$

we may replace $\mathbf{v}$ by $\mathbf{v}^D$ if necessary to assume

$$\frac{k}{r} \geq -1.$$

Furthermore, if $\frac{k}{r} = -1$ we may additionally assume $\frac{l}{r} \geq -1 - \frac{e}{2}$. The Bogomolov inequality gives a further restriction on $\nu(\mathbf{v})$.

LEMMA 3.8. *With* $\mathbf{v}$ *as above, if* $M(\mathbf{v})$ *is nonempty, then* $\frac{l}{r} \geq -1 + \frac{ke}{2r}$.

PROOF. There is nothing to prove in the case $\frac{k}{r} = -1$, since then we are already assuming $\frac{l}{r} \geq -1 - \frac{e}{2}$. Assume $\frac{k}{r} > -1$ and $\frac{l}{r} < -1 + \frac{ke}{2r}$. We show $M(\mathbf{v})$ is empty. Let

$$P(\nu) = \chi(\mathcal{O}_{\mathbb{F}_e}) + \frac{1}{2}(\nu^2 - \nu \cdot K_{\mathbb{F}_e}),$$

so that the Riemann-Roch formula takes the form

$$\chi(\mathbf{v}) = r(P(\nu(\mathbf{v})) - \Delta(\mathbf{v})).$$

Then since $\chi(\mathbf{v}) = 0$, we have

$$\begin{aligned}\Delta(\mathbf{v}) = P(\nu(\mathbf{v})) &= \left(\frac{k}{r}+1\right)\left(\frac{l}{r}+1\right) - \frac{1}{2}e\left(\frac{k}{r}+1\right)\left(\frac{k}{r}\right)\\ &= \left(\frac{k}{r}+1\right)\left(\frac{l}{r}+1-\frac{ek}{2r}\right) < 0.\end{aligned}$$

Thus, by the Bogomolov inequality, $M(\mathbf{v})$ is empty. □

Our assumptions on $\mathbf{v}$ now give some simple H^2-vanishing results for semistable sheaves.

LEMMA 3.9. *With* $\mathbf{v}$ *as above, if* $\mathcal{E}$ *is an* H*-semistable sheaf of character* $\mathbf{v}$*, then*

$$H^2(\mathbb{F}_e, \mathcal{E}) = H^2(\mathbb{F}_e, \mathcal{E}(-E)) = 0.$$

PROOF. Since $\frac{k}{r} \geq -1$, Lemma 3.8 in particular implies $\frac{l}{r} \geq -1 - \frac{e}{2}$. In fact, we prove the stronger result that if $\mathbf{v} \in K(\mathbb{F}_e)$ is any character such that $\frac{k}{r} \geq -2$, $\frac{l}{r} \geq -2 - e$, and at least one of the inequalities is strict, then any H-semistable sheaf $\mathcal{F}$ of character $\mathbf{v}$ has $H^2(\mathbb{F}_e, \mathcal{F}) = 0$. Taking $\mathcal{F} = \mathcal{E}$ or $\mathcal{F} = \mathcal{E}(-E)$ proves the original statement.

We use Serre duality to write

$$H^2(\mathbb{F}_e, \mathcal{F}) = \operatorname{Ext}^2(\mathcal{O}_{\mathbb{F}_e}, \mathcal{F}) = \operatorname{Hom}(\mathcal{F}, K_{\mathbb{F}_e})^*.$$

Since $\mathcal{F}$ and $K_{\mathbb{F}_e}$ are both H-semistable, the vanishing will follow if $\mu_H(\mathcal{F}) > \mu_H(K_{\mathbb{F}_e})$, which is equivalent to $H \cdot (\nu(\mathbf{v}) - K_{\mathbb{F}_e}) > 0$. The nef cone of $\mathbb{F}_e$ is spanned by F and $E + eF$. We compute

$$\begin{aligned}F \cdot (\nu(\mathbf{v}) - K_{\mathbb{F}_e}) &= \frac{k}{r} + 2\\ (E+eF) \cdot (\nu(\mathbf{v}) - K_{\mathbb{F}_e}) &= \frac{l}{r} + 2 + e.\end{aligned}$$

By our assumption on $\mathbf{v}$, both of these intersection numbers are nonnegative, and at least one of them is positive. Since H is ample, it is a positive combination of the extremal nef classes, and it follows that $H \cdot (\nu(\mathbf{v}) - K_{\mathbb{F}_e}) > 0$. □

Next we describe characters $\mathbf{v}$ such that weak Brill-Noether fails for $M(\mathbf{v})$ if the moduli space is nonempty. It will turn out that these are the only characters where weak Brill-Noether fails.

PROPOSITION 3.10. *Let $\mathbf{v}$ be a character as above, and suppose $\mathcal{E}$ is an H-semistable sheaf of character $\mathbf{v}$. If $\chi(\mathcal{E}(-E)) > 0$, then $H^0(\mathbb{F}_e, \mathcal{E}) \neq 0$. More explicitly, if*

$$\frac{l - ke}{r} = \nu(\mathbf{v}) \cdot E < -1,$$

then $H^0(\mathbb{F}_e, \mathcal{E}) \neq 0$.

PROOF. The restriction sequence

$$0 \to \mathcal{E}(-E) \to \mathcal{E} \to \mathcal{E}|_E \to 0$$

shows that if $H^0(\mathbb{F}_e, \mathcal{E}(-E)) \neq 0$, then $H^0(\mathbb{F}_e, \mathcal{E}) \neq 0$. By Lemma 3.9, we have $H^2(\mathbb{F}_e, \mathcal{E}(-E)) = 0$. Therefore, $H^0(\mathbb{F}_e, \mathcal{E}(-E)) \neq 0$ since $\chi(\mathcal{E}(-E)) > 0$. The second statement follows immediately from Riemann-Roch. □

Conversely, we have the main result of this section.

THEOREM 3.11. *Let $\mathbf{v}$ be a character as above, and suppose $\mathcal{E}$ is a general H-semistable sheaf of character $\mathbf{v}$. If $\chi(\mathcal{E}(-E)) \leq 0$, then $\mathcal{E}$ has no cohomology.*

Moreover, unless $\mathcal{E}$ is a direct sum of copies of $\mathcal{O}_{\mathbb{P}^1 \times \mathbb{P}^1}(-1,-1)$, $\mathcal{E}$ admits a resolution of the form

$$0 \to \mathcal{O}_{\mathbb{F}_e}(-E-(e+1)F)^a \to \mathcal{O}_{\mathbb{F}_e}(-E-eF)^b \oplus \mathcal{O}_{\mathbb{F}_e}(-F)^c \to \mathcal{E} \to 0$$

for some nonnegative numbers a, b, c.

PROOF. We may assume $\mathcal{E}$ is not a direct sum of copies of $\mathcal{O}_{\mathbb{P}^1 \times \mathbb{P}^1}(-1,-1)$, since there is nothing to prove in that case. By Rudakov's classification, $\mathcal{E}$ is a direct sum of copies of $\mathcal{O}_{\mathbb{P}^1 \times \mathbb{P}^1}(-1,-1)$ if and only if $e = 0$ and $k = l = -r$, so we will assume these equalities do not all hold.

By Example 3.2, the collection

$$A_1 = \mathcal{O}_{\mathbb{F}_e}(-E-(e+1)F), \quad A_2 = \mathcal{O}_{\mathbb{F}_e}(-E-eF), \quad A_3 = \mathcal{O}_{\mathbb{F}_e}(-F), \quad \mathcal{O}_{\mathbb{F}_e}$$

is a strong exceptional collection. These line bundles give a basis for the vector space $K(\mathbb{F}_e) \otimes \mathbb{Q}$. Therefore, the character $\mathbf{v}$ can be expressed uniquely as a formal linear combination

$$\mathbf{v} = -a \operatorname{ch} A_1 + b \operatorname{ch} A_2 + c \operatorname{ch} A_3 + d \operatorname{ch} \mathcal{O}_X$$

for some $a, b, c, d \in \mathbb{Q}$. These numbers can be conveniently computed using the Euler pairing. Since $\chi(\mathbf{v}) = 0$, we have $d = 0$. We view the above decomposition of the Chern character as corresponding formally to an exact sequence

$$0 \to \mathcal{O}_{\mathbb{F}_e}(-E-(e+1)F)^a \xrightarrow{\phi} \mathcal{O}_{\mathbb{F}_e}(-E-eF)^b \oplus \mathcal{O}_X(-F)^c \to \mathcal{E} \to 0$$

where $\mathcal{E}$ has character $\mathbf{v}$. We now compute the exponents a, b, c using the formal properties of the Euler pairing and verify that, under our assumptions on $\mathbf{v}$, they are all nonnegative integers.

Step 1: Nonnegativity of the exponents. By Lemma 3.4, $c = \chi(\mathcal{E}(F))$. By Lemma 2.7, we get

$$c = \chi(\mathcal{E}(F)) = c_1(\mathcal{E}) \cdot F + r = k + r \geq 0$$

since $\frac{k}{r} \geq -1$. Hence, $c \geq 0$.

Next, applying $\chi(\mathcal{O}_X(E), -)$, we get

$$b = -\chi(\mathcal{O}_X(E), \mathcal{E}) = -\chi(\mathcal{E}(-E)).$$

Our assumption $\chi(\mathcal{E}(-E)) \leq 0$ yields $b \geq 0$.

The nonnegativity of the exponent a is the most challenging. By Lemmas 2.7 and 3.4,

$$a = -\chi(\mathcal{E}(-E - F)) = l - ke + k + r.$$

By assumption,

$$l - ke + r \geq 0 \quad \text{and} \quad l - \frac{ke}{2} + r \geq 0.$$

Hence, if either $k \geq 0$, or $k < 0$ and $e \geq 2$, then $a \geq 0$.

It remains to consider the cases when $k < 0$ and $e = 0$ or 1. Suppose $a < 0$ to get a contradiction. Since we are assuming $k \geq -r$, we must have $l < 0$. We claim that

$$\operatorname{Ext}^2(\mathcal{O}(E + F), \mathcal{E}) = \operatorname{Hom}(\mathcal{E}, \mathcal{O}(-E - (e + 1)F))^* = 0$$

by stability. To see this compare the $H = E + \alpha F$ slopes (with $\alpha > e$)

$$\mu_H(\mathcal{E}) = \frac{k}{r}\alpha + \frac{l}{r} - \frac{ke}{r} \geq \frac{k}{r}\alpha + \frac{l}{r} - \frac{ke}{2r} \geq -\alpha - 1 = \mu_H(\mathcal{O}(-E - (e + 1)F)).$$

The first inequality is strict unless $e = 0$. If $e = 0$, then the second inequality is strict unless $k = l = -r$. Since we are assuming $\mathcal{E}$ is not a direct sum of copies of $\mathcal{O}_{\mathbb{P}^1 \times \mathbb{P}^1}(-1, -1)$, we conclude that the inequality is always strict, and the vanishing $\operatorname{Ext}^2(\mathcal{O}(E + F), \mathcal{E}) = 0$ holds. Then since a is negative, we deduce $\operatorname{Hom}(\mathcal{O}(E + F), \mathcal{E}) \neq 0$. This contradicts the H-semistability of $\mathcal{E}$, since

$$\mu_H(\mathcal{O}(E + F)) > 0 > \mu_H(\mathcal{E}).$$

Therefore $a \geq 0$ in every case.

Step 2: A general sheaf with the specified resolution is locally free of class $\mathbf{v}$. Since

$$\mathcal{H}om(\mathcal{O}_{\mathbb{F}_e}(-E - (e + 1)F), \mathcal{O}_{\mathbb{F}_e}(-E - eF)) \cong \mathcal{O}_{\mathbb{F}_e}(F) \quad \text{and}$$
$$\mathcal{H}om(\mathcal{O}_{\mathbb{F}_e}(-E - (e + 1)F), \mathcal{O}_{\mathbb{F}_e}(-F)) \cong \mathcal{O}_{\mathbb{F}_e}(E + eF)$$

are globally generated, by Lemma 3.7, the cokernel of a general map ϕ is locally free. By the paragraph preceding Step 1 and the choice of the exponents a, b, c, the bundle $\mathcal{E}$ must have character $\mathbf{v}$.

Step 3: Any locally free sheaf $\mathcal{E}$ with the specified resolution is prioritary. By Theorem 2.1,

$$\operatorname{Ext}^1(A_1, A_2 \otimes \mathcal{O}_{F_e}(-F)) = H^1(\mathbb{F}_e, \mathcal{O}_{\mathbb{F}_e}) = 0,$$
$$\operatorname{Ext}^1(A_1, A_3 \otimes \mathcal{O}_{F_e}(-F)) = H^1(\mathbb{F}_e, \mathcal{O}_{\mathbb{F}_e}(E + (e - 1)F)) = 0$$

and

$$\operatorname{Ext}^2(A_2, A_2 \otimes \mathcal{O}_{F_e}(-F)) = \operatorname{Ext}^2(A_3, A_3 \otimes \mathcal{O}_{F_e}(-F)) = H^2(\mathbb{F}_e, \mathcal{O}_{\mathbb{F}_e}(-F)) = 0,$$
$$\operatorname{Ext}^2(A_2, A_3 \otimes \mathcal{O}_{F_e}(-F)) = H^2(\mathbb{F}_e, \mathcal{O}_{\mathbb{F}_e}(E + (e - 2)F)) = 0.$$

By Lemma 3.5, we conclude that $\mathcal{E}$ is F-prioritary.

Step 4: Conclusion of the proof. By Proposition 3.6, the open subset $S \subset \operatorname{Hom}(A_1^a, A_2^b \oplus A_3^c)$ parameterizing locally free quotients parameterizes a complete family of prioritary sheaves. Any sheaf parameterized by S has no cohomology by Lemma 3.4. Since the stack $\mathcal{P}_{\mathbb{F}_e,F}(\mathbf{v})$ of prioritary sheaves of character $\mathbf{v}$ is irreducible and $\mathcal{M}(\mathbf{v})$ is a dense open substack of $\mathcal{P}_{\mathbb{F}_e,F}(\mathbf{v})$, we conclude that weak Brill-Noether holds for $M(\mathbf{v})$. Furthermore, the general sheaf in $M(\mathbf{v})$ has the required resolution. In particular, note that the moduli space is unirational since it is dominated by an open set in $\operatorname{Hom}(A_1^a, A_2^b \oplus A_3^c)$. □

3.2. Applications to blowups of $\mathbb{P}^2$. In this subsection, let Γ be a set of k distinct points $p_1, \dots, p_k$ on $\mathbb{P}^2$ and let X denote the blowup of $\mathbb{P}^2$ along Γ. Let L denote the pullback of the hyperplane class and let E_i denote the exceptional divisor lying over p_i. Our methods have the following consequence.

THEOREM 3.12. *Let* $\mathbf{v}$ *be a Chern character on* X *such that*

$$r(\mathbf{v}) \geq 2, \quad \chi(\mathbf{v}) = 0, \quad \nu(\mathbf{v}) = \delta L - \sum_{i=1}^{k} \alpha_i E_i, \text{ where } \delta, \alpha_i \geq 0 \text{ and } \delta - \sum_{i=1}^{k} \alpha_i \geq -1.$$

Then the stack $\mathcal{P}_{X,L-E_1}(\mathbf{v})$ *is nonempty, and a general sheaf parameterized by* $\mathcal{P}_{X,L-E_1}(\mathbf{v})$ *has no cohomology.*

If $\delta - 2\sum_{i=1}^k \alpha_i + 2 \geq 0$, *then the general sheaf* $\mathcal{E}$ *in* $\mathcal{P}_{X,L-E_1}(\mathbf{v})$ *has a resolution of the form*

$$0 \longrightarrow \mathcal{O}_X(-2L)^a \xrightarrow{\phi} \mathcal{O}_X(-L)^b \oplus \bigoplus_{i=1}^{k} \mathcal{O}(-E_i)^{c_i} \longrightarrow \mathcal{E} \longrightarrow 0, \tag{2}$$

Otherwise, $\mathcal{E}$ *has a resolution of the form*

$$0 \longrightarrow \mathcal{O}_X(-2L)^a \oplus \mathcal{O}_X(-L)^b \xrightarrow{\phi} \bigoplus_{i=1}^{k} \mathcal{O}(-E_i)^{c_i} \longrightarrow \mathcal{E} \longrightarrow 0. \tag{3}$$

The exponents are given by

$$a = r(\mathbf{v})(\delta - \sum_{i=1}^{k} \alpha_i + 1), \quad c_i = r(\mathbf{v})\alpha_i, \quad b = \left| r(\mathbf{v}) + a - \sum_{i=1}^{k} c_i \right|.$$

PROOF. The linear system $|L - E_1|$ defines a map from X to $\mathbb{P}^1$ giving X the structure of a birationally ruled surface. By [**Wal98**, Proposition 2], the stack of prioritary sheaves $\mathcal{P}_{X,L-E_1}(\mathbf{v})$ is smooth and irreducible. By Example 3.3,

$$A_1 = \mathcal{O}(-2L), \quad A_2 = \mathcal{O}(-L), \quad A_3 = \mathcal{O}(-E_1), \dots, A_{k+2} = \mathcal{O}(-E_k), \quad \mathcal{O}$$

is a strong exceptional collection on X. Lemma 3.4 computes the exponents of a sheaf with Chern character $\mathbf{v}$ with the given resolutions (2) or (3). Our assumptions on $\mathbf{v}$ imply that these exponents are positive; the requirement on the sign of $\delta - 2\sum_{i=1}^k \alpha_i + 2$ which is used to determine the form of the resolution ensures that the exponent b is positive. Lemma 3.4 also implies that sheaves with these resolutions have no cohomology.

Since

$$\mathcal{H}om(\mathcal{O}_X(-2L), \mathcal{O}_X(-L)) \cong \mathcal{O}_X(L), \quad \mathcal{H}om(\mathcal{O}_X(-2L), \mathcal{O}_X(-E_l)) \cong \mathcal{O}_X(2L - E_i)$$

and

$$\mathcal{H}om(\mathcal{O}_X(-L), \mathcal{O}_X(-E_i)) = \mathcal{O}_X(L - E_i)$$

are globally generated and the rank of $\mathbf{v}$ is at least 2, by Lemma 3.7, the cokernel $\mathcal{E}$ of a general map ϕ is locally free in both cases.

By Lemma 2.3, $H^1(X, \mathcal{O}_X(D)) = 0$ for D a divisor of the form

$$E_1, \quad L, \quad L + E_1 - E_i, \quad \emptyset, \quad E_1 - E_i.$$

Hence,

$$\mathrm{Ext}^1(A_1, A_l \otimes \mathcal{O}_X(-L + E_1)) = 0 \text{ for } 2 \le l \le k+2$$

and

$$\mathrm{Ext}^1(A_2, A_l \otimes \mathcal{O}_X(-L + E_1)) = 0 \text{ for } 3 \le l \le k+2.$$

Similarly, $H^2(X, \mathcal{O}_X(D)) = 0$ when D is an integral divisor of the form $dL - \sum a_i E_i$ with $d \ge -2$. Hence,

$$\mathrm{Ext}^2(A_i, A_l \otimes \mathcal{O}_X(-L + E_i)) = 0 \text{ for } 2 \le i, l \le k+2.$$

By Lemma 3.5, all locally-free sheaves with resolutions given by (2) or (3) are prioritary with respect to $\mathcal{O}(L - E_1)$. We conclude that the stack of prioritary sheaves $\mathcal{P}_{X,L-E_1}(\mathbf{v})$ is nonempty. Finally, by Proposition 3.6, this is a complete family of prioritary sheaves. Since the stack $\mathcal{P}_{X,L-E_1}(\mathbf{v})$ is irreducible, the general prioritary sheaf with Chern character $\mathbf{v}$ has no cohomology and has a resolution of the form (2) or (3) depending on the sign of the exponent of $\mathcal{O}_X(-L)$. □

COROLLARY 3.13. *Let H be an ample divisor on X such that $H \cdot (-2L + \sum_{i=2}^k E_i) < 0$. Let $\mathbf{v}$ be a stable Chern character on X satisfying the hypotheses of Theorem* 3.12. *Then the moduli space $M_{X,H}(\mathbf{v})$ is unirational and satisfies weak Brill-Noether.*

PROOF. By Walter [**Wal98**], an H-semistable sheaf is prioritary with respect to $L - E_1$. Consequently, $\mathcal{M}_{X,H}(\mathbf{v})$ is an open substack of $\mathcal{P}_{X,L-E_1}(\mathbf{v})$. The corollary follows from Theorem 3.12. □

REMARK 3.14. Let X be the blowup of $\mathbb{P}^2$ along k collinear points. Let $\mathcal{E}$ be a coherent sheaf with Chern character $\mathbf{v}$ with

$$\nu(\mathbf{v}) = \delta L - \sum_{i=1}^k \alpha_i E_i, \quad \delta - \sum_{i=1}^k \alpha_i < -1.$$

Assume $\mathcal{E}$ is semistable with respect to an ample divisor H such that

$$\nu(\mathcal{E}) \cdot H > -2L \cdot H.$$

(For example, these assumptions hold for the line bundle $\mathcal{O}_X(L - E_1 - E_2 - E_3)$ and any ample class $H = L - \epsilon \sum_{i=1}^k E_i$ for sufficiently small $\epsilon > 0$.) Then we claim that the cohomology of $\mathcal{E}$ does not vanish and $M_{X,H}(\mathbf{v})$ does not satisfy weak Brill-Noether. By stability,

$$\mathrm{Ext}^2(\mathcal{O}(L - \sum_{i=1}^k E_i), \mathcal{E})) = \mathrm{Hom}(\mathcal{E}, \mathcal{O}(-2L))^* = 0.$$

By Lemma 2.7,

$$\chi(\mathcal{O}(L - \sum_{i=1}^k E_i), \mathcal{E}) = \mathrm{rk}(\mathcal{E})(-\delta + \sum_{i=1}^k \alpha_i - 1) > 0.$$

We conclude that

$$\mathrm{Hom}(\mathcal{O}(L - \sum_{i=1}^k E_i), \mathcal{E}) \neq 0.$$

However, since the k points are collinear, composing the natural map $\mathrm{Hom}(\mathcal{O}, \mathcal{O}(L - \sum_{i=1}^k E_i))$ with a nonzero morphism to $\mathcal{E}$, we see that $H^0(X, \mathcal{E}) \neq 0$. Hence, without further assumptions on the positions of the points, Theorem 3.12 is sharp. However, if we assume that the points are general, we can use different strong exceptional collections to extend the range where weak Brill-Noether holds.

3.3. Applications to blowups of Hirzebruch surfaces. A very similar theorem holds for blowups of Hirzebruch surfaces. Let X be the blowup of a Hirzebruch surface $\mathbb{F}_e$, $e \geq 1$, along k distinct points not lying on the exceptional curve E. Denote by E and F the pullbacks of the corresponding classes on $\mathbb{F}_e$ and let $E_1, \dots, E_k$ denote the exceptional divisors lying over the points $p_1, \dots, p_k$. Let $\mathcal{E}$ be a locally sheaf of rank at least 2 with total slope

$$\nu(\mathcal{E}) := \frac{\mathrm{ch}_1(\mathcal{E})}{\mathrm{rk}(\mathcal{E})} = \alpha E + \beta F - \sum_{i=1}^k \alpha_i E_i.$$

THEOREM 3.15. *Let* $\mathbf{v}$ *be a Chern character on* X *such that* $\nu(\mathbf{v})$ *satisfies*

$$\alpha_i \geq 0, \quad \alpha - \sum_{i=1}^k \alpha_i \geq -1, \quad \text{and} \quad \beta - \sum_{i=1}^k \alpha_i + 1 \geq \max((e-1)\alpha, e\alpha).$$

Then the stack $\mathcal{P}_{X,F}(\mathbf{v})$ *is nonempty and the general sheaf in* $\mathcal{P}_{X,F}(\mathbf{v})$ *has no cohomology. Furthermore, the general sheaf in* $\mathcal{P}_{X,F}(\mathbf{v})$ *admits a resolution of the form*

$$0 \to \mathcal{O}_X(-E-(e+1)F)^a \xrightarrow{\phi} \mathcal{O}_X(-E-eF)^b \oplus \mathcal{O}_X(-F)^c \oplus \bigoplus_{i=1}^k \mathcal{O}_X(-E_i)^{d_i} \to \mathcal{E} \to 0,$$

where

$$a = r(\mathbf{v})(\beta - (e-1)\alpha - \sum_{i=1}^k \alpha_i + 1), \quad b = r(\mathbf{v})(\beta - e\alpha - \sum_{i=1}^k \alpha_i + 1),$$

$$c = r(\mathbf{v})(\alpha - \sum_{i=1}^k \alpha_i + 1) \quad d_i = r(\mathbf{v})\alpha_i.$$

PROOF. Since the proof of this theorem is similar to the proof of Theorems 3.11 and 3.12, we leave the routine verifications to the reader. By Lemma 2.4, the sequence

$$\mathcal{O}_X(-E-(e+1)F), \quad \mathcal{O}_X(-E-eF) \quad \mathcal{O}_X(-F), \quad \mathcal{O}_X(-E_1), \dots, \mathcal{O}_X(-E_k), \quad \mathcal{O}_X$$

is a strong exceptional collection. By Lemma 3.4, the exponents are as claimed and are positive by assumption. Lemma 3.7 applies to show that the cokernel of a general map ϕ is locally free. By Lemma 2.4, Lemma 3.5 applies. By Lemma 3.5 and Proposition 3.6, one obtains a complete family of F-prioritary sheaves. Since the stack of F-prioritary sheaves is irreducible, the theorem follows. □

As usual, we obtain the following corollary.

COROLLARY 3.16. *Let* H *be an ample divisor on* X *such that* $H \cdot (-2E - (e+1)F + \sum_{i=1}^k E_i) < 0$. *Let* $\mathbf{v}$ *be a stable Chern character satisfying the assumptions of Theorem* 3.15. *Then* $M_{X,H}(\mathbf{v})$ *is unirational and satisfies weak Brill-Noether.*

4. Blowups of $\mathbb{P}^2$ revisited

Let $X = \mathrm{Bl}_{p_1,\ldots,p_k} \mathbb{P}^2$ be the blowup of $\mathbb{P}^2$ at k distinct points $p_1, \ldots, p_k \in \mathbb{P}^2$. In this section we study the weak Brill-Noether problem for X. In particular, we give some sufficient conditions on a character $\mathbf{v}$ of Euler characteristic 0 for weak Brill-Noether to hold for certain moduli spaces of sheaves of character $\mathbf{v}$. We will prove sharper results when X is a del Pezzo surface of degree at least 4 in the next section, using the basic tools developed in this section as a starting point. Fix $F = L - E_1$, so that the complete series $|F|$ induces a map $X \to \mathbb{P}^1$ with fiber class F, exhibiting X as a birationally ruled surface.

If H is an ample divisor such that $H \cdot (K_X + F) < 0$, then any H-semistable sheaf is automatically F-prioritary, so that $\mathcal{P}_{X,F}(\mathbf{v})$ contains the stack $\mathcal{M}_{X,H}(\mathbf{v})$ of H-semistable sheaves as a dense open substack whenever H-semistable sheaves of character $\mathbf{v}$ exist. Note that such polarizations H always exist; since $F \cdot (K_X + F) = -2$, any ample divisor H spanning a ray sufficiently close to the ray spanned by the nef divisor F will do the trick. In what follows we work primarily with prioritary sheaves instead of semistable sheaves. Prioritary sheaves have the advantage that they are much easier to construct than semistable sheaves.

4.1. Prioritary direct sums of line bundles. By Walter's irreducibility theorem, in order to prove a general prioritary sheaf of some character $\mathbf{v}$ has no cohomology, it suffices to construct a particular such sheaf. It is often possible to do this by considering elementary modifications of direct sums of line bundles. First we give a criterion for determining when a direct sum of line bundles is prioritary.

LEMMA 4.1. *Let $\mathcal{E} = L_1 \oplus \cdots \oplus L_r$ be a direct sum of line bundles. Suppose that N is a nef divisor such that*

$$N \cdot (L_i - L_j) < -N \cdot (F + K_X)$$

for all i, j. Then $\mathcal{E}$ is F-prioritary.

PROOF. The vanishing $\mathrm{Ext}^2(\mathcal{E}, \mathcal{E}(-F)) = 0$ will follow if $\mathrm{Ext}^2(L_i, L_j(-F)) = 0$ for all i, j. We have

$$\mathrm{Ext}^2(L_i, L_j(-F)) \cong H^2(-L_i + L_j - F) \cong H^0(L_i - L_j + F + K_X)^*.$$

Then

$$N \cdot (L_i - L_j + F + K_X) < 0,$$

and since N is nef we conclude $L_i - L_j + F + K_X$ is not effective. Therefore $\mathrm{Ext}^2(L_i, L_j(-F)) = 0$. □

Further prioritary sheaves of lower Euler characteristic can be constructed by elementary modifications.

LEMMA 4.2. *Let $\mathcal{E}$ be an F-prioritary sheaf on X, and let $p \in X$ be a point where $\mathcal{E}$ is locally free. Pick a surjection $\mathcal{E} \to \mathcal{O}_p$, and consider the elementary modification $\mathcal{E}'$ defined by the sequence*

$$0 \to \mathcal{E}' \to \mathcal{E} \to \mathcal{O}_p \to 0.$$

Then $\mathcal{E}'$ is F-prioritary with $\chi(\mathcal{E}') = \chi(\mathcal{E}) - 1$. If furthermore $h^0(\mathcal{E}) > 0$ and p and the map $\mathcal{E} \to \mathcal{O}_p$ are general, then additionally $h^0(\mathcal{E}') = h^0(\mathcal{E}) - 1$.

PROOF. Clearly $\mathcal{E}'$ is torsion-free since $\mathcal{E}$ is. Then $\operatorname{Ext}^2(\mathcal{E}', \mathcal{E}'(-F))$ is a quotient of $\operatorname{Ext}^2(\mathcal{E}, \mathcal{E}'(-F))$, so it suffices to prove the latter group vanishes. Tensoring the exact sequence by $\mathcal{O}_X(-F)$ and applying $\operatorname{Hom}(\mathcal{E}, -)$, we get an exact sequence

$$\operatorname{Ext}^1(\mathcal{E}, \mathcal{O}_p) \to \operatorname{Ext}^2(\mathcal{E}, \mathcal{E}'(-F)) \to \operatorname{Ext}^2(\mathcal{E}, \mathcal{E}(-F)).$$

Then $\operatorname{Ext}^2(\mathcal{E}, \mathcal{E}(-F)) = 0$ since $\mathcal{E}$ is prioritary and $\operatorname{Ext}^1(\mathcal{E}, \mathcal{O}_p) = 0$ since $\mathcal{E}$ is locally free at p. If $h^0(\mathcal{E}) > 0$ and p is general, then $\mathcal{E}$ has a section which does not vanish at p. Therefore a general surjection $\mathcal{E} \to \mathcal{O}_p$ induces a surjective map $H^0(\mathcal{E}) \to H^0(\mathcal{O}_p)$, and we conclude $h^0(\mathcal{E}') = h^0(\mathcal{E}) - 1$. □

These two lemmas motivate the next definition.

DEFINITION 4.3. Fix a nef divisor N such that $-N \cdot (F + K_X) \geq 2$. We call a direct sum $\mathcal{E} = L_1 \oplus \cdots \oplus L_r$ of line bundles (N-)*good* if the following properties are satisfied.

(1) $\mathcal{E}$ has no higher cohomology: $h^i(\mathcal{E}) = 0$ for $i > 0$.
(2) For all i, j we have $N \cdot (L_i - L_j) \leq 1$. In particular, $\mathcal{E}$ is F-prioritary by Lemma 4.1.

Fix an integer $r \geq 1$. We let

$$\Lambda_{r,N} = \{c_1(\mathcal{E}) : \mathcal{E} \text{ is a rank } r \text{ good direct sum of line bundles}\} \subset N^1(X)_{\mathbb{Z}}.$$

We then have the next result.

COROLLARY 4.4. *Suppose* $\mathbf{v} \in K(X)$ *has* $\chi(\mathbf{v}) = 0$ *and* $r = r(\mathbf{v}) > 0$, *and fix a nef divisor* N *as in Definition 4.3. If* $c_1(\mathbf{v}) \in \Lambda_{r,N}$, *then* $\mathcal{P}_{X,F}(\mathbf{v})$ *is nonempty and a general sheaf* $\mathcal{E} \in \mathcal{P}_{X,F}(\mathbf{v})$ *has no cohomology.*

PROOF. Let $\mathcal{E}$ be a rank r good direct sum of line bundles with $c_1(\mathcal{E}) = c_1(\mathbf{v})$. Then $\chi := \chi(\mathcal{E}) \geq 0$, $\mathcal{E}$ is prioritary, and $\mathcal{E}$ has no higher cohomology. Therefore, by Lemma 4.2, performing χ general elementary modifications to $\mathcal{E}$ produces a prioritary sheaf $\mathcal{E}'$ with no cohomology and $\operatorname{ch} \mathcal{E}' = \mathbf{v}$. □

Our next result is our strongest result on the weak Brill-Noether problem for an arbitrary blowup of $\mathbb{P}^2$.

THEOREM 4.5. *Let* $X = \operatorname{Bl}_{p_1,\ldots,p_k} \mathbb{P}^2$ *be a blowup of* $\mathbb{P}^2$ *at* k *distinct points. Let* $\mathbf{v} \in K(X)$ *with* $r = r(\mathbf{v}) > 0$, *and write*

$$\nu(\mathbf{v}) := \frac{c_1(\mathbf{v})}{r(\mathbf{v})} = \delta L - \alpha_1 E_1 - \cdots - \alpha_k E_k,$$

so that the coefficients $\delta, \alpha_i \in \mathbb{Q}$. *Assume that* $\delta \geq 0$ *and* $\alpha_i \geq 0$ *for all* i. *Suppose that the line bundle*

$$\lfloor \delta \rfloor L - \lceil \alpha_1 \rceil E_1 - \cdots - \lceil \alpha_k \rceil E_k$$

has no higher cohomology. Then choosing $N = L$, *we have* $c_1(\mathbf{v}) \in \Lambda_{r,L}$. *In particular, if* $\chi(\mathbf{v}) = 0$, *then* $\mathcal{P}_{X,F}(\mathbf{v})$ *is nonempty and the general* $\mathcal{E} \in \mathcal{P}_{X,F}(\mathbf{v})$ *has no cohomology.*

PROOF. In more detail, write

$$\nu(\mathbf{v}) = \left(d + \frac{p}{r}\right) L - \left(a_1 + \frac{p_1}{r}\right) E_1 - \cdots - \left(a_k + \frac{p_k}{r}\right) E_k$$

where $d = \lfloor \delta \rfloor$ and $a_i = \lfloor \alpha_i \rfloor$. Suppose $M = eL - b_1 E_1 - \cdots - b_k E_k$ is a line bundle such that $e \in \{d, d+1\}$ and $b_i \in \{a_i, a_i + 1\}$ for all i. Our assumptions imply that

M has no higher cohomology. Now we can construct a direct sum $\mathcal{E} = L_1 \oplus \cdots \oplus L_r$ of line bundles of this form such that $c_1(\mathcal{E}) = c_1(\mathbf{v})$. Indeed, we only need to ensure that exactly p of the L_i's have a coefficient of $d+1$ on L, and similarly for the other coefficients. By construction, $\mathcal{E}$ is L-good. □

5. Del Pezzo surfaces

In this section we improve on Theorem 4.5 in the special case of a smooth del Pezzo surface X of degree $4 \leq d \leq 7$. Thus $X = \mathrm{Bl}_{p_1,\ldots,p_k} \mathbb{P}^2$ is a blowup of $\mathbb{P}^2$ at $2 \leq k = 9-d \leq 5$ points with no three lying on a line, and $-K_X = 3L - E_1 - \cdots - E_k$ is ample. As in the previous section we fix a fiber class $F = L - E_1$ and study F-prioritary sheaves. Our main theorem is the following.

THEOREM 5.1. *Let X be a del Pezzo surface of degree $4 \leq d \leq 7$. Let $\mathbf{v} \in K(X)$ with $\chi(\mathbf{v}) = 0$, and suppose $c_1(\mathbf{v})$ is nef. Then the stack $\mathcal{P}_{X,F}(\mathbf{v})$ of prioritary sheaves is nonempty and a general $\mathcal{E} \in \mathcal{P}_{X,F}(\mathbf{v})$ has no cohomology.*

The cone of curves $\mathrm{NE}(X)$ is spanned by the classes of the (-1)-curves on X; the (-1)-curves are an exceptional divisor, a line through two points, or a conic through 5 points if $d = 4$ (review Example 2.2). Dually, the nef cone $\mathrm{Nef}(X) \subset N^1(X)$ is the subcone of classes ν such that $\nu \cdot C \geq 0$ for every (-1)-curve C.

The main additional ingredient that goes into the proof of Theorem 5.1 is the action of the Weyl group on $\mathrm{Pic}(X)$. The Weyl group preserves the intersection pairing, so it preserves the nef cone, the Euler characteristic, and the dimensions of cohomology groups of line bundles. The Weyl group acts transitively on length $\ell \leq k$ configurations $(C_1, \ldots, C_\ell)$ of disjoint (-1)-curves unless $\ell = k - 1$, in which case the action has two orbits [**Man97**].

The Weyl group additionally preserves the canonical bundle K_X. For this reason, we will take $N = -K_X$ in Definition 4.3 and study $(-K_X)$-good direct sums of line bundles—note that the needed inequality $K_X.(F + K_X) \geq 2$ follows from our assumption that $d \geq 4$. Then the Weyl group additionally preserves the set $\Lambda_{r,-K_X}$ of first Chern classes of rank r $(-K_X)$-good direct sums of line bundles. Theorem 5.1 is a direct consequence of the next result and Corollary 4.4.

PROPOSITION 5.2. *If X is a del Pezzo surface of degree $4 \leq d \leq 7$ we have $\mathrm{Nef}(X)_{\mathbb{Z}} \subset \Lambda_{r,-K_X}$.*

Since the Weyl group preserves both the nef cone and $\Lambda_{r,-K_X}$, we only need to show that if $D \in \mathrm{Nef}(X)_{\mathbb{Z}}$, then some translate of D by a Weyl group element is in $\Lambda_{r,-K_X}$. Our next lemma further allows us to repeatedly replace D with a different divisor D' until D' is on the boundary of the nef cone.

LEMMA 5.3. *Let $D \in \mathrm{Nef}(X)_{\mathbb{Z}}$ be a nef divisor, and suppose $D \cdot E_i \geq D \cdot E_j$. Suppose $D' = D - E_i + E_j$ is nef and that $D' \in \Lambda_{r,-K_X}$. Then $D \in \Lambda_{r,-K_X}$.*

PROOF. Since $D' \in \Lambda_{r,-K_X}$, there is a $(-K_X)$-good direct sum of line bundles $\mathcal{E}'$ such that $c_1(\mathcal{E}') = D'$. Since $D \cdot E_i \geq D \cdot E_j$, we have $D' \cdot E_i > D' \cdot E_j$, and at least one of the line bundles L' in $\mathcal{E}'$ must have $L' \cdot E_i > L' \cdot E_j \geq -1$. If we put $M = L' + E_i - E_j$, then since L' has no higher cohomology it follows that M has no higher cohomology. Additionally, $M \cdot (-K_X) = L' \cdot (-K_X)$. Then if we define $\mathcal{E}$ by taking $\mathcal{E}'$ and replacing L' with M, it follows that $\mathcal{E}$ is $(-K_X)$-good and $c_1(\mathcal{E}) = D$. □

Now let $D \in \mathrm{Nef}(X)_{\mathbb{Z}}$ be arbitrary. If there are i, j such that Lemma 5.3 can be applied, then we replace D with D'. We iterate this process until we arrive at a divisor D such that the lemma cannot be applied for any pair of indices i, j. Note that if C is any (-1)-curve on X, then $C \cdot (-E_i + E_j) \geq -1$. This implies that D must be orthogonal to some (-1)-curve. If $k \geq 3$, then applying an element of the Weyl group we may assume D is orthogonal to E_k. In this case, we reduce to studying a divisor class on the blowup at $k-1$ points.

Likewise, if $k = 2$ then by inspection D is orthogonal to either E_1 or E_2. That is, up to the Weyl group action, D takes the form $D = dL - aE_1$ for some $0 \leq a \leq d$. In the end, we have reduced the proof of Proposition 5.2 to the following statement.

PROPOSITION 5.4. *Let $D = dL - aE_1$ with $0 \leq a \leq d$ be a nef divisor on $X = \mathrm{Bl}_{p_1,p_2}\mathbb{P}^2$, and let $r \geq 1$. Then $D \in \Lambda_{r,-K_X}$.*

REMARK 5.5. It is crucial to regard D as a divisor on the blowup at 2 points instead of as a divisor on the blowup at 1 point. For example, consider $D = 2L$ and $r = 3$. Then $D \cdot (-K_X) = 6$, so we need to express $2L$ as a sum of 3 line bundles of $(-K_X)$-degree 2 with no higher cohomology. The only such line bundle on the blowup at 1 point is $L - E_1$, so there is no way to achieve this. On the other hand, on the blowup at 2 points, there is the additional line bundle $E_1 + E_2$ of $(-K_X)$-degree 2, and

$$2L = (L - E_1) + (L - E_2) + (E_1 + E_2).$$

PROOF OF PROPOSITION 5.4. The proof is by induction on r. The result is clear for $r = 1$. Let $r \geq 2$. If $r = 2$ and $D = L - E_1$, then $D = (L - 2E_1) + E_1$, so we may ignore this case. We will prove the following claim.

Claim: Suppose it is not the case that $r = 2$ and $D = L - E_1$. Write

$$\frac{D \cdot (-K_X)}{r} = \frac{3d - a}{r} = m + \frac{p}{r} \qquad (0 \leq p < r).$$

(We clearly have $m \geq 0$.) Then there is a line bundle M with $M \cdot (-K_X) = m$ such that M has no higher cohomology and $D - M$ is nef.

First observe that the claim implies the result. Indeed, by induction and Lemma 5.3 it follows that $D - M \in \Lambda_{r-1,-K_X}$. We have

$$\frac{(D - M) \cdot (-K_X)}{r - 1} = m + \frac{p}{r-1},$$

so $D - M$ can be written as $c_1(\mathcal{E})$ for a good sum $\mathcal{E}$ of p line bundles of $(-K_X)$-degree $m+1$ and $(r-1)-p$ line bundles of $(-K_X)$-degree m. Then $D = c_1(\mathcal{E} \oplus M)$, and $\mathcal{E} \oplus M$ is a good sum.

Next we prove the claim by constructing the line bundle M. Note that the claim is trivial if $m = 0$, since then we can take $M = 0$. In what follows we assume $m > 0$. Write

$$m = 3s + t \qquad (0 \leq t < 3).$$

For $0 \leq t < 3$ we define line bundles M_t as follows:

$$\begin{aligned} M_0 &= 0 \\ M_1 &= E_1 \\ M_2 &= E_1 + E_2 \end{aligned}$$

noting that $M_t.(-K_X) = t$. Then we put

$$B = sL + M_t,$$

and observe $B \cdot (-K_X) = m$. The line bundle B is the *basic line bundle of* $(-K_X)$-*degree* m. Note that B has no higher cohomology.

We now modify B by adding some number α of copies of $L - 3E_1$ and some number β of copies of $L - 2E_1 - E_2$ to it; call such a line bundle

$$B' = B + \alpha(L - 3E_1) + \beta(L - 2E_1 - E_2) = d'L - a'E_1 - b'E_2.$$

Then B' still has $B' \cdot (-K_X) = m$ since $L - 3E_1$ and $L - 2E_1 - E_2$ are both orthogonal to K_X. For B' to be the desired line bundle M that proves the claim, we need to choose α and β so that B' has no higher cohomology and $D - B'$ is nef.

For $D - B'$ to be nef we must have $(D - B') \cdot E_2 \geq 0$, so we have $-1 \leq b' \leq 0$. Then either $\beta = 0$, or $t = 2$ and $\beta = 1$. Similarly, considering E_1 shows $-1 \leq a' \leq a$. Now suppose we first make β as large as possible, then make α as large as possible, without violating the inequalities $b' \leq 0$ or $a' \leq a$. We claim that then $D - B'$ is nef, i.e. that $(D - B') \cdot (L - E_1 - E_2) \geq 0$. There are a couple cases to consider.

Case 1: $t = 2$ and $a = 0$. In this case $\alpha = \beta = 0$. We have $D \cdot (L - E_1 - E_2) = d$ and $B' = sL + E_1 + E_2$, so $B' \cdot (L - E_1 - E_2) = s + 2$. Then

$$3(d - (s + 2)) = 3d - m + t - 6 = mr + p - m - 4 = m(r - 1) + p - 4.$$

Now $m \equiv 2 \pmod 3$ and $m > 0$, so $d \geq s + 2$ unless $m = 2$, $s = 0$, and $d = 1$. But then $m = \lfloor 3/r \rfloor$, so this is impossible since $r \geq 2$. Therefore $d \geq s + 2$ and $D - B'$ is nef. Note that also B' has no higher cohomology, so we may take $M = B'$ in this case to complete the proof.

Case 2: $t \neq 2$ or $a > 0$. If $t = 2$ but $a > 0$ then we have $\beta = 1$, and therefore $b' = 0$. We also have $b' = 0$ if $t \neq 2$, so $b' = 0$ in every case. Then B' takes one of the following three forms:

$$d'L - (a - 2)E_1 \qquad d'L - (a - 1)E_1 \qquad d'L - aE_1.$$

In order to have $(D - B') \cdot (L - E_1 - E_2) \geq 0$, we will need to compare d' with d. To do this we first compute d'. Observe that the total number of line bundles added to the basic line bundle is

$$\alpha + \beta = \left\lfloor \frac{a + t}{3} \right\rfloor;$$

write

$$a + t = 3(\alpha + \beta) + \ell \qquad (0 \leq \ell < 3).$$

Then

$$d' = s + \alpha + \beta,$$

and

$$3(d - d') = 3d - 3s - 3(\alpha + \beta) = (mr + a + p) - (m - t) - (a + t - \ell) = m(r - 1) + p + \ell.$$

Therefore $d' < d$ in any case, which implies $D - B'$ is nef unless $B' = d'L - (a - 2)E_1$. In this case we need the stronger inequality $d' \leq d - 2$, or equivalently we must show

$$m(r - 1) + p + \ell > 3.$$

But if $B' = d'L - (a - 2)E_1$ then $\ell = 2$, and the only way the inequality can fail is if $m = 1$, $r = 2$, and $p = 0$. In this case $3d - a = 2$, and since $a \leq d$ we have

$d = a = 1$. Thus this case is the special case $D = L - E_1$, $r = 2$, which we have already excluded. Therefore $D - B'$ is nef.

Finally we are ready to construct the line bundle M that proves the claim in case $t \neq 2$ or $a > 0$. Starting from the basic line bundle B, add a single copy of $L - 2E_1 - E_2$ if it won't violate the inequality $b' \leq 0$ (this won't violate the inequality $a' \leq a$ since $a > 0$ if $t \neq 2$). After that, repeatedly add copies of $L - 3E_1$. Sometime before the inequality $a' \leq a$ is violated, we will have $B' \cdot (L - E_1 - E_2) \leq D \cdot (L - E_1 - E_2)$. Let M be the first B' where this inequality holds. Each copy of $L - 2E_1 - E_2$ or $M - 3E_1$ that is added decreases the intersection number with $M - E_1 - E_2$ by 2. Thus we will additionally have $M \cdot (L - E_1 - E_2) \geq D \cdot (L - E_1 - E_2) - 1 \geq -1$. Since $b' = 0$, this implies M has no higher cohomology and $D - M$ is nef. □

References

[Bea83] Arnaud Beauville, *Complex algebraic surfaces*, London Mathematical Society Lecture Note Series, vol. 68, Cambridge University Press, Cambridge, 1983. Translated from the French by R. Barlow, N. I. Shepherd-Barron and M. Reid. MR732439

[Bon89] A. I. Bondal, *Representations of associative algebras and coherent sheaves* (Russian), Izv. Akad. Nauk SSSR Ser. Mat. **53** (1989), no. 1, 25–44; English transl., Math. USSR-Izv. **34** (1990), no. 1, 23–42. MR992977

[Cos06a] Izzet Coskun, *Degenerations of surface scrolls and the Gromov-Witten invariants of Grassmannians*, J. Algebraic Geom. **15** (2006), no. 2, 223–284. MR2199064

[Cos06b] Izzet Coskun, *The enumerative geometry of Del Pezzo surfaces via degenerations*, Amer. J. Math. **128** (2006), no. 3, 751–786. MR2230924

[CH15] Izzet Coskun and Jack Huizenga, *The birational geometry of the moduli spaces of sheaves on $\mathbb{P}^2$*, Proceedings of the Gökova Geometry-Topology Conference 2014, Gökova Geometry/Topology Conference (GGT), Gökova, 2015, pp. 114–155. MR3381441

[Gim87] Alessandro Gimigliano, *ON LINEAR SYSTEMS OF PLANE CURVES*, ProQuest LLC, Ann Arbor, MI, 1987. Thesis (Ph.D.)–Queen's University (Canada). MR2635606

[GHi94] Lothar Göttsche and André Hirschowitz, *Weak Brill-Noether for vector bundles on the projective plane*, Algebraic geometry (Catania, 1993/Barcelona, 1994), Lecture Notes in Pure and Appl. Math., vol. 200, Dekker, New York, 1998, pp. 63–74. MR1651090

[Harb84] Brian Harbourne, *The geometry of rational surfaces and Hilbert functions of points in the plane*, Proceedings of the 1984 Vancouver conference in algebraic geometry, CMS Conf. Proc., vol. 6, Amer. Math. Soc., Providence, RI, 1986, pp. 95–111. MR846019

[Hart77] Robin Hartshorne, *Algebraic geometry*, Springer-Verlag, New York-Heidelberg, 1977. Graduate Texts in Mathematics, No. 52. MR0463157

[Hir89] André Hirschowitz, *Une conjecture pour la cohomologie des diviseurs sur les surfaces rationnelles génériques* (French), J. Reine Angew. Math. **397** (1989), 208–213. MR993223

[Hui16] Jack Huizenga, *Effective divisors on the Hilbert scheme of points in the plane and interpolation for stable bundles*, J. Algebraic Geom. **25** (2016), no. 1, 19–75. MR3419956

[Hui17] Jack Huizenga, *Birational geometry of moduli spaces of sheaves and Bridgeland stability*, Surveys on recent developments in algebraic geometry, Proc. Sympos. Pure Math., vol. 95, Amer. Math. Soc., Providence, RI, 2017, pp. 101–148. MR3727498

[HuL10] Daniel Huybrechts and Manfred Lehn, *The geometry of moduli spaces of sheaves*, 2nd ed., Cambridge Mathematical Library, Cambridge University Press, Cambridge, 2010. MR2665168

[KO95] S. A. Kuleshov and D. O. Orlov, *Exceptional sheaves on Del Pezzo surfaces* (Russian, with Russian summary), Izv. Ross. Akad. Nauk Ser. Mat. **58** (1994), no. 3, 53–87; English transl., Russian Acad. Sci. Izv. Math. **44** (1995), no. 3, 479–513. MR1286839

[LeP97] J. Le Potier, *Lectures on vector bundles*, Cambridge Studies in Advanced Mathematics, vol. 54, Cambridge University Press, Cambridge, 1997. Translated by A. Maciocia. MR1428426

[Man97] Yu. I. Manin and M. Hazewinkel, *Cubic forms: algebra, geometry, arithmetic*, North-Holland Publishing Co., Amsterdam-London; American Elsevier Publishing Co., New

York, 1974. Translated from the Russian by M. Hazewinkel; North-Holland Mathematical Library, Vol. 4. MR0460349

[Rud88] A. N. Rudakov, *Exceptional vector bundles on a quadric* (Russian), Izv. Akad. Nauk SSSR Ser. Mat. **52** (1988), no. 4, 788–812, 896; English transl., Math. USSR-Izv. **33** (1989), no. 1, 115–138. MR966985

[Seg60] Beniamino Segre, *Alcune questioni su insiemi finiti di punti in geometria algebrica* (Italian), Univ. e Politec. Torino Rend. Sem. Mat. **20** (1960/1961), 67–85. MR0169119

[Wal98] Charles Walter, *Irreducibility of moduli spaces of vector bundles on birationally ruled surfaces*, Algebraic geometry (Catania, 1993/Barcelona, 1994), Lecture Notes in Pure and Appl. Math., vol. 200, Dekker, New York, 1998, pp. 201–211. MR1651095

Department of Mathematics, Statistics and CS, University of Illinois at Chicago, Chicago, Illinois 60607

Email address: coskun@math.uic.edu

Department of Mathematics, The Pennsylvania State University, University Park, Pennsylvania 16802

Email address: huizenga@psu.edu

Contemporary Mathematics
Volume **712**, 2018
http://dx.doi.org/10.1090/conm/712/14344

Excellence in prime characteristic

Rankeya Datta and Karen E. Smith

To Professor Lawrence Ein on the occasion of his sixtieth birthday.

ABSTRACT. Fix any field K of characteristic p such that $[K : K^p]$ is finite. We discuss excellence for Noetherian domains whose fraction field is K, showing for example, that R is excellent if and only if the Frobenius map is finite on R. Furthermore, we show R is excellent if and only if it admits some non-zero p^{-e}-linear map for R (in the language of Blickle and Böckle), or equivalently, that the Frobenius map $R \to F_*R$ defines a **solid** R-algebra structure on F_*R (in the language of Hochster). In particular, this means that generically F-finite, Frobenius split Noetherian domains are always excellent. We also show that non-excellent rings are abundant and easy to construct in prime characteristic, even within the world of regular local rings of dimension one inside function fields.

1. Introduction

The notion of **excellence** for a commutative ring was introduced by Grothendieck. A Noetherian ring is excellent, essentially, if the ring satisfies a list of axioms that ensures it behaves much like a finitely generated algebra over a field; see Definition 2.1. An arbitrary Noetherian ring can be quite pathological, but the class of excellent rings is supposed to be the most general setting to which one can expect the deep ideas of algebraic geometry, such as resolution of singularities, to extend.

Although a common hypothesis in the literature, excellent rings are not widely understood. They are often dismissed with sentences like the following quoted from Wikipedia: *"Essentially all Noetherian rings that occur naturally in algebraic geometry or number theory are excellent; in fact it is quite hard to construct examples of Noetherian rings that are not excellent"* [**29**]. In this paper we show that on the contrary, non-excellent rings are quite easy to construct and are abundant, even among regular local rings of dimension one. Our setting is prime characteristic since Dedekind domains of characteristic zero are always excellent [**15**, Cor 8.2.40]. The examples we construct, moreover, are *generically F-finite*, unlike other known examples such as Nagata's $k \otimes_{k^p} k^p[[t]]$ (whenever $[k : k^p] = \infty$), whose completion map is purely inseparable.

The first author was partially supported by a Department of Mathematics Graduate Fellowship at the University of Michigan.

The second author was partially supported by NSF grant DMS-1501625.

Excellence in prime characteristic is closely connected to another common hypothesis, that of *F-finiteness.* A ring of characteristic p is F-finite if the Frobenius (or p-th power) map is finite. A well-known theorem of Kunz ensures that F-finite rings of characteristic p are excellent [**14**, Thm 2.5]. The converse is also true under the additional hypothesis that the fraction field is F-finite. Put differently, a domain is F-finite if and only if it is excellent and generically F-finite. For example, for any domain R whose fraction field is the function field of some algebraic variety over a perfect (e.g. algebraically closed) field, R is F-finite if and only if R is excellent. Because this fact does not seem to be well-known, we show in Section 2 how this statement follows easily from known facts in the literature.

In Section 3, we turn toward the issue of p^{-e}-linear maps. For a ring R of prime characteristic p, a p^{-e}-linear map is a map $R \xrightarrow{\phi} R$ of the underlying abelian group that satisfies $\phi(r^{p^e}s) = r\phi(s)$ for all $r, s \in S$. Any splitting of the Frobenius map is such a p^{-1}-linear map. In Section 3, we show that for Noetherian domains with F-finite fraction field, there are no non-zero p^{-e}-linear maps at all unless R is excellent! Put differently using the language of Hochster [**10**], we show that a generically F-finite Noetherian domain is a solid algebra via Frobenius if and only if it is excellent. In particular, if a generically F-finite Noetherian domain is Frobenius split, it must be excellent.

The study of p^{-e}-linear maps, or equivalently, elements of $\mathrm{Hom}_R(F^e_* R, R)$, was formalized by Manuel Blickle and later used by Karl Schwede to give an alternate and more global approach to *test ideals,* an important topic in characteristic p commutative algebra. Our results show that Schwede's approach to test ideals relies much more heavily on F-finiteness than previously understood. Test ideals can be viewed as "prime characteristic analogs" of multiplier ideals due to the results in [**27**] and [**7**] (see also [**9**] and [**8**]). While they have attracted great interest in birational algebraic geometry, our results in Section 4 offer a cautionary tale about the limits of this approach.

In Section 4, we consider excellence in the setting of *discrete valuation rings* of a function field of characteristic p. Excellence in this case is equivalent to the DVR being *divisorial,* a topic explored in [**4**]. We show here how this makes it is easy to write down explicit examples of non-excellent discrete valuation rings. Moreover, a simple countability argument shows that among domains whose fraction field is, say the function field of $\mathbb{P}^2$, **non-excellent** regular local rings of dimension one are far more abundant than the excellent ones.

This paper is largely expository, drawing heavily on the work in [**4**] (and the corrections in [**5**]) where most of the harder proofs of the results discussed here appear. However, we are emphasizing a different aspect of the subject than in that paper, drawing conclusions not explicit there.

Acknowledgments: We are honored to help celebrate the birthday of Lawrence Ein, who has been a tremendous inspiration and support for the second author, both mathematically and personally. We thank the referee for their careful comments, and, in particular, for suggesting a generalization of an earlier version of Proposition 3.7.

2. Excellence

An arbitrary Noetherian ring can exhibit pathological behavior. For instance, the integral closure of a Noetherian domain in a finite extension of its fraction field can fail to be Noetherian, and Noetherian rings can have saturated chains of prime ideals of different lengths. Excellent rings were introduced by Grothendieck in [**6**, exposé IV] to rule out such pathologies. We review Grothendieck's definition and some other relevant properties of excellent rings. Another good source is [**17**, Chapter 32].

DEFINITION 2.1. [**6**, Déf 7.8.2] A Noetherian ring A is **excellent** if it satisfies the following properties:

(1) A is *universally catenary*. This means that every finitely generated A-algebra has the property that for any two prime ideals $p \subseteq q$, all saturated chains of prime ideals from p to q have the same length.[1]
(2) All *formal fibers* of A are *geometrically regular*. This means that for every $p \in \operatorname{Spec}(A)$, the fibers of the natural map $\operatorname{Spec}(\widehat{A_p}) \to \operatorname{Spec}(A_p)$ induced by completion along p are geometrically regular in the sense that for each $x \in \operatorname{Spec}(A_p)$, the ring $\widehat{A_p} \otimes_{A_p} K$ is regular for any finite field extension K of the residue field $\kappa(x)$.
(3) For every finitely generated A algebra B, the regular locus of $\operatorname{Spec}(B)$ is open; that is, the set
$$\{q \in \operatorname{Spec}(B) \ : B_q \text{ is a regular local ring}\}$$
is open in $\operatorname{Spec}(B)$.

The class of excellent rings is closed under homomorphic image, localization, and finite-type algebra extensions. Since every field is excellent, it follows that nearly every ring one is likely to encounter in classical algebraic geometry is excellent. Likewise, because the ring of integers is excellent and all complete local rings are excellent, familiar rings likely to arise in number theory are excellent as well.

Among the many properties of excellent rings, the following, sometimes called the **Japanese** or **N2** property, will be important for us later.

PROPOSITION 2.2. [**6**, exposé IV, 7.8.3 (vi)]. *Let A be a Noetherian excellent domain. The integral closure of A in any finite extension of its fraction field is finite as an A-module.*

We construct discrete valuation rings of characteristic p which fail to be Japanese in Subsection 4.1.

2.1. Excellence in prime characteristic.

Fix a commutative ring R of prime characteristic p. The Frobenius map is the ring homomorphism $R \xrightarrow{F} R$ sending each element to its p-th power. It is convenient to denote the target copy of R by F_*R. Thus the notation F_*R denotes the ring R but viewed as an R-module via the Frobenius map: an element $r \in R$ acts on $x \in F_*R$ to produce $r^p x$. Similarly, iterating the Frobenius map, F^e_*R denotes the R-algebra defined by the e-th iterate of Frobenius $R \xrightarrow{F^e} F^e_*R$ sending $r \mapsto r^{p^e}$.

DEFINITION 2.3. The ring R is **F-finite** if the Frobenius map is finite; that is, R is F-finite if F_*R is a finitely generated R-module.

[1] The first example of a non-catenary Noetherian ring was given by Nagata in [**19**].

F-finite rings are ubiquitous. For example, it is easy to check that every perfect field is F-finite, and that a finitely generated algebra over an F-finite ring is F-finite. Furthermore, F-finiteness is preserved under homomorphic images, localization and completion, similar to excellence. Indeed, the two notions are closely related:

THEOREM 2.4. *Let R be a Noetherian domain whose fraction field K satisfies $[K : K^p] < \infty$. Then R is excellent if and only if R is F-finite.*

PROOF. One direction of Theorem 2.4 is the following famous result of Kunz:

THEOREM 2.5. [**14**, Thm 2.5] *Let R be any Noetherian ring of prime characteristic. If the Frobenius map $R \to F_*R$ is finite, then R is excellent.*

So to prove Theorem 2.4, we only need to prove the converse under the hypothesis that R is a domain with F-finite fraction field. The ring R^p, with fraction field K^p, is excellent because it is isomorphic to R via the Frobenius map. Since $K^p \hookrightarrow K$ is finite by assumption, the integral closure S of R^p in K is a finitely generated R^p-module by Proposition 2.2. But clearly every element of R is integral over R^p, as each $r \in R$ satisfies the integral polynomial $x^p - r^p$ over R^p. This means that R is an R^p-submodule of the Noetherian R^p-module S, hence R itself is a Noetherian R^p-module. That is, R is finitely generated as an R^p-module, and the Frobenius map is finite. In other words, R is F-finite. □

COROLLARY 2.6. *Let R be a reduced, Noetherian ring of characteristic p whose total quotient ring K is F-finite. Then R is excellent if and only if R is F-finite.*

PROOF. The backward implication is again a consequence of Kunz's Theorem 2.5. So assume that R is excellent. Let $q_1, \ldots, q_n$ be the minimal primes of R. We denote the corresponding minimal primes of R^p by q_i^p. Let K_i be the fraction field of R/q_i, so that K_i^p is the fraction field of R^p/q_i^p. Then we have a commutative diagram

$$\begin{array}{ccccc} R & \hookrightarrow & R/q_1 \times \cdots \times R/q_n & \hookrightarrow & K_1 \times \cdots \times K_n \cong K \\ \uparrow & & \uparrow & & \uparrow \\ R^p & \hookrightarrow & R^p/q_1^p \times \cdots \times R^p/q_n^p & \hookrightarrow & K_1^p \times \cdots \times K_n^p \cong K^p \end{array}$$

where all rings involved are R^p-modules, and the horizontal maps are injections because R is reduced. Since R is excellent, so is each quotient R/q_i, and F-finiteness of K implies that each K_i is also a finitely generated K_i^p-module. Thus, Theorem 2.4 implies that each R/q_i is F-finite, that is, R/q_i a finitely generated $(R/q_i)^p = R^p/q_i^p$-module. As a consequence,

$$R^p/q_1^p \times \cdots \times R^p/q_n^p \hookrightarrow R/q_1 \times \cdots \times R/q_n$$

is a finite map, and so is the map $R^p \hookrightarrow R^p/q_1^p \times \cdots \times R^p/q_n^p$. This shows that $R/q_1 \times \cdots \times R/q_n$ is a finitely generated R^p-module, and being a submodule of the Noetherian R^p-module $R/q_1 \times \cdots \times R/q_n$, R is also a finitely generated R^p-module. Thus, R is F-finite. □

Theorem 2.4 offers a simple way to think about excellence in prime characteristic, at least for domains in function fields. In Section 4, we use Theorem 2.4 to easily construct many nice examples of non-excellent rings.

In the spirit of Theorem 2.4, there is also an equivalence of excellence and F-finiteness in a slightly different context:

THEOREM 2.7. [**14**, Corollary 2.6] *Let $(R, \mathfrak{m})$ be a Noetherian local ring with F-finite residue field. Then R is excellent if and only if R is F-finite.*

It is worth pointing out that Theorem 2.7 is closely related to Theorem 2.4. Indeed the backward implication follows from Theorem 2.4. Moreover, the hypothesis of Theorem 2.7 ensures that the completion $\widehat{R}$ is F-finite, because Cohen's structure theorem shows that a complete Noetherian local ring of equal characteristic $p > 0$ is F-finite if and only if the residue field is F-finite. The new implication in Theorem 2.7 then says that if R is excellent, that is, when the completion map $R \to \widehat{R}$ is well-behaved, F-finiteness descends from $\widehat{R}$ to R.

2.2. Frobenius splitting vs. F-purity. The hypothesis of F-finiteness is often seen in contexts where Frobenius splitting is discussed. We recall the definitions of Frobenius splitting and the closely related notion of F-purity, which is sometimes confused with it. These notions were originally defined in [**18**] and [**12**], respectively.

DEFINITION 2.8. Let R be an arbitrary commutative ring of prime characteristic p.

(a) The ring R is **Frobenius split** if the map $R \xrightarrow{F} F_*R$ splits as a map of R-modules, that is, there exists an R-module map $F_*R \to R$ such that the composition
$$R \xrightarrow{F} F_*R \to R$$
is the identity map.

(b) The ring R is **F-pure** if $R \xrightarrow{F} F_*R$ is a pure map of R-modules; this means that the map remains injective after tensoring with any R-module M.

It is easy to see that Frobenius split rings are always F-pure. It is also well-known that in the presence of F-finiteness, a Noetherian ring is Frobenius split if and only if it is F-pure [**12**, Corollary 5.2]. However, the relationship between F-purity and Frobenius splitting for a general excellent ring is less understood. Corollary 2.6 clarifies that, at least in a large and important setting, there is little difference between the F-finite and excellent settings for the question of comparing splitting versus purity:

COROLLARY 2.9. *For an excellent Noetherian reduced ring whose total quotient ring is F-finite, Frobenius splitting is equivalent to F-purity. For an excellent local Noetherian ring whose residue field F-finite, Frobenius splitting is equivalent to F-purity.*

PROOF OF COROLLARY. It easily follows from the definitions that a split map is pure, so Frobenius splitting always implies F-purity. Our hypotheses in both statements imply F-finiteness (from Corollary 2.6 and Theorem 2.7, respectively), so splitting and purity are equivalent by [**12**, Corollary 5.2]. □

REMARK 2.10. We do not know any example of an *excellent* F-pure ring that is not F-split. As we see in Section 4, there are plenty of non-excellent examples even among regular local rings.

3. Maps inverse to Frobenius

Test ideals are an important technical tool in both commutative algebra and birational geometry. The original test ideal of Hochster and Huneke is the ideal generated by all the *test elements* for tight closure; they show such test elements exist for excellent local rings in [**11**]. Many recent authors have taken the point of view that a slightly smaller ideal, sometimes called the non-finitistic test ideal, is the more natural object; this ideal is known to be the same as Hochster and Huneke's test ideal in many cases and conjectured to be the same quite generally. See the surveys [**23**] or [**28**] for more information on this history.

An important insight of Schwede is that (under appropriate hypothesis) the test ideal can be defined independently of tight closure.

DEFINITION 3.1. Fix an F-finite ring R. An ideal J is said to be uniformly F-compatible if for all e and all $\phi \in \mathrm{Hom}_R(F_*^e R, R)$, we have $\phi(F_*^e(J)) \subset J$.

It is not at all obvious that non-trivial uniformly F-compatible ideals exist. Schwede shows, however, using a deep theorem of Hochster and Huneke [**11**, Theorem 5.10], that there is in fact a unique smallest non-zero such ideal [**22**]. This is the (non-finitistic) **test ideal**.

The point we want to emphasize is that the modules $\mathrm{Hom}_R(F_*^e R, R)$ play a crucial role in this approach to test ideals. Note also that a splitting of Frobenius is a particular element of $\mathrm{Hom}_R(F_*^1 R, R)$, namely a map $F_*R \xrightarrow{\phi} R$ satisfying $\phi(1) = 1$.

Our next theorem shows, however, that there is little hope to use this approach beyond the F-finite case.

THEOREM 3.2. *Let R be a Noetherian domain of characteristic p whose fraction field is F-finite. Then the following are equivalent:*

(1) *R is excellent.*
(2) *The Frobenius endomorphism $R \xrightarrow{F} F_*R$ is finite.*
(3) *The module $\mathrm{Hom}_R(F_*R, R)$ is non-zero.*
(4) *For all $e > 0$, the module $\mathrm{Hom}_R(F_*^e R, R)$ is non-zero.*
(5) *There exists $e > 0$ such that $\mathrm{Hom}_R(F_*^e R, R)$ is not the trivial module.*

Conditions (3)-(5) in Theorem 3.2 can be stated using Hochster's notion of a **solid algebra**.

DEFINITION 3.3. An R-algebra A is **solid** if there exists a non-trivial R-module map $A \to R$.

Thus condition (3) above precisely states that F_*R is a solid R-algebra via Frobenius, or equivalently, that R is a solid R^p-algebra. Similarly conditions (4) and (5) deal with the solidity of R over R^{p^e}. The theorem states that if R is a domain whose fraction field is F-finite, then R is a solid algebra via Frobenius if and only if R is excellent.

REMARK 3.4. It is worth emphasizing that the generic F-finite assumption in Theorem 3.2 is essential. Fix a field k of characteristic p such that $[k : k^p] = \infty$. Then $R = k[x_1, ..., x_n]$ is an excellent domain that is not F-finite; in this case $F_*^e R$ is a free R-module so there are many non-zero maps in $Hom_R(F_*^e R, R)$.

REMARK 3.5. There are many applications of the module $\mathrm{Hom}_R(F^e_*R, R)$ which motivate its study more generally. Schwede was the first to apply it to the test ideal in [**20**] and [**21**], but the R-module $\mathrm{Hom}_R(F^e_*R, R)$ plays a role in many related stories in birational geometry in characteristic p. For example, under suitable hypothesis including F-finiteness, the module $\mathrm{Hom}_R(F^e_*R, R)$ can be identified with the global sections of the sheaf $F^e_*\mathcal{O}_X((1-p^e)K_X)$ on $X = \mathrm{Spec}\ R$. Each section of this sheaf can be identified with a $\mathbb{Q}$-divisor Δ on Spec R such that $K_X + \Delta$ is $\mathbb{Q}$-Cartier. This idea is applied to understanding log-Fano varieties in prime characteristic in [**25**]. Many other applications are described in [**23**] and [**3**].

It is also worth pointing out for the experts in tight closure that for a local F-finite ring R, the uniformly F-compatible ideals defined in terms of the module $\mathrm{Hom}_R(F^e_*R, R)$ can be interpreted as dual to the $\mathcal{F}(E)$-submodules of E (where E is an injective hull of the residue field and $\mathcal{F}(E)$ is the ring of all Frobenius operators on it) studied in [**16**], the largest of which is the tight closure of zero. The dual characterization used by Schwede to define the test ideal was first carried out in the Gorenstein case in [**24**] and [**26**].

The proof of Theorem 3.2 requires the following lemma, which is independent of the characteristic.

LEMMA 3.6. *Let* $R \xrightarrow{f} S$ *be an injective ring homomorphism of Noetherian domains such that the induced map of fraction fields* $\mathrm{Frac}(R) \hookrightarrow \mathrm{Frac}(S)$ *is finite. If the canonical map*
$$S \to \mathrm{Hom}_R(\mathrm{Hom}_R(S, R), R)$$
is injective, then f *is also a finite map.*

PROOF. Note that if M is a finitely generated R-module, then also $\mathrm{Hom}_R(M, R)$ is finitely generated. Thus, the lemma follows by Noetherianity if we can show that $\mathrm{Hom}_R(S, R)$ is a finitely generated R-module. Let n be the degree of the field extension $\mathrm{Frac}(S)/\mathrm{Frac}(R)$. Then there exists a basis $x_1, \ldots, x_n$ of $\mathrm{Frac}(S)$ over $\mathrm{Frac}(R)$ such that $x_i \in S$ [**1**, 5.1.7].

Let T be the free R-submodule of S generated by the x_i. It is clear that S/T is a torsion R-module. Then applying $\mathrm{Hom}_R(-, R)$ to the short exact sequence
$$0 \to T \to S \to S/T \to 0$$
we get the exact sequence
$$0 \to \mathrm{Hom}_R(S/T, R) \to \mathrm{Hom}_R(S, R) \to \mathrm{Hom}_R(T, R).$$
Since S/T is a torsion R-module and R is a domain, $\mathrm{Hom}_R(S/T, R) = 0$. Thus, $\mathrm{Hom}_R(S, R)$ is a submodule of $\mathrm{Hom}_R(T, R)$, which is free of rank n. But R is a Noetherian ring, and so $\mathrm{Hom}_R(S, R)$ is also finitely generated. □

A necessary condition for the injectivity of $S \to \mathrm{Hom}_R(\mathrm{Hom}_R(S, R), R)$ in the situation of the previous lemma is for the module $\mathrm{Hom}_R(S, R)$ to be non-trivial. If only the non-triviality of this module is assumed, injectivity of $S \to \mathrm{Hom}_R(\mathrm{Hom}_R(S, R), R)$ follows for a large class of examples as shown in the following result:

PROPOSITION 3.7. *Let* $R \overset{f}{\hookrightarrow} S$ *be an injective ring homomorphism of arbitrary domains such that the induced map* $\mathrm{Frac}(R) \hookrightarrow \mathrm{Frac}(S)$ *is algebraic. If* S *is a solid* R*-algebra, then the canonical map* $S \to \mathrm{Hom}_R(\mathrm{Hom}_R(S, R), R)$ *is injective. If, in addition,* R *and* S *are Noetherian and* f *is generically finite, then* f *is a finite map.*

PROOF. By non-triviality of $\mathrm{Hom}_R(S,R)$, there exists an R-linear map $S \xrightarrow{\phi} R$ such that $\phi(1) \neq 0$, and so, for all non-zero $r \in R$, $\phi(r) = r\phi(1) \neq 0$. For the injectivity of

$$S \to \mathrm{Hom}_R(\mathrm{Hom}_R(S,R),R),$$

it suffices to show that for each non-zero $s \in S$, there exists $\varphi \in \mathrm{Hom}_R(S,R)$ such that $\varphi(s) \neq 0$. Now since s is algebraic over $\mathrm{Frac}(R)$, there exists $\sum_{i=0}^n a_i T^i \in R[T]$ such that $a_0 \neq 0$, and

$$a_n s^n + a_{n-1}s^{n-1} + \dots a_1 s + a_0 = (a_n s^{n-1} + a_{n-1}s^{n-2} + \dots + a_1)s + a_0 = 0.$$

Suppose ℓ_λ is left multiplication by λ, where $\lambda := a_n s^{n-1} + a_{n-1}s^{n-2} + \dots + a_1 \in S$. Then $\phi \circ \ell_\lambda \in \mathrm{Hom}_R(S,R)$, and

$$\phi \circ \ell_\lambda(s) = \phi(-a_0) = -a_0\phi(1) \neq 0,$$

which proves injectivity of $S \to \mathrm{Hom}_R(\mathrm{Hom}_R(S,R),R)$.

If $R \xrightarrow{f} S$ is a generically finite map of Noetherian domains, then f is a finite map by Lemma 3.6 and what we just proved. □

REMARK 3.8. As a special case of Proposition 3.7, we obtain the following result: Let R be any domain and K be any field containing R. If the integral closure $\overline{R}$ of R in K is a solid R-algebra, then the canonical map $\overline{R} \to \mathrm{Hom}_R(\mathrm{Hom}_R(\overline{R},R),R)$ is injective. In particular, a Noetherian domain R is Japanese precisely when the integral closure of R in any finite extension of its fraction field is a solid R-algebra.

PROOF OF THEOREM 3.2. We already know (1) and (2) are equivalent from Theorem 2.4.

For (2) implies (3), assume F_*R is a finitely generated R-module. Let K be the fraction field of R, and denote by F_*K the fraction field of F_*R, again emphasizing the K-vector space structure via Frobenius. Note $F_*K = F_*R \otimes_R K$. Since

$$\mathrm{Hom}_R(F_*R,R) \otimes_R K \cong \mathrm{Hom}_K(F_*K,K) \neq 0,$$

it follows that $\mathrm{Hom}_R(F_*R,R) \neq 0$.

We now show (3) implies (4). If $\mathrm{Hom}_R(F_*R,R)$ is non-trivial, then there exists $\phi : F_*R \to R$ such that

$$\phi(1) = c \neq 0.$$

By induction, suppose there exists $\varphi \in \mathrm{Hom}_R(F_*^{e-1}R,R)$ such that $\varphi(1) \neq 0$. Then the p^{-e}-linear map

$$F_*^e R \xrightarrow{F_*^{e-1}(\phi)} F_*^{e-1}R \xrightarrow{\varphi} R$$

maps $c^{(p^{e-1}-1)p} \mapsto c\varphi(1) \neq 0$, showing that $\mathrm{Hom}_R(F_*^e R,R)$ is non-trivial.

Obviously, (4) implies (5). We finish the proof by proving that (5) implies (2). By assumption, $F_*^e K$ is a finite extension of K. We now apply Proposition 3.7, taking taking $S = F_*^e R$ and $f = F^e$. The proposition implies that F^e is a finite map. Thus, also F is a finite map, and we have proved (5) implies (2). □

COROLLARY 3.9. *If R is a non-excellent domain of characteristic $p > 0$ which is generically F-finite, then $Hom(F_*^e R,R) = 0$ for all $e \in \mathbb{N}$.*

Basically, this corollary means that we can not expect to develop a theory of test ideals for non-excellent rings, at least, not a theory that uses the ideas of uniform F-compatibility.

4. Examples of non-excellent rings

Given that the class of excellent rings is so large, it is natural to wonder how one can possibly find natural classes of examples of non-excellent rings. The next theorem gives one source.

THEOREM 4.1. *Let K be a field of characteristic p such that $[K : K^p] < \infty$. For any discrete valuation ring V of K, the following are equivalent:*

(1) *V is excellent;*
(2) *V is F-finite;*
(3) *V is Frobenius split.*

Moreover, if K is a function field over a ground field k, and V is a discrete valuation ring of K/k, then (1)-(3) are equivalent to V being a divisorial valuation ring of K/k.

Recall that a **divisorial valuation ring** of K is one that obtained as the local ring along some prime divisor of a normal model of K/k. In particular, if K/k is a function field of transcendence degree d over k, then any divisorial valuation ring of K has residue field of transcendence degree $d-1$ over k.

PROOF. The equivalence of (1) and (2) is a straightforward consequence of Theorem 2.4. For the proof of (2) $\Rightarrow$ (3), we use the fact that Frobenius is flat for a regular local ring [**13**, Theorem 2.1]. In particular, when V is F-finite, flatness implies that F_*V is a free V-module, which gives a splitting of the Frobenius map. Conversely, a splitting of Frobenius gives the non-triviality of $\mathrm{Hom}_V(F_*V, V)$. Then V is F-finite by Theorem 3.2.

Finally, (1)-(3) is equivalent to V being divisorial when it is a discrete valuation ring of a function field K/k by [**4**, Corollary 6.6.3]. □

REMARK 4.2. The paper [**4**], with corrections in [**5**], shows more generally that a (not necessarily Noetherian) valuation ring with F-finite function field will *always* be divisorial if it is F-finite; see [**5**, Thm 0.1]. Thus, in the class of valuation rings of an F-finite function field, F-finiteness implies Noetherian.

4.1. Some Non-excellent DVRs. Let k be the algebraic closure of the finite field $\mathbb{F}_p$, and fix $K = k(x, y)$, the function field of $\mathbb{P}^2_k$. For concreteness, we consider discrete valuations of K centered at the origin (the point defined by the ideal (x, y)). The reader will immediately observe that our technique generalizes to any function field over k.

Choose any non-unit power series $p(t) \in k[[t]]$ which is *not algebraic* over the subfield $k(t)$ of $k((t))$. Note that such $p(t)$ are abundant: the field $k(t)$ is countable, hence so its algebraic closure, whereas $k[[t]]$ is uncountable (consisting of all infinite sequences of elements in k). Thus, there are uncountably many different choices of non-unit power series $p(t)$ non-algebraic over $k(t)$.

Consider the ring homomorphism

$$k[x, y] \hookrightarrow k[[t]]$$

obtained by sending $x \mapsto t$ and $y \mapsto p(t)$. Our assumption on $p(t)$ ensures this map is injective. Consider the induced inclusion of fraction fields

$$k(x, y) \hookrightarrow k((t)).$$

The standard t-adic valuation on $k((t))$ restricts to some discrete valuation on $k(x, y)$ which takes the value 1 on x. Its valuation ring is $V_p = k[[t]] \cap k(x, y)$, whose maximal ideal is generated by any element of minimal non-zero value, such as x. We have a local map of local rings

$$V_p \hookrightarrow k[[t]]$$

in which the maximal ideal of $k[[t]]$ obviously contracts to the maximal ideal of V_p. In particular, the residue field of V_p satisfies

$$k \hookrightarrow V_p/\mathfrak{m}_{V_p} \hookrightarrow k[[t]]/(t) \cong k.$$

Hence, the residue field of V_p is k, which has transcendence degree zero over k. This means that the discrete valuation ring V_p of $k(x, y)/k$ is *not divisorial.*

Moreover, by Theorem 4.1, because V_p is not divisorial, it is neither excellent nor F-finite. This gives us examples of non-excellent regular local rings of dimension 1, whose fraction field is $k(x, y)$.

REMARK 4.3. Since one can similarly embed $k(x_1, \ldots, x_n)$ in $k((t))$ for any $n \geq 2$, our method easily generalizes to produce examples of non-divisorial DVRs in the function field of $\mathbb{P}^n_k$, for all $n \geq 2$ and k of characteristic p. These can be extended to non-divisorial DVRs on the function field of any variety over k.

The above construction shows that there are many more non-excellent DVRs than excellent ones. For example, among DVRs of $\mathbb{P}^2_k$, we have:

COROLLARY 4.4. *Let $K = k(x, y)$, where k is the algebraic closure of $\mathbb{F}_p$. The set of all discrete valuation rings of K/k is an uncountable set, with the excellent ones among them forming a countable subset.*

PROOF. We first show that our construction above already gives *uncountably many* non-excellent valuation rings in $k(x, y)$ over k. We have already observed that there are uncountably many different choices of the power series $p(t)$ giving a homorphism of fields

$$k(x, y) \hookrightarrow k((t))$$

that maps $x \mapsto t$ and $y \mapsto p(t)$. Each such homomorphism then gives a discrete valuation

$$v_{p(t)} : k(x, y)^{\times} \hookrightarrow k((t))^{\times} \xrightarrow{t-adic} \mathbb{Z}, \tag{4.1}$$

whose associated valuation ring is a non-excellent discrete valuation ring. We now claim that each choice of $p(t)$ yields a *different* valuation ring of $k(x, y)/k$.

Let $p(t) = \Sigma_{n \geq 0} a_n t^n$ and $q(t) = \Sigma_{n \geq 0} b_n t^n$ be two different power series, and let $i \in \mathbb{N} \cup \{0\}$ be the smallest integer such that $a_i \neq b_i$. From the definitions of $v_{p(t)}$ and $v_{q(t)}$ (see 4.1) we get

$$v_{p(t)}(y - (a_0 + a_1 x + \cdots + a_i x^i)) > i \quad \text{and} \quad v_{q(t)}(y - (a_0 + a_1 x + \cdots + a_i x^i)) = i.$$

Thus the fraction

$$\frac{x^i}{y - (a_0 + a_1 x + \cdots + a_i x^i)}$$

is in the valuation ring for the valuation $v_{q(t)}$ but not the one for $v_{p(t)}$, showing that each choice of power series $p(t)$ gives rise to a distinct valuation ring of $k(x, y)$. This completes the proof that $k(x, y)/k$ has uncountably many non-excellent DVRs.

On the other hand, let us show more generally that for any countable algebraically closed field k of characteristic p, and any function field K of k, the set divisorial valuation rings of K/k is countable. Note that any such valuation ring is the localization of a finitely generated, normal k-subalgebra R of K at a height 1 prime. Observe that being a finitely generated field extension of a countable field, K itself is also countable. Thus the collection of all finitely generated k-subalgebras R of K is countable. Any such R is itself countable, and since every ideal of R is finitely generated, the set of ideals of R is countable. This clearly implies countability of the collection S of pairs (R, p), where R is a finitely generated, normal k-subalgebra of K with fraction field K and height one prime p, completing the proof. □

To summarize: randomly choosing a discrete valuation ring in $k(x, y)/k$, we expect it to be *non-excellent* since there are only countably many excellent valuation rings. Equivalently, there are only countably many F-finite discrete valuation rings in $k(x, y)/k$, namely the same ones which are excellent.

REMARK 4.5. See also [**30**, Chapter VI], [**2**] and [**15**, Example 8.2.31] where these types of rings are discussed. In particular, [**15**] gives a different argument for the failure of the Japanese property in a specific case.

References

[1] M. F. Atiyah and I. G. Macdonald, *Introduction to commutative algebra*, Addison-Wesley Publishing Co., Reading, Mass.-London-Don Mills, Ont., 1969. MR0242802

[2] Bruce Bennett, *On the structure of non-excellent curve singularities in characteristic p*, Inst. Hautes Études Sci. Publ. Math. **42** (1973), 129–170. MR0318144

[3] Manuel Blickle and Karl Schwede, *p^{-1}-linear maps in algebra and geometry*, Commutative algebra, Springer, New York, 2013, pp. 123–205. MR3051373

[4] Rankeya Datta and Karen E. Smith, *Frobenius and valuation rings*, Algebra Number Theory **10** (2016), no. 5, 1057–1090. MR3531362

[5] Rankeya Datta and Karen E. Smith, *Correction to the article "Frobenius and valuation rings" [MR3531362]*, Algebra Number Theory **11** (2017), no. 4, 1003–1007. MR3665644

[6] A. Grothendieck and J. A. Dieudonné, *Eléments de géométrie algébrique. I* (French), Grundlehren der Mathematischen Wissenschaften [Fundamental Principles of Mathematical Sciences], vol. 166, Springer-Verlag, Berlin, 1971. MR3075000

[7] Nobuo Hara, *Geometric interpretation of tight closure and test ideals*, Trans. Amer. Math. Soc. **353** (2001), no. 5, 1885–1906. MR1813597

[8] Nobuo Hara and Shunsuke Takagi, *On a generalization of test ideals*, Nagoya Math. J. **175** (2004), 59–74. MR2085311

[9] Nobuo Hara and Ken-Ichi Yoshida, *A generalization of tight closure and multiplier ideals*, Trans. Amer. Math. Soc. **355** (2003), no. 8, 3143–3174. MR1974679

[10] Melvin Hochster, *Solid closure*, Commutative algebra: syzygies, multiplicities, and birational algebra (South Hadley, MA, 1992), Contemp. Math., vol. 159, Amer. Math. Soc., Providence, RI, 1994, pp. 103–172. MR1266182

[11] Melvin Hochster and Craig Huneke, *F-regularity, test elements, and smooth base change*, Trans. Amer. Math. Soc. **346** (1994), no. 1, 1–62. MR1273534

[12] Melvin Hochster and Joel L. Roberts, *The purity of the Frobenius and local cohomology*, Advances in Math. **21** (1976), no. 2, 117–172. MR0417172

[13] Ernst Kunz, *Characterizations of regular local rings for characteristic p*, Amer. J. Math. **91** (1969), 772–784. MR0252389

[14] Ernst Kunz, *On Noetherian rings of characteristic p*, Amer. J. Math. **98** (1976), no. 4, 999–1013. MR0432625

[15] Qing Liu, *Algebraic geometry and arithmetic curves*, Oxford Graduate Texts in Mathematics, vol. 6, Oxford University Press, Oxford, 2002. Translated from the French by Reinie Erné; Oxford Science Publications. MR1917232

[16] Gennady Lyubeznik and Karen E. Smith, *On the commutation of the test ideal with localization and completion*, Trans. Amer. Math. Soc. **353** (2001), no. 8, 3149–3180. MR1828602

[17] Hideyuki Matsumura, *Commutative algebra*, 2nd ed., Mathematics Lecture Note Series, vol. 56, Benjamin/Cummings Publishing Co., Inc., Reading, Mass., 1980. MR575344

[18] V. B. Mehta and A. Ramanathan, *Frobenius splitting and cohomology vanishing for Schubert varieties*, Ann. of Math. (2) **122** (1985), no. 1, 27–40. MR799251

[19] Masayoshi Nagata, *On the chain problem of prime ideals*, Nagoya Math. J. **10** (1956), 51–64. MR0078974

[20] Karl Schwede, *F-adjunction*, Algebra Number Theory **3** (2009), no. 8, 907–950. MR2587408

[21] Karl Schwede, *Centers of F-purity*, Math. Z. **265** (2010), no. 3, 687–714. MR2644316

[22] Karl Schwede, *Test ideals in non-$\mathbb{Q}$-Gorenstein rings*, Trans. Amer. Math. Soc. **363** (2011), no. 11, 5925–5941. MR2817415

[23] Karl Schwede and Kevin Tucker, *A survey of test ideals*, Progress in commutative algebra 2, Walter de Gruyter, Berlin, 2012, pp. 39–99. MR2932591

[24] Karen E. Smith, *Test ideals in local rings*, Trans. Amer. Math. Soc. **347** (1995), no. 9, 3453–3472. MR1311917

[25] Karl Schwede and Karen E. Smith, *Globally F-regular and log Fano varieties*, Adv. Math. **224** (2010), no. 3, 863–894. MR2628797

[26] Karen E. Smith, *F-rational rings have rational singularities*, Amer. J. Math. **119** (1997), no. 1, 159–180. MR1428062

[27] Karen E. Smith, *The multiplier ideal is a universal test ideal*, Comm. Algebra **28** (2000), no. 12, 5915–5929. Special issue in honor of Robin Hartshorne. MR1808611

[28] Karen E. Smith and Wenliang Zhang, *Frobenius splitting in commutative algebra*, Commutative algebra and noncommutative algebraic geometry. Vol. I, Math. Sci. Res. Inst. Publ., vol. 67, Cambridge Univ. Press, New York, 2015, pp. 291–345. MR3525475

[29] Wikipedia, *Excellent ring — Wikipedia, the free encyclopedia*, `https://en.m.wikipedia.org/wiki/Excellent_ring`, 2017, [Online; accessed 8-February-2017].

[30] Oscar Zariski and Pierre Samuel, *Commutative algebra. Vol. II*, Springer-Verlag, New York-Heidelberg, 1975. Reprint of the 1960 edition; Graduate Texts in Mathematics, Vol. 29. MR0389876

Department of Mathematics, University of Michigan, 530 Church Street, Ann Arbor, Michigan 48109

Email address: `rankeya@umich.edu`

Department of Mathematics, University of Michigan, 530 Church Street, Ann Arbor, Michigan 48109

Email address: `kesmith@umich.edu`

Contemporary Mathematics
Volume **712**, 2018
http://dx.doi.org/10.1090/conm/712/14345

Motivic zeta functions and infinite cyclic covers

Manuel González Villa, Anatoly Libgober, and Laurenţiu Maxim

To Lawrence Ein on the occasion of his 60th birthday

ABSTRACT. We associate with an infinite cyclic cover of a punctured neighborhood of a simple normal crossing divisor on a complex quasi-projective manifold (assuming certain finiteness conditions are satisfied) a rational function in $K_0(\mathrm{Var}_{\mathbb{C}}^{\hat{\mu}})[\mathbb{L}^{-1}]$, which we call *motivic infinite cyclic zeta function*, and show its birational invariance. Our construction is a natural extension of the notion of *motivic infinite cyclic cover* introduced by the authors, and as such, it generalizes the Denef-Loeser motivic Milnor zeta function of a complex hypersurface singularity germ.

1. Introduction

Motivated by the importance in topology (e.g., in knot theory [**Ro76**], but see also [**Mi67**]) and algebraic geometry (e.g., for the study of Alexander-type invariants of complex hypersurface complements, see [**DL92, DN04, Li82, Li94, Ma06**]) of infinite cyclic covers, and by the work of Denef and Loeser on motivic invariants of singularities of complex hypersurfaces, in [**GVLM**] we attached to an infinite cyclic cover associated to a punctured neighborhood of a simple normal crossing divisor E on a complex quasi-projective manifold X and a holonomy map Δ, an element $S_{X,E,\Delta}$ in the Grothendieck ring $K_0(\mathrm{Var}_{\mathbb{C}}^{\hat{\mu}})$ of algebraic $\mathbb{C}$-varieties endowed with a good action of the pro-finite group $\hat{\mu} = \lim \mu_n$ of roots of unity, which we called a *motivic infinite cyclic cover*.

In this note, we extend the above-mentioned construction to attach to an infinite cyclic cover associated to a punctured neighborhood of a simple normal crossing divisor E on a complex quasi-projective manifold X, together with holonomy Δ and log discrepancy ν, a rational function $Z_{X,E,\Delta,\nu}(T)$ in $K_0(\mathrm{Var}_{\mathbb{C}}^{\hat{\mu}})[\mathbb{L}^{-1}][[T]]$, which we call a *motivic infinite cyclic zeta function*. It is worthwhile to notice that, although in [**GVLM**] we allowed the holonomy map to take arbitrary integer values, here we restrict the holonomy to always take strictly positive values. This technical

2010 *Mathematics Subject Classification.* Primary 32S55, 14J17, 14J70, 14H30, 32S45.

Key words and phrases. Infinite cyclic cover of finite type, motivic infinite cyclic cover, motivic infinite cyclic zeta function, Milnor fiber, motivic Milnor fiber, motivic zeta function.

M. González Villa was supported in part by the grant MTM2013-45710-C2-2-P from Ministerio de Economía y Competitividad of the Goverment of Spain.

A. Libgober was supported in part by a grant from the Simons Foundation.

L. Maxim was supported in part by a grant from the National Science Foundation.

condition is preserved under blow-ups (hence under birational maps), and it allows us to relate $Z_{X,E,\Delta,\nu}(T)$ with $S_{X,E,\Delta}$.

Our construction of the motivic infinite cyclic zeta function is topological in the sense that it does not make use of arc spaces. One of our main results, Theorem 3.14, shows that our notion of motivic infinite cyclic zeta function is a birational invariant, or equivalently, $Z_{X,E,\Delta,\nu}(T)$ is an invariant of the punctured neighborhood of E in X. The main results of this note are enhanced versions of (and rely on) similar facts about motivic infinite cyclic covers obtained in our earlier paper [**GVLM**].

Finally, in the last section we discuss the significance of motivic infinite cyclic zeta functions for the monodromy conjecture.

2. Infinite cyclic cover of finite type

Let X be a smooth complex quasi-projective variety of dimension $r+1$ and E a (reduced) simple normal crossing divisor on X which shall be called a *deletion* (or *deleted) divisor*. Let $E = \sum_{i\in J} E_i$ be a decomposition of E into irreducible components E_i, where we assume that all divisors E_i are smooth. We use the following natural stratification of X given by the intersections of the irreducible components of E: for each $I \subseteq J$ consider

$$E_I = \bigcap_{i\in I} E_i \quad \text{and} \quad E_I^\circ = E_I \setminus \bigcup_{j\notin I} E_j. \tag{2.1}$$

Clearly, $X = \bigcup_{I\subseteq J} E_I^\circ$, $X \setminus E = E_\emptyset^\circ$ and $E = \bigcup_{\emptyset\neq I\subseteq J} E_I^\circ$.

Let $T_{X,E}^*$ be a punctured regular neighborhood of E on X, e.g., see [**GVLM**, Section 2] and the references therein. Note that $T_{X,E}^*$ is homotopy equivalent to the boundary of a regular neighborhood of E in X, which sometimes is called the *link* of E. In the following, we may discard the subscript X and just write T_E^*. Note that T_E^* is a union of locally trivial fibrations $T_{E_I^\circ}^* \to E_I^\circ$ over the strata E_I° (with $\emptyset \neq I \subseteq J$) of E, the fiber of the latter fibration being diffeomorphic to $(\mathbb{C}^*)^{|I|}$, where $|I|$ denotes the number of elements in the set I.

DEFINITION 2.1 ([**GVLM**]). *Infinite cyclic cover of finite type.*
Let $\Delta : \pi_1(T_E^*) \to \mathbb{Z}$ be an epimorphism[1], and for any $i \in J$ let δ_i denote the boundary of a small oriented disk transversal to the irreducible component E_i. Let $\widetilde{T}_{X,E,\Delta}^*$ be the corresponding infinite cyclic cover (with Galois group $\mathbb{Z}$) of the punctured regular neighborhood T_E^* of a simple normal crossing divisor $E \subset X$. We call such an infinite cyclic cover $\widetilde{T}_{X,E,\Delta}^*$ *of finite type* if $m_i = \Delta(\delta_i) \neq 0$ for all $i \in J$. Sometimes we omit Δ and X in the notation and write simply $\widetilde{T}_E^*$. The map Δ is also referred to as the *holonomy* of this infinite cyclic cover.

REMARK 2.2. Proposition 2.4 of [**GVLM**] shows that if $\widetilde{T}_E^*$ is an infinite cyclic cover of finite type, then, for any $i \in \mathbb{Z}$, the rational vector spaces $H_c^i(\widetilde{T}_E^*)$ and $H^i(\widetilde{T}_E^*)$ are finite dimensional. This fact justifies the terminology used in Definition 2.1. In this note, we deal only with infinite cyclic covers of finite type.

[1]The surjectivity assumption is made here solely for convenience (in which case the corresponding infinite cyclic cover is connected), all results in this paper being valid for arbitrary homomorphisms to $\mathbb{Z}$.

REMARK 2.3. The infinite cyclic cover $\widetilde{T}^*_{X,E,\Delta}$ has the structure of a complex manifold, but it is not an algebraic variety. An algebro-geometric (motivic) realization of $\widetilde{T}^*_{X,E,\Delta}$ was given in [**GVLM**] via the associated motivic infinite cyclic cover $S_{X,E,\Delta}$. Moreover, it was proved in [**GVLM**, Theorem 3.7] that $S_{X,E,\Delta}$ is an invariant of the punctured neighborhood of E in X, i.e., it is invariant under the following equivalence relation.

DEFINITION 2.4 ([**GVLM**]). Let $T_1 = T_{X_1,E_1}$ and $T_2 = T_{X_2,E_2}$ be two regular neighborhoods of normal crossing divisors, and $\Delta_i : \pi_1(T_i^*) \to \mathbb{Z}$, $i = 1, 2$, be epimorphisms defined on the fundamental groups of the corresponding punctured neighborhoods.

We say that the punctured neighborhoods T_1^* and T_2^* of E_1 and resp. E_2 are *equivalent* if there exists a birational map $\Phi : X_1 \to X_2$, which is biregular on $T_1^* \subset X_1$ (and, respectively, Φ^{-1} is biregular on $T_2^* \subset X_2$) and which moreover induces a map $\Phi|_{T_1^*} : T_1^* \to T_2^*$ such that $\Phi(T_1^*)$ and T_2^* are deformation retracts of each other.

We say that the pairs (T_1^*, Δ_1) and (T_2^*, Δ_2) are *equivalent* if the punctured neighborhoods T_1^* and T_2^* are equivalent in the above sense, and the following diagram commutes:

(2.2)
$$\begin{array}{ccc} \pi_1(T_1^*) & \xrightarrow{(\Phi|_{T_1^*})_*} & \pi_1(T_2^*) \\ & {\scriptstyle \Delta_1}\searrow \quad \swarrow{\scriptstyle \Delta_2} & \\ & \mathbb{Z} & \end{array}$$

Here $(\Phi|_{T_1^*})_*$ is the homomorphism induced by $\Phi|_{T_1^*}$ on the fundamental groups.

The basic example of equivalent pairs (T_1^*, Δ_1) and (T_2^*, Δ_2), which is crucial for the proofs of Theorem 3.14 and [**GVLM**, Theorem 3.7] is the following.

EXAMPLE 2.5. Let (X, E) be as above, and consider the blow-up $p : X' := Bl_Z X \to X$ of X along the smooth center $Z \subset X$ of codimension ≥ 2 in X. Assume moreover that Z is contained in E and has normal crossings with the components of E. Denote by E_* the exceptional divisor of the blow-up p, which is isomorphic to the projectivized normal bundle over Z, i.e., $E_* \cong \mathbb{P}(\nu_Z)$. Let us denote the preimage of the divisor E_i in X' by E'_i. Denote by E' the normal crossing divisor in X' formed by the E'_i together with E_*. Assume that there is $I \subseteq J$ maximal such that Z is contained in E_I. We consider the (surjective) homomorphism given by the composition

$$\Delta' : \pi_1(T^*_{X',E'}) \to \pi_1(T^*_{X,E}) \xrightarrow{\Delta} \mathbb{Z},$$

resulting from the identification $T^*_{X'E'} \xrightarrow{\cong} T^*_{X,E}$ induced by the blow-down map and the fact that the center Z is contained in E. We then have:

$$\Delta'(\delta'_i) = \Delta(\delta_i) = m_i \quad (i \in J) \quad \text{and} \quad m_* := \Delta'(\delta_*) = \sum_{i \in I} m_i, \tag{2.3}$$

where δ'_i and δ_* are the meridians about the components E'_i and E_* of E'. Indeed, the blow-down map takes the 2-disk transversal to E_* (at a generic point) and bounded by δ_*, to the disk in X transversal to the components $E_i, i \in I$, containing Z and disjoint from the remaining components of E, i.e., one has the relation $\delta_* = \sum_{i \in I} \delta_i$ in $H_1(T_E^*)$.

3. Motivic infinite cyclic zeta functions

Most of our calculations will be done in the ring of power series with coefficients in the localized Grothendieck ring $\mathcal{M}_{\mathbb{C}}^{\hat{\mu}} = K_0(\mathrm{Var}_{\mathbb{C}}^{\hat{\mu}})[\mathbb{L}^{-1}]$, where $\mathrm{Var}_{\mathbb{C}}^{\hat{\mu}}$ is the category of complex algebraic varieties endowed with good $\hat{\mu}$-actions. Let us recall some definitions, e.g., see [**DL01**].

For a positive integer n, we denote by μ_n the group of all n-th roots of unity (in $\mathbb{C}$). The groups μ_n form a projective system with respect to the maps $\mu_{d\cdot n} \to \mu_n$ defined by $\alpha \mapsto \alpha^d$, and we denote by $\hat{\mu} := \lim \mu_n$ the projective limit of the μ_n.

Let X be a complex algebraic variety. A good μ_n-action on X is an algebraic action $\mu_n \times X \to X$, such that each orbit is contained in an affine subvariety of X. (This last condition is automatically satisfied if X is quasi-projective.) A good $\hat{\mu}$-action on X is a $\hat{\mu}$-action which factors through a good μ_n-action, for some n.

The Grothendieck ring $K_0(\mathrm{Var}_{\mathbb{C}}^{\hat{\mu}})$ of the category $\mathrm{Var}_{\mathbb{C}}^{\hat{\mu}}$ of complex algebraic varieties endowed with a good $\hat{\mu}$-action is generated by classes $[Y, \sigma]$ of isomorphic varieties endowed with good $\hat{\mu}$-actions, modulo the following relations:

- $[Y, \sigma] = [Y \setminus Y', \sigma_{|_{Y\setminus Y'}}] + [Y', \sigma_{|_{Y'}}]$, if $Y' \subseteq Y$ is a closed σ-invariant subset.
- $[Y \times Y', (\sigma, \sigma')] = [Y, \sigma][Y', \sigma']$.
- $[Y \times \mathbb{A}_{\mathbb{C}}^1, \sigma] = [Y \times \mathbb{A}_{\mathbb{C}}^1, \sigma']$, if σ and σ' are two affine liftings of the $\mathbb{C}^*$-action on Y.

The third relation above is included for completeness. However, it is not needed in this paper.

We let $\mathbb{L}$ be the class in $K_0(\mathrm{Var}_{\mathbb{C}}^{\hat{\mu}})$ of $\mathbb{A}_{\mathbb{C}}^1$, with the trivial $\hat{\mu}$-action. We denote by $\mathcal{M}_{\mathbb{C}}^{\hat{\mu}}$ the ring obtained from $K_0(\mathrm{Var}_{\mathbb{C}}^{\hat{\mu}})$ by inverting $\mathbb{L}$.

We now introduce the main objects of the paper. Firstly, recall that J denotes the index set for the irreducible components of the deletion divisor E. For $I \subseteq J$, we can use the exact sequence

$$\mathbb{Z}^{|I|} \to \pi_1(T_{E_I^\circ}^*) \to \pi_1(E_I^\circ) \to 0$$

of the locally trivial fibration $T_{E_I^\circ}^* \to E_I^\circ$ together with the (restricted) holomony $\Delta : \pi_1(T_{E_I^\circ}^*) \to \mathbb{Z}$ to define a map

$$\Delta_I : \pi_1(E_I^\circ) \to \mathbb{Z}/m_I\mathbb{Z} \tag{3.1}$$

where $m_I := \gcd(m_i | i \in I)$ is the index of the image of $\mathbb{Z}^{|I|}$ in $\mathbb{Z}$. For complete details, see Lemma 3.1 and 3.2 in [**GVLM**].

Definition 3.1. We denote by $\widetilde{E_I^\circ}$ the unramified cover of E_I° with Galois group $\mathbb{Z}/m_I\mathbb{Z}$ induced by the map Δ_I from (3.1).

Remark that the cover $\widetilde{E_I^\circ}$ is an algebraic variety endowed with a good μ_{m_I}-action. The cover $\widetilde{E_I^\circ}$ has n connected components, where n is the index of $\Delta(\pi_1(T_{E_I^\circ}^*))$ in $\mathbb{Z}$. Each of the connected components is the cyclic cover of E_I° with the covering group $n\mathbb{Z}/m_I\mathbb{Z}$.

Remark 3.2. As shown in [**GVLM**], the locally trivial $(\mathbb{C}^*)^{|I|}$-fibration $T_{E_I^\circ}^* \to E_I^\circ$ induces a fibration

$$\widetilde{T}_{E_I^\circ}^* \to \widetilde{E}_I^\circ$$

on the infinite cyclic cover $\widetilde{T}^*_{E_I^\circ}$, with connected fiber $\widetilde{(\mathbb{C}^*)^{|I|}} \cong (\mathbb{C}^*)^{|I|-1}$, the infinite cyclic cover of $(\mathbb{C}^*)^{|I|}$ defined by $\ker(\mathbb{Z}^{|I|} \to m_I\mathbb{Z})$.

In order to define the motivic zeta function of an infinite cyclic cover, we also need the notion of *log discrepancy* with respect to a pair in the sense of, for instance, Kollar and Mori [**KM98**]. The actual definition of a log discrepancy is not needed here, but the reader may consult [**KM98**, Section 2.2 and Definition 2.25] for complete details. Let us only recall that for a pair (X, E) as in our setup, a log discrepancy ν with respect to (X, E) is determined by a collection of integers $\{\nu_i\}_{i\in J}$, one for each irreducible component E_i of E, $i \in J$, and moreover it satisfies the following blow-up relation (3.2), which can be derived from [**KM98**, Lemma 2.29] under the additional hypothesis listed below.

Let X' be the blow-up $p : X' := Bl_Z X \to X$ of X along a smooth center $Z \subset X$ of codimension ≥ 2. Assume moreover that the center Z is contained in E and that it has normal crossings with the components of E, as in Example 2.5. Denote by E_* the exceptional divisor of the blow-up, and by E'_i the proper transform of E_i $(i \in J)$, with $E' = E_* + \sum_{i\in J} E'_i$. Then the log discrepancies ν'_i (resp. ν_*) of E'_i (resp. E_*) satisfy

$$\nu'_i = \nu_i \quad \text{for all } i \in J, \quad \text{and} \quad \nu_* = \sum_{i\in I} \nu_i + c, \tag{3.2}$$

where I is the maximal subset of J such that Z is contained in E_I, and c denotes the codimension of Z in E_I.

EXAMPLE 3.3. (i) A global holomorphic form ω on X supported along a simple normal crossing divisor E defines a log-discrepancy ν, with ν_i given by the multiplicity of E_i in the divisor of the form ω plus one, i.e., $\nu_i - 1 = \mathrm{ord}_{E_i}\omega$.

(ii) Let $\{f = 0\}$ be a germ of holomorphic function at the origin in $\mathbb{C}^n$, let $\pi : X \to \mathbb{C}^n$ be an embedded resolution of the singularities of f, and let E be the total transform of $\{f = 0\}$. The standard volume form $\omega = dz_1 \wedge ... \wedge dz_n$ gives a log-discrepancy of the pair (X, E), see [**Ig88**]. See also Section 4.

REMARK 3.4. A blow-up $p : X' := Bl_Z X \to X$ as above identifies the set of irreducible components, labelled by the set of indices J, of the normal crossing divisor E on X with a subset of irreducible components of the divisor E' in X'. Moreover, the punctured neighborhood of E in X and that of E' in X' are homeomorphic. For birational maps $X \to Y$, i.e., for compositions of blow-ups and blow-downs, there are similar identifications of certain components of normal crossing divisors in X and resp. Y which have homeomorphic punctured neighborhoods. The blow-up condition (3.2) may impose in this case additional relations on the values ν_i of a log-discrepancy with respect to a pair (X, E). For example, suppose that the exceptional divisor in a 4-manifold X' is $E_* = \mathbb{P}^2 \times \mathbb{P}^1$ and can be contracted to either $Z_1 = \mathbb{P}^2$ in X_1 or $Z_2 = \mathbb{P}^1$ in X_2. Suppose a divisor F intersects E_* in $\mathbb{P}^2 \times P$, with $P \in \mathbb{P}^1$, and a divisor G intersects E_* along $\mathbb{P}^1 \times \mathbb{P}^1$. Then, since the codimension of Z_1 in F is 1 and the codimension of Z_2 in G is 2, and the corresponding values ν_F and ν_G of the log-discrepancies are invariant because of the first relation in (3.2), it follows that $\nu_* = \nu_F + 1 = \nu_G + 2$.

The previous remark motivates the following definition.

DEFINITION 3.5. Let $T_1 = T_{X_1,E_1}$ and $T_2 = T_{X_2,E_2}$ be regular neighborhoods of normal crossing divisors as above, with holonomies $\Delta_i : \pi_1(T_i^*) \to \mathbb{Z}$, $i = 1, 2$, and log discrepancies ν^i, $i = 1, 2$ defined with respect to (X_1, E_1) (resp. (X_2, E_2)). We say that the triples (T_1^*, Δ_1, ν^1) and (T_2^*, Δ_2, ν^2) are *equivalent* if (T_1^*, Δ_1) and (T_2^*, Δ_2) are equivalent in the sense of Definition 2.4 and the log-discrepancies ν^1 and ν^2 are connected by a finite sequence of blow-up relations as in (3.2).

The following Definition and Lemma provide a natural situation where log-discrepancies exist.

DEFINITION 3.6. Let T_E^* be a punctured neighborhood of a normal crossing divisor E on a smooth projective manifold X. A holomorphic volume form on T_E^* is a meromorphic form ω on X such that the irreducible components of the divisor associated to ω are either irreducible components of E or have empty intersection with E.

LEMMA 3.7. *Let T_E^* be a punctured neighborhood of a normal crossing divisor E on a smooth projective manifold X and let ω be a holomorphic volume form on T_E^*. Then $\nu(E_i) = \mathrm{ord}_{E_i}(\omega) - 1$ defines a log-discrepancy, i.e., satisfies the blow-up relations in (3.2).*

EXAMPLE 3.8. Let $\{f = 0\}$ be a germ of holomorphic function at the orgin in $\mathbb{C}^n$ as in Example 3.3 (ii). The standard volume form $\omega = dz_1 \wedge \dots \wedge dz_n$ defines a holomorphic volume form on a tubular neighborhood of the strict transform of $\{f = 0\}$ in the compactification of X, where $\pi : X \to \mathbb{C}^n$ is an embedded resolution of the singularities of f. The corresponding log-discrepancies coincide with those used by Igusa in [**Ig88**].

We introduce now the main object of this paper, the motivic infinite cyclic zeta function. In order to do so we will assume from now that the holonomy map Δ takes only strictly positive values. The reason for this assumption is two-fold. This condition is enough for our definition and it is stable under blow-ups, see Remarks 3.10 and 3.15. It also allows to relate motivic infinite cyclic zeta functions and the motivic infinite cyclic covers from [**GVLM**], see Remark 3.13.

DEFINITION 3.9. *Motivic zeta function associated to an infinite cyclic cover of finite type and a choice of log-discrepancy*
Let T_E^* be the punctured neighborhood of a normal crossing divisor E in a quasi-projective manifold X as in Section 2, let $\Delta : \pi_1(T_E^*) \to \mathbb{Z}$ be an epimorphism such that the corresponding infinite cyclic cover $\widetilde{T}_{X,E}^*$ is of finite type, and let ν be a log-discrepancy with respect to (X, E). Assume there is a choice of orientations of all δ_i with $i \in J$ such that $m_i > 0$. For each fixed subset $A \subseteq J$, we define the corresponding *motivic infinite cyclic zeta function* (of finite type) of T_E^* as

$$Z_{X,E,\Delta,\nu}^A(T) := \sum_{\substack{\emptyset \neq I \subseteq J \\ A \cap I \neq \emptyset}} [\widetilde{E_I^\circ}, \sigma_I](\mathbb{L} - 1)^{|I|-1} \prod_{i \in I} \frac{\mathbb{L}^{-\nu_i} T^{m_i}}{1 - \mathbb{L}^{-\nu_i} T^{m_i}} \in \mathcal{M}_{\mathbb{C}}^{\hat{\mu}}[[T]]_{\mathrm{sr}}, \tag{3.3}$$

where $\mathcal{M}_{\mathbb{C}}^{\hat{\mu}}[[T]]_{\mathrm{sr}}$ is the $\mathcal{M}_{\mathbb{C}}^{\hat{\mu}}$-submodule of *rational series* of $\mathcal{M}_{\mathbb{C}}^{\hat{\mu}}[[T]]$ generated by 1 and finite products of terms $\frac{\mathbb{L}^{-\nu_i} T^{m_i}}{1 - \mathbb{L}^{-\nu_i} T^{m_i}}$ with $(m_i, \nu_i) \in \mathbb{Z}_{>0} \times \mathbb{Z}$.

Sometimes we omit Δ and ν from the notation. In the case $A = J$, we use the simpler notation $Z_{X,E,\Delta,\nu}(T)$ or $Z_{X,E}(T)$.

In what follows, we will also use the notation

$$A_i := \mathbb{L}^{-\nu_i} T^{m_i}$$

for any $i \in J$.

REMARK 3.10. The factors $A_i(1-A_i)^{-1}$ in (3.3) are well-defined since $m_i > 0$ for all i.

Before stating and proving the main result of this section, let us discuss the relation between the motivic zeta function associated to an infinite cyclic cover of finite type and the motivic infinite cyclic cover introduced in [**GVLM**]. First, let us recall the definition of the motivic infinite cyclic cover:

DEFINITION 3.11 ([**GVLM**]). For each fixed subset $A \subseteq J$, we define the corresponding *motivic infinite cyclic cover* (of finite type) of T_E^* by

$$S^A_{X,E,\Delta} := \sum_{\substack{\emptyset \neq I \subseteq J \\ A \cap I \neq \emptyset}} (-1)^{|I|-1} [\widetilde{E_I^\circ}, \sigma_I] (\mathbb{L}-1)^{|I|-1} \in K_0(\mathrm{Var}_{\mathbb{C}}^{\hat{\mu}}). \tag{3.4}$$

As before, when $A = J$, we use the notation $S_{X,E,\Delta}$ or $S_{X,E}$.

REMARK 3.12. By constrast with Definition 3.9, the definition of motivic infinite cyclic cover does not make any extra assumption on the holonomy map. The reason we make such restrictions in Definition 3.9 is to be able to recover the motivic infinite cyclic cover $S^A_{X,E,\Delta}$ from $Z^A_{X,E,\Delta,\nu}(T)$.

There exists a unique $\mathcal{M}^{\hat{\mu}}_{\mathbb{C}}$-linear morphism

$$\lim_{T \to +\infty} : \mathcal{M}^{\hat{\mu}}_{\mathbb{C}}[[T]]_{\mathrm{sr}} \to \mathcal{M}^{\hat{\mu}}_{\mathbb{C}} \tag{3.5}$$

such that

$$\lim_{T \to +\infty} \prod_{i \in I} \frac{\mathbb{L}^{-\nu_i} T^{m_i}}{1 - \mathbb{L}^{-\nu_i} T^{m_i}} = (-1)^{|I|}, \tag{3.6}$$

when $(m_i, \nu_i) \in \mathbb{Z}_{>0} \times \mathbb{Z}$ (see [**GLM**, Section 2.8] or [**DL98**, Lemma 4.1.1]). This morphism allows to relate the infinite cyclic zeta function and the motivic infinite cyclic cover as follows

$$S^A_{X,E,\Delta} = - \lim_{T \to +\infty} Z^A_{X,E,\Delta,\nu}(T) \in \mathcal{M}^{\hat{\mu}}_{\mathbb{C}}. \tag{3.7}$$

REMARK 3.13. Note that the motivic infinite cyclic cover forgets all information about the log-discrepancy ν. However, equation (3.7) does not hold if one allows the holomony map to take non-positive values. Indeed, if $m_i < 0$ then we have that $\lim_{T \to +\infty} A_i(1-A_i)^{-1} = 0$, and if $m_i = 0$ then $\lim_{T \to +\infty} A_i(1-A_i)^{-1} = A_i(1-A_i)^{-1}$.

The main result of this section is the following.

THEOREM 3.14. *The above notion of motivic infinite cyclic zeta function is an invariant of* $(T^*_{X,E}, \Delta, \nu)$, *i.e., it is preserved under the equivalence relation of Definition* 3.5.

The proof of Theorem 3.14 is similar to (and relies on) that of [**GVLM**, Theorem 3.7], and we will use freely facts proved in loc. cit. In fact, since any birational map $X_1 \to X_2$ providing an equivalence between punctured neighborhoods (cf. Definition 2.4) is, by the Weak Factorization Theorem [**AKMW**] (see also [**Bo02**] for the non-complete case), a composition of blow-ups and blow-downs, each inducing an equivalence between the corresponding punctured neighborhoods, *it suffices to show that the above expression (3.3) is invariant under blowing up along a smooth center in E*. Let us consider

$$p : X' := Bl_Z X \to X$$

the blow-up of X along the smooth center $Z \subset E$ of codimension ≥ 2 in X. Denote by E_* the exceptional divisor of the blow-up p, which is isomorphic to the projectivized normal bundle over Z, i.e., $E_* \cong \mathbb{P}(\nu_Z)$. We may also assume that the center Z of the blow-up is contained in E and has normal crossings with the components of E (cf. [**AKMW**, Theorem 0.3.1,(6)]). Let us denote as before the preimage of the divisor E_i in X' by E_i'. Denote by E' the normal crossing divisor in X' formed by the E_i' together with E_*. Let $J' = J \cup \{*\}$ be the family of indices of the divisor E'. For $I \subseteq J$ we denote by $I' \subseteq J'$ the family $I \cup \{*\}$. Finally, let $A' = A \cup \{*\}$.

By the above reduction to the normal crossing situation, we may assume that there is $I \subseteq J$ maximal such that Z is contained in E_I. We consider the (surjective) homomorphism given by the composition

$$\Delta' : \pi_1(T^*_{X',E'}) \to \pi_1(T^*_{X,E}) \stackrel{\Delta}{\to} \mathbb{Z},$$

resulting from the identification $T^*_{X'E'} \stackrel{\cong}{\to} T^*_{X,E}$ induced by the blowing down map, as in Example 2.5. We also consider the log discrepancy ν' with respect to (X', E') obtained from ν by the blow-up relation (3.2). Note that $\widetilde{T}^*_{X',E',\Delta'}$ is of finite type since $\widetilde{T}^*_{X,E,\Delta}$ is so and $T^*_{X',E'} \cong T^*_{X,E}$. Moreover, by Lemma 3.2 in [**GVLM**] and Definition 3.9 applied to (X', E', Δ', ν'), we can define the corresponding motivic infinite cyclic zeta function by:

$$Z^{A'}_{X',E',\Delta',\nu'}(T) := \sum_{\substack{\emptyset \neq K \subset J' \\ K \cap A' \neq \emptyset}} [\widetilde{E^{\circ}_K}, \sigma_K](\mathbb{L}-1)^{|K|-1} \prod_{k \in K} \frac{\mathbb{L}^{-\nu_k} T^{m_k}}{1 - \mathbb{L}^{-\nu_k} T^{m_k}}. \tag{3.8}$$

We will also sometimes denote $\mathbb{L}^{-\nu_*} T^{m_*}$ by A_*.

REMARK 3.15. The factor $A_*(1-A_*)^{-1}$ in (3.8) is well-defined because $m_i > 0$ for all i, and therefore $m_* = \sum_{i \in I} m_i > 0$.

Theorem 3.14 follows now from the following proposition.

PROPOSITION 3.16. *With the above notations, we have the following identification:*

$$Z^A_{X,E,\Delta,\nu}(T) = Z^{A'}_{X',E',\Delta',\nu'}(T) \in \mathcal{M}^{\hat{\mu}}_{\mathbb{C}}[[T]]_{\rm sr}. \tag{3.9}$$

Note that we can always restrict the comparison of zeta functions in Proposition 3.16 to strata in the center of blowup and in the exceptional divisor, respectively. Indeed, the blow-up map induces an isomorphism outside the center Z, so the strata in $E \setminus Z$ and $E' \setminus E_*$ are in one-to-one isomorphic correspondence; moreover, these isomorphisms can be lifted (e.g., by Lemma 3.2 in [**GVLM**]) to the corresponding unramified covers. It also suffices to prove the above result only in the case $A = J$.

The proof of Proposition 3.16 is by induction on the dimension of the center of blow-up.

3.1. Beginning of induction. Let us consider the following examples in relation to the starting case of induction, i.e., when the center Z is a point.

EXAMPLE 3.17. Let X be a surface and let E_1 and E_2 be two smooth curves intersecting transversally at a point P. Let us consider the blow-up $X' = Bl_Z X$ of X at the center $Z = P$. The exceptional divisor is $E_* \cong \mathbb{P}^1$ and we have $E_*^\circ \cong \mathbb{C}^*$. Let $\delta_i \in H_1(T^*_{E_i^\circ}, \mathbb{Z})$ $(i = 1, 2)$ be the class of the fiber of the projection of punctured neighborhood $T^*_{E_i^\circ}$ onto the stratum E_i°. If $\Delta(\delta_1) = m_1$, $\Delta(\delta_2) = m_2$, $\nu(\delta_1) = \nu_1$, $\nu(\delta_2) = \nu_2$, and we let $m = \gcd(m_1, m_2)$, then the contribution of P to $Z_{X,E}(T)$ is

$$[\mu_m](\mathbb{L} - 1)\frac{A_1}{1 - A_1}\frac{A_2}{1 - A_2},$$

and the contributions of the exceptional divisor E_* to $Z_{X',E'}(T)$ are

$$\frac{A_*}{1 - A_*}\left((\mathbb{L} - 1)\left([\widetilde{E_1' \cap E_*}]\frac{A_1}{1 - A_1} + [\widetilde{E_2' \cap E_*}]\frac{A_2}{1 - A_2}\right) + [\widetilde{E_*^\circ}]\right).$$

Because $m = \gcd(m_1, m_1+m_2) = \gcd(m_2, m_1+m_2)$, we get the following equalities: $[\widetilde{E_1' \cap E_*}, \sigma_{\Delta'}] = [\widetilde{E_2' \cap E_*}, \sigma_{\Delta'}] = [\mu_m]$. Finally, it follows from Lemma 3.2 in [**GVLM**] that $[\widetilde{E_*^\circ}, \sigma_{\Delta'}] = [\mu_m](\mathbb{L}-1)$, see [**GVLM**, Example 3.9] for more details. Hence, after factoring out $[\mu_m](\mathbb{L}-1)$, it remains to show that the two contributions

$$\frac{A_1}{1 - A_1}\frac{A_2}{1 - A_2} \quad \text{and} \quad \frac{A_*}{1 - A_*}\left[\left(\frac{A_1}{1 - A_1} + \frac{A_2}{1 - A_2}\right) + 1\right]$$

coincide. After reducing to the common denominator $(1 - A_1)(1 - A_2)(1 - A_*)$, we just need to check that

$$A_1A_2(1 - A_*) = A_1(1 - A_2)A_* + (1 - A_1)A_2A_* + A_*(1 - A_1)(1 - A_2),$$

which further simplifies to $A_1A_2 = A_*$. The latter claim holds since in this case $m_* = m_1 + m_2$ and $\nu_* = \nu_1 + \nu_2$ (here $\operatorname{codim}_{E_I} Z = 0$).

Note that in the case when P belongs to only one irreducible component, say E_1, we have $[\widetilde{E_1^\circ}|_P] = [\mu_{m_1}]$ and $[\widetilde{E_*^\circ}, \sigma_\Delta] = [\mu_{m_1}]\mathbb{L}$. In this case, the contributions of P and resp. E_* to $Z_{X,E}(T)$ and resp. $Z_{X',E'}(T)$ are

$$[\mu_{m_1}]\frac{A_1}{1 - A_1} \quad \text{and resp.} \quad [\mu_{m_1}](\mathbb{L} - 1)\frac{A_1}{1 - A_1}\frac{A_*}{1 - A_*} + [\mu_{m_1}]\mathbb{L}\frac{A_*}{1 - A_*}.$$

Again, after factoring out $[\mu_{m_1}]$ and reducing to common denominator, we need to check that

$$A_1(1 - A_*) = (\mathbb{L} - 1)A_1A_* + \mathbb{L}(1 - A_1)A_*,$$

which simplifies to $A_1 = \mathbb{L}A_*$. The latter claim holds since $m_* = m_1$ and $\nu_* = \nu_1 + 1$ (here $\operatorname{codim}_{E_I} Z = 1$). □

EXAMPLE 3.18. Let X be a threefold and E_1, E_2, E_3 be three divisors intersecting transversally at a point P. Consider the blow-up X' of X at the center $Z = P$, so here $I = J = \{1, 2, 3\}$. The divisor $E = \sum_{i=1}^3 E_i$ of X transforms into the divisor E' in X' consisting of the proper transforms E_i' of the irreducible components E_i of E $(i = 1, 2, 3)$, together with the exceptional component $E_* \cong \mathbb{P}^2$. As already mentioned, it suffices to restrict the comparison of $Z_{X,E}(T)$ and $Z_{X',E'}(T)$ only to contributions coming from the strata in the center of blow-up and the exceptional divisor, respectively. Let $m_I := \gcd(m_1, m_2, m_3)$. Then the contribution of P to $Z_{X,E}(T)$ is

$$[\mu_{m_I}](\mathbb{L}-1)^2 \frac{A_1}{1-A_1}\frac{A_2}{1-A_2}\frac{A_3}{1-A_3}.$$

The exceptional divisor E_* acquires seven strata induced from the stratification of E'. These strata are:

- $L_{\{i,j\}} = E_* \cap E'_i \cap E'_j$, for $i, j \in \{1,2,3\}$ with $i \neq j$,
- $L_{\{i\}} = (E_* \cap E'_i) \setminus (L_{\{i,j\}} \cup L_{\{i,k\}})$, with $\{i,j,k\} = \{1,2,3\}$,
- $E_*^\circ = E_* \setminus \bigcup_{i=1}^3 E'_i$.

Note that the strata E_*° and $L_{\{i\}}$ are complex tori of dimension 2 and 1, respectively, while the strata $L_{\{i,j\}}$ are points.

It follows from Lemma 3.2 of [**GVLM**] that for each of seven strata of E_*, the corresponding unbranched covers appearing in (3.3) have $m_I = \gcd(m_1, m_2, m_3)$ components, each of which is biregular to the stratum itself (since all these strata are tori), see Example 3.6 in [**GVLM**] for more details. Hence the contribution of E_* to $Z_{X',E'}(T)$ is:

$$[\mu_{m_I}](\mathbb{L}-1)^2 \frac{A_*}{1-A_*}\left(1 + \frac{A_1}{1-A_1} + \frac{A_2}{1-A_2} + \frac{A_3}{1-A_3}\right.$$
$$\left.+\frac{A_1}{1-A_1}\frac{A_2}{1-A_2} + \frac{A_1}{1-A_1}\frac{A_3}{1-A_3} + \frac{A_2}{1-A_2}\frac{A_3}{1-A_3}\right).$$

After we factor out $[\mu_{m_I}](\mathbb{L}-1)^2$ and reduce to common denominator, we need to check that $A_1A_2A_3(1-A_*)$ is equal to

$$A_*\Big[(1-A_1)(1-A_2)(1-A_3) + A_1(1-A_2)(1-A_3) + (1-A_1)A_2(1-A_3)$$
$$+ (1-A_1)(1-A_2)A_3 + A_1A_2(1-A_3) + A_1(1-A_2)A_3 + (1-A_1)A_2A_3\Big],$$

which further reduces to showing that $A_1A_2A_3 = A_*$. This last equality holds because $m_* = m_1 + m_2 + m_3$ and $\nu_* = \nu_1 + \nu_2 + \nu_3$ (here $\operatorname{codim}_{E_I} Z = 0$). □

EXAMPLE 3.19. Let X be a threefold, and $E = E_1 + E_2$ be a simple normal crossing divisor on X, with holonomy values m_1 and resp. m_2 on the meridians about its irreducible components. Let $m = \gcd(m_1, m_2)$. Choose a point Z contained in the (one-dimensional) intersection $E_J := E_1 \cap E_2$, for $J = I = \{1,2\}$, and consider the blow-up $X' = Bl_Z X$ of X along the center Z. We denote the exceptional divisor $\mathbb{P}(\nu_Z)$ by E_*. The divisor E is transformed under the blow-up into the divisor E' in X' consisting of the proper transforms E'_i $(i \in J)$ of the irreducible components E_i of E, together with the exceptional divisor $E_* \cong \mathbb{P}^2$.

Let us explicitly describe the contribution of the center Z and that of the exceptional divisor E_* to the zeta functions $Z_{X,E}(T)$ and $Z_{X',E'}(T)$, respectively. Clearly, the class $[\widetilde{E_J}|_Z, \sigma_\Delta]$ equals $[\mu_m]$. So the contribution to $Z_{X,E}(T)$ consists of

$$[\mu_m](\mathbb{L}-1)\frac{A_1}{1-A_1}\frac{A_2}{1-A_2}.$$

On the other hand, the exceptional divisor E_* acquires four strata induced from the stratification of E', namely,

- $L_J = E'_1 \cap E'_2 \cap E_*$, which is just a point.
- $L_{\{i\}} = E_* \cap E'_i \setminus L_J \cong \mathbb{C}$, for $i \in J$.
- $E^\circ_* = E_* \setminus (E'_1 \cup E'_2) \cong \mathbb{C} \times \mathbb{C}^*$.

Note that, since any of the four strata in E_* is either simply-connected or a product of a simply-connected space with a torus, any finite connected unbranched cover of such a stratum is biregular to the stratum itself. Moreover, as shown in [**GVLM**, Example 3.11], for each of the four strata of E_*, the corresponding unbranched cover appearing in (3.3) has exactly m connected components. So the contribution of E_* to $Z_{X',E'}(T)$ is given by:

$$[\mu_m](\mathbb{L}-1)\frac{A_*}{1-A_*}\left[\mathbb{L}+\mathbb{L}\left(\frac{A_1}{1-A_1}+\frac{A_2}{1-A_2}\right)+(\mathbb{L}-1)\frac{A_1}{1-A_1}\frac{A_2}{1-A_2}\right]. \tag{3.10}$$

After factoring out $[\mu_m](\mathbb{L}-1)$ out and reducing to common denominator, it remains to check the equality

$$\begin{aligned}&A_1A_2(1-A_*)\\ &\quad = A_*\left[\mathbb{L}(1-A_1)(1-A_2)+\mathbb{L}\big(A_1(1-A_2)+(1-A_1)A_2\big)+(\mathbb{L}-1)A_1A_2\right],\end{aligned}$$

which simplifies to $A_1A_2 = \mathbb{L}A_*$. The latter equality holds because $m_* = m_1 + m_2$ and $\nu_* = \nu_1 + \nu_2 + 1$ (here $\operatorname{codim}_{E_I} Z = 1$). □

EXAMPLE 3.20. Let X be a threefold, and $E = E_1$ a (simple normal crossing) divisor on X, with holonomy value $m := m_1$. Choose Z a point contained in E and consider the blow-up $X' = Bl_Z X$ of X along Z. Denote by E_* the exceptional divisor which is isomorphic to $\mathbb{P}^2$ and by E' the proper transform of E. In this case, the exceptional divisor E_* has two strata, namely, $E^\circ_* = E_* \setminus E' \cong \mathbb{L}^2$ and $L_1 = E_* \cap E' \cong \mathbb{P}^1$. In particular, any finite connected unbranched cover of such a stratum is biregular to the stratum itself. And it can be easily seen that for each of the two strata of E_*, the corresponding branched covers have exactly m connected components. Thus, the contributions of Z to $Z_{X,E}(T)$, and resp. of E_* to $Z_{(X',E')}(T)$ are

$$[\mu_m]\frac{A_1}{1-A_1}, \quad \text{resp.} \quad [\mu_m]\mathbb{L}^2\frac{A_*}{1-A_*}+[\mu_m](\mathbb{L}^2-1)\frac{A_1}{1-A_1}\frac{A_*}{1-A_*}.$$

After we factor out $[\mu_m]$ and reduce to common denominator, it remains to show that

$$A_1(1-A_*) = \mathbb{L}^2(1-A_1)A_* + (\mathbb{L}^2-1)A_1A_*,$$

or equivalently $A_1 = \mathbb{L}^2 A_*$, which follows from $m_* = m$ and $\nu_* = \nu_1 + 2$ (here $\operatorname{codim}_{E_I} Z = 2$).

Let us now prove the beginning case of induction for Proposition 3.16.

PROPOSITION 3.21. *The assertion of Proposition* 3.16 *holds in the case when the center of blow-up Z is zero-dimensional.*

It suffices to prove Proposition 3.21 in the case when the center of blow-up is a single point. Indeed, the blow-up at a finite number of points can be regarded as a finite number of single point blow-ups.

We can thus assume that Z is a point. Let $r+1 = \operatorname{codim}_X Z$, which, by our assumption, equals $\dim X$. Then the exceptional divisor is $E_* \cong \mathbb{P}^r$. The divisor

$E = \sum_{i\in J} E_i$ of X transforms under the blow-up into the divisor E' in X' consisting of the proper transforms E_i' of the irreducible components E_i of E, together with the exceptional component E_*. It suffices to restrict the comparison of zeta functions $Z_{X,E}(T)$ and $Z_{X',E'}(T)$ only to contributions coming from the strata in the center of blow-up and the exceptional divisor, respectively.

As in the above examples, we need to describe the stratification of $E_* \cong \mathbb{P}^r$ induced from that of E' (see (2.1) for the latter). Assume that

$$Z \subseteq \bigcap_{i=1}^{k} E_i.$$

We recall here the following result from [**GVLM**].

LEMMA 3.22. [**GVLM**, Lemma 3.13] *For each k with $1 \leq k \leq r+1$ we have the following identity in $K_0(\mathrm{Var}_{\mathbb{C}})$:*

$$[\mathbb{P}^r] = \sum_{l=0}^{k-1} \binom{k}{l} \mathbb{L}^{r-k+1}(\mathbb{L}-1)^{k-l-1} + [\mathbb{P}^{r-k}]. \tag{3.11}$$

The right-hand side describes the stratification of the exceptional divisor $E_* \cong \mathbb{P}^r$ induced by the divisor $\sum_{i=1}^{k} E_i'$ consisting of the proper transforms of components of E containing the center of blow-up. More precisely, by setting $I := \{1, \cdots, k\}$, the strata of E_* are:

- *$L_I := (\bigcap_{i=1}^{k} E_i') \cap E_*$, which is isomorphic to $\mathbb{P}^{r-k}$.*
- *$\binom{k}{l}$ strata of dimension $r-l$ and of the form*

$$L_G := (\bigcap_{i\in G} E_i') \cap E_* \setminus \bigcup_{i\in I\setminus G} E_i',$$

  with $G \subset I$ and $1 \leq |G| = l \leq k-1$, each of which is isomorphic to $\mathbb{C}^{r-k+1} \times (\mathbb{C}^*)^{k-l-1}$. The class of each such stratum in $K_0(\mathrm{Var}_{\mathbb{C}})$ is equal to $\mathbb{L}^{r-k+1}(\mathbb{L}-1)^{k-l-1}$.
- *$E_*^\circ := E_* \setminus \bigcup_{i=1}^{k} E_i'$, of dimension r, which is isomorphic to $\mathbb{C}^{r-k+1} \times (\mathbb{C}^*)^{k-1}$, and whose class in $K_0(\mathrm{Var}_{\mathbb{C}})$ is $\mathbb{L}^{r-k+1}(\mathbb{L}-1)^{k-1}$.*

We also need the following easy fact:

LEMMA 3.23. *For a ring R with unit containing the elements $A_1, \ldots, A_k$ and $1 + A_1, \ldots, 1 + A_k$, for any $k \in \mathbb{N}$ we have that*

$$\prod_{i=1}^{k}(1 - A_i) + \sum_{l=1}^{k} \sum_{\substack{G\subseteq\{1,2,\ldots,k\}\\ |G|=l}} \prod_{i\in G} A_i \prod_{j\in\{1,2,\ldots,k\}\setminus G} (1 - A_j) = 1 \in R \tag{3.12}$$

PROOF. This follows iteratively from the fact that $(1 - A_1) + A_1 = 1$ and, if we denote by $B(k)$ the left-hand side of (3.12) then we have that

$$B(l)(1 - A_{l+1}) + B(l)A_{l+1} = B(l+1).$$

□

PROOF. (of Proposition 3.21)
As already pointed out, it suffices to check the invariance (3.9) of motivic infinite cyclic zeta function under blow-up in the case when Z is a single point.

Assume $Z \subseteq \bigcap_{i=1}^k E_i$. Let $I = \{1, 2, \dots, k\}$ and set $m = \gcd(m_1, \cdots, m_k)$, where the m_i are the values of the holonomy on the meridians δ_i about the components E_i. Clearly, the class $[\widetilde{E}^\circ_I{}_{|Z}, \sigma_\Delta]$ equals $[\mu_m]$. Therefore, the corresponding contribution of Z to the left hand-side of (3.9) is

$$[\mu_m](\mathbb{L}-1)^{k-1}\prod_{i=1}^k \frac{\mathbb{L}^{-\nu_i}T^{m_i}}{1-\mathbb{L}^{-\nu_i}T^{m_i}}.$$

On the other hand, for any stratum S of the exceptional divisor E_* (as described in Lemma 3.22), the motive of the corresponding unbranched cover of Definition 3.1 can be computed by $[\widetilde{S}, \sigma_{\Delta'}] = [\mu_m][S, \sigma_\Delta]$, see the proof of Proposition 3.12 in [**GVLM**] for more details.

Taking into account the description of the stratification in Lemma 3.22 the contribution of E_* to the right hand-side of (3.9) is

$$[\mu_m](\mathbb{L}-1)^{k-1}\frac{A_*}{1-A_*}$$
$$\cdot\left(\mathbb{L}^{r-k+1} + \sum_{l=1}^{k-1}\mathbb{L}^{r-k+1}\sum_{\substack{G\subseteq I\\ |G|=l}}\prod_{i\in G}\frac{A_i}{1-A_i} + [\mathbb{P}^{r-k}](\mathbb{L}-1)\prod_{i=1}^k\frac{A_i}{1-A_i}\right).$$

The first summand in the above expression corresponds to the stratum E_*° of dimension r, the summands indexed by l correspond to the $\binom{k}{l}$ strata L_G of dimension $r-l$ (with $G \subset I$, $1 \le |G| = l \le k-1$), while the last summand corresponds to the stratum L_I of dimension $r-k$.

After we factor out $[\mu_m](\mathbb{L}-1)^{k-1}$ and we reduce to the common denominator $(1-A_*)\prod_{i=1}^k(1-A_i)$, we are left to verify that $\prod_{i=1}^k A_i(1-A_*)$ equals

$$A_*\Big[\mathbb{L}^{r-k+1}\prod_{i=1}^k(1-A_i)+\sum_{l=1}^{k-1}\mathbb{L}^{r-k+1}\sum_{\substack{I\subseteq K\\ |I|=l}}\prod_{i\in I}A_i\prod_{j\in K\setminus I}(1-A_j)+(\mathbb{L}^{r-k+1}-1)\prod_{i=1}^k A_i\Big].$$

The last claim follows from Lemma 3.23. Indeed, it can be reduced to to $\prod_{i=1}^k A_i = \mathbb{L}^{r-k+1}A_*$, which holds because $m_* = \sum_{i=1}^k m_i$ and $\nu_* = \sum_{i=1}^k \nu_i + r - k + 1$. Recall that here $\operatorname{codim}_{E_I} Z = r - k + 1$. □

3.2. Invariance of the motivic infinite cyclic covers under blowups: general case. The main ideas of the proof of the induction step are already visible in the following example.

EXAMPLE 3.24. Let X be a threefold and $E = \sum_{i\in J} E_i$, with $J = \{1, 2, 3\}$, be a simple normal crossing divisor. Let Z be the intersection of E_1 and E_2. Set $I = \{1, 2\}$, so in the notations from the introduction, we have that $Z = E_I$. The component E_3 is transversal to Z. Let us consider the blow-up $X' = Bl_Z X$ of X along the center Z. As before, we denote the exceptional divisor $\mathbb{P}(\nu_Z)$ by E_*.

The strata in Z are $E_I^\circ = E_1 \cap E_2 \setminus E_3$ and the point $E_J = \cap_{i\in J} E_i$, so the contribution of the center Z to the motivic infinite cyclic zeta function $Z_{X,E}(T)$ is:

$$[\widetilde{E_I^\circ}](\mathbb{L}-1)\frac{A_1}{1-A_1}\frac{A_2}{1-A_2} + [\widetilde{E_J}](\mathbb{L}-1)^2\frac{A_1}{1-A_1}\frac{A_2}{1-A_2}\frac{A_3}{1-A_3}.$$

The exceptional divisor E_* acquires a stratification with strata of the form:

$$L_H := (\bigcap_{i\in H} E'_i) \cap E_* \setminus \bigcup_{i\in J\setminus H} E'_i,$$

with $H \subset J$, where the dense open stratum E_*° in E_* is identified with $L_\emptyset$. More precisely, the strata of E_* are in this case the following:

- $L_{\{1,3\}} = E_* \cap E'_1 \cap E'_3$, $L_{\{2,3\}} = E_* \cap E'_2 \cap E'_3$.
- $L_{\{1\}} = (E_* \cap E'_1) \setminus E'_3$, $L_{\{2\}} = (E_* \cap E'_2) \setminus E'_3$, $L_{\{3\}} = (E_* \cap E'_3) \setminus (E'_1 \cup E'_2)$.
- $E_*^\circ = E_* \setminus \bigcup_{i=1}^3 E'_i$.

So the contribution of the exceptional divisor E_* to $Z_{X',E'}(T)$ is:

$$\frac{A_*}{1-A_*}\Bigg([\widetilde{E_*^\circ}] + [\widetilde{L_{\{1\}}}](\mathbb{L}-1)\frac{A_1}{1-A_1} + [\widetilde{L_{\{2\}}}](\mathbb{L}-1)\frac{A_2}{1-A_2}$$
$$+ [\widetilde{L_{\{3\}}}](\mathbb{L}-1)\frac{A_3}{1-A_3} + [\widetilde{L_{\{1,3\}}}](\mathbb{L}-1)^2\frac{A_1}{1-A_1}\frac{A_3}{1-A_3}$$
$$+[\widetilde{L_{\{2,3\}}}]\mathbb{L}-1)^2\frac{A_2}{1-A_2}\frac{A_3}{1-A_3}\Bigg).$$

Note that by Example 3.17, applied to the blow-up of the point E_J of intersection of transversal curves $E_1 \cap E_3$ and $E_2 \cap E_3$ inside the surface E_3, we have that:

$$[\widetilde{E_J}](\mathbb{L}-1)\frac{A_1}{1-A_1}\frac{A_2}{1-A_2}$$
$$= \left([\widetilde{L_{\{3\}}}] + ([\widetilde{L_{\{1,3\}}}]\frac{A_1}{1-A_1} + [\widetilde{L_{\{2,3\}}}]\frac{A_2}{1-A_2})(\mathbb{L}-1)\right)\frac{A_*}{1-A_*}.$$

In particular, this relation is consistent with the fact that $m_* = m_1 + m_2$ and $\nu_* = \nu_1 + \nu_2$.

So in order to show that the contributions of Z and E_* to the zeta functions $Z_{X,E}(T)$ and respectively $Z_{X',E'}(T)$ coincide, it suffices to prove the equality:

$$[\widetilde{E_I^\circ}](\mathbb{L}-1)\frac{A_1}{1-A_1}\frac{A_2}{1-A_2}$$
$$= \left([\widetilde{E_*^\circ}] + ([\widetilde{L_{\{1\}}}]\frac{A_1}{1-A_1} + [\widetilde{L_{\{2\}}}]\frac{A_2}{1-A_2})(\mathbb{L}-1)\right)\frac{A_*}{1-A_*}.$$

Next, note that by the definition of blow-up, we have isomorphisms $L_{\{1\}} \cong E_I^\circ \cong L_{\{2\}}$ which, moreover, extend (by Lemma 3.2 in [**GVLM**]) to isomorphisms between the corresponding unbranched covers appearing in the definition of our motivic zeta function. Also, by Lemma 3.1 of [**GVLM**], the (Zariski) locally trivial fibration $E_*^\circ \to E_I^\circ$ (with fiber $\mathbb{P}^1 \setminus \{2 \text{ points}\} = \mathbb{C}^*$) can be lifted to a $\mathbb{C}^*$-fibration $\widetilde{E_*^\circ} \to \widetilde{E_I^\circ}$. Finally, the Zariski triviality implies that $[\widetilde{E_*^\circ}, \sigma_{\Delta'}] = [\widetilde{E_I^\circ}, \sigma_\Delta](\mathbb{L}-1)$, which reduces the above claim to checking that

$$[\widetilde{E_I^\circ}](\mathbb{L}-1)\frac{A_1}{1-A_1}\frac{A_2}{1-A_2} = [\widetilde{E_I^\circ}](\mathbb{L}-1)\frac{A_*}{1-A_*}\left(1 + \frac{A_1}{1-A_1} + \frac{A_2}{1-A_2}\right). \tag{3.13}$$

After we factor out $[\widetilde{E_I^\circ}, \sigma_\Delta](\mathbb{L}-1)$ and reduce to common denominator, this amounts to

$$A_1A_2(1-A_*) = A_*(1-A_1)(1-A_2) + A_*A_1(1-A_2) + A_*(1-A_1)A_2,$$

or equivalently, $A_1A_2 = A_*$. The latter equality follows from $m_* = m_1 + m_2$ and $\nu_* = \nu_1 + \nu_2$. □

PROOF. (of Proposition 3.16)
The proof below is an enhanced version of (and relies on) that of [**GVLM**, Proposition 3.8], and we will use freely facts proved in loc.cit.

As already mentioned, the proof of Proposition 3.16 is by induction on the dimension of the center of blow-up. The beginning of induction (i.e., the case of one point) is proved in Proposition 3.21. Note that, in general, the center of blow-up Z is either contained in a component E_i of E, or it is transversal to it, or it doesn't intersect it at all. We refer to components of the second kind as *transversal* components of E (with respect to Z). By collecting all indices i of components of E containing Z, we note that the center Z is contained in a set E_I (for some $I \subseteq J$) given by intersections of components of the deleted divisor. In particular, Z gets an induced stratification from that of E_I. So, there is a dense open stratum $Z \cap E_I^\circ$ in Z, together with positive codimension strata obtained by intersecting Z with collections of transversal components.

We begin the proof by first studying the case when the center of blowup is of type E_I, for some $I \subseteq J$. Let X' be the blowup of X along the center Z defined as the intersection $E_I := \bigcap_{i=1}^k E_i$ of some of the irreducible components of the deleted divisor E (so $I = \{1, 2, \cdots, k\}$), and also assume that the irreducible components E_j for $j = k+1, \ldots, \ell$ of E intersect the center Z transversally, and no other components of E intersect Z. In this case, Z is stratified by a top dimensional open dense stratum E_I°, and by positive codimension strata obtained by intersecting Z with some of the transversal components E_j (with $j = k+1, \ldots, \ell$), i.e., strata of the form $E_{I\cup K}^\circ$, where $K \neq \emptyset$ and $K \subseteq \{k+1, \ldots, \ell\}$. Therefore, the contributions to the motivic infinite cyclic zeta function $Z_{X,E}(T)$ supported on Z are
(3.14)

$$\prod_{i=1}^{k} \frac{A_i}{1-A_i} \left([\widetilde{E_I^\circ}](\mathbb{L}-1)^{k-1} + \sum_{\substack{K \subseteq \{k+1,\ldots,\ell\} \\ K \neq \emptyset}} [\widetilde{E_{I\cup K}^\circ}](\mathbb{L}-1)^{k+|K|-1} \prod_{h\in K} \frac{A_h}{1-A_h} \right).$$

After blowing up X along Z, we get the deleted divisor $E' = (\bigcup_{j\in J} E_j') \cup E_*$ of $X' = Bl_Z X$, where E_* is the exceptional locus of the blowup and E_j' is the proper transform of E_j (for $j \in J$). Note that, by the choice of the center Z of blow-up, the k-fold intersection of the proper transforms of components E_i with $i = 1, \ldots, k$ is empty, i.e., $\bigcap_{i=1}^k E_i' = \emptyset$. The exceptional divisor E_* is stratified by the top dimensional open stratum $L_\emptyset = E_*^\circ$, by the codimension s (for $s < k$) strata obtained by intersecting E_* with s-fold intersections of the components $E_1', \ldots, E_k'$ of E', i.e., by the strata L_G with $G \subset I$ is a proper subset of I, and by strata contained in intersections of the proper transforms E_j' for $j = k+1, \ldots, \ell$ of the transversal components, i.e., strata of the type $L_{G\cup K}$ where $G \subset I$ $(G \neq I)$ a (possibly empty) subset and K is a nonempty subset of $\{k+1, \ldots, \ell\}$. Therefore the contributions to the motivic infinite cyclic zeta function $Z_{X',E'}(T)$ supported on E_* are:

$$(3.15)\quad \frac{A_*}{1-A_*}\left([\widetilde{E_*^\circ}] + \sum_{\substack{G\subset I,\\ G\neq\emptyset,I}} [\widetilde{L_G}](\mathbb{L}-1)^{|G|}\prod_{g\in G}\frac{A_g}{1-A_g} + \right.$$
$$\left.\sum_{\substack{G\subset I,\\ G\neq I}}\ \sum_{\substack{K\subseteq\{k+1,\dots,\ell\}\\ K\neq\emptyset}} [\widetilde{L_{G\cup K}}](\mathbb{L}-1)^{|K|+|G|}\prod_{h\in G\cup K}\frac{A_h}{1-A_h}\right).$$

We can now apply induction on the dimension of the center of blowup, and the exclusion-inclusion principle, to show that strata of the center Z which are contained in intersections of the transversal components E_j, for $j=k+1,\cdots,\ell$, give equal contributions to the zeta functions $Z_{X,E}(T)$ and $Z_{X',E'}(T)$. More precisely, for each positive codimension stratum $E^\circ_{I\cup K}$ (with $K\subseteq\{k+1,\dots,\ell\}, K\neq\emptyset$) of Z, we get by induction for the blow-up of E_K along the center $Z\cap E_K=E_{I\cup K}$, and with deletion divisor $E_K\cap(\sum_{i=1}^k E_i)$, a relation of the type

$$(*_K)\quad (\mathbb{L}-1)^{k-1}\prod_{i=1}^k\frac{A_i}{1-A_i}$$
$$\cdot\left([\widetilde{E^\circ_{I\cup K}}] + \sum_{\substack{K\subset K'\subseteq\{k+1,\dots,\ell\}\\ K'\setminus K\neq\emptyset}} [\widetilde{E^\circ_{I\cup K'}}](\mathbb{L}-1)^{|K'\setminus K|}\prod_{h\in K'\setminus K}\frac{A_h}{1-A_h}\right)$$
$$=\frac{A_*}{1-A_*}\left([\widetilde{L_K}] + \sum_{\substack{K\subset K'\subset\{k+1,\dots,\ell\}\\ K'\setminus K\neq\emptyset}} [\widetilde{L_{K'}}](\mathbb{L}-1)^{|K'\setminus K|}\prod_{h\in K'\setminus K}\frac{A_h}{1-A_h}\right.$$
$$+\sum_{\substack{G\subset I,\\ G\neq\emptyset,I}} [\widetilde{L_{G\cup K}}](\mathbb{L}-1)^{|G|}\prod_{g\in G}\frac{A_g}{1-A_g}$$
$$\left.+\sum_{\substack{G\subset I,\\ G\neq\emptyset,I}}\ \sum_{\substack{K\subset K'\subset\{k+1,\dots,\ell\}\\ K'\setminus K\neq\emptyset}} [\widetilde{L_{G\cup K'}}](\mathbb{L}-1)^{|G\cup(K'\setminus K)|}\prod_{g\in G}\frac{A_g}{1-A_g}\prod_{h\in K'\setminus K}\frac{A_h}{1-A_h}\right).$$

By summing up all the products $(*_K)\cdot(\mathbb{L}-1)^{|K|}\prod_{h\in K}\frac{A_h}{1-A_h}$ for the positive codimension strata $E^\circ_{I\cup K}$ of Z, we reduce the comparison of (3.14) and (3.15) to proving the identity:
(3.16)

$$[\widetilde{E_I^\circ}](\mathbb{L}-1)^{k-1}\prod_{i=1}^k\frac{A_i}{1-A_i}=\frac{A_*}{1-A_*}\left([\widetilde{E_*^\circ}] + \sum_{\substack{G\subset I,\\ G\neq\emptyset,I}} [\widetilde{L_G}](\mathbb{L}-1)^{|G|}\prod_{g\in G}\frac{A_g}{1-A_g}\right),$$

i.e., it remains to show that the contribution of the dense open stratum of the center Z to the zeta function $Z_{X,E}(T)$ coincides with the contribution to $Z_{X',E'}(T)$ of the strata supported on the exceptional divisor E_* and which are not contained in the proper transforms of the transversal components of E.

Note that in the proof of Proposition 3.8 in [**GVLM**] we have shown that, for any subset $G \subset I = \{1, \ldots, k\}$, $G \neq I$ (but including the empty set corresponding to $L_\emptyset = E_*^\circ$), we have the following identification:

$$[\widetilde{L_G}] = [\widetilde{E_I^\circ}](\mathbb{L} - 1)^{k-|G|-1}.$$

Therefore we can factor out $[\widetilde{E_I^\circ}](\mathbb{L} - 1)^{k-1}$ in (3.16) and we are left to check that

$$\prod_{i=1}^{k} \frac{A_i}{1 - A_i} = \frac{A_*}{1 - A_*} \left(1 + \sum_{\substack{G \subset I, \\ G \neq \emptyset, I}} \prod_{g \in G} \frac{A_g}{1 - A_g} \right).$$

The last claim follows from Lemma 3.23. Indeed, it reduces to $\prod_{i=1}^{k} A_i = A_*$, which holds since $m_* = \sum_{i \in I} m_i$ and $\nu_* = \sum_{i \in I} \nu_i$. (Here $\operatorname{codim}_{E_I} Z = 0$.)

Let us now explain the proof in the general case, i.e., when the center Z is strictly contained in some set E_I, for $I \subseteq J$, and let $I = \{1, \cdots, k\}$. Assume that the codimension of Z in X is $d + 1 \geq k$. Therefore the codimension of Z in E_I is $d + 1 - k$. Again, by induction, it suffices to show that the contribution of the dense open stratum $Z^\circ := Z \cap E_I^\circ$ of the center Z to the zeta function $Z_{X,E}(T)$ coincides with the contribution to $Z_{X',E'}(T)$ of the strata supported on the exceptional divisor $E_* = \mathbb{P}(\nu_Z)$ which are not contained in the proper transforms of the transversal components components of E (with respect to Z), that is,

(3.17)
$$[\widetilde{Z^\circ}](\mathbb{L} - 1)^{k-1} \prod_{i=1}^{k} \frac{A_i}{1 - A_i} = \frac{A_*}{1 - A_*} \left([\widetilde{E_*^\circ}] + \sum_{\substack{G \subseteq I, \\ G \neq \emptyset}} [\widetilde{L_G}](\mathbb{L} - 1)^{|G|} \prod_{g \in G} \frac{A_g}{1 - A_g} \right).$$

On the right hand side of (3.17), we use the same notation as before for the stratification of the exceptional divisor E_*. Note that in this case we have to also allow $G = I$ in the sum of the right hand side term of (3.17) because Z is strictly contained in E_I, and therefore $\cap_{i=1}^{k} E_i' \neq \emptyset$.

Recall that in the proof of Proposition 3.8 in [**GVLM**] we have showed that, for any subset $G \subset I = \{1, \ldots, k\}$, $G \neq I$ (but including the empty set corresponding to $L_\emptyset = E_*^\circ$), we have:

$$[\widetilde{L_G}] = [\widetilde{Z^\circ}]\mathbb{L}^{d-k+1}(\mathbb{L} - 1)^{k-|G|-1}. \tag{3.18}$$

Moreover, for $G = I$, we have proved that

$$[\widetilde{L_I}] = [\widetilde{Z^\circ}](\mathbb{L}^{d-k} + \mathbb{L}^{d-k-1} + \cdots + \mathbb{L} + 1). \tag{3.19}$$

By substituting the equalities (3.18) and (3.19) into (3.17), and factoring out $[\widetilde{Z}^\circ](\mathbb{L}-1)^{k-1}$, it remains to show that:

$$(3.20)\quad \prod_{i=1}^{k}\frac{A_i}{1-A_i}$$
$$=\frac{A_*}{1-A_*}\left(\mathbb{L}^{d-k+1}\Big(1+\sum_{\substack{G\subset I,\\ G\neq\emptyset,I}}\prod_{g\in G}\frac{A_g}{1-A_g}\Big)+(\mathbb{L}^{d-k+1}-1)\prod_{i=1}^{k}\frac{A_i}{1-A_i}\right).$$

Reducing to the common denominator $\prod_{i=1}^{k}(1-A_i)(1-A_*)$, it is then enough to check that

$$\prod_{i=1}^{k}A_i(1-A_*)$$
$$=A_*\mathbb{L}^{d-k+1}\left(\prod_{i=1}^{k}(1-A_i)+\sum_{\substack{G\subseteq I\\ G\neq\emptyset}}\prod_{g\in G}A_g\prod_{j\in I\setminus G}(1-A_j)\right)-A_*\prod_{i=1}^{k}A_i.$$

By using Lemma 3.23, this can be further reduced to showing that $\prod_{i=1}^{k}A_i=A_*\mathbb{L}^{d-k+1}$, which holds since $m_*=m_1+\cdots+m_k$ and $\nu_*=\nu_1+\cdots+\nu_k+d-k+1$. (Here $\operatorname{codim}_{E_I}Z=d-k+1$). □

REMARK 3.25. The proof of Proposition 3.16 works whenever (3.3) and (3.8) are defined. Remarks 3.10 and 3.15 explain that it is sufficient (although not necessary) to assume that the holonomy takes strictly positive values. An alternative sufficient hypothesis is that either $\nu_i\neq 0$ or $m_i\neq 0$ for all $i\in J\cup\{*\}$. The rest of the proof does not assume that the holonomy map takes always strictly positive values. In fact, the proof shows that the expression defining $Z^A_{X,E,\Delta,\nu}(T)$ in (3.3) is well-defined in the sense of the Definition 3.5 if and only if the holomony and log-discrepancy values satisfy the correct blow-up relations given in (2.3) and (3.2) and never vanish simultaneously.

As a consequence, Theorem 3.14 holds as long as either $\nu_i\neq 0$ or $m_i\neq 0$ for all irreducible components of E and their transforms along any sequence of blow-ups and blow-downs.

In [**Ve03**], Veys has used the condition that either $\nu_i\neq 0$ or $m_i\neq 0$ to introduce *naive motivic zeta functions* in the cases of a Kawamata log terminal and divisorial log terminal (dlt) pair (X,E) and in the case that the pair (X,E) has no strictly log canonical singularities assuming the existence of a log relative minimal model of (X,E) which is dlt. See Section 5, in particular (5.2) and (5.1), for an explanation of the adjective "naive".

4. Relation with the Denef-Loeser motivic zeta function

Denef and Loeser introduced the *local motivic zeta function* at a point x for a non-constant morphism $f : \mathbb{C}^{d+1} \to \mathbb{C}$ with $f(x) = 0$ (e.g., see [**DL01**, Definition 3.2.1] and the references therein) as

$$Z_{f,x}(T) := \sum_{n\geq 1} [\mathcal{X}_{n,1}]\mathbb{L}^{-(d+1)n}T^n \in \mathcal{M}_{\mathbb{C}}^{\hat{\mu}}[[T]], \tag{4.1}$$

where $\mathcal{X}_{n,1}$ denotes the set of $(n+1)$-jets φ of $\mathbb{C}^{d+1}$ such that $f \circ \varphi = t^n + \dots$. Note that there is a good action of the group μ_n (hence of $\hat{\mu}$) on $\mathcal{X}_{n,1}$ by $\lambda \times \varphi \mapsto \varphi(\lambda \cdot t)$. Moreover, Denef and Loeser [**DL01**, Theorem 3.3.1] proved with the help of an embedded resolution of the singularities of f that the motivic zeta function $Z_{f,x}(T)$ belongs to $\mathcal{M}_{\mathbb{C}}^{\hat{\mu}}[[T]]_{\mathrm{sr}}$. Then they also defined the *local motivic Milnor fibre* $\mathcal{S}_{f,x}$ (e.g., see [**DL01**, Definition 3.5.3] and the references therein) as a limit in the sense of (3.5):

$$\mathcal{S}_{f,x} := -\lim_{T\to+\infty} Z_{f,x}(T) \in \mathcal{M}_{\mathbb{C}}^{\hat{\mu}}, \tag{4.2}$$

see also [**GLM**, Section 2.8] and [**DL98**, Lemma 4.1.1].

In [**GVLM**, Theorem 5.1] we related the concepts of local motivic Milnor fibre $\mathcal{S}_{f,x}$ and the motivic infinite cyclic cover $S_{X,E,\Delta}$. The following result extends this relation to the realm of motivic zeta function and the motivic infinite cyclic zeta function, respectively.

THEOREM 4.1. *Let $f : \mathbb{C}^{d+1} \to \mathbb{C}$ be a non constant morphism with $f(x) = 0$, and $p : X \to \mathbb{C}^{d+1}$ be a log-resolution of the singularities of the pair $(\mathbb{C}^{d+1}, f^{-1}(0))$. Choose p so that $(p^{-1}(x))_{\mathrm{red}}$ is a union of components of $(p^{-1}(f^{-1}(0)))_{\mathrm{red}}$. Let $E = \sum_{j\in J} E_j$ be the decomposition of $(p^{-1}(f^{-1}(0)))_{\mathrm{red}}$ into irreducible components, and let $A = \{i \in J \mid E_i \subset p^{-1}(x)\}$. Then the following hold:*

(1) *For $\epsilon > 0$ small enough, and $B(x, \epsilon)$ a ball of radius ϵ centered at $x \in \mathbb{C}^{d+1}$, the resolution map p provides a biholomorpic identification between $B(x, \epsilon) \setminus \{f = 0\}$ and $T^*_{E^A}$, the punctured regular neighborhood of the divisor $E^A := \sum_{i\in A} E_i$. In particular, the map $\gamma \to \int_\gamma \frac{df}{f}$ can be viewed as a holonomy homomorphism $\Delta : \pi_1(T^*_{E^A}) \to \mathbb{Z}$ of the punctured neighborhood of E^A. This holonomy map takes the boundary δ_i of any small disk transversal to the irreducible component E_i of E^A to the multiplicity m_i of E_i in the divisor of $f \circ p$, i.e., $\Delta(\delta_i) = m_i$ for all $i \in A$.*

(2) *The map taking the boundary δ_i of any small disk transversal to the irreducible component E_i of E^A, to the multiplicity of E_i in the divisor of the pullback of the standard volume form $\omega = dx_1 \cdots dx_{d+1}$ plus one, defines a log-discrepancy ν, i.e., $\nu_i - 1 = \mathrm{ord}_{E_i} p^*\omega$.*

(3) *One has the identity in $\mathcal{M}_{\mathbb{C}}^{\hat{\mu}}[[T]]_{\mathrm{sr}}$:*

$$Z_{f,x}(T) = Z^A_{X,E,\Delta,\nu}(T).$$

PROOF. The result follows from Theorem 5.1 in [**GVLM**] and [**GLM**, Section 2.8], see also [**DL98**, Lemma 4.1.1]. □

REMARK 4.2. Note that Theorems 4.1 and 3.14 give a direct argument of the fact that the right-hand side of the formula in [**DL01**, Theorem 3.3.1] expressing the motivic zeta function in terms of the data of a log resolution is in fact independent of the choice of log resolution. This was a priori known only because of the relation (4.1) defining the motivic zeta function in terms of arc spaces (which are intrinsic invariants), see also the discussion in [**DL01**, Section 3.5]. Moreover, we do not make use of the third relation in the definition of the equivariant Grothendieck ring for proving our result.

5. Further specializations of the motivic infinite cyclic zeta function

In the previous section, we discussed the specialization of motivic infinite cyclic zeta function to the motivic infinite cyclic cover $S^A_{X,E,\Delta}$. The remarks after Definition 3.11 explain how to recover $S^A_{X,E,\Delta}$ from $Z^A_{X,E,\Delta,\nu}(T)$. In our previous work we proved that in general the Betti realization of $S^A_{X,E,\Delta}$ is given (up to sign) by the cohomology with compact support of the infinite cyclic cover of finite type $\widetilde{T}^*_{X,E,\Delta}$ of the punctured neighborhood $T^*_{X,E}$, see [**GVLM**, Proposition 4.2]. We also hinted that there should be a stronger version of this result involving the Hodge realization of $S^A_{X,E,\Delta}$, see [**GVLM**, Remark 4.3]. In this section, we discuss further specializations of our motivic infinite zeta functions.

For any strictly positive integer e and any character α of order e we can use the equivariant Euler characteristic $\chi_{\mathrm{top}}(-,\alpha)$ to introduce a *twisted topological zeta function* as

$$\chi_{\mathrm{top}}\left((\mathbb{L}-1)Z^A_{X,E,\Delta,\nu}(\mathbb{L}^{-s}),\alpha\right).$$

One can also consider p-adic realizations as in [**DL98**, Section 2.4], see also [**Ni10**, Section 5.3].

One can also introduce a *naive motivic infinite cyclic zeta function* $Z^{\mathrm{naive},A}_{X,E,\Delta,\nu}(T)$ as

$$\sum_{\substack{\emptyset\neq I\subseteq J\\ A\cap I\neq\emptyset}}[E_I^\circ](\mathbb{L}-1)^{|I|}\prod_{i\in I}\frac{\mathbb{L}^{-\nu_i}T^{m_i}}{1-\mathbb{L}^{-\nu_i}T^{m_i}}\in\mathcal{M}_{\mathbb{C}}[[T]]_{\mathrm{sr}}. \tag{5.1}$$

It is easy to check that under the hypothesis of Theorem 4.1 the zeta function $Z^{\mathrm{naive},A}_{X,E,\Delta,\nu}(T)$ coincides with the naive motivic zeta function of Denef and Loeser, which is given by

$$Z^{\mathrm{naive}}_{f,x}(T):=\sum_{n\geq 1}[\mathcal{X}_n]\mathbb{L}^{-(d+1)n}T^n\in\mathcal{M}_{\mathbb{C}}[[T]], \tag{5.2}$$

where $\mathcal{X}_n$ denotes the set of $(n+1)$-jets φ of $\mathbb{C}^{d+1}$ such that $f\circ\varphi=at^n+\dots$, for some $a\neq 0$.

Hartmann [**Ha16**, Corollaries 8.5 and 8.6] has introduced a $\mathcal{M}_{\mathbb{C}}$-linear quotient map $\mathcal{M}^{\mu}_{\mathbb{C}}\to\mathcal{M}_{\mathbb{C}}$ that allows to recover $Z^{\mathrm{naive},A}_{X,E,\Delta,\nu}(T)$ from $Z^A_{X,E,\Delta,\nu}(T)$. The naive motivic infinite cyclic cover equals

$$(\mathbb{L}-1)Z^A_{X,E,\Delta,\nu}(T)/\hat{\mu}\in\mathcal{M}_{\mathbb{C}},$$

where $Z^A_{X,E,\Delta,\nu}(T)/\hat{\mu}$ denotes the image under the extension to $\mathcal{M}^{\hat{\mu}}_{\mathbb{C}}[[T]]\to\mathcal{M}_{\mathbb{C}}[[T]]$ of the above mentioned map. See also [**Ha15**, Remarks 7.1 and 7.2].

There are two further specializations of the naive infinite cyclic motivic zeta function worth mentioning. Firstly, there is a *Hodge infinite cyclic cover zeta function* defined as

$$\mathcal{H}^A_{X,E,\Delta,\nu}(T) := \sum_{\substack{\emptyset\neq I\subseteq J\\ A\cap I\neq\emptyset}} H(E_I^\circ;u,v)\prod_{i\in I}\frac{(uv-1)T^{m_i}}{(uv)^{\nu_i}-T^{m_i}} \in \mathbb{Q}(u,v)[[T]],$$

where $H(-;u,v)$ is the Hodge-Deligne polynomial. Compare with [**SV09**, Section 2.2]. Secondly, there is the topological Euler realization $Z^{\mathrm{top},A}_{X,E,\Delta,\nu}(s)$ given by

$$\chi_{\mathrm{top}}(Z^{\mathrm{naive},A}_{X,E,\Delta,\nu}(\mathbb{L}^{-s})) = \sum_{\substack{\emptyset\neq I\subseteq J\\ A\cap I\neq\emptyset}} \chi_{\mathrm{top}}(E_I^\circ)\prod_{i\in I}\frac{1}{\nu_i+sm_i} \in \mathbb{Q}(s),$$

where $\chi_{\mathrm{top}}(-)$ denotes the topological Euler characteristic. It is again immediate to check that in the case of Theorem 4.1 this coincides with the topological zeta function introduced by Denef and Loeser.

Finally, Veys [**Ve03**] has investigated how to use the theory of motivic zeta functions to generalize the theory of E-stringy invariants of Batyrev beyond the log terminal case. Schepers and Veys showed in [**SV09**, Proposition 2.3], as a consequence of the inversion of the adjunction of Ein, Mustaţa and Yasuda [**EMY**, Theorem 1.6], that under the hypothesis of Theorem 4.1 and assuming that $f^{-1}(0)$ is irreducible normal (and canonical), the local stringy E-function of the hypersurface $f^{-1}(0)$ can be recovered as a *residue* of the local Hodge zeta function, i.e., the local stringy E-function can be obtained as:

$$-\frac{1}{uv(uv-1)}\mathcal{H}^A_{X,E,\Delta,\nu}(T)(T-uv)|_{T=uv}.$$

6. Motivic zeta functions for arbitrary divisors. Relation to the monodromy conjecture

The framework of the zeta functions associated to infinite cyclic covers seems to be useful for the study of the monodromy conjecture and some of its generalizations (like for example the generalized monodromy conjecture proposed by Némethi and Veys in [**Ve07, NV10, NV12**]).

The monodromy conjecture predicts a relation between the poles of the local zeta function $Z_{f,x}(T)$ and the eigenvalues of local monodromies (i.e., monodromies of local Milnor fibers at points) of the hypersurface defined by f. However, it is worth to notice here that the Grothendieck ring is not a domain and therefore one should be careful when speaking of poles. The *Igusa-Denef-Loeser motivic monodromy conjecture* can be formulated as follows (see for example [**DL98**, Remark at the end of Section 2.4] and [**Lo09**, Section 3.4.1]).

CONJECTURE 6.1. *There exists a finite subset S of $\mathbb{Z}_{>0}\times\mathbb{Z}_{>0}$ such that*

$$Z_{f,x}(T)\in\mathcal{M}^{\hat{\mu}}_{\mathbb{C}}\left[T,\frac{1}{1-\mathbb{L}^{-a}T^b}\right]_{(a,b)\in S}$$

and such that for each $(a,b)\in S$, the value $\exp(-2\pi ib/a)$ is a monodromy eigenvalue of f at some point of $f^{-1}(0)$.

There are other, naive (see [**Ve06**, Section 6.8]), topological (see [**DL92**, Conjecture 3.3.2]) or p-adic (see [**Ig88**]) versions of the monodromy conjecture that come from considering different specializations of the zeta function, like the ones mentioned in the previous section. Naive, topological and/or p-adic versions of the monodromy conjecture have been proven in, among others, the case of curves by Loeser, certain surfaces by Veys, Rodrigues, Lemahieu, Van Proeyen, Bories, superisolated singularities and quasi-ordinary singularities by Artal-Bartolo, Cassou-Nogués, Luengo and Melle-Hernández, and hyperplane arrangements by Budur, Mustaţa and Teitler. The Igusa-Denef-Loeser motivic monodromy conjecture is different from the naive one as explained in [**Ca16**, Section 4 and Subsection 5.2]. Nicaise and Veys have informed us [**NV15**] that the Igusa-Denef-Loeser motivic monodromy conjecture can be checked in the case of plane curves (though this fact is not available in the literature).

One of the first difficulties in the study of the monodromy conjecture is the computation of the set S, whose elements are called *poles*. Explicit formulae for the zeta function in terms of an embedded resolution of singularities (see for instance [**DL01**, Theorem 3.3.1]) show that any exceptional divisor gives a candidate pole (a, b). But concrete computations also show the existence of many cancellations and that most irreducible exceptional divisors do not contribute to S.

On the other hand, typically only a few eigenvalues are recovered from the set S. Moreover, those eigenvalues coming from S cannot be determined before-hand.

Némethi and Veys have proposed an alternative approach in [**Ve07, NV10, NV12**] to study the topological Monodromy Conjecture in the case of singularities of plane curves. By considering a finite family of *allowed* differential forms associated to f (see [**NV12**, Section 1.5.3 and Definition 4.1.1] for more details), they obtain a finite family of topological zeta functions such that their corresponding sets of poles recover all (and only) the eigenvalues of f at x. This works because, while the first coordinate of each element of S is totally determined by f, the second coordinate depends on the differential form. Moreover, Némethi and Veys show that arbitrary (i.e., not allowed) differential forms may produce poles (a, b') such that $\exp(-2\pi i b'/a)$ is not an eigenvalue of the local monodromies of $f^{-1}(0)$.

Cauwbergs has extended the approach of Némethi and Veys in the case of naive motivic zeta function [**Ca16**, Corollary 5.2]. However, he has also shown in [**Ca16**, Section 5.2], using twisted topological zeta functions as specializations of the (non naive) motivic zeta functions, that there is no set of allowed forms able to recover all eigenvalues from the poles of the (non naive) motivic zeta function.

Let us conclude this note by suggesting a version of the monodromy conjecture in the case of the infinite cyclic covers, as considered in this paper. Let D be any divisor on a smooth quasi-projective variety Y, with $T^*_{Y,D}$ a punctured neighborhood of D in Y. Under suitable assumptions, we can associate a motivic zeta function to the pair (Y, D) as follows. Fix, as before, a holonomy map $\Delta : \pi_1(T^*_{Y,D}) \to \mathbb{Z}$, and let $\widetilde{T}^*_{Y,D}$ be the corresponding infinite cyclic cover. Let (X, E) be a log resolution of (Y, D), with log discrepancies ν_i, $i \in J$, defined as usual in terms of the relative canonical divisor of the resolution map. Assume moreover that the induced holonomy (denoted also by Δ) on the punctured neighborhood $T^*_{X,E}$ of E in X makes the infinite cyclic cover $\widetilde{T}^*_{X,E}$ (and hence $\widetilde{T}^*_{Y,D}$) of finite type. Then the Alexander modules $H^i(\widetilde{T}^*_{Y,D}; \mathbb{C})$ $(i \geq 0)$ are torsion over the Laurent polynomial ring $\mathbb{C}[t^{\pm 1}]$, with the t-multiplication corresponding to the action of a generator of the covering group $\mathbb{Z}$ (e.g., see [**GVLM**, Proposition 2.4 and Remark 2.5]). The order of the torsion module $H^i(\widetilde{T}^*_{Y,D};\mathbb{C})$ is called the i-th Alexander polynomial of $T^*_{Y,D}$, and will be denoted by $\delta_i(t)$.

The monodromy zeta function $Z_{Y,D,\Delta}(T)$ of the infinite cyclic cover $\widetilde{T}^*_{Y,D}$ can then be defined by formula (3.3) of Definition 3.9. According to Theorem 3.14, this is a well-defined invariant of the punctured neighborhood of D in Y.

It is then interesting to compare the poles of the motivic zeta function $Z_{Y,D,\Delta}(T)$ with the zeros of the Alexander polynomials $\delta_i(t)$ $(i \geq 0)$ of the punctured neighborhood $T^*_{Y,D}$ of D in Y. It was already noted in [**GVLM**] that the zeros of $\delta_i(t)$ $(i \geq 0)$ are roots of unity. In fact, one would be tempted to believe that the following claim is true:

If (a,b) *is a pole of the motivic zeta function* $Z_{Y,D,\Delta}(T)$ *in the sense of Conjecture* 6.1, *then* $\exp(-2\pi ib/a)$ *is a zero of some Alexander polynomial* $\delta_i(t)$ $(i \geq 0)$ *of the punctured neighborhood of* D.

The motivation behind such a claim goes as follows. Let us assume that a linking number can be defined on the complement of D, e.g., D is homologically trivial in a simply-connected algebraic manifold Y, and let Δ be the holonomy defined by the linking number homomorphism. If, moreover, D is a Cartier divisor, then an argument involving a Mayer-Vietoris spectral sequence can be used to show that any root of an Alexander polynomial $\delta_i(t)$ of the punctured neighborhood of D in Y is also an eigenvalue of some monodromy operator acting on the local Milnor fiber (with respect to a local defining equation of D) at some point on D. When coupled with the above claim (assuming it holds true), this fact would then imply a version of the Igusa-Denef-Loeser monodromy Conjecture 6.1. For a different variant of Conjecture 6.1, see [**Xu16**].

We should also point out that in the case of plane curves the above claim is actually equivalent to the Conjecture 6.1, as can be easily seen by using a Mayer-Vietoris long exact sequence. Hence our claim is true in this case by [**NV15**], as already indicated above. Nevertheless, in general, the collection of zeros of the Alexander polynomials $\delta_i(t)$ $(i \geq 0)$ of the punctured neighborhood is (stricly) contained in the set of eigenvalues of the local monodromies at points along the divisor D. So, in some sense, the above claim complements the Némethi-Veys approach mentioned earlier.

Acknowledgment. The authors are grateful to the anonymous referee for carefully reading the manuscript, and for his valuable comments and constructive suggestions.

References

[AKMW] Dan Abramovich, Kalle Karu, Kenji Matsuki, and Jarosław Włodarczyk, *Torification and factorization of birational maps*, J. Amer. Math. Soc. **15** (2002), no. 3, 531–572. MR1896232

[Bo02] Laurent Bonavero, *Factorisation faible des applications birationnelles (d'après Abramovich, Karu, Matsuki, Włodarczyk et Morelli)* (French, with French summary), Astérisque **282** (2002), Exp. No. 880, vii, 1–37. Séminaire Bourbaki, Vol. 2000/2001. MR1975173

[Ca16] Thomas Cauwbergs, *Splicing motivic zeta functions*, Rev. Mat. Complut. **29** (2016), no. 2, 455–483. MR3490596

[DL92] J. Denef and F. Loeser, *Caractéristiques d'Euler-Poincaré, fonctions zêta locales et modifications analytiques* (French), J. Amer. Math. Soc. **5** (1992), no. 4, 705–720. MR1151541

[DL98] Jan Denef and François Loeser, *Motivic Igusa zeta functions*, J. Algebraic Geom. **7** (1998), no. 3, 505–537. MR1618144

[DL01] Jan Denef and François Loeser, *Geometry on arc spaces of algebraic varieties*, European Congress of Mathematics, Vol. I (Barcelona, 2000), Progr. Math., vol. 201, Birkhäuser, Basel, 2001, pp. 327–348. MR1905328

[DiL06] Alexandru Dimca and Anatoly Libgober, *Regular functions transversal at infinity*, Tohoku Math. J. (2) **58** (2006), no. 4, 549–564. MR2297199

[DN04] Alexandru Dimca and András Némethi, *Hypersurface complements, Alexander modules and monodromy*, Real and complex singularities, Contemp. Math., vol. 354, Amer. Math. Soc., Providence, RI, 2004, pp. 19–43. MR2087802

[EMY] Lawrence Ein, Mircea Mustaţă, and Takehiko Yasuda, *Jet schemes, log discrepancies and inversion of adjunction*, Invent. Math. **153** (2003), no. 3, 519–535. MR2000468

[GVLM] Manuel González Villa, Anatoly Libgober, and Laurenţiu Maxim, *Motivic infinite cyclic covers*, Adv. Math. **298** (2016), 413–447. MR3505746

[GLM] Gil Guibert, François Loeser, and Michel Merle, *Iterated vanishing cycles, convolution, and a motivic analogue of a conjecture of Steenbrink*, Duke Math. J. **132** (2006), no. 3, 409–457. MR2219263

[Ha15] A. Hartmann, *Equivariant motivic integration on formal schemes and the motivic zeta function*, arXiv:1511.08656.

[Ha16] Annabelle Hartmann, *The quotient map on the equivariant Grothendieck ring of varieties*, Manuscripta Math. **151** (2016), no. 3-4, 419–451. MR3556826

[Ig88] Jun-ichi Igusa, *b-functions and p-adic integrals*, Algebraic analysis, Vol. I, Academic Press, Boston, MA, 1988, pp. 231–241. MR992457

[KM98] János Kollár and Shigefumi Mori, *Birational geometry of algebraic varieties*, Cambridge Tracts in Mathematics, vol. 134, Cambridge University Press, Cambridge, 1998. With the collaboration of C. H. Clemens and A. Corti; Translated from the 1998 Japanese original. MR1658959

[Li82] A. Libgober, *Alexander polynomial of plane algebraic curves and cyclic multiple planes*, Duke Math. J. **49** (1982), no. 4, 833–851. MR683005

[Li94] A. Libgober, *Homotopy groups of the complements to singular hypersurfaces. II*, Ann. of Math. (2) **139** (1994), no. 1, 117–144. MR1259366

[Lo09] François Loeser, *Seattle lectures on motivic integration*, Algebraic geometry—Seattle 2005. Part 2, Proc. Sympos. Pure Math., vol. 80, Amer. Math. Soc., Providence, RI, 2009, pp. 745–784. MR2483954

[Ma06] Laurentiu Maxim, *Intersection homology and Alexander modules of hypersurface complements*, Comment. Math. Helv. **81** (2006), no. 1, 123–155. MR2208801

[Mi67] John W. Milnor, *Infinite cyclic coverings*, Conference on the Topology of Manifolds (Michigan State Univ., E. Lansing, Mich., 1967), Prindle, Weber & Schmidt, Boston, Mass., 1968, pp. 115–133. MR0242163

[NV10] András Némethi and Willem Veys, *Monodromy eigenvalues are induced by poles of zeta functions: the irreducible curve case*, Bull. Lond. Math. Soc. **42** (2010), no. 2, 312–322. MR2601558

[NV12] András Némethi and Willem Veys, *Generalized monodromy conjecture in dimension two*, Geom. Topol. **16** (2012), no. 1, 155–217. MR2872581

[Ni10] Johannes Nicaise, *An introduction to p-adic and motivic zeta functions and the monodromy conjecture*, Algebraic and analytic aspects of zeta functions and L-functions, MSJ Mem., vol. 21, Math. Soc. Japan, Tokyo, 2010, pp. 141–166. MR2647606

[NV15] J. Nicaise, W. Veys, Private communication, 2015.

[Ro76] Dale Rolfsen, *Knots and links*, Publish or Perish, Inc., Berkeley, Calif., 1976. Mathematics Lecture Series, No. 7. MR0515288

[SV09] Jan Schepers and Willem Veys, *Stringy E-functions of hypersurfaces and of Brieskorn singularities*, Adv. Geom. **9** (2009), no. 2, 199–217. MR2523840

[Ve03] Willem Veys, *Stringy zeta functions for $\mathbb{Q}$-Gorenstein varieties*, Duke Math. J. **120** (2003), no. 3, 469–514. MR2030094

[Ve06] Willem Veys, *Arc spaces, motivic integration and stringy invariants*, Singularity theory and its applications, Adv. Stud. Pure Math., vol. 43, Math. Soc. Japan, Tokyo, 2006, pp. 529–572. MR2325153

[Ve07] Willem Veys, *Monodromy eigenvalues and zeta functions with differential forms*, Adv. Math. **213** (2007), no. 1, 341–357. MR2331246

[Xu16] Chenyang Xu, *Motivic zeta function via dlt modification*, Michigan Math. J. **65** (2016), no. 1, 89–103. MR3466817

Department of Mathematics, University of Wisconsin-Madison, 480 Lincoln Drive, Madison, Wisconsin 53706

Email address: villa@math.wisc.edu

Department of Mathematics, University of Illinois at Chicago, 851 S Morgan Street, Chicago, Illinois 60607

Email address: libgober@uic.edu

Department of Mathematics, University of Wisconsin-Madison, 480 Lincoln Drive, Madison, Wisconsin 53706

Email address: maxim@math.wisc.edu

Contemporary Mathematics
Volume **712**, 2018
http://dx.doi.org/10.1090/conm/712/14346

Algebraic fiber spaces over abelian varieties: Around a recent theorem by Cao and Păun

Christopher Hacon, Mihnea Popa, and Christian Schnell

To Lawrence Ein, on his 60*th birthday.*

ABSTRACT. We present a simplified proof for a recent theorem by Junyan Cao and Mihai Păun, confirming a special case of Iitaka's $C_{n,m}$ conjecture: if $f\colon X \to Y$ is an algebraic fiber space, and if the Albanese mapping of Y is generically finite over its image, then we have the inequality of Kodaira dimensions $\kappa(X) \geq \kappa(Y) + \kappa(F)$, where F denotes a general fiber of f. We include a detailed survey of the main algebraic and analytic techniques, especially the construction of singular hermitian metrics on pushforwards of adjoint bundles (due to Berndtsson, Păun, and Takayama).

A. Introduction

1. Main result. In the classification of algebraic varieties up to birational equivalence, the most fundamental invariants of a smooth projective variety X are the spaces of global sections of the pluricanonical bundles $\omega_X^{\otimes m}$. The rate of growth of the plurigenera $P_m(X) = \dim H^0(X, \omega_X^{\otimes m})$ is measured by the *Kodaira dimension*

$$\kappa(X) = \limsup_{m \to +\infty} \frac{\log P_m(X)}{\log m} \in \{-\infty, 0, 1, \dots, \dim X\}.$$

The following conjecture by Iitaka and Viehweg predicts the behavior of the Kodaira dimension in families. Recall that an *algebraic fiber space* is a surjective morphism with connected fibers between two smooth projective varieties.

CONJECTURE. *Let $f\colon X \to Y$ be an algebraic fiber space with general fiber F. Provided that $\kappa(Y) \geq 0$, the Kodaira dimension of X satisfies the inequality*

$$\kappa(X) \geq \kappa(F) + \max\{\kappa(Y), \operatorname{var}(f)\},$$

where $\operatorname{var}(f)$ *measures the birational variation in moduli of the fibers.*

Using analytic techniques, Cao and Păun [**6**] have recently proved the conjectured subadditivity of the Kodaira dimension in the case where Y is an abelian variety; as $\kappa(Y) = 0$, this amounts to the inequality

$$\kappa(X) \geq \kappa(F).$$

With very little extra work, one can deduce the subadditivity of the Kodaira dimension in any algebraic fiber space whose base Y has maximal Albanese dimension, meaning that the Albanese mapping $Y \to \mathrm{Alb}(Y)$ is generically finite over its image. This includes of course the case where Y is a curve of genus ≥ 1, where the following result was first proved by Kawamata [**22**, Theorem 2].

THEOREM 1.1. *Let $f\colon X \to Y$ be an algebraic fiber space with general fiber F. Assume that Y has maximal Albanese dimension, then $\kappa(X) \geq \kappa(F) + \kappa(Y)$.*

REMARK. A proof of this result is also claimed in [**11**], however the proof given there is incomplete because of a serious mistake in §4.

The purpose of this paper is to explain a simplified proof of the Cao-Păun theorem that combines both analytic and algebraic techniques. We first reduce to the case when $\kappa(X) = 0$ and Y is an abelian variety, where we then prove a more precise statement (Theorem 5.2). This is done in Chapter B. We then take the opportunity to provide a detailed survey of the results that are used in the proof, for the benefit of those readers who are more familiar with one or the other side of the story.

In Chapter C, we discuss the main algebraic tools, contained mostly in the papers [**9, 18, 25, 31, 36**], namely results from generic vanishing theory. The upshot of the discussion is that when $f\colon X \to A$ is a fiber space over an abelian variety, with $\kappa(X) = 0$, then for all m sufficiently large and divisible, $f_*\omega_X^{\otimes m}$ is a unipotent vector bundle on A, meaning a successive extension of copies of $\mathscr{O}_A$. This is as far as the algebraic techniques seem to go at present. While we recall the basic generic vanishing and Fourier-Mukai tools involved, as this topic is well-established in the literature, we provide fewer background details. Sources where a comprehensive treatment can be found include the lecture notes [**39**], as well as [**30, 35**].

In Chapter D and Chapter E we discuss the main analytic tools, contained mostly in the papers [**3, 6, 37**], namely the existence of singular metrics with semi-positive curvature (in a suitable sense) on pushforwards of pluricanonical bundles, and a very surprising criterion for such a metric to be smooth and flat. This time, the upshot is that when $f\colon X \to A$ is a fiber space onto an abelian variety, with $\kappa(X) = 0$, then $f_*\omega_X^{\otimes m}$ is a vector bundle with a flat hermitian metric. Because a unipotent vector bundle with a flat hermitian metric must be trivial, the algebraic and analytic results together lead to the conclusion in Theorem 1.1. Since the analytic results are still new, and are likely to be less familiar to algebraic geometers, we decided to include as many details as possible. For another survey of these and related results, we recommend [**34**].

REMARK. For the sake of exposition, we present only the simplest version of the result by Cao and Păun. One can tweak the proof of Theorem 1.1 to show that the inequality in Theorem 1.1 still holds when X is replaced by a klt pair (X, Δ), and F by the pair (F, Δ_F); this is done in [**6**, Theorem 4.22].

2. What is new? The presentation in Chapter D contains various small improvements compared to the original papers [**3, 6, 37**]. We briefly summarize the main points here. Let $f\colon X \to Y$ be a projective and surjective holomorphic mapping between two complex manifolds. Given a holomorphic line bundle L on X, and a singular hermitian metric h on L with semi-positive curvature, we construct a singular hermitian metric on the torsion-free coherent sheaf

$$\mathscr{F} = f_*\big(\omega_{X/Y} \otimes L \otimes \mathcal{I}(h)\big),$$

and show that this metric has semi-positive curvature (in the sense that the logarithm of the norm of every local section of the dual sheaf is plurisubharmonic). In [**37**], Păun and Takayama constructed a singular hermitian metric with semi-positive curvature on the larger sheaf $f_*(\omega_{X/Y} \otimes L)$, under the additional assumption that the restriction of (L, h) to a general fiber of f has trivial multiplier ideal. Another difference with [**37**] is that we do not use approximation by smooth metrics or results about Stein manifolds; instead, both the construction of the metric, and the proof that it has semi-positive curvature, rely on the Ohsawa-Takegoshi extension theorem with sharp estimates, recently proved by Błocki and Guan-Zhou [**4**, **16**]. This approach was suggested to us by Mihai Păun.

NOTE. Berndtsson and Lempert [**2**] explain how one can use the curvature properties of pushforwards of adjoint bundles to get a relatively short proof of (one version of) the Ohsawa-Takegoshi theorem with sharp estimates. This suggests that the two results are not so far from each other. That said, we hope that using the Ohsawa-Takegoshi theorem as a black box will make the proof of the main result more accessible to algebraic geometers than it would otherwise be.

We introduce what we call the "minimal extension property" for singular hermitian metrics (see §20), and show that, as a consequence of the Ohsawa-Takegoshi theorem with sharp estimates, the singular hermitian metric on $\mathscr{F}$ always has this property. We then use the minimal extension property, together with some basic inequalities from analysis, to give an alternative proof for the following result by Cao and Păun: when Y is projective, $\mathscr{F}$ is a hermitian flat bundle if and only if the line bundle $\det \mathscr{F}$ has trivial first Chern class in $H^2(Y, \mathbb{R})$. The original argument in [**6**] relied on some results by Raufi about curvature tensors for singular hermitian metrics [**38**]. We also show that when Y is projective, every nontrivial morphism of sheaves $\mathscr{F} \to \mathscr{O}_Y$ is split surjective; this result is new.

In Chapter E, we apply these results to construct canonical singular hermitian metrics with semi-positive curvature on the sheaves $f_*\omega_{X/Y}^{\otimes m}$ for $m \geq 2$. Here, one small improvement over [**37**] is the observation that these metrics are continuous on the Zariski-open subset of Y where $f \colon X \to Y$ is submersive.

Our discussion of generic vanishing theory in Chapter C is fairly standard, but includes (in §11) a new result relating the structure of the cohomological support loci $V^0(\omega_X^{\otimes m})$ for $m \geq 2$ to the Iitaka fibration of X. Here the main Theorem 5.2 is one of the crucial ingredients.

3. Acknowledgments. We thank Mihai Păun for encouraging us to write this paper, and for many useful discussions and advice about its contents. We also thank Dano Kim and Luigi Lombardi for reading and commenting on a draft version. During the preparation of the paper, CH was partially supported by NSF grants DMS-1300750 and DMS-1265285 and by a grant from the Simons Foundation (Award #256202). MP was partially supported by NSF grant DMS-1405516 and by a Simons Fellowship. CS was partially supported by NSF grants DMS-1404947 and DMS-1551677, and by a Centennial Fellowship from the American Mathematical Society.

B. Proof of the main statement

4. Main analytic and algebraic input. In this section we lay out the tools neeed to prove the main result. We also give a brief sketch of the proof, which is presented in more detail in §5. The rest of the paper will be devoted to a detailed survey of the results stated here.

We first note that one can reduce Theorem 1.1 to the special case when $\kappa(X) = 0$ and $Y = A$ is an abelian variety, with the help of the Iitaka fibration; the argument for this is recalled in §5 below. We will therefore make these assumptions in the remainder of this section. The condition $\kappa(X) = 0$ is equivalent to saying that $P_m(X) \leq 1$ for all $m \in \mathbb{N}$, with equality for m sufficiently large and divisible. Let F be the general fiber of $f\colon X \to A$. Our goal is to prove that $\kappa(F) = 0$. What we will actually show is that $P_m(F) = 1$ whenever $P_m(X) = 1$; this is enough to conclude that $\kappa(F) = 0$.

Fix now an integer $m \in \mathbb{N}$ such that $P_m(X) = 1$, and consider the pushforward of the m-th pluricanonical bundle

$$\mathscr{F}_m = f_*\omega_X^{\otimes m}.$$

This is a torsion-free coherent sheaf on A, whose rank at the generic point of A is equal to $P_m(F)$. (In fact, this holds for every smooth fiber of f, by invariance of plurigenera.) The space of global sections of $\mathscr{F}_m$ has dimension

$$h^0\bigl(A, f_*\omega_X^{\otimes m}\bigr) = \dim H^0(X, \omega_X^{\otimes m}) = P_m(X) = 1.$$

To obtain the conclusion, it is enough to show that $\mathscr{F}_m$ has rank 1 generically; we will in fact prove the stronger statement that $\mathscr{F}_m \simeq \mathscr{O}_A$. This uses both algebraic and analytic properties of $\mathscr{F}_m$.

Generic vanishing and unipotency. We first explain the algebraic input. We borrow an idea from generic vanishing theory, initiated in [**14, 15**], namely to consider the locus

$$\begin{aligned} V^0(A, \mathscr{F}_m) &= \bigl\{\, P \in \operatorname{Pic}^0(A) \bigm| H^0(A, \mathscr{F}_m \otimes P) \neq 0 \,\bigr\} \\ &= \bigl\{\, P \in \operatorname{Pic}^0(A) \bigm| H^0\bigl(X, \omega_X^{\otimes m} \otimes f^*P\bigr) \neq 0 \,\bigr\} \subseteq \operatorname{Pic}^0(A). \end{aligned}$$

The following result by Chen-Hacon [**9**, §3], Lai [**25**, Theorem 3.5] and Siu [**42**, Theorem 2.2] describes the structure of $V^0(A, \mathscr{F}_m)$; it is a generalization of a famous theorem by Simpson [**40**]. The proof by Simpson (which applies when $m = 0, 1$) relies on Hodge theory and the Gelfond-Schneider theorem in transcendence theory; the cited works use a construction with cyclic coverings, originally due to Viehweg, to reduce the general case to the case $m = 1$. We review the argument in §10.

THEOREM 4.1. *Let X be a smooth projective variety. For each $m \in \mathbb{N}$, the locus*

$$\bigl\{\, P \in \operatorname{Pic}^0(X) \bigm| H^0\bigl(X, \omega_X^{\otimes m} \otimes P\bigr) \neq 0 \,\bigr\} \subseteq \operatorname{Pic}^0(X)$$

is a finite union of abelian subvarieties translated by points of finite order.

This theorem implies that $V^0(A, \mathscr{F}_m)$ is also a finite union of abelian subvarieties translated by points of finite order. The reason is that, as $f\colon X \to A$ has connected fibers, the pullback morphism $f^*\colon \operatorname{Pic}^0(A) \to \operatorname{Pic}^0(X)$ is injective. Since we are assuming that $P_m(X) = 1$, we have $\mathscr{O}_A \in V^0(A, \mathscr{F}_m)$; let $s_0 \in H^0(X, \omega_X^{\otimes m})$ be any nontrivial section. Now we observe that $\kappa(X) = 0$ forces

$$V^0(A, \mathscr{F}_m) = \{\mathscr{O}_A\}.$$

To see why, suppose that we had $P \in V^0(A, \mathscr{F}_m)$ for some nontrivial line bundle $P \in \operatorname{Pic}^0(A)$. By Theorem 4.1, we can assume that P has finite order $d \neq 1$. Let

$$s_1 \in H^0(A, \mathscr{F}_m \otimes P) = H^0\big(X, \omega_X^{\otimes m} \otimes f^*P\big)$$

be any nontrivial section; then $s_0^{\otimes d}$ and $s_1^{\otimes d}$ are two linearly independent sections of $\omega_X^{\otimes dm}$, contradicting the fact that $P_{dm}(X) = 1$.

Knowing the locus $V^0(A, \mathscr{F}_m)$ gives us a lot of information about $\mathscr{F}_m$, due to the following result [**36**, Theorem 1.10]. It is based on a vanishing theorem for pushforwards of pluricanonical bundles, which is again proved using Viehweg's cyclic covering construction; we review the argument in §8.

THEOREM 4.2. *Let $f\colon X \to A$ be a morphism from a smooth projective variety to an abelian variety. For every $m \in \mathbb{N}$, the sheaf $\mathscr{F}_m = f_*\omega_X^{\otimes m}$ is a GV-sheaf on A.*

Recall that a coherent sheaf $\mathscr{F}$ on an abelian variety A is called a *GV-sheaf* if its cohomology support loci

$$V^i(A, \mathscr{F}) = \big\{\, P \in \operatorname{Pic}^0(A) \;\big|\; H^i(A, \mathscr{F} \otimes P) \neq 0 \,\big\}$$

satisfy the inequalities $\operatorname{codim} V^i(A, \mathscr{F}) \geq i$ for every $i \in \mathbb{N}$. This property can be seen as a variant of (semi-)positivity on abelian varieties; in fact every GV-sheaf on A is nef, see [**32**, Theorem 4.1].

NOTE. A more conceptual description involves the Fourier-Mukai transform

$$\mathbf{R}\Phi_P\colon D^b_{coh}(\mathscr{O}_A) \to D^b_{coh}\big(\mathscr{O}_{\operatorname{Pic}^0(A)}\big),$$

which is an equivalence between the bounded derived categories of coherent sheaves on A and the dual abelian variety $\operatorname{Pic}^0(A)$. In terms of the Fourier-Mukai transform, $\mathscr{F}$ is a GV-sheaf if and only if the complex of sheaves

$$\mathbf{R}\mathcal{H}om\Big(\mathbf{R}\Phi_P(\mathscr{F}), \mathscr{O}_{\hat{A}}\Big) \in D^b_{coh}\big(\mathscr{O}_{\operatorname{Pic}^0(A)}\big)$$

is concentrated in degree 0, and is therefore again a coherent sheaf $\hat{\mathscr{F}}$ on $\operatorname{Pic}^0(A)$. By the base change theorem, the support of $\hat{\mathscr{F}}$ is precisely the locus $V^0(A, \mathscr{F})$.

In the case at hand, we have $V^0(A, \mathscr{F}_m) = \{\mathscr{O}_A\}$; consequently, $\hat{\mathscr{F}}_m$ is a successive extension of skyscraper sheaves supported at the origin in $\operatorname{Pic}^0(A)$. We will use this via the following elementary consequence; see §9 for details. Recall first from [**28**] that a vector bundle U on A is called *unipotent* if it has a filtration

$$0 = U_0 \subset U_1 \subset \cdots \subset U_n = U$$

such that $U_i/U_{i-1} \simeq \mathscr{O}_A$ for all $i = 1, \ldots, n$. Note in particular that $\det U \simeq \mathscr{O}_A$. More generally, U is called *homogeneous* if it has a filtration

$$0 = U_0 \subset U_1 \subset \cdots \subset U_n = U$$

such that U_i/U_{i-1} is isomorphic to a line bundle in $\operatorname{Pic}^0(A)$ for all $i = 1, \ldots, n$. A homogeneous vector bundle U is called *decomposable* if $U = U_1 \oplus U_2$, where the U_i are non-zero vector bundles, and *indecomposable* if this is not the case.

COROLLARY 4.3. *Let X be a smooth projective variety with $\kappa(X) = 0$, and let $f\colon X \to A$ be an algebraic fiber space over an abelian variety.*

(a) *If $\mathscr{F}_m \neq 0$ for some $m \in \mathbb{N}$, then the coherent sheaf $\mathscr{F}_m$ is an indecomposable homogeneous vector bundle.*

(b) *If* $H^0(X, \omega_X^{\otimes m}) \neq 0$ *for some* $m \in \mathbb{N}$, *then the coherent sheaf* $\mathscr{F}_m$ *is an indecomposable unipotent vector bundle.*

Singular hermitian metrics on pushforwards of pluricanonical bundles. We now come to the analytic input. To motivate it, recall that the space of global sections of $\mathscr{F}_m$ has dimension $P_m(X) = 1$. In order to show that $P_m(F) = 1$, we therefore need to argue that the unipotent vector bundle $\mathscr{F}_m$ is actually the trivial bundle $\mathscr{O}_A$. For the moment this seems quite hopeless with algebraic methods, so it is at this point that the analytic methods take over.

The crucial development that allows us to proceed is recent work on the notion of a singular hermitian metric on a torsion-free sheaf; the highlight of this study is the following remarkable result by Păun and Takayama [**37**, Theorem 3.3.5]. In order to state it, recall that to a singular hermitian metric h on a line bundle L, one associates the multiplier ideal sheaf $\mathcal{I}(h) \subseteq \mathscr{O}_X$, consisting of those functions that are locally square-integrable with respect to h.

THEOREM 4.4. *Let* $f\colon X \to Y$ *be a projective morphism of smooth varieties, and let* (L, h) *be a line bundle on* X *with a singular hermitian metric of semi-positive curvature. Then the torsion-free sheaf* $f_*\big(\omega_{X/Y} \otimes L \otimes \mathcal{I}(h)\big)$ *has a canonical singular hermitian metric with semi-positive curvature.*

The relevant definitions and the proof are described in Chapter D and Chapter E, where we also present another key statement. Indeed, Cao and Păun [**6**, Corollary 2.9 and Theorem 5.23] show that their singular hermitian metrics behave much like smooth metrics with Griffiths semi-positive curvature: if the determinant line bundle $\det \mathscr{F}$ has trivial first Chern class, then $\mathscr{F}$ is actually a hermitian flat bundle. This is (a) below; part (b) is new.

THEOREM 4.5. *Let* $f\colon X \to Y$ *be a surjective morphism of smooth projective varieties. Let* (L, h) *be a line bundle on* X *with a singular hermitian metric of semi-positive curvature, and define* $\mathscr{F} = f_*\big(\omega_{X/Y} \otimes L \otimes \mathcal{I}(h)\big)$.

(a) *If* $c_1(\det \mathscr{F}) = 0$ *in* $H^2(Y, \mathbb{R})$, *then the torsion-free sheaf* $\mathscr{F}$ *is locally free, and the singular hermitian metric in Theorem 4.4 is smooth and flat.*

(b) *Every nonzero morphism* $\mathscr{F} \to \mathscr{O}_Y$ *is split surjective.*

The application of these results to Theorem 1.1 stems from the fact that the bundles $\mathscr{F}_m$ with $m \geq 2$ naturally fit into this framework. Let us briefly summarize how this works when $f\colon X \to Y$ is an algebraic fiber space with general fiber F. For every $m \in \mathbb{N}$ such that $P_m(F) \neq 0$, the spaces of m-canonical forms on the smooth fibers of f induce in a canonical way a singular hermitian metric with semi-positive curvature on $\omega_{X/Y}$, called the *m-th Narasimhan-Simha metric.* (For $m = 1$, the Narasimhan-Simha metric is of course just the Hodge metric.) Denote by h the induced singular hermitian metric on the line bundle $L = \omega_{X/Y}^{\otimes(m-1)}$. Pretty much by construction, the inclusion

$$f_*\big(\omega_{X/Y} \otimes L \otimes \mathcal{I}(h)\big) \subseteq f_*\big(\omega_{X/Y} \otimes L\big) = f_*\omega_{X/Y}^{\otimes m}$$

is generically an isomorphism, and so Theorem 4.4 and Theorem 4.5 apply.

COROLLARY 4.6. *Let* $f\colon X \to Y$ *be an algebraic fiber space.*

(a) *For any* $m \in \mathbb{N}$, *the torsion-free sheaf* $f_*\omega_{X/Y}^{\otimes m}$ *has a canonical singular hermitian metric with semi-positive curvature.*

(b) *If* $c_1(\det f_*\omega_{X/Y}^{\otimes m}) = 0$ *in* $H^2(Y, \mathbb{R})$, *then* $f_*\omega_{X/Y}^{\otimes m}$ *is locally free, and the singular hermitian metric on it is smooth and flat.*

(c) *Every nonzero morphism* $f_*\omega_{X/Y}^{\otimes m} \to \mathscr{O}_Y$ *is split surjective.*

In our case, $\mathscr{F}_m = f_*\omega_X^{\otimes m}$ is a unipotent vector bundle by Corollary 4.3, and so the hypothesis in (b) is satisfied; after this point, the proof of Theorem 1.1 becomes straightforward.

5. Proof of Theorem 1.1. We now explain how Theorem 1.1 follows quickly by combining the results outlined in the previous section. Recall that we are starting with an algebraic fiber space $f\colon X \to Y$, where X is a smooth projective variety, and Y is of maximal Albanese dimension. Let us note right away that one can perform a useful reduction, following in part the argument in [**8**, Theorem 4.9].

LEMMA 5.1. *To prove Theorem 1.1, it is enough to assume that* $\kappa(X) = 0$ *and that* Y *is an abelian variety.*

PROOF. We begin by showing that if $\kappa(X) = -\infty$, then $\kappa(F) = -\infty$. If this were not the case, then we could pick some $m > 0$ such that $P_m(F) > 0$ and hence $f_*\omega_X^{\otimes m} \neq 0$. Let $Y \to A$ be the Albanese morphism of Y, and $g\colon X \to A$ the induced morphism. Since F is an irreducible component of the general fiber of $X \to g(X) \subseteq A$, it follows that $g_*\omega_X^{\otimes m} \neq 0$. By Theorem 4.2, $g_*\omega_X^{\otimes m}$ is a GV-sheaf, and in particular by the general Lemma 7.4 below, the set

$$V^0(g_*\omega_X^{\otimes m}) = \bigl\{ P \in \operatorname{Pic}^0(A) \bigm| H^0(A, g_*\omega_X^{\otimes m} \otimes P) \neq 0 \bigr\}$$

is non-empty. Now by Theorem 4.1 and the comments immediately after, $V^0(g_*\omega_X^{\otimes m})$ contains a torsion point $P \in \operatorname{Pic}^0(A)$, i.e. there is an integer $k > 0$ such that $P^{\otimes k} \simeq \mathscr{O}_A$. But then $h^0(X, \omega_X^{\otimes m} \otimes g^*P) = h^0(A, g_*\omega_X^{\otimes m} \otimes P) \neq 0$ and so

$$h^0(X, \omega_X^{\otimes km}) = h^0\bigl(X, (\omega_X^{\otimes m} \otimes P)^{\otimes k}\bigr) \neq 0.$$

This contradicts the assumption $\kappa(X) = -\infty$.

Assume now that $\kappa(X) \geq 0$. We will first prove the statement in the case that $\kappa(Y) = 0$. By Kawamata's theorem [**21**, Theorem 1], since Y is of maximal Albanese dimension, it is in fact birational to its Albanese variety and so we may assume that Y is an abelian variety. Let $h\colon X \to Z$ the Iitaka fibration of X. Since we are allowed to work birationally, we can assume that Z is smooth. We denote by G its general fiber, so that in particular $\kappa(G) = 0$. By the same result of Kawamata, the Albanese map of G is surjective, so we deduce that $B = f(G) \subseteq Y$ is an abelian subvariety. If $G \to B' \to B$ is the Stein factorization, then $B' \to B$ is an étale map of abelian varieties. We thus have an induced fiber space

$$G \longrightarrow B'$$

over an abelian variety, with $\kappa(G) = 0$, and whose general fiber is $H = F \cap G$. Assuming that Theorem 1.1 holds for algebraic fibers spaces of Kodaira dimension zero over abelian varieties, we obtain $\kappa(H) = 0$. Note however that H is also an irreducible component of the general fiber of

$$h_{|F}\colon F \longrightarrow h(F).$$

Considering the Stein factorization of this morphism, the easy addition formula [**27**, Corollary 2.3], implies that

$$\kappa(F) \leq \kappa(H) + \dim h(F) = \dim h(F).$$

(Note that we can assume that $g(F)$ is smooth, by passing to a birational model.) Since $\dim h(F) \leq \dim Z = \kappa(X)$, we obtain the required inequality $\kappa(F) \leq \kappa(X)$.

Finally we prove the general case. Since Y has maximal Albanese dimension, after replacing it by a resolution of singularities of an étale cover of its Stein factorization, and X by a resolution of the corresponding fiber product, by [**21**, Theorem 13] we may assume that $Y = Z \times K$ where Z is of general type and K is an abelian variety. In particular $\kappa(Y) = \dim Z = \kappa(Z)$. If E is the general fiber of the induced morphism $X \to Z$, then the induced morphism $E \to K$ has general fiber isomorphic to F. By what we have proven above, we deduce that $\kappa(E) \geq \kappa(F)$. We then have the required inequality

$$\kappa(X) = \kappa(Z) + \kappa(E) \geq \kappa(Y) + \kappa(F),$$

where the first equality is [**21**, Theorem 3], since Z is of general type. □

We may therefore assume that $f\colon X \to A$ is a fiber space onto an abelian variety, and $\kappa(X) = 0$. Note that this last condition means that we have $h^0(X, \omega_X^{\otimes m}) = 1$ for all sufficiently divisible integers $m > 0$. The task at hand is to show that $\kappa(F) = 0$. (It is a well known consequence of the easy addition formula [**27**, Corollary 2.3] that if $\kappa(F) = -\infty$, then $\kappa(X) = -\infty$.) We show in fact the following more precise statement:

THEOREM 5.2. *If $f\colon X \to A$ is an algebraic fiber space over an abelian variety, with $\kappa(X) = 0$, then we have*

$$\mathscr{F}_m = f_*\omega_X^{\otimes m} \simeq \mathscr{O}_A$$

for every $m \in \mathbb{N}$ such that $H^0(X, \omega_X^{\otimes m}) \neq 0$.

PROOF. From Corollary 4.3, we know that $\mathscr{F}_m$ is an indecomposable unipotent vector bundle on A. In particular,

$$\det \mathscr{F}_m \simeq \mathscr{O}_A.$$

Corollary 4.6 implies then that $\mathscr{F}_m$ has a smooth hermitian metric that is flat. Thus $\mathscr{F}_m$ is a successive extension of trivial bundles $\mathscr{O}_A$ that can be split off as direct summands with the help of the flat metric. It follows that in fact $\mathscr{F}_m \simeq \mathscr{O}_A^{\oplus r}$, the trivial bundle of some rank $r \geq 1$. But then, since

$$h^0(A, f_*\omega_X^{\otimes m}) = h^0(X, \omega_X^{\otimes m}) = 1,$$

we obtain that $r = 1$, which is the statement of the theorem. □

In the remaining chapters, we explain the material in §4 in more detail.

C. Generic vanishing

6. Canonical bundles and their pushforwards. As explained above, the algebraic tools used in this paper revolve around the topic of generic vanishing. This study was initiated by Green and Lazarsfeld [**14**, **15**], in part as an attempt to provide a useful weaker version of Kodaira Vanishing for the canonical bundle, in the absence of twists by positive line bundles. An important addition was provided in work of Simpson [**40**]. The results of Green-Lazarsfeld were extended to include higher direct images of canonical bundles in [**18**]. From the point of view of this paper, the main statements to keep in mind are summarized in the following theorem. Recall that for any coherent sheaf $\mathscr{F}$ on an abelian variety A, we consider for all $k \geq 0$ the *cohomological support loci*

$$V^k(\mathscr{F}) = \{ P \in \mathrm{Pic}^0(A) \mid H^k(X, \mathscr{F} \otimes P) \neq 0 \}$$

They are closed subsets of $\mathrm{Pic}^0(A)$, by the semi-continuity theorem for cohomology.

THEOREM 6.1. *If $f \colon X \to A$ is a morphism from a smooth projective variety to an abelian variety, then for any $j, k \geq 0$ we have*

(1) [**18**] $\mathrm{codim}_{\mathrm{Pic}^0(A)} V^k(R^j f_* \omega_X) \geq k$.
(2) [**15, 40**] *Every irreducible component of $V^k(R^j f_* \omega_X)$ is a translate of an abelian subvariety of A by a point of finite order.*

What we use in this paper are (partial) extensions of these results to push-forwards of pluricanonical bundles $f_* \omega_X^{\otimes m}$, for $m \geq 2$. We describe these in the following sections, beginning with an abstract study in the next.

7. The *GV* property and unipotency. Let A be an abelian variety of dimension g. The generic vanishing property (1) in the theorem above can be formalized into the following:

DEFINITION 7.1. The sheaf $\mathscr{F}$ is called a *GV-sheaf* on A if it satisfies

$$\mathrm{codim}_{\mathrm{Pic}^0(A)} V^k(\mathscr{F}) \geq k \qquad \text{for all } k \geq 0.$$

We will identify $\mathrm{Pic}^0(A)$ with the dual abelian variety $\widehat{A}$, and denote by P the normalized Poincaré bundle on $A \times \widehat{A}$. It induces the integral transforms

$$\mathbf{R}\Phi_P \colon D^b_{coh}(\mathscr{O}_A) \longrightarrow D^b_{coh}(\mathscr{O}_{\widehat{A}}), \quad \mathbf{R}\Phi_P \mathscr{F} = \mathbf{R}p_{2*}(p_1^* \mathscr{F} \otimes P).$$

and

$$\mathbf{R}\Psi_P \colon D^b_{coh}(\mathscr{O}_{\widehat{A}}) \longrightarrow D^b_{coh}(\mathscr{O}_A), \quad \mathbf{R}\Psi_P \mathscr{G} = \mathbf{R}p_{1*}(p_2^* \mathscr{G} \otimes P).$$

These functors are known from [**28**, Theorem 2.2] to be equivalences of derived categories, usually called the Fourier-Mukai transforms; moreover,

$$\mathbf{R}\Psi_P \circ \mathbf{R}\Phi_P = (-1_A)^*[-g] \quad \text{and} \quad \mathbf{R}\Phi_P \circ \mathbf{R}\Psi_P = (-1_{\widehat{A}})^*[-g], \tag{7.2}$$

where $[-g]$ denotes shifting g places to the right.

Standard applications of base change (see e.g. [**32**, Lemma 2.1] and [**31**, Proposition 3.14]) lead to the following basic properties of GV-sheaves:

LEMMA 7.3. *Let $\mathscr{F}$ be a coherent sheaf on A. Then:*

(1) *$\mathscr{F}$ is a GV-sheaf if and only if*

$$\mathrm{codim}_{\widehat{A}} \mathrm{Supp}\, R^k \Phi_P \mathscr{F} \geq k \quad \text{for all} \quad k \geq 0.$$

(2) *If $\mathscr{F}$ is a GV-sheaf, then*

$$V^g(\mathscr{F}) \subseteq \cdots \subseteq V^1(\mathscr{F}) \subseteq V^0(\mathscr{F}).$$

To give a sense of what is going on, here is a sketch of the proof of part (1): note that the restriction of ${p_1}^* \mathscr{F} \otimes P$ to a fiber $A \times \{\alpha\}$ of p_2 is isomorphic to the sheaf $\mathscr{F} \otimes \alpha$ on A, and so fiberwise we are looking at the cohomology groups $H^k(A, \mathscr{F} \otimes \alpha)$. A simple application of the theorem on cohomology and base change then shows for every $m \geq 0$ that

$$\bigcup_{k \geq m} \mathrm{Supp}\, R^k \Phi_P \mathscr{F} = \bigcup_{k \geq m} V^k(\mathscr{F}).$$

This implies the result by descending induction on k.

LEMMA 7.4. *If $\mathscr{F}$ is a GV-sheaf on A, then $\mathscr{F} = 0$ if and only if $V^0(\mathscr{F}) = \emptyset$.*

PROOF. By Lemma 7.3, we see that $V^0(\mathscr{F}) = \emptyset$ is equivalent to $V^k(\mathscr{F}) = \emptyset$ for all $k \geq 0$, which by base change is in turn equivalent to $\mathbf{R}\Phi_P\mathscr{F} = 0$. By Mukai's derived equivalence, this is equivalent to $\mathscr{F} = 0$. □

The following proposition is the same as [**18**, Corollary 3.2(4)], since it can be seen that the assumption on $\mathscr{F}$ imposed there is equivalent to that of being a GV-sheaf. This is the main way in which generic vanishing is used in this paper; for the definition of a unipotent vector bundle see §4.

PROPOSITION 7.5. *Let $\mathscr{F}$ be a GV-sheaf on an abelian variety A. If $V^0(\mathscr{F}) = \{0\}$, then $\mathscr{F}$ is a unipotent vector bundle.*

PROOF. By [**28**, Example 2.9], if $g = \dim A$, then $\mathscr{F}$ is a unipotent vector bundle if and only if

$$R^i\Phi_P\mathscr{F} = 0 \quad \text{for all } i \neq g, \qquad \text{and} \qquad R^g\Phi_P\mathscr{F} = \mathscr{G}, \tag{7.6}$$

where $\mathscr{G}$ is a coherent sheaf supported at the origin $0 \in \widehat{A}$. To review the argument, notice that if this is the case, then if $l = \text{length}(\mathscr{G}) > 0$, we have $h^0(\widehat{A}, \mathscr{G}) \neq 0$ and so there is a short exact sequence

$$0 \longrightarrow k(0) \longrightarrow \mathscr{G} \longrightarrow \mathscr{G}' \longrightarrow 0$$

where $\mathscr{G}'$ is a coherent sheaf supported at the origin $0 \in \widehat{A}$, with $\text{length}(\mathscr{G}') = l - 1$. Applying $\mathbf{R}\Psi_P$ we obtain a short exact sequence of vector bundles on A

$$0 \longrightarrow \mathscr{O}_A \longrightarrow R^0\Psi_P\mathscr{G} \longrightarrow R^0\Psi_P\mathscr{G}' \longrightarrow 0,$$

and by (7.2) we have $R^0\Psi_P\mathscr{G} = (-1_A)^*\mathscr{F}$. It is then not hard to see that $\mathscr{F}' = R^0\Psi_P\mathscr{G}'$ also satisfies the hypotheses in (7.6) and so, proceeding by induction on l, we may assume that $\mathscr{F}'$ is a unipotent vector bundle. It follows that $\mathscr{F}$ is also a unipotent vector bundle as well (since it is an extension of a unipotent vector bundle by $\mathscr{O}_A$).

We now check that the two conditions in (7.6) are satisfied. By Lemma 7.3 the hypothesis implies that

$$V^i(\mathscr{F}) \subseteq \{0\} \qquad \text{for all } i \geq 0.$$

By base change one obtains that $R^i\Phi_P\mathscr{F}$ is supported at most at $0 \in \widehat{A}$ for $0 \leq i \leq g$. It remains to show that $R^i\Phi_P\mathscr{F} = 0$ for $i \neq g$. Note that

$$H^j(\widehat{A}, R^i\Phi_P\mathscr{F} \otimes \alpha) = 0 \qquad \text{for all } j > 0,\ 0 \leq i \leq g, \text{ and } \alpha \in \text{Pic}^0(\widehat{A}),$$

and so by base change we have

$$R^j\Psi_P(R^i\Phi_P\mathscr{F}) = 0 \qquad \text{for all } j > 0 \text{ and } 0 \leq i \leq g.$$

By an easy argument involving the spectral sequence of the composition of two functors, since $\mathbf{R}\Psi_P \circ \mathbf{R}\Phi_P = (-1_A)^*[-g]$, it then follows that $R^0\Psi_P(R^i\Phi_P\mathscr{F}) = \mathcal{H}^i\big((-1_A)^*\mathscr{F}[-g]\big)$, and so in particular

$$R^0\Psi_P(R^i\Phi_P\mathscr{F}) = 0 \ \text{ for } \ i < g.$$

But then $\mathbf{R}\Psi_P(R^i\Phi_P\mathscr{F}) = 0$ for $i < g$, and hence $R^i\Phi_P\mathscr{F} = 0$ for $i < g$.

□

For later use, we note that a very useful tool for detecting generic vanishing is a cohomological criterion introduced in [**18**, Corollary 3.1]. Before stating it, we recall that an ample line bundle N on an abelian variety B induces an isogeny

$$\varphi_N : B \longrightarrow \widehat{B}, \quad x \to t_x^* N \otimes N^{-1},$$

where t_x denotes translation by $x \in B$.

THEOREM 7.7. *A coherent sheaf $\mathscr{F}$ on A is a GV-sheaf if and only if given any sufficiently large power M of an ample line bundle on $\widehat{A}$, one has*

$$H^i\big(A, \mathscr{F} \otimes R^g\Psi_P(M^{-1})\big) = 0 \quad \text{for all } \ i > 0.$$

If $\varphi_M \colon \widehat{A} \to A$ is the isogeny induced by M, this is also equivalent to

$$H^i(\widehat{A}, \varphi_M^* \mathscr{F} \otimes M) = 0 \quad \text{for all } \ i > 0.$$

REMARK. Note that since M is ample, $H^i(\widehat{A}, M^{-1} \otimes \alpha) = 0$ for all $i < g$ and $\alpha \in \operatorname{Pic}^0(\widehat{A}) \simeq A$, and so $R^i\Psi_P(M^{-1}) = 0$ for $i \neq g$. If we denote $R^g\Psi_P(M^{-1}) = \widehat{M^{-1}}$, then by [**28**, Proposition 3.11] we have $\varphi_M^* \widehat{M^{-1}} \simeq M^{\oplus h^0(M)}$, hence the second assertion.

8. Pushforwards of pluricanonical bundles. In this section we explain the proof of Theorem 4.2, following [**36**, §5]. In *loc. cit.* we noted that a very quick proof can be given based on the general effective vanishing theorem for pushforwards of pluricanonical bundles proved there. However, another more self-contained, if less efficient, proof using cyclic covering constructions is also given; we choose to explain this here, as cyclic covering constructions are in the background of many arguments in this article. We first recall Kollár's vanishing theorem [**23**, Theorem 2.1].

THEOREM 8.1. *Let $f : X \to Y$ be a morphism of projective varieties, with X smooth. If L is an ample line bundle on Y, then*

$$H^j(Y, R^i f_* \omega_X \otimes L) = 0 \ \text{ for all } i \text{ and all } j > 0.$$

PROOF OF THEOREM 4.2. Let $M = L^{\otimes d}$, where L is an ample and globally generated line bundle on $\widehat{A}$, and d is an integer that can be chosen arbitrarily large. Let $\varphi_M \colon \widehat{A} \to A$ be the isogeny induced by M. According to Theorem 7.7, it is enough to show that

$$H^i(\widehat{A}, \varphi_M^* f_* \omega_X^{\otimes m} \otimes M) = 0 \quad \text{for all } i > 0.$$

Equivalently, we need to show that

$$H^i(\widehat{A}, h_* \omega_{X_1}^{\otimes m} \otimes L^{\otimes d}) = 0 \quad \text{for all } i > 0,$$

where $h \colon X_1 \to \widehat{A}$ is the base change of $f \colon X \to A$ via φ_M, as in the diagram

$$\begin{array}{ccc} X_1 & \longrightarrow & X \\ \big\downarrow{\scriptstyle h} & & \big\downarrow{\scriptstyle f} \\ \widehat{A} & \xrightarrow{\ \varphi_M\ } & A \end{array}$$

We can conclude immediately if we know that there exists a bound $d = d(g, m)$, i.e. depending only on $g = \dim A$ and m, such that the vanishing in question holds for any morphism h. (Note that we cannot apply Serre Vanishing here, as construction depends on the original choice of M.) But Proposition 8.2 below shows that there exists a morphism $\varphi : Z \to \widehat{A}$ with Z smooth projective, and $k \leq g+m$, such that $h_*\omega_{X_1}^{\otimes m} \otimes L^{\otimes k(m-1)}$ is a direct summand of $\varphi_*\omega_Z$. Applying Kollár vanishing, Theorem 8.1, we deduce that

$$H^i(\widehat{A}, h_*\omega_{X_1}^{\otimes m} \otimes L^{\otimes d}) = 0 \quad \text{for all} \quad i > 0 \quad \text{and all} \quad d \geq (g+m)(m-1)+1,$$

which finishes the proof. (The main result of [**36**] shows that one can in fact take $d \geq m(g+1) - g$.) □

PROPOSITION 8.2. *Let $f : X \to Y$ be a morphism of projective varieties, with X smooth and Y of dimension n. Let L be an ample and globally generated line bundle on Y, and $m \geq 1$ an integer. Then there exists a smooth projective variety Z with a morphism $\varphi : Z \to Y$, and an integer $0 \leq k \leq n+m$, such that $f_*\omega_X^{\otimes m} \otimes L^{\otimes k(m-1)}$ is a direct summand in $\varphi_*\omega_Z$.*

PROOF. The sheaf $f_*\omega_X^{\otimes m} \otimes L^{\otimes pm}$ is globally generated for some sufficiently large p. Denote by k the minimal $p \geq 0$ for which this is satisfied.

Consider now the adjunction morphism

$$f^* f_*\omega_X^{\otimes m} \to \omega_X^{\otimes m}.$$

After blowing up X, if necessary, we can assume that the image sheaf is of the form $\omega_X^{\otimes m} \otimes \mathscr{O}_X(-E)$ for a divisor E with normal crossing support. As $f_*\omega_X^{\otimes m} \otimes L^{\otimes km}$ is globally generated, the line bundle

$$\omega_X^{\otimes m} \otimes f^*L^{\otimes km} \otimes \mathscr{O}_X(-E)$$

is globally generated as well. It is therefore isomorphic to $\mathscr{O}_X(D)$, where D is an irreducible smooth divisor, not contained in the support of E, such that $D+E$ also has normal crossings. We have thus arranged that

$$(\omega_X \otimes f^*L^{\otimes k})^{\otimes m} \simeq \mathscr{O}_X(D+E).$$

We can now take the associated covering of X of degree m, branched along $D+E$, and resolve its singularities. This gives us a generically finite morphism $g\colon Z \to X$ of degree m, and we denote $\varphi = f \circ g : Z \to Y$.

By a well-known calculation of Esnault and Viehweg, see e.g. [**13**, Lemma 2.3], the direct image $g_*\omega_Z$ contains the sheaf

$$\begin{aligned}\omega_X \otimes \left(\omega_X \otimes f^*L^{\otimes k}\right)^{\otimes m-1} &\otimes \mathscr{O}_X\left(-\left\lfloor \tfrac{m-1}{m}(D+E)\right\rfloor\right) \\ \simeq \omega_X^{\otimes m} \otimes f^*L^{\otimes k(m-1)} &\otimes \mathscr{O}_X\left(-\left\lfloor \tfrac{m-1}{m}E\right\rfloor\right)\end{aligned}$$

as a direct summand. If we now apply f_*, we find that

$$f_*\Big(\omega_X^{\otimes m} \otimes \mathscr{O}_X\left(-\left\lfloor \tfrac{m-1}{m}E\right\rfloor\right)\Big) \otimes L^{\otimes k(m-1)}$$

is a direct summand of $\varphi_*\omega_Z$. Finally, E is the relative base locus of $\omega_X^{\otimes m}$, and so

$$f_*\Big(\omega_X^{\otimes m} \otimes \mathscr{O}_X\left(-\left\lfloor \tfrac{m-1}{m}E\right\rfloor\right)\Big) \simeq f_*\omega_X^{\otimes m}.$$

In other words, $f_*\omega_X^{\otimes m} \otimes L^{\otimes k(m-1)}$ is a direct summand in $\varphi_*\omega_Z$. By Theorem 8.1, the sheaf $f_*\omega_X^{\otimes m} \otimes L^{\otimes k(m-1)+n+1}$ is 0-regular in the sense of Castelnuovo-Mumford, and hence globally generated.[1] By our minimal choice of k, this is only possible if

$$k(m-1)+n+1 \geq (k-1)m+1,$$

which is equivalent to $k \leq n+m$. □

9. Fiber spaces over abelian varieties. Let $f\colon X \to A$ be a fiber space over an abelian variety. For simplicity, for each $m \geq 0$ we denote

$$\mathscr{F}_m = f_*\omega_X^{\otimes m}.$$

Note that $\mathscr{F}_0 = \mathscr{O}_A$. Though this is not really necessary for the argument, we first remark that we can be precise about the values of $m \geq 1$ for which $\mathscr{F}_m \neq 0$.

LEMMA 9.1. *We have* $\mathscr{F}_m \neq 0$ *if and only if there exists* $P \in \mathrm{Pic}^0(A)$ *such that* $H^0(X, \omega_X^{\otimes m} \otimes f^*P) \neq 0$.

PROOF. By Theorem 4.2 we know that $\mathscr{F}_m$ is a GV-sheaf on A for all $m \geq 1$. We conclude from Lemma 7.4 that $\mathscr{F}_m \neq 0$ if and only if $V^0(\mathscr{F}_m) \neq \emptyset$, which by the projection formula is precisely the statement of the lemma. □

The purpose of this subsection is to give the

PROOF OF COROLLARY 4.3. We will only prove the second statement, since the first one is similar. We fix an m such that $H^0(A, \mathscr{F}_m) = H^0(X, \omega_X^{\otimes m}) \neq 0$. In particular $\mathscr{F}_m$ is a non-trivial GV-sheaf on A. Since $\kappa(X) = 0$, we have $h^0(A, \mathscr{F}_m) = 1$, and in particular $0 \in V^0(\mathscr{F}_m)$. We claim that

$$V^0(\mathscr{F}_m) = \{0\},$$

which implies that $\mathscr{F}_m$ is unipotent by Proposition 7.5.

To see this, note first that by Theorem 4.1 and the comments immediately after, $V^0(\mathscr{F}_m)$ is a union of torsion translates of abelian subvarieties of $\mathrm{Pic}^0(A)$. Then, proceeding as in [**7**, Lemma 2.1], if there were two distinct points $P, Q \in V^0(\mathscr{F}_m)$ we could assume that they are both torsion of the same order k. Since f is a fiber space, the mapping

$$f^*\colon \mathrm{Pic}^0(A) \longrightarrow \mathrm{Pic}^0(X)$$

is injective, and so f^*P and f^*Q are distinct as well. Now if $P \in V^0(\mathscr{F}_m)$, then

$$H^0(X, \omega_X^{\otimes m} \otimes f^*P) \simeq H^0(A, \mathscr{F}_m \otimes P) \neq 0,$$

and similarly for Q. Let $D \in |mK_X + f^*P|$ and $G \in |mK_X + f^*Q|$, so that $kD, kG \in |mkK_X|$. Since $h^0(X, \omega_X^{\otimes mk}) = 1$, it follows that $kD = kF$, and hence $f^*P = f^*Q$. This is the required contradiction.

Finally, since $h^0(A, \mathscr{F}_m) = 1$, it is clear that $\mathscr{F}_m$ is indecomposable. □

---

[1] Recall that a sheaf $\mathscr{F}$ on Y is 0-*regular* with respect to an ample and globally generated line bundle L if

$$H^i(Y, \mathscr{F} \otimes L^{\otimes -i}) = 0 \quad \text{for all } i > 0.$$

The Castelnuovo-Mumford Lemma says that every 0-regular sheaf is globally generated.

10. Cohomological support loci for pluricanonical bundles. In this section we explain an important ingredient used in Corollary 4.3, namely Theorem 4.1, the analogue of Simpson's theorem for the 0-th cohomological support locus of a pluricanonical bundle. We give a slight generalization, emphasizing again the ubiquitous cyclic covering trick.

For a coherent sheaf $\mathscr{F}$ on an abelian variety A, for each $k \geq 1$ we consider the following refinement of $V^0(\mathscr{F})$, namely

$$V_k^0(\mathscr{F}) = \{\, P \in \operatorname{Pic}^0(A) \mid h^0(X, \mathscr{F} \otimes P) \geq k \,\}.$$

THEOREM 10.1. *Let $f : X \to A$ be a morphism from a smooth projective variety to an abelian variety, and fix integers $m, k \geq 1$. Then every irreducible component of $V_k^0(f_*\omega_X^{\otimes m})$ is a torsion subvariety, i.e. a translate of an abelian subvariety of $\operatorname{Pic}^0(A)$ by a torsion point.*

To prove Theorem 10.1, we first collect a few lemmas. Given a smooth projective variety X, and a line bundle L on X with $\kappa(L) \geq 0$, recall that the *asymptotic multiplier ideal* of L is defined as

$$\mathcal{I}(\|L\|) = \mathcal{I}\Big(\tfrac{1}{p}D\Big) \subseteq \mathscr{O}_X,$$

where p is any sufficiently large and divisible integer, D is the divisor of a general section in $H^0(X, L^{\otimes p})$, and the ideal sheaf on the right is the multiplier ideal of the $\mathbb{Q}$-divisor $\frac{1}{p}D$; see [**26**, Ch. 11]. It is easy to see that the ideal sheaf $\mathcal{I}(\|L\|)$ is independent of the choice of p and D. Further properties of asymptotic multiplier ideals appear in the proof of Lemma 10.4 below.

LEMMA 10.2. *There exists a morphism $g \colon Y \to X$ with Y smooth and projective, such that the sheaf $\omega_X \otimes L \otimes \mathcal{I}(\|L\|)$ is a direct summand of $g_*\omega_Y$.*

PROOF. Take D as above, and let $\mu \colon X' \to X$ be a log resolution of (X, D) such that X' is smooth and μ^*D plus the exceptional divisor of μ is a divisor with simple normal crossing support. Then

$$\mu^*L^{\otimes p} = \mathscr{O}_{X'}(\mu^*D),$$

and we let $f \colon Y \to X'$ be a resolution of singularities of the degree p branched covering of X' defined by μ^*D. According to the calculation of Esnault and Viehweg recalled in the proof of Proposition 8.2, $f_*\omega_Y$ contains as a direct summand the sheaf

$$\omega_{X'} \otimes \mu^*L \otimes \mathscr{O}_{X'}\Big(-\big\lfloor \tfrac{1}{p}\mu^*D \big\rfloor\Big) \simeq \mu^*(\omega_X \otimes L) \otimes \mathscr{O}_{X'}\Big(K_{X'/X} - \big\lfloor \tfrac{1}{p}\mu^*D \big\rfloor\Big),$$

and so $\mu_*f_*\omega_Y$ contains as a direct summand the sheaf

$$\omega_X \otimes L \otimes \mu_*\mathscr{O}_{X'}\Big(K_{X'/X} - \big\lfloor \tfrac{1}{p}\mu^*D \big\rfloor\Big) = \omega_X \otimes L \otimes \mathcal{I}\big(\|L\|\big).$$

We can therefore take $g = \mu \circ f$. □

LEMMA 10.3. *If $\mathscr{F}$ and $\mathscr{G}$ are two coherent sheaves on an abelian variety A, and $\mathscr{F}$ is a direct summand of $\mathscr{G}$, then every irreducible component of $V_k^0(\mathscr{F})$ is also an irreducible component of $V_\ell^0(\mathscr{G})$ for some $\ell \geq k$.*

PROOF. Let $Z \subseteq V_k^0(\mathscr{F})$ be an irreducible component. We can assume without loss of generality that $k = \min\{\, h^0(X, \mathscr{F} \otimes \alpha) \mid \alpha \in Z \,\}$. By assumption, we have a decomposition $\mathscr{G} \simeq \mathscr{F} \oplus \mathscr{F}'$. We define

$$\ell = k + \min\{\, h^0(X, \mathscr{F}' \otimes \alpha) \mid \alpha \in Z \,\}.$$

By the semicontinuity of $h^0(A, \mathscr{F}' \otimes \alpha)$ and $h^0(A, \mathscr{F} \otimes \alpha)$, it follows that there is a neighborhood U of the generic point of Z such that $h^0(\mathscr{F}' \otimes \alpha) \leq \ell - k$ and $h^0(\mathscr{F} \otimes \alpha) \leq k$ for any $\alpha \in U$. Since $h^0(\mathscr{F} \otimes \alpha) < k$ for any $\alpha \in U \setminus (U \cap Z)$ it is easy to see that Z is an irreducible component of $V_\ell^0(\mathscr{G})$. □

LEMMA 10.4. *If $V_k^0(f_*\omega_X^{\otimes m})$ contains a point, then it also contains a torsion subvariety through that point.*

PROOF. Take any point in $V_k^0(f_*\omega_X^{\otimes m})$. Since $\mathrm{Pic}^0(X)$ is divisible, we may assume that our point is of the form $L_0^{\otimes m}$ for some $L_0 \in \mathrm{Pic}^0(X)$. This means that

$$h^0\big(X, \omega_X^{\otimes m} \otimes f^*L_0^{\otimes m}\big) \geq k.$$

For $r \geq 0$, set $\mathcal{I}_r = \mathcal{I}\big(\|\omega_X^{\otimes r} \otimes f^*L_0^{\otimes r}\|\big)$. According to Lemma 10.2, there exists a morphism $g\colon Y \to X$ such that

$$\omega_X \otimes (\omega_X \otimes f^*L_0)^{\otimes(m-1)} \otimes \mathcal{I}_{m-1} = \omega_X^{\otimes m} \otimes L_0^{\otimes(m-1)} \otimes \mathcal{I}_{m-1}$$

is a direct summand of $g_*\omega_Y$. Consequently, $f_*(\omega_X^{\otimes m} \otimes L_0^{m-1} \otimes \mathcal{I}_{m-1})$ is a direct summand of $h_*\omega_Y$, where $h = f \circ g\colon Y \to A$. By Simpson's theorem we know that, for any ℓ, every irreducible component of $V_\ell^0(h_*\omega_Y)$ is a torsion subvariety. Together with Lemma 10.3, this shows that every irreducible component of

$$V_k^0\big(f_*(\omega_X^{\otimes m} \otimes L_0^{\otimes(m-1)} \otimes \mathcal{I}_{m-1})\big)$$

is a torsion subvariety. We observe that this set contains L_0: the reason is that since $\mathcal{I}_m \subseteq \mathcal{I}_{m-1}$ (see [**26**, Theorem 11.1.8]), we have

$$\begin{aligned} H^0\big(X, (\omega_X \otimes f^*L_0)^{\otimes m} \otimes \mathcal{I}_m\big) &\subseteq H^0\big(X, (\omega_X \otimes f^*L_0)^{\otimes m} \otimes \mathcal{I}_{m-1}\big) \\ &\subseteq H^0\big(X, (\omega_X \otimes f^*L_0)^{\otimes m}\big), \end{aligned}$$

and the two spaces on the outside are equal because the subscheme defined by $\mathcal{I}_m$ is contained in the base locus of $(\omega_X \otimes f^*L_0)^{\otimes m}$ by [**26**, Theorem 11.1.8].

Now let W be an irreducible component of $V_k^0\big(f_*(\omega_X^{\otimes m} \otimes L_0^{\otimes(m-1)} \otimes \mathcal{I}_{m-1})\big)$ passing through the point L_0. For every $L \in W$, we have

$$h^0\big(X, \omega_X^{\otimes m} \otimes L_0^{\otimes(m-1)} \otimes L\big) \geq h^0\big(X, \omega_X^{\otimes m} \otimes L_0^{\otimes(m-1)} \otimes \mathcal{I}_{m-1} \otimes L\big) \geq k,$$

and so $L_0^{\otimes(m-1)} \otimes W \subseteq V_k^0(f_*\omega_X^{\otimes m})$. As noted above, $L_0^{\otimes(m-1)} \otimes W$ contains the point $L_0^{\otimes m}$; it is also a torsion subvariety, because W is a torsion subvariety and $L_0 \in W$. □

PROOF OF THEOREM 10.1. Let $Z \subseteq V_k^0(f_*\omega_X^{\otimes m})$ be an irreducible component; we have to show that Z is a torsion subvariety. In case Z is a point, this follows directly from Lemma 10.4, so let us assume that $\dim Z \geq 1$. Let $Z_0 \subseteq Z$ denote the Zariski-open subset obtained by removing the intersection with the other irreducible components of $V_k^0(f_*\omega_X^{\otimes m})$. Then again by Lemma 10.4, every point of Z_0 lies on a torsion subvariety that is contained in Z. Because there are only countably many torsion subvarieties in $\mathrm{Pic}^0(X)$, Baire's theorem implies that Z itself must be a torsion subvariety. □

11. Iitaka fibration and cohomological support loci. In this section, we use Theorem 5.2 to give a precise description of the cohomological support loci

$$V^0(\omega_X^{\otimes m}) = \{ P \in \operatorname{Pic}^0(X) \mid H^0(X, \omega_X^{\otimes m} \otimes P) \neq 0 \}$$

for all $m \geq 2$, in terms of the Iitaka fibration of X. After a birational modification of X, the Iitaka fibration can be realized as a morphism $f \colon X \to Y$, where Y is smooth projective of dimension $\kappa(X)$. By the universal property of the Albanese mapping, we obtain a commutative diagram

$$\begin{array}{ccc} X & \xrightarrow{a_X} & A_X \\ \downarrow{\scriptstyle f} & & \downarrow{\scriptstyle a_f} \\ Y & \xrightarrow{a_Y} & A_Y \end{array}$$

where $A_X = \operatorname{Alb}(X)$ and $A_Y = \operatorname{Alb}(Y)$ are the two Albanese varieties. The following simple lemma appears in [**9**, Lemma 2.6].

LEMMA 11.1. *With notation as above, the following things are true:*

(a) *The homomorphism a_f is surjective with connected fibers.*

(b) *Setting $K = \ker(a_f)$, we have a short exact sequence*

$$0 \to \operatorname{Pic}^0(Y) \to \operatorname{Pic}^0(X) \to \operatorname{Pic}^0(K) \to 0.$$

(c) *If F is a general fiber of f, then the kernel of the natural homomorphism $\operatorname{Pic}^0(X) \to \operatorname{Pic}^0(F)$ is a finite union of torsion translates of* $\operatorname{Pic}^0(Y)$.

Using this lemma and the results of Green-Lazarsfeld [**14, 15**] and Simpson [**40**], one can prove the following results about the locus $V^0(\omega_X)$:

(1) There are finitely many quotient abelian varieties $\operatorname{Alb}(X) \to B_i$ and finitely many torsion points $\alpha_i \in \operatorname{Pic}^0(X)$ such that

$$V^0(\omega_X) = \bigcup_{i=1}^{n} \big(\alpha_i + \operatorname{Pic}^0(B_i)\big).$$

This is proved in [**15**, Theorem 0.1] and [**40**]. Note that $V^0(\omega_X)$ may be empty; in that case, we take $n = 0$.

(2) We have $\operatorname{Pic}^0(B_i) \subseteq \operatorname{Pic}^0(Y)$, where $f \colon X \to Y$ is the Iitaka fibration; when X is of maximal Albanese dimension, then the union of the $\operatorname{Pic}^0(B_i)$ generates $\operatorname{Pic}^0(Y)$. This is proved in [**9, 10**].

(3) At a general point P of the i-th irreducible component $\alpha_i + \operatorname{Pic}^0(B_i)$, one has $s \cup v = 0$ for all $s \in H^0(X, \omega_X \otimes P)$ and all $v \in H^1(B_i, \mathscr{O}_{B_i})$; conversely, if $s \in H^0(X, \omega_X \otimes P)$ is nonzero and $s \cup v = 0$ for some $v \in H^1(X, \mathscr{O}_X)$, then necessarily $v \in H^1(B_i, \mathscr{O}_{B_i})$.

NOTE. One can interpret property (3) as follows. If $P \in \operatorname{Pic}^0(X)$ is a general point of a component of $V^0(\omega_X)$, and we identify the tangent space to $\operatorname{Pic}^0(X)$ at the point P with the vector space $H^1(X, \mathscr{O}_X)$, then $s \cup v = 0 \in H^1(X, \omega_X \otimes P)$ if and only if s deforms to first order in the direction of v. Property (3) then says that if s deforms to first order in the direction of v, then it deforms to arbitrary order.

It turns out that the cohomology support loci $V^0(\omega_X^{\otimes m})$ for $m \geq 2$ are governed by the Iitaka fibration $f \colon X \to Y$: in contrast to the case $m = 1$, every irreducible component is now simply a torsion translate of $\operatorname{Pic}^0(Y)$.

THEOREM 11.2. *Let X be a smooth complex projective variety, and let F be a general fiber of the Iitaka fibration $f\colon X \to Y$. Let $m \geq 2$.*

(a) *For every torsion point $\alpha \in \operatorname{Pic}^0(X)$, and every $\beta \in \operatorname{Pic}^0(Y)$, we have*
$$h^0(X, \omega_X^{\otimes m} \otimes \alpha) = h^0(X, \omega_X^{\otimes m} \otimes \alpha \otimes f^*\beta).$$

(b) *There exist finitely many torsion points $\alpha_1, \dots, \alpha_n \in \operatorname{Pic}^0(X)$ such that*
$$V^0(\omega_X^{\otimes m}) = \bigcup_{i=1}^{n} \bigl(\alpha_i + \operatorname{Pic}^0(Y)\bigr).$$

(c) *At every point $\alpha \in V^0(\omega_X^{\otimes m})$, one has $s \cup v = 0$ for all $s \in H^0(X, \omega_X^{\otimes m} \otimes \alpha)$ and all $v \in H^1(Y, \mathscr{O}_Y)$; conversely, if $s \in H^0(X, \omega_X^{\otimes m} \otimes \alpha)$ is nonzero and $s \cup v = 0$ for some $v \in H^1(X, \mathscr{O}_X)$, then necessarily $v \in H^1(Y, \mathscr{O}_Y)$.*

PROOF. We begin by proving (a), following Jiang's version [**20**, Lemma 3.2] of the original argument in [**19**, Proposition 2.12]. Let $g\colon X \to A_Y$ be the morphism induced by composing f with the Albanese map of Y.

$$\begin{array}{ccc} X & \xrightarrow{a_X} & A_X \\ \downarrow{\scriptstyle f} & \searrow^{g} & \downarrow{\scriptstyle a_f} \\ Y & \xrightarrow{a_Y} & A_Y \end{array}$$

Let H be an ample divisor on A_Y. By construction, g factors through the Iitaka fibration of X, and so there is an integer $d \gg 0$ such that
$$dK_X \sim g^*H + E \tag{11.3}$$
for some effective divisor E on X. In particular, all sufficiently large and divisible powers of ω_X have nontrivial global sections.

Now consider the torsion-free coherent sheaf
$$\mathscr{F} = g_*\Bigl(\omega_X^{\otimes m} \otimes \alpha \otimes \mathcal{I}\bigl(\|\omega_X^{\otimes(m-1)}\|\bigr)\Bigr)$$
on the abelian variety A_Y. From our discussion of asymptotic multiplier ideals in §10, it is easy to see that we have inclusions
$$\mathcal{I}\bigl(\|\omega_X^{\otimes m} \otimes \alpha\|\bigr) = \mathcal{I}\bigl(\|\omega_X^{\otimes m}\|\bigr) \subseteq \mathcal{I}\bigl(\|\omega_X^{\otimes(m-1)}\|\bigr),$$
by choosing the integer $p \in \mathbb{N}$ in the definition of the asymptotic multiplier ideal as a multiple of the order of the torsion point $\alpha \in \operatorname{Pic}^0(X)$. Since
$$H^0\Bigl(X, \omega_X^{\otimes m} \otimes \alpha \otimes \mathcal{I}\bigl(\|\omega_X^{\otimes m} \otimes \alpha\|\bigr)\Bigr) = H^0\bigl(X, \omega_X^{\otimes m} \otimes \alpha\bigr),$$
this shows that $H^0(A_Y, \mathscr{F}) = H^0(X, \omega_X^{\otimes m} \otimes \alpha)$. For $p \in \mathbb{N}$ sufficiently large and divisible, we have
$$\mathcal{I}\bigl(\|\omega_X^{\otimes(m-1)}\|\bigr) = \mu_*\mathscr{O}_{X'}\Bigl(K_{X'/X} - \bigl\lfloor \tfrac{1}{p} F \bigr\rfloor\Bigr).$$
Here $\mu\colon X' \to X$ is a log resolution of the complete linear system of $\omega_X^{\otimes p(m-1)}$: the divisor $F + D$ has simple normal crossing support, the linear system $|D|$ is base point free, and
$$p(m-1)\mu^*K_X \sim F + D. \tag{11.4}$$
(see [**26**, 9.2.10]). We may also assume that the larger divisor $F + D + \mu^*E$ has simple normal crossing support.

Now $\mathscr{F}$ is the pushforward, via the mapping $g \circ \mu \colon X' \to A_Y$, of the line bundle

$$\mu^*\alpha \otimes \mathscr{O}_{X'}\Big(K_{X'} + (m-1)\mu^* K_X - \big\lfloor \tfrac{1}{p} F \big\rfloor\Big),$$

and for any $\varepsilon \in \mathbb{Q}$, we have the $\mathbb{Q}$-linear equivalence of $\mathbb{Q}$-divisors

$$(m-1)\mu^* K_X - \big\lfloor \tfrac{1}{p} F \big\rfloor \sim_{\mathbb{Q}} \tfrac{1-\varepsilon}{p}(F + D) + \tfrac{\varepsilon(m-1)}{d}(\mu^* g^* H + \mu^* E) - \big\lfloor \tfrac{1}{p} F \big\rfloor$$

by combining (11.3) and (11.4). This allows us to write

$$K_{X'} + (m-1)\mu^* K_X - \big\lfloor \tfrac{1}{p} F \big\rfloor \sim_{\mathbb{Q}} K_{X'} + \Delta + \tfrac{\varepsilon(m-1)}{d}\mu^* g^* H,$$

where Δ is the $\mathbb{Q}$-divisor on X' given by the formula

$$\Delta = \tfrac{1-\varepsilon}{p} D + \tfrac{\varepsilon(m-1)}{d}\mu^* E + \tfrac{1-\varepsilon}{p} F - \big\lfloor \tfrac{1}{p} F \big\rfloor.$$

By construction, the support of Δ is a divisor with simple normal crossings; and if we choose $\varepsilon > 0$ sufficiently small, then Δ is a boundary divisor, meaning that the coefficient of every irreducible component belongs to the interval $[0, 1)$. To see that $\Delta \geq 0$ it suffices to observe that $p(m-1)\mu^* E \geq dF$ and to check that $\lfloor \Delta \rfloor = 0$ it suffices to observe that the coefficients of $\frac{1}{p} D + \{\frac{1}{p} F\}$ are < 1 and apply continuity. In particular, the pair (X', Δ) is klt. We can now apply the version for $\mathbb{Q}$-divisors of Kollár's vanishing theorem [**24**, §10] and conclude that the pushforward of

$$\mu^*(\alpha \otimes g^*\beta) \otimes \mathscr{O}_{X'}\Big(K_{X'} + (m-1)\mu^* K_X - \big\lfloor \tfrac{1}{p} F \big\rfloor\Big)$$

under the map $g \circ \mu \colon X' \to A_Y$ has vanishing higher cohomology for all $\beta \in \operatorname{Pic}^0(Y)$. Together with the projection formula, this shows that

$$H^i\big(A_Y, \mathscr{F} \otimes \beta\big) = 0$$

for every $i > 0$ and every $\beta \in \operatorname{Pic}^0(Y)$. It follows that

$$h^0\big(A_Y, \mathscr{F} \otimes \beta\big) = \chi\big(A_Y, \mathscr{F} \otimes \beta\big)$$

has the same value for every $\beta \in \operatorname{Pic}^0(Y)$. But then

$$\begin{aligned} h^0\big(X, \omega_X^{\otimes m} \otimes \alpha\big) &= h^0\big(A_Y, \mathscr{F}\big) = h^0\big(A_Y, \mathscr{F} \otimes g^*\beta\big) \\ &= h^0\Big(X, \omega_X^{\otimes m} \otimes \alpha \otimes g^*\beta \otimes \mathcal{I}\big(\|\omega_X^{\otimes(m-1)}\|\big)\Big) \\ &\leq h^0\big(X, \omega_X^{\otimes m} \otimes \alpha \otimes g^*\beta\big), \end{aligned}$$

and by semicontinuity, we conclude that in fact

$$h^0\big(X, \omega_X^{\otimes m} \otimes \alpha\big) = h^0\big(X, \omega_X^{\otimes m} \otimes \alpha \otimes g^*\beta\big).$$

We next prove (b). Assuming that $V^0(\omega_X^{\otimes m})$ is nonempty, there are by Theorem 10.1 finitely many distinct torsion elements $\alpha_1, \dots, \alpha_s \in \operatorname{Pic}^0(X)$, and abelian subvarieties $B_i \subset \operatorname{Pic}^0(X)$, such that

$$V^0(\omega_X^{\otimes m}) = \bigcup_{i=1}^{s} (\alpha_i + B_i).$$

By (a) we know that $\operatorname{Pic}^0(Y) \subseteq B_i$; indeed, (a) implies that for every torsion point $\alpha \in V^0(\omega_X^{\otimes m})$ we have $\alpha + \operatorname{Pic}^0(Y) \subseteq V^0(\omega_X^{\otimes m})$, and this applies of course to α_i. To prove (b), it is therefore enough to show that $B_i \subseteq \operatorname{Pic}^0(Y)$. Take an arbitrary element $\alpha \in V^0(\omega_X^{\otimes m})$, and let $s \in H^0(X, \omega_X^{\otimes m} \otimes \alpha)$ be a nonzero global section.

The restriction of s to a general fiber F of the Iitaka fibration $f\colon X \to Y$ is then a nonzero global section of

$$\omega_F^{\otimes m} \otimes \alpha|_F,$$

and because $\kappa(F) = 0$, it follows that $\alpha|_F$ is torsion in $\mathrm{Pic}^0(F)$. According to Lemma 11.1, a nonzero multiple of α therefore belongs to $\mathrm{Pic}^0(Y)$. This is enough to conclude that $B_i \subseteq \mathrm{Pic}^0(Y)$, and so (b) is proved.

To prove (c), note first that standard arguments (see for instance [**13**, Lemma 12.6]) imply the first half, namely that if $v \in H^1(Y, \mathscr{O}_Y)$ then

$$s \cup v = 0 \qquad \text{for all } s \in H^0\big(X, \omega_X^{\otimes m} \otimes \alpha\big).$$

For the second half, suppose that $s \in H^0(X, \omega_X^{\otimes m} \otimes \alpha)$ is a nonzero global section such that $s \cup v = 0$ for some $v \in H^1(X, \mathscr{O}_X)$. Restricting to a general fiber F of the Iitaka fibration $f\colon X \to Y$, we get

$$0 = s|_F \cup v|_F \in H^1\big(F, \omega_F^{\otimes m} \otimes \alpha|_F\big),$$

where $s|_F \in H^0\big(F, \omega_F^{\otimes m} \otimes \alpha|_F\big)$ is nonzero, and $v|_F \in H^1(F, \mathscr{O}_F)$. Since $\alpha \in V^0(\omega_X^{\otimes m})$, we have $\alpha|_F = \alpha_i|_F$ for some $i = 1, \dots, s$, as a consequence of Lemma 11.1 and (b). In particular, $\alpha|_F$ is torsion, say of order k, and so

$$s^{\otimes k}|_F \in H^0\big(F, \omega_F^{\otimes km}\big).$$

Let $a_F\colon F \to A_F$ denote the Albanese mapping of F. Recalling that $\kappa(F) = 0$, we get from Theorem 5.2 that $(a_F)_*\omega_F^{\otimes km} = \mathscr{O}_{A_F}$. Under the isomorphism

$$H^0\big(F, \omega_F^{\otimes km}\big) = H^0\big(A_F, \mathscr{O}_{A_F}\big),$$

our nonzero section $s^{\otimes k}|_F$ therefore corresponds to a nonzero constant $\sigma \in \mathbb{C}$; likewise, under the isomorphism

$$H^1(F, \mathscr{O}_F) = H^1\big(A_F, \mathscr{O}_{A_F}\big),$$

the vector $v|_F$ corresponds to a vector $u \in H^1\big(A_F, \mathscr{O}_{A_F}\big)$. It is not hard to see that the two isomorphisms are compatible with cup product; consequently, $s^{\otimes k} \cup v = 0$ implies that $\sigma u = 0$, and hence that $u = 0$. By the infinitesimal version of Lemma 11.1, this means that $v \in H^1(Y, \mathscr{O}_Y)$, as asserted. □

D. Singular metrics on pushforwards of adjoint line bundles

12. Plurisubharmonic functions. Let X be a complex manifold. We begin our survey of the analytic techniques by recalling the following important definition; see for example [**12**, I.5] for more details.

DEFINITION 12.1. A function $\varphi\colon X \to [-\infty, +\infty)$ is called *plurisubharmonic* if it is upper semi-continuous, locally integrable, and satisfies the mean-value inequality

$$(\varphi \circ \gamma)(0) \leq \frac{1}{\pi} \int_\Delta (\varphi \circ \gamma)\, d\mu \tag{12.2}$$

for every holomorphic mapping $\gamma\colon \Delta \to X$ from the open unit disk $\Delta \subseteq \mathbb{C}$.

Suppose that φ is plurisubharmonic. From (12.2) one can deduce, by integrating over the space of lines through a given point, that the mean-value inequality

$$(\varphi \circ \iota)(0) \leq \frac{1}{\mu(B)} \int_B (\varphi \circ \iota)\, d\mu$$

also holds for any open embedding $\iota\colon B \hookrightarrow X$ of the open unit ball $B \subseteq \mathbb{C}^n$; here n is the local dimension of X at the point $\iota(0)$. In other words, every plurisubharmonic function is also *subharmonic*. Together with local integrability, this implies that φ is locally bounded from above.

LEMMA 12.3. *Every plurisubharmonic function on a compact complex manifold is locally constant.*

PROOF. Let φ be a plurisubharmonic function on a compact complex manifold X. As φ is upper semi-continuous and locally bounded from above, it achieves a maximum on every connected component of X. The mean-value inequality then forces φ to be locally constant. □

Observe that a plurisubharmonic function is uniquely determined by its values on any subset whose complement has measure zero. Indeed, the mean-value inequality provides an upper bound on the value at any point $x \in X$, and the upper semi-continuity a lower bound. One also has the following analogue of the Riemann and Hartogs extension theorems for holomorphic functions [**12**, Theorem I.5.24]; by what we have just said, there can be at most one extension in each case.

LEMMA 12.4. *Let $Z \subseteq X$ be a closed analytic subset, and let φ be a plurisubharmonic function on $X \setminus Z$.*

(a) *If* $\operatorname{codim} Z \geq 2$, *then φ extends to a plurisubharmonic function on X.*
(b) *If* $\operatorname{codim} Z = 1$, *then φ extends to a plurisubharmonic function on X if and only if it is locally bounded near every point of Z.*

A plurisubharmonic function φ determines a coherent sheaf of ideals $\mathcal{I}(\varphi) \subseteq \mathscr{O}_X$, called the *multiplier ideal sheaf*, whose sections over any open subset $U \subseteq X$ consist of those holomorphic functions $f \in H^0(U, \mathscr{O}_X)$ for which the function $|f|^2 e^{-\varphi}$ is locally integrable. We use the convention that the value of the product is 0 at points $x \in X$ where $f(x) = 0$ and $\varphi(x) = -\infty$.

Since plurisubharmonic functions are locally bounded from above, Montel's theorem in several variables implies the following compactness property.

PROPOSITION 12.5. *Let $\varphi\colon B \to [-\infty, +\infty)$ be a plurisubharmonic function on the open unit ball $B \subseteq \mathbb{C}^n$. Consider the collection of holomorphic functions*

$$H_K(\varphi) = \left\{ f \in H^0(B, \mathscr{O}_B) \;\middle|\; \int_B |f|^2 e^{-\varphi}\, d\mu \leq K \right\}.$$

Any sequence of functions in $H_K(\varphi)$ has a subsequence that converges uniformly on compact subsets to an element of $H_K(\varphi)$.

PROOF. The mean-value inequality for holomorphic functions implies that all functions in $H_K(\varphi)$ are uniformly bounded on every closed ball of radius $R < 1$. Let us briefly review the argument. Because φ is locally bounded from above, there is a constant $C \geq 0$ such that $\varphi \leq C$ on the closed ball of radius $(R+1)/2$. Fix a point $z \in \overline{B}_R(0)$ in the closed ball of radius R, and a holomorphic function $f \in H_K(\varphi)$. By the mean-value inequality,

$$|f(z)|^2 \leq \frac{1}{r^n \mu(B)} \int_{B_r(z)} |f|^2\, d\mu \leq \frac{e^C}{r^n \mu(B)} \int_{B_r(z)} |f|^2 e^{-\varphi}\, d\mu \leq \frac{K \cdot e^C}{r^n \mu(B)},$$

where $r = (1-R)/2$. By the n-dimensional version of Montel's theorem [**17**, Theorem I.A.12], this uniform bound implies that any sequence $f_0, f_1, f_2, \ldots \in H_K(\varphi)$

has a subsequence that converges uniformly on compact subsets to a holomorphic function $f \in H^0(B, \mathscr{O}_B)$. By Fatou's lemma,

$$\int_B |f|^2 e^{-\varphi} \, d\mu \leq \liminf_{k \to +\infty} \int_B |f_k|^2 e^{-\varphi} \, d\mu \leq K,$$

which means that $f \in H_K(\varphi)$. □

NOTE. The example of an orthonormal sequence in $H_K(\varphi)$ shows that the convergence need not be with respect to the L^2-norm.

13. Singular hermitian metrics on line bundles. Many of the newer applications of analytic techniques in algebraic geometry – such as Siu's proof of the invariance of plurigenera – rely on the notion of *singular hermitian metrics* on holomorphic line bundles. The word "singular" here means two things at once: first, that the metric is not necessarily C^∞; second, that certain vectors in the fibers of the line bundle are allowed to have either infinite length or length zero.

Let X be a complex manifold, and let L be a holomorphic line bundle on X with a singular hermitian metric h. In any local trivialization of L, such a metric is represented by a "weight function" of the form $e^{-\varphi}$, where φ is a measurable function with values in $[-\infty, +\infty]$. More precisely, suppose that the restriction of L to an open subset $U \subseteq X$ is trivial, and that $s_0 \in H^0(U, L)$ is a nowhere vanishing holomorphic section. Then any other holomorphic section $s \in H^0(U, L)$ can be written as $s = f s_0$ for a unique holomorphic function f on U, and the length squared of s with respect to the singular hermitian metric h is

$$|s|_h^2 = |f|^2 e^{-\varphi}. \tag{13.1}$$

The points where φ is not finite correspond to singularities of the metric: at points where $\varphi(x) = -\infty$, the metric becomes infinite; at points where $\varphi(x) = +\infty$, the metric stops being positive definite.

NOTE. At points $x \in U$ where $\varphi(x) = -\infty$, we use the following convention: the product in (13.1) equals 0 if $f(x) = 0$; otherwise, it equals $+\infty$. With this rule in place, $|s|_h$ is a well-defined measurable function on U with values in $[0, +\infty]$.

We say that a singular hermitian metric h is *continuous* if the local weight functions φ are continuous functions with values in $[-\infty, +\infty]$. This is equivalent to asking that, for every open subset $U \subseteq X$ and every section $s \in H^0(U, L)$, the function $|s|_h \colon U \to [0, +\infty]$ should be continuous.

We say that the pair (L, h) has *semi-positive curvature* if the local weight functions φ are plurisubharmonic. In that case, φ is locally integrable, and the curvature current of (L, h) can be defined, in the sense of distributions, by the formula

$$\Theta_h = \frac{\sqrt{-1}}{2\pi} \partial \bar{\partial} \varphi.$$

It is easy to see that Θ_h is a well-defined closed positive $(1,1)$-current on X; its cohomology class in $H^2(X, \mathbb{R})$ equals the first Chern class $c_1(L)$. Conversely, if the current Θ_h is positive, then one can make the local weight functions φ plurisubharmonic by modifying them on a set of measure zero [**12**, Theorem I.5.8].

NOTE. Most authors include the condition of local integrability into the definition of a singular hermitian metric. We use a different convention, so as to be consistent with the definition of singular hermitian metrics on vector bundles later on.

A singular hermitian metric of semi-positive curvature is automatically positive definite at every point. Indeed, φ is locally bounded from above, and so the factor $e^{-\varphi}$ in the local expression for h may equal $+\infty$ at certain points, but has to be locally bounded from below by a positive constant. Moreover, φ is upper semi-continuous, and so the function $|s|_h \colon U \to [0, +\infty]$ is not just measurable, but even lower semi-continuous, for every holomorphic section $s \in H^0(U, L)$ on some open subset $U \subseteq X$.

LEMMA 13.2. *Suppose that X is compact, and that h is a singular hermitian metric with semi-positive curvature on a holomorphic line bundle L. If $c_1(L) = 0$ in $H^2(X, \mathbb{R})$, then h is actually a smooth metric with zero curvature.*

PROOF. The cohomology class of the closed positive $(1,1)$-current Θ_h equals zero in $H^2(X, \mathbb{R})$, and so there is a globally defined plurisubharmonic function ψ in X such that $\Theta_h = \frac{\sqrt{-1}}{2\pi} \partial\bar{\partial}\psi$. By Lemma 12.3, ψ is locally constant, and so $\Theta_h = 0$. Now all the local weight functions φ satisfy $\partial\bar{\partial}\varphi = 0$, and are therefore smooth functions; but this means exactly that h is a smooth metric. □

The curvature assumption implies that the *multiplier ideal sheaf* $\mathcal{I}(h) \subseteq \mathscr{O}_X$ is a coherent sheaf of ideals on X; in the notation from above, a holomorphic function $f \colon U \to \mathbb{C}$ is a section of $\mathcal{I}(h)$ if and only if the function $|f|^2 e^{-\varphi}$ is locally integrable. Consequently, the subspace

$$H^0\big(X, L \otimes \mathcal{I}(h)\big) \subseteq H^0(X, L)$$

consists of all global holomorphic sections of L for which the lower semi-continuous function $|s|_h^2 \colon X \to [0, +\infty]$ is locally integrable.

14. The Ohsawa-Takegoshi extension theorem. It is known that a line bundle on a projective complex manifold admits a singular hermitian metric with semi-positive curvature if and only if it is pseudo-effective. The power of the metric approach to positivity comes from fact that one can extend holomorphic sections from submanifolds with precise bounds on the norm of the extension. The most important result in this direction is the famous *Ohsawa-Takegoshi theorem.*[2]

Let X be a complex manifold of dimension n, and let (L, h) be a holomorphic line bundle with a singular hermitian metric of semi-positive curvature. What we actually need is the "adjoint version" of the Ohsawa-Takegoshi theorem, which is about extending sections of the adjoint bundle $\omega_X \otimes L$, or equivalently, holomorphic n-forms with coefficients in L. Before we can state the theorem, we first have to introduce some notation.

Given $\beta \in H^0(X, \omega_X \otimes L)$, we define a nonnegative measurable (n,n)-form $|\beta|_h^2$ as follows: view $\beta \wedge \overline{\beta}$ as a smooth (n,n)-form with coefficients in $L \otimes \overline{L}$, compose with the singular hermitian metric h, and then multiply by the factor $c_n = 2^{-n}(-1)^{n^2/2}$. Locally, we can write $\beta = f s_0 \otimes dz_1 \wedge \cdots \wedge dz_n$, and then

$$|\beta|_h^2 = |f|^2 e^{-\varphi} (dx_1 \wedge dy_1) \wedge \cdots \wedge (dx_n \wedge dy_n), \tag{14.1}$$

[2] We use the name "Ohsawa-Takegoshi theorem" for convenience only; in reality, there is a large collection of different L^2-extension theorems in complex analysis, of which Theorem 14.4 below is an important but nevertheless special case. For more on this topic, see for example [**12**].

where $z_1 = x_1 + y_1\sqrt{-1}, \dots, z_n = x_n + y_n\sqrt{-1}$ are local holomorphic coordinates on U. Using this notation, we have

$$H^0\bigl(X, \omega_X \otimes L \otimes \mathcal{I}(h)\bigr) \\ = \bigl\{ \beta \in H^0(X, \omega_X \otimes L) \bigm| |\beta|_h^2 \text{ is locally integrable} \bigr\}.$$

We also define the L^2-norm of the element $\beta \in H^0(X, \omega_X \otimes L)$ to be

$$\|\beta\|_h^2 = \int_X |\beta|_h^2 \in [0, +\infty]. \tag{14.2}$$

When X is compact, $\beta \in H^0\bigl(X, \omega_X \otimes L \otimes \mathcal{I}(h)\bigr)$ is equivalent to having $\|\beta\|_h^2 < +\infty$; in general, finiteness of the L^2-norm is a much stronger requirement.

Now suppose that $f\colon X \to B$ is a holomorphic mapping to the open unit ball $B \subseteq \mathbb{C}^r$. We assume that f is projective and that $0 \in B$ is a regular value of f; the central fiber $X_0 = f^{-1}(0)$ is therefore a projective complex manifold of dimension $n - r = \dim X - \dim B$. We denote by (L_0, h_0) the restriction of (L, h) to X_0. As long as h_0 is not identically equal to $+\infty$, it defines a singular hermitian metric with semi-positive curvature on L_0, and we have the space

$$H^0\bigl(X_0, \omega_{X_0} \otimes L_0 \otimes \mathcal{I}(h_0)\bigr) \tag{14.3}$$

of holomorphic $(n-r)$-forms with coefficients in L_0 that are square-integrable with respect to h_0; as before, the defining condition is that the integral

$$\|\alpha\|_{h_0}^2 = \int_{X_0} |\alpha|_{h_0}^2$$

should be finite; note that the definition of $|\alpha|_{h_0}^2$ involves the constant c_{n-r}.

The Ohsawa-Takegoshi theorem says that every section of $\omega_{X_0} \otimes L_0 \otimes \mathcal{I}(h_0)$ can be extended to a section of $\omega_X \otimes L \otimes \mathcal{I}(h)$ with finite L^2-norm – and, crucially, it provides a universal upper bound on the L^2-norm of the extension. (If $h_0 \equiv +\infty$, then the space in (14.3) is trivial and the extension problem is not interesting.) Here $\beta \in H^0(X, \omega_X \otimes L)$ is an extension of $\alpha \in H^0(X_0, \omega_{X_0} \otimes L_0)$ if

$$\beta|_{X_0} = \alpha \wedge df = \alpha \wedge (df_1 \wedge \cdots \wedge df_r),$$

where $f = (f_1, \dots, f_r)$. That said, the precise statement of the Ohsawa-Takegoshi extension theorem is the following.

THEOREM 14.4. *Let $f\colon X \to B$ be a projective morphism such that $0 \in B$ is a regular value. Let (L, h) be a holomorphic line bundle with a singular hermitian metric of semi-positive curvature. Denote by (L_0, h_0) the restriction to the central fiber $X_0 = f^{-1}(0)$, and suppose that $h_0 \not\equiv +\infty$. Then for every $\alpha \in H^0\bigl(X_0, \omega_{X_0} \otimes L_0 \otimes \mathcal{I}(h_0)\bigr)$, there exists at least one $\beta \in H^0\bigl(X, \omega_X \otimes L \otimes \mathcal{I}(h)\bigr)$ with*

$$\beta|_{X_0} = \alpha \wedge df \quad \text{and} \quad \|\beta\|_h^2 \le \mu(B) \cdot \|\alpha\|_{h_0}^2.$$

The special thing about this form of the extension theorem is that the constant $\mu(B) = \pi^r/r!$ in the estimate is the volume of the unit ball $B \subseteq \mathbb{C}^r$; the example of a product $X = B \times X_0$ shows that this is optimal. Earlier proofs of the Ohsawa-Takegoshi theorem, for example by Siu or Păun [**33, 41**], only gave a weaker estimate, in which $\mu(B)$ had to be replaced by a certain constant $C_0 \le 200$. The proof of the sharp estimate is due to Błocki and Guan-Zhou [**4, 16**]. There is also a (weaker) version of the Ohsawa-Takegoshi theorem for the case where the fibers are compact Kähler manifolds, proved by Cao [**5**].

PROOF OF THEOREM 14.4. In [**16**, §3.12], the result is stated only for "projective families", meaning in the case where $f\colon X \to B$ is smooth and everywhere submersive, but the same proof works as long as $0 \in B$ is a regular value. Guan and Zhou have $(2\pi)^r/r!$ as the constant, but the extra factor of 2^r goes away because our definition of $\|\beta\|_h^2$ and $\|\alpha\|_h^2$ involves dividing by 2^n and 2^{n-r}, respectively.

For the reader who wants to look up the result in [**16**], we briefly explain how to deduce Theorem 14.4 from Guan and Zhou's main theorem. Choose an embedding $X \hookrightarrow B \times \mathbb{P}^N$, and let $H \subseteq X$ be the preimage of a sufficiently general hyperplane in $\mathbb{P}^N$. Then $X \setminus H$ is a Stein manifold and $X_0 \setminus X_0 \cap H$ a closed submanifold. We can now apply [**16**, Theorem 2.2] to the pair (X, X_0), taking $A = 0$, $c_A(t) \equiv 1$, and $\Psi = r \log|f|^2 = r \log\bigl(|f_1|^2 + \cdots + |f_r|^2\bigr)$. □

NOTE. Observe that if we write the inequality in Theorem 14.4 in the form

$$\frac{1}{\mu(B)} \|\beta\|_h^2 \leq \|\alpha\|_{h_0}^2,$$

then it looks like a mean-value inequality; this fact will play a crucial role later or, when we construct singular hermitian metrics on pushforwards of adjoint bundles.

15. Coherent sheaves and Fréchet spaces. In this section, we briefly review some fundamental results about section spaces of coherent sheaves on complex manifolds. Recall that a *Fréchet space* is a Hausdorff topological vector space, whose topology is induced by a countable family of semi-norms, and which is complete with respect to this family of semi-norms. Most of the familiar theorems about Banach spaces, such as the open mapping theorem or the closed graph theorem, remain true for Fréchet spaces.

EXAMPLE 15.1. On a complex manifold X, the vector space $H^0(X, \mathscr{O}_X)$ of all holomorphic functions on X is a Fréchet space, under the topology of uniform convergence on compact subsets. More precisely, each compact subset $K \subseteq X$ gives rise to a semi-norm

$$\|f\|_K = \sup_{x \in K} |f(x)|$$

on the space $H^0(X, \mathscr{O}_X)$; to get a countable family, write X as a countable union of compact subsets. The same construction works for any open subset $U \subseteq X$, and when $U \subseteq V$, the restriction mapping $H^0(V, \mathscr{O}_X) \to H^0(U, \mathscr{O}_X)$ is continuous.

In fact, the section spaces of *all* coherent sheaves on a given complex manifold can be made into Fréchet spaces in a consistent way; the construction is explained for example in [**17**, Ch. VIII, §A]. Let $\mathscr{F}$ be a coherent sheaf on a complex manifold X. Then for every open subset $U \subseteq X$, the space of sections $H^0(U, \mathscr{F})$ has the structure of a Fréchet space, in such a way that if $U \subseteq V$, the restriction mapping $H^0(V, \mathscr{F}) \to H^0(U, \mathscr{F})$ is continuous. Moreover, if $\phi\colon \mathscr{F} \to \mathscr{G}$ is any morphism between two coherent sheaves, then the induced mappings $\phi_U\colon H^0(U, \mathscr{F}) \to H^0(U, \mathscr{G})$ are all continuous. The Fréchet space topology has several other good properties, such as the following [**17**, Proposition VIII.A.2].

PROPOSITION 15.2. *If $\mathscr{F} \subseteq \mathscr{G}$, then $H^0(U, \mathscr{F})$ is a closed subspace of $H^0(U, \mathscr{G})$.*

Let $f\colon X \to Y$ be a proper holomorphic mapping between complex manifolds, and let $\mathscr{F}$ be a coherent sheaf on X. By Grauert's coherence theorem, the pushforward sheaf $f_*\mathscr{F}$ is a coherent sheaf on Y. The vector space

$$H^0(Y, f_*\mathscr{F}) = H^0(X, \mathscr{F})$$

therefore has two (a priori different) Fréchet space topologies, one coming from Y, the other from X.

PROPOSITION 15.3. *In the situation just described, the two Fréchet space topologies on $H^0(Y, f_*\mathscr{F}) = H^0(X, \mathscr{F})$ are equal.*

PROOF. Since the problem is local, we may replace Y by a Stein open subset and assume that we have a surjective morphism

$$\mathscr{O}_Y^{\oplus m} \to f_*\mathscr{F}.$$

The induced mapping $H^0(Y, \mathscr{O}_Y)^{\oplus m} \to H^0(Y, f_*\mathscr{F})$ is continuous and surjective; by the open mapping theorem, the topology on $H^0(Y, f_*\mathscr{F})$ must be the quotient topology. We also get a morphism $\mathscr{O}_X^{\oplus m} \to \mathscr{F}$, and therefore a factorization

$$H^0(Y, \mathscr{O}_Y)^{\oplus m} \to H^0(X, \mathscr{O}_X)^{\oplus m} \to H^0(X, \mathscr{F}).$$

Both mappings are continuous: the first because, f being proper, uniform convergence on compact subsets of Y implies uniform convergence on compact subsets of X; the second because $\mathscr{O}_X^{\oplus m} \to \mathscr{F}$ is a morphism. It follows that the identity mapping

$$H^0(Y, f_*\mathscr{F}) \to H^0(X, \mathscr{F})$$

is continuous; by the open mapping theorem, it must be a homeomorphism. □

16. Singular hermitian inner products. Before we can talk about singular hermitian metrics on vector bundles, we first have to be clear about what we mean by a "singular" hermitian inner product on a vector space. The purpose of this section is to define this notion with some care. Throughout, we let V be a finite-dimensional complex vector space. There are two ways in which a hermitian inner product can be singular: there may be vectors whose length is $+\infty$, and others whose length is 0. The best way to formalize this is to work not with the inner product itself, but with the associated length function [**3**, §3].

DEFINITION 16.1. A *singular hermitian inner product* on a finite-dimensional complex vector space V is a function $|-|_h \colon V \to [0, +\infty]$ with the following properties:

(1) $|\lambda v|_h = |\lambda| \cdot |v|_h$ for every $\lambda \in \mathbb{C} \setminus \{0\}$ and every $v \in V$, and $|0|_h = 0$.
(2) $|v + w|_h \leq |v|_h + |w|_h$ for every $v, w \in V$.
(3) $|v + w|_h^2 + |v - w|_h^2 = 2|v|_h^2 + 2|w|_h^2$ for every $v, w \in V$.

Our convention is that an inequality is satisfied if both sides are equal to $+\infty$. It is easy to deduce from the axioms that both

$$V_0 = \{ v \in V \mid |v|_h = 0 \} \quad \text{and} \quad V_{fin} = \{ v \in V \mid |v|_h < +\infty \}$$

are linear subspaces of V. We say that h is *positive definite* if $V_0 = 0$; we say that h is *finite* if $V_{fin} = V$. Clearly, $|-|_h$ is a semi-norm on V_{fin}; it is a norm if and only if $V_0 = 0$. The third axiom is the parallelogram law for this semi-norm. The formula

$$\langle v, w \rangle_h = \frac{1}{4} \sum_{k=0}^{3} (\sqrt{-1})^k \cdot \left| v + (\sqrt{-1})^k w \right|_h$$

therefore defines a semi-definite hermitian inner product on the subspace V_{fin}; it is positive definite if and only if $V_0 = 0$. We use the same notation for the induced hermitian inner product on the quotient space V_{fin}/V_0.

Given a singular hermitian inner product h on V, we obtain a singular hermitian inner product h^* on the dual space $V^* = \operatorname{Hom}_{\mathbb{C}}(V, \mathbb{C})$ by setting

$$|f|_{h^*} = \sup \left\{ \frac{|f(v)|}{|v|_h} \;\middle|\; v \in V \text{ with } |v|_h \neq 0 \right\}$$

for any linear functional $f \in V^*$, with the understanding that a fraction with denominator $+\infty$ is equal to 0. (If $V_0 = V$, then we define $|f|_{h^*} = 0$ for $f = 0$, and $|f|_{h^*} = +\infty$ otherwise.) It is easy to see that $|f|_{h^*} = 0$ if and only if f annihilates the subspace V_{fin}, and that $|f|_{h^*} < +\infty$ if and only if f annihilates the subspace V_0. One then checks that h^* is again a singular hermitian inner product on V^*, and that the resulting hermitian inner product $\langle -, - \rangle_{h^*}$ on

$$\frac{\{ f \in V^* \mid |f|_{h^*} < +\infty \}}{\{ f \in V^* \mid |f|_{h^*} = 0 \}} \simeq \operatorname{Hom}_{\mathbb{C}}\big(V_{fin}/V_0, \mathbb{C}\big)$$

agrees with the one naturally induced by $\langle -, - \rangle_h$. Here is another way to think about h^*. From a nonzero linear functional $f \colon V \to \mathbb{C}$, we get an induced singular hermitian inner product on $\mathbb{C}$ by setting

$$|\lambda|_{h,f} = \inf\{ |v|_h \mid v \in V \text{ satisfies } f(v) = \lambda \}$$

If $\lambda \neq 0$, this quantity is $+\infty$ unless the restriction of f to the subspace V_{fin} is nonzero; if $V_0 = V$, then $|\lambda|_{h,f} = 0$ for every $\lambda \in \mathbb{C}$. Taking into account various special cases, the following result is immediate from the definition.

LEMMA 16.2. *Let $f \colon V \to \mathbb{C}$ be a nonzero linear functional. Then*

$$|\lambda|_{h,f} = \frac{|\lambda|}{|f|_{h^*}}$$

for every nonzero $\lambda \in \mathbb{C}$.

Let $r = \dim V$. Since the product of 0 and $+\infty$ is undefined, we do not get a singular hermitian inner product on

$$\det V = \bigwedge^r V$$

unless $V_0 = 0$ or $V_{fin} = V$. But when h is either positive definite or finite, there is a well-defined singular hermitian inner product $\det h$ on the one-dimensional vector space $\det V$. If $V_{fin} = V$, we declare that

$$|v_1 \wedge \cdots \wedge v_r|_{\det h} = \det \begin{pmatrix} \langle v_1, v_1 \rangle_h & \cdots & \langle v_1, v_r \rangle_h \\ \vdots & \ddots & \vdots \\ \langle v_r, v_1 \rangle_h & \cdots & \langle v_r, v_r \rangle_h \end{pmatrix}.$$

If $V_{fin} \neq V$ and $V_0 = 0$, we let $|-|_{\det h}$ equal $+\infty$ on all nonzero elements of $\det V$.

17. Singular hermitian metrics on vector bundles. The purpose of this section is to extend the concept of singular hermitian metrics from holomorphic line bundles to holomorphic vector bundles of arbitrary rank. Let X be a complex manifold, and let E be a holomorphic vector bundle on X of some rank $r \geq 1$.

DEFINITION 17.1. A *singular hermitian metric* on E is a function h that associates to every point $x \in X$ a singular hermitian inner product $|-|_{h,x} \colon E_x \to [0, +\infty]$ on the complex vector space E_x, subject to the following two conditions:

(1) h is finite and positive definite almost everywhere, meaning that for all x outside a set of measure zero, $|-|_{h,x}$ is a hermitian inner product on E_x.
(2) h is measurable, meaning that the function

$$|s|_h \colon U \to [0, +\infty], \quad x \mapsto \big|s(x)\big|_{h,x},$$

is measurable whenever $U \subseteq X$ is open and $s \in H^0(U, E)$.

In the case $r = 1$, this specializes to the definition of singular hermitian metrics on holomorphic line bundles. The requirement that h be measurable is extremely weak: the singular hermitian metrics that we will actually encounter below are at least semi-continuous. The advantage of the above definition is that it behaves well under duality. By applying the general construction from the previous section, we obtain on each fiber

$$E_x^* = \operatorname{Hom}_{\mathbb{C}}(E_x, \mathbb{C})$$

of the dual bundle E^* a singular hermitian inner product $|-|_{h^*,x}$. The following result shows that these form a singular hermitian metric on E^*.

PROPOSITION 17.2. *A singular hermitian metric h on a holomorphic vector bundle E induces a singular hermitian metric h^* on the dual bundle E^*.*

PROOF. If $|-|_{h,x}$ is finite and positive definite, then $|-|_{h^*,x}$ is also finite and positive definite, and so the first condition in the definition is clearly satisfied. The second condition is of a local nature, and so we may assume without loss of generality that E is the trivial bundle of rank r. Denote by $s_1, \dots, s_r \in H^0(X, E)$ the global sections corresponding to a choice of trivialization. The expression

$$H_{i,j}(x) = \langle s_i(x), s_j(x) \rangle_{h,x}$$

is well-defined outside a set of measure zero, and the resulting function is measurable. Denote by $H \in \operatorname{Mat}_{r \times r}(\mathbb{C})$ the $r \times r$-matrix with entries $H_{i,j}$. Then h^* is represented by the transpose of the matrix H^{-1}, in the natural trivialization of E^*; the usual formula for the inverse of a matrix shows that all entries of this matrix are again measurable functions. □

NOTE. In more sheaf-theoretic terms, a singular hermitian metric on a holomorphic vector bundle E is a morphism of sheaves of sets

$$|-|_h \colon E \to \mathscr{M}_X$$

from E to the sheaf of measurable functions on X with values in $[0, +\infty]$. The following conditions need to be satisfied:

(1) One has $|fs|_h = |f| \cdot |s|_h$ for every $s \in H^0(U, E)$ and every $f \in H^0(U, \mathscr{O})$.
(2) If $s \in H^0(U, E)$ and $|s|_h = 0$ almost everywhere, then $s = 0$.
(3) For almost every point $x \in X$, the function $|-|_{h,x} \colon E_x \to [0, +\infty]$ is a singular hermitian inner product (in the sense of Definition 17.1).

Again, we use the convention that $|f| \cdot |s|_h = 0$ at points where f is zero.

18. Semi-positive curvature. Let h be a singular hermitian metric on a holomorphic vector bundle E, and denote by h^* the induced singular hermitian metric on the dual bundle E^*. Suppose for a moment that h is smooth, and denote by Θ_h the curvature tensor of the Chern connection; it is a $(1,1)$-form with coefficients in the bundle $\operatorname{End}(E)$. One says that (E,h) has *semi-positive curvature in the sense of Griffiths* if, for every choice of holomorphic tangent vector $\xi \in T_x X$, the matrix $\Theta_h(\xi,\bar{\xi})$ is positive semi-definite [**12**, VII.6]. This is known to be equivalent to the condition that the function $\log|f|_{h^*}$ is plurisubharmonic for every local section $f \in H^0(U,E^*)$. In the singular case, we use this condition as the definition.

DEFINITION 18.1. We say that the pair (E,h) has *semi-positive curvature* if the function $\log|f|_{h^*}$ is plurisubharmonic for every $f \in H^0(U,E^*)$.

The point of this definition is that it allows us to talk about the curvature of a singular hermitian metric without mentioning the curvature tensor: unlike in the case of line bundles, the curvature tensor of h does not in general make sense even as a distribution [**38**, Theorem 1.3]. The following lemma gives an equivalent formulation of the definition.

LEMMA 18.2. *Let h be a singular hermitian metric on E. Then (E,h) has semi-positive curvature if, and only if, for every open subset $U \subseteq X$ and every nonzero morphism $E|_U \to L$ to a line bundle, the induced singular hermitian metric on L has semi-positive curvature.*

PROOF. The construction of the induced singular hermitian metric on L works as in Lemma 16.2. At each point $x \in U$, the linear mapping $E_x \to L_x$ between fibers induces a singular hermitian inner product on the one-dimensional complex vector space L_x: the length of a vector $\lambda \in L_x$ is the infimum of $|e|_{h,x}$ over all $e \in E_x$ that map to λ. (If $E_x \to L_x$ is zero, then the infimum equals $+\infty$ whenever $\lambda \neq 0$.)

Let us compute the curvature of the induced metric. After replacing X by an open neighborhood of a given point in U, we may assume that L is trivial; our morphism $E \to \mathscr{O}_X$ is then given by a linear functional $f \in H^0(X,E^*)$. Let $e^{-\varphi}$ be the weight function of the induced metric. The formula in Lemma 16.2 says that

$$e^{-\varphi(x)} = \frac{1}{|f(x)|^2_{h^*,x}}$$

for every $x \in X$. Taking logarithms, we get $\varphi = 2\log|f|_{h^*}$, which is plurisubharmonic because the pair (E,h) has semi-positive curvature. □

Suppose that (E,h) has semi-positive curvature. Since plurisubharmonic functions are locally bounded from above, the singular hermitian inner product $|-|_{h^*,x}$ on E_x^* must be finite for every $x \in X$; dually, every $|-|_{h,x}$ must be positive definite. The determinant line bundle $\det E$ therefore has a well-defined singular hermitian metric that we denote by the symbol $\det h$. We will prove later (in Proposition 25.1) that the pair $(\det E, \det h)$ again has semi-positive curvature.

When (E,h) has semi-positive curvature, the pointwise length of any holomorphic section of E^* is an upper semi-continuous function. Likewise, the pointwise length of any holomorphic section of E is a lower semi-continuous function.

LEMMA 18.3. *If (E, h) has semi-positive curvature, then for any $s \in H^0(X, E)$, the function $|s|_h : X \to [0, +\infty]$ is lower semi-continuous.*

PROOF. Since the question is local, we may assume without loss of generality that X is the open unit ball in $\mathbb{C}^n$, and E the trivial bundle of rank $r \geq 1$. We have

$$|s|_h \geq \frac{|f(s)|}{|f|_{h^*}}$$

for every $f \in H^0(X, E^*)$, and it is easy to see that $|s|_h$ is the pointwise supremum of the collection of functions on the right-hand side. Because $\log|f|_{h^*}$ is upper semi-continuous, each

$$\frac{|f(s)|}{|f|_{h^*}} = |f(s)| \cdot e^{-\log|f|_{h^*}}$$

is a lower semi-continuous function from X to $[0, +\infty]$; their pointwise supremum is therefore also lower semi-continuous. □

EXAMPLE 18.4. The following example, due to Raufi [**38**, Theorem 1.3], shows that the function $|s|_h$ can indeed have jumps. Let E be the trivial bundle of rank 2 on $\mathbb{C}$. We first define a singular hermitian metric h^* on the dual bundle E^*: at each point $z \in \mathbb{C}$, it is represented by the matrix

$$\begin{pmatrix} 1+|z|^2 & z \\ \bar{z} & |z|^2 \end{pmatrix}.$$

From this, one computes that the singular hermitian metric h on E is given by

$$\frac{1}{|z|^4} \begin{pmatrix} |z|^2 & -z \\ -\bar{z} & 1+|z|^2 \end{pmatrix}$$

as long as $z \neq 0$. Contrary to what this formula might suggest, one has

$$\big|(1,0)\big|_{h,0} = 1;$$

the length of the vector $(1, 0)$ is thus $|z|^{-2}$ for $z \neq 0$, but 1 for $z = 0$.

19. Singular hermitian metrics on torsion-free sheaves. Let X be a complex manifold, and let $\mathscr{F}$ be a torsion-free coherent sheaf on X. Let $X(\mathscr{F}) \subseteq X$ denote the maximal open subset where $\mathscr{F}$ is locally free; then $X \setminus X(\mathscr{F})$ is a closed analytic subset of codimension ≥ 2. If $\mathscr{F} \neq 0$, then the restriction of $\mathscr{F}$ to the open subset $X(\mathscr{F})$ is a holomorphic vector bundle E of some rank $r \geq 1$.

DEFINITION 19.1. A *singular hermitian metric* on $\mathscr{F}$ is a singular hermitian metric h on the holomorphic vector bundle E. We say that such a metric has *semi-positive curvature* if the pair (E, h) has semi-positive curvature.

Suppose that $\mathscr{F}$ has a singular hermitian metric with semi-positive curvature. Since $X \setminus X(\mathscr{F})$ has codimension ≥ 2, every holomorphic section of the dual bundle E^* extends to a holomorphic section of the reflexive coherent sheaf

$$\mathscr{F}^* = \mathcal{H}om(\mathscr{F}, \mathscr{O}_X),$$

and every plurisubharmonic function on $X(\mathscr{F})$ extends to a plurisubharmonic function on X (see Lemma 12.4). For every open subset $U \subseteq X$ and every holomorphic section $f \in H^0(U, \mathscr{F}^*)$, we thus obtain a well-defined plurisubharmonic function

$$\log|f|_{h^*} : U \to [-\infty, +\infty).$$

Note that the function $|f|_{h^*}$ is upper semi-continuous.

What about holomorphic sections of the sheaf $\mathscr{F}$ itself? For any $s \in H^0(U, \mathscr{F})$, the function $|s|_h$ is lower semi-continuous on $U \cap X(\mathscr{F})$. In a suitable neighborhood of every point in U, we can imitate the proof of Lemma 18.3 and take the pointwise supremum of the functions

$$|f(s)| \cdot e^{-\log|f|_{h^*}},$$

where f runs over all sections of $\mathscr{F}^*$. Since the pointwise supremum of a family of lower semi-continuous functions is again lower semi-continuous, we obtain in this manner a distinguished extension

$$|s|_h \colon U \to [0, +\infty]$$

to a lower semi-continuous function on U.

Definition 19.2. We say that a singular hermitian metric on $\mathscr{F}$ is *continuous* if, for every open subset $U \subseteq X$ and every holomorphic section $s \in H^0(U, \mathscr{F})$, the function $|s|_h \colon U \to [0, +\infty]$ is continuous.

Proposition 19.3. *Let $\phi\colon \mathscr{F} \to \mathscr{G}$ be a morphism between two torsion-free coherent sheaves that is generically an isomorphism. If $\mathscr{F}$ has a singular hermitian metric with semi-positive curvature, then so does $\mathscr{G}$.*

Proof. Let h denote the singular hermitian metric on $\mathscr{F}$. On the open subset of $X(\mathscr{F}) \cap X(\mathscr{G})$ where ϕ is an isomorphism, $\mathscr{G}$ clearly acquires a singular hermitian metric that we also denote by h for simplicity. Because the dual morphism $\phi^* \colon \mathscr{G}^* \to \mathscr{F}^*$ is injective, the function $\log|f|_{h^*}$ is plurisubharmonic for every $f \in H^0(U, \mathscr{G}^*) \subseteq H^0(U, \mathscr{F}^*)$. Consequently, h extends to a singular hermitian metric with semi-positive curvature on all of $X(\mathscr{G})$. □

Example 19.4. If $\mathscr{F}$ has a singular hermitian metric of semi-positive curvature, then the same is true for the reflexive hull $\mathscr{F}^{**}$.

20. The minimal extension property. The Ohsawa-Takegoshi theorem leads us to consider the following "minimal extension property" for singular hermitian metrics. To keep the statement simple, let us assume that X is a connected complex manifold of dimension n, and denote by $B \subseteq \mathbb{C}^n$ the open unit ball.

Definition 20.1. We say that a singular hermitian metric on $\mathscr{F}$ has the *minimal extension property* if there exists a nowhere dense closed analytic subset $Z \subseteq X$ with the following two properties:

(1) $\mathscr{F}$ is locally free on $X \setminus Z$, or equivalently, $X \setminus Z \subseteq X(\mathscr{F})$.
(2) For every embedding $\iota\colon B \hookrightarrow X$ with $x = \iota(0) \in X \setminus Z$, and every $v \in E_x$ with $|v|_{h,x} = 1$, there is a holomorphic section $s \in H^0\big(B, \iota^*\mathscr{F}\big)$ such that

$$s(0) = v \quad \text{and} \quad \frac{1}{\mu(B)} \int_B |s|_h^2 \, d\mu \leq 1;$$

here (E, h) denotes the restriction to the open subset $X(\mathscr{F})$.

The point of the minimal extension property is the ability to extend sections over the "bad" locus Z, with good control on the norm of the extension. We will see later that pushforwards of adjoint line bundles always have this property, as a consequence of the Ohsawa-Takegoshi theorem.

EXAMPLE 20.2. The minimal extension property rules out certain undesirable examples like the following. Let $Z \subseteq X$ be a closed analytic subset of codimension ≥ 2, and let $\mathcal{I}_Z \subseteq \mathscr{O}_X$ denote the ideal sheaf of Z. Then $\mathcal{I}_Z$ is trivial on $X \setminus Z$, and the constant hermitian metric on this trivial bundle is a singular hermitian metric with semi-positive curvature on $\mathcal{I}_Z$. But this metric does not have the minimal extension property, because a holomorphic function $f: B \to \mathbb{C}$ with $f(0) = 1$ and

$$\frac{1}{\mu(B)} \int_B |f|^2 d\mu \leq 1$$

must be constant.

21. Pushforwards of adjoint line bundles. Let X be a complex manifold of dimension n, and let (L, h) be a holomorphic line bundle with a singular hermitian metric of semi-positive curvature. If X is compact, the space $H^0(X, \omega_X \otimes L)$ is finite-dimensional, and the formula

$$\|\beta\|_h^2 = \int_X |\beta|_h^2$$

endows it with a positive definite singular hermitian inner product that is finite on the subspace $H^0\big(X, \omega_X \otimes L \otimes \mathcal{I}(h)\big)$. We are now going to analyze how this construction behaves in families.

Suppose then that $f: X \to Y$ is a projective surjective morphism between two connected complex manifolds, with $\dim X = n$ and $\dim Y = r$; the general fiber of f is a projective complex manifold of dimension $n - r$, but there may be singular fibers. Let (L, h) be a holomorphic line bundle with a singular hermitian metric of semi-positive curvature on X. The following important theorem was essentially proved by Păun and Takayama [**37**, Theorem 3.3.5], building on earlier results for smooth morphisms by Berndtsson and Păun [**1**, **3**].

THEOREM 21.1. *Let $f: X \to Y$ be a projective surjective morphism between two connected complex manifolds. If (L, h) is a holomorphic line bundle with a singular hermitian metric of semi-positive curvature on X, then the pushforward sheaf*

$$\mathscr{F} = f_*\big(\omega_{X/Y} \otimes L \otimes \mathcal{I}(h)\big)$$

has a canonical singular hermitian metric H. This metric has semi-positive curvature and satisfies the minimal extension property.

The metric in the theorem is uniquely characterized by a simple property that we now describe. Recall from (14.2) and (14.1) that any $\beta \in H^0\big(X, \omega_X \otimes L \otimes \mathcal{I}(h)\big)$ gives rise to a locally integrable (n, n)-form $|\beta|_h^2$. Any such form can be integrated against compactly supported smooth functions, and therefore defines a current of bidegree (n, n) on X. If we use brackets to denote the evaluation pairing between (n, n)-currents and compactly supported smooth functions, then

$$\langle |\beta|_h^2, \phi \rangle = \int_X \phi \cdot |\beta|_h^2.$$

By the same token, any section $\beta \in H^0(Y, \omega_Y \otimes \mathscr{F})$ defines a current of bidegree (r, r) on Y that we denote by the symbol $|\beta|_H^2$. Now suppose that

$$\beta \in H^0(U, \omega_Y \otimes \mathscr{F}) \simeq H^0\big(f^{-1}(U), \omega_X \otimes L \otimes \mathcal{I}(h)\big).$$

The singular hermitian metric H is uniquely characterized by the condition that

$$\langle |\beta|_H^2, \phi \rangle = \langle |\beta|_h^2, f^*\phi \rangle$$

for every compactly supported smooth function $\phi \in A_c(U)$. Said differently, $|\beta|_H^2$ is the pushforward of the current $|\beta|_h^2$ under the proper mapping f.

COROLLARY 21.2. *In the situation of Theorem 21.1, suppose that the inclusion*

$$f_*\big(\omega_{X/Y} \otimes L \otimes \mathcal{I}(h)\big) \hookrightarrow f_*(\omega_{X/Y} \otimes L)$$

is generically an isomorphism. Then $f_*(\omega_{X/Y} \otimes L)$ also has a singular hermitian metric with semi-positive curvature and the minimal extension property.

PROOF. The existence of the metric follows from Proposition 19.3. The minimal extension property continues to hold because every section of $f_*\big(\omega_{X/Y} \otimes L \otimes \mathcal{I}(h)\big)$ is of course also a section of $f_*(\omega_{X/Y} \otimes L)$. □

EXAMPLE 21.3. If we apply Theorem 21.1 to the identity morphism $\mathrm{id}\colon X \to X$, we only get a singular hermitian metric on $L \otimes \mathcal{I}(h)$. To recover the singular hermitian metric on L that we started from, we can use Corollary 21.2.

The proof of Theorem 21.1 gives the following additional information about the singular hermitian metric on $\mathscr{F} = f_*\big(\omega_{X/Y} \otimes L \otimes \mathcal{I}(h)\big)$ (see the end of §24).

COROLLARY 21.4. *In the situation of Theorem 21.1, suppose that $f\colon X \to Y$ is submersive and that the singular hermitian metric h on the line bundle L is continuous. Then the singular hermitian metric H on $\mathscr{F}$ is also continuous.*

The following three sections explain the proof of Theorem 21.1. In a nutshell, it is an application of the Ohsawa-Takegoshi extension theorem. We present the argument in three parts that rely on successively stronger versions of the extension theorem: first the ability to extend sections from a fiber; then the fact that there is a universal bound on the norm of the extension; and finally the optimal bound in Theorem 14.4.

22. Proof of the pushforward theorem, Part I. Our first goal is to define the singular hermitian metric on

$$\mathscr{F} = f_*\big(\omega_{X/Y} \otimes L \otimes \mathcal{I}(h)\big),$$

and to establish a few basic facts about it. In this part of the proof, we only use the weakest version of the Ohsawa-Takegoshi extension theorem, namely the ability to extend sections from a fiber.

The idea is to construct the metric first over a Zariski-open subset $Y \setminus Z$ where everything is nice, and then to extend it over the bad locus Z. To begin with, choose a nowhere dense closed analytic subset $Z \subseteq Y$ with the following three properties:

(1) The morphism f is submersive over $Y \setminus Z$.
(2) Both $\mathscr{F}$ and the quotient sheaf $f_*(\omega_{X/Y} \otimes L)/\mathscr{F}$ are locally free on $Y \setminus Z$.
(3) On $Y \setminus Z$, the locally free sheaf $f_*(\omega_{X/Y} \otimes L)$ has the base change property.

By the base change theorem, the third condition will hold as long as the coherent sheaves $R^i f_*(\omega_{X/Y} \otimes L)$ are locally free on $Y \setminus Z$. The restriction of $\mathscr{F}$ to the open subset $Y \setminus Z$ is a holomorphic vector bundle E of some rank $r \geq 1$. The second and third condition together guarantee that

$$E_y = \mathscr{F}|_y \subseteq f_*(\omega_{X/Y} \otimes L)|_y = H^0\big(X_y, \omega_{X_y} \otimes L_y\big)$$

whenever $y \in Y \setminus Z$. As before, (L_y, h_y) denotes the restriction of (L, h) to the fiber $X_y = f^{-1}(y)$; it may happen that $h_y \equiv +\infty$. The Ohsawa-Takegoshi theorem gives us the following additional information about E_y.

LEMMA 22.1. *For any $y \in Y \setminus Z$, we have inclusions*

$$H^0\big(X_y, \omega_{X_y} \otimes L_y \otimes \mathcal{I}(h_y)\big) \subseteq E_y \subseteq H^0\big(X_y, \omega_{X_y} \otimes L_y\big).$$

PROOF. If $h_y \equiv +\infty$, then the subspace of the left is trivial, which means that the asserted inclusion is true by default. If h_y is not identically equal to $+\infty$, then given $\alpha \in H^0\big(X_y, \omega_{X_y} \otimes L_y \otimes \mathcal{I}(h_y)\big)$ and a suitable open neighborhood U of the point y, there is by Theorem 14.4 some

$$\beta \in H^0\big(U, \omega_Y \otimes \mathscr{F}\big) \simeq H^0\big(f^{-1}(U), \omega_X \otimes L \otimes \mathcal{I}(h)\big)$$

such that $\beta|_{X_y} = \alpha \wedge df$. Since ω_Y is trivial on U, this gives us a section of $\mathscr{F}$ in a neighborhood of the fiber X_y whose restriction to X_y agrees with α. $\square$

NOTE. We will see in a moment that the two subspaces

$$H^0\big(X_y, \omega_{X_y} \otimes L_y \otimes \mathcal{I}(h_y)\big) \subseteq E_y$$

are equal for almost every $y \in Y \setminus Z$. But unless $\mathscr{F} = 0$, the two subspaces are different for example at points where h_y is identically equal to $+\infty$.

We can now define on each E_y with $y \in Y \setminus Z$ a singular hermitian inner product in the following manner. Given an element

$$\alpha \in E_y \subseteq H^0\big(X_y, \omega_{X_y} \otimes L_y\big),$$

we can integrate over the compact complex manifold X_y and define

$$|\alpha|_{H,y}^2 = \int_{X_y} |\alpha|_{h_y}^2 \in [0, +\infty].$$

It is easy to see that $|-|_{H,y}$ is a positive definite singular hermitian inner product. Clearly $|\alpha|_{H,y} < +\infty$ if and only if $\alpha \in H^0\big(X_y, \omega_{X_y} \otimes L_y \otimes \mathcal{I}(h_y)\big)$; in light of Lemma 22.1, our singular hermitian inner product $|-|_{H,y}$ is therefore finite precisely on the subspace $H^0\big(X_y, \omega_{X_y} \otimes L_y \otimes \mathcal{I}(h_y)\big) \subseteq E_y$.

Let us now analyze how the individual singular hermitian inner products $|-|_{H,y}$ fit together on $Y \setminus Z$. Fix a point $y \in Y \setminus Z$ and an open neighborhood $U \subseteq Y \setminus Z$ biholomorphic to the open unit ball $B \subseteq \mathbb{C}^r$; after pulling everything back to U, we may assume without loss of generality that $Y = B$ and $Z = \emptyset$ and $y = 0$. Denote by $t_1, \dots, t_r$ the standard coordinate system on B; then the canonical bundle ω_B is trivialized by the global section $dt_1 \wedge \cdots \wedge dt_r$, and the volume form on B is

$$d\mu = c_r(dt_1 \wedge \cdots \wedge dt_r) \wedge (d\bar{t}_1 \wedge \cdots \wedge d\bar{t}_r).$$

Fix a holomorphic section $s \in H^0(B, E)$, and denote by

$$\beta = s \wedge (dt_1 \wedge \cdots \wedge dt_r) \in H^0\big(B, \omega_B \otimes E\big) \simeq H^0\big(X, \omega_X \otimes L \otimes \mathcal{I}(h)\big)$$

the corresponding holomorphic n-form on X with coefficients in L. Since $f \colon X \to B$ is smooth, Ehresmann's fibration theorem shows that X is diffeomorphic to the product $B \times X_0$. After choosing a Kähler metric ω_0 on X_0, we can write

$$|\beta|_h^2 = F \cdot d\mu \wedge \frac{\omega_0^{n-r}}{(n-r)!}, \tag{22.2}$$

where $F\colon B \times X_0 \to [0, +\infty]$ is lower semi-continuous and locally integrable; the reason is of course that the local weight functions for (L, h) are upper semi-continuous functions. At every point $y \in B$, we then have by construction

$$|s(y)|_{H,y}^2 = \int_{X_0} F(y, -) \frac{\omega_0^{n-r}}{(n-r)!}. \tag{22.3}$$

By Fubini's theorem, the function $y \mapsto |s(y)|_{H,y}$ is measurable; moreover, since F is locally integrable and X_0 is compact, we must have $|s(y)|_{H,y} < +\infty$ for almost every $y \in B$. Being coherent, E is generated over B by a finite number of global sections; the singular hermitian inner product $|-|_{H,y}$ is therefore finite and positive-definite for almost every $y \in B$, hence for almost every $y \in Y \setminus Z$. In particular, the first inclusion in Lemma 22.1 is an equality for almost every $y \in Y \setminus Z$. We may summarize the conclusion as follows.

PROPOSITION 22.4. *On $Y \setminus Z$, the singular hermitian inner products $|-|_{H,y}$ determine a singular hermitian metric on the holomorphic vector bundle E.*

While we are not yet ready to show that (E, H) has semi-positive curvature, we can already show that the function $|s|_H$ is always lower semi-continuous.

PROPOSITION 22.5. *For any open subset $U \subseteq Y \setminus Z$ and any section $s \in H^0(U, E)$, the function $|s|_H\colon U \to [0, +\infty]$ is lower semi-continuous.*

PROOF. As before, we may assume that $U = B$ is the open unit ball in $\mathbb{C}^m$; it is clearly sufficient to show that $|s|_H$ is lower semi-continuous at the origin. In other words, we need to argue that

$$|s(0)|_{H,0} \leq \liminf_{k \to +\infty} |s(y_k)|_{H,y_k}$$

holds for every sequence $y_0, y_1, y_2, \ldots \in B$ that converges to the origin. As in (22.2), the given section $s \in H^0(B, E)$ determines a lower semi-continuous function $F\colon B \times X_0 \to [0, +\infty]$ such that (22.3) is satisfied. By the lower semi-continuity of F and Fatou's lemma, we obtain

$$\begin{aligned} \int_{X_0} F(0, -) \frac{\omega_0^{n-r}}{(n-r)!} &\leq \int_{X_0} \liminf_{k \to +\infty} F(y_k, -) \frac{\omega_0^{n-r}}{(n-r)!} \\ &\leq \liminf_{k \to +\infty} \int_{X_0} F(y_k, -) \frac{\omega_0^{n-r}}{(n-r)!}, \end{aligned}$$

which is the desired inequality up to taking square roots. □

23. Proof of the pushforward theorem, Part II. Having defined (E, H) on the open subset $Y \setminus Z$, our next task is to say something about the induced singular hermitian metric H^* on the dual vector bundle E^*. In particular, we need to prove that the norm of any local section of $\mathscr{F}^*$ is uniformly bounded in the neighborhood of any point in Z, and that its logarithm is an upper semi-continuous function. This part of the argument relies on the existence of a uniform bound in the Ohsawa-Takegoshi theorem, but not on the precise value of the constant. Let us start by reformulating the statement of the Ohsawa-Takegoshi in terms of the pair (E, H).

LEMMA 23.1. *For every embedding* $\iota\colon B \hookrightarrow Y$ *with* $y = \iota(0) \in Y \setminus Z$, *and for every* $\alpha \in E_y$ *with* $|\alpha|_{H,y} = 1$, *there is a holomorphic section* $s \in H^0(B, \iota^*\mathscr{F})$ *with*

$$s(0) = \alpha \quad \textit{and} \quad \int_B |s|_H^2 \, d\mu \leq C_0,$$

where C_0 *is the same constant as in the Ohsawa-Takegoshi theorem.*

PROOF. After pulling everything back to B, we may assume that $Y = B$ and $y = 0$. Since $|\alpha|_{H,0} = 1$, by Theorem 14.4 there exists an element $\beta \in H^0\big(X, \omega_X \otimes L \otimes \mathcal{I}(h)\big)$ with

$$\beta|_{X_0} = \alpha \wedge df \quad \text{and} \quad \|\beta\|_h^2 = \int_X |\beta|_h^2 \leq C_0.$$

In fact, one can take $C_0 = \mu(B)$, but the exact value of the constant is not important here. Using $dt_1 \wedge \cdots \wedge dt_r$ as a trivialization of the canonical bundle ω_B, we may consider β as a holomorphic section $s \in H^0(B, \mathscr{F})$; the two conditions from above then turn into

$$s(0) = \alpha \quad \text{and} \quad \int_B |s|_H^2 \, d\mu \leq C_0,$$

due to the fact that $d\mu = c_r(dt_1 \wedge \cdots \wedge dt_r) \wedge (d\bar{t}_1 \wedge \cdots \wedge d\bar{t}_r)$. □

Fix an open subset $U \subseteq Y$ and a holomorphic section $g \in H^0(U, \mathscr{F}^*)$; after replacing Y by the open subset U, we may assume without loss of generality that $g \in H^0(Y, \mathscr{F}^*)$. Consider the measurable function

$$\psi = \log|g|_{H^*} \colon Y \setminus Z \to [-\infty, +\infty]. \tag{23.2}$$

Ultimately, our goal is to show that ψ extends to a plurisubharmonic function on all of Y. The following boundedness result is the crucial step in this direction.

PROPOSITION 23.3. *Every point in* Y *has an open neighborhood* $U \subseteq Y$ *such that* $\psi = \log|g|_{H^*}$ *is bounded from above by a constant on* $U \setminus U \cap Z$.

PROOF. Choose two sufficiently small open neighborhoods $U \subseteq V \subseteq Y$ of the given point, such that $\overline{V}$ is compact, $\overline{U} \subseteq V$, and for every point $y \in \overline{U}$, there is an embedding $\iota\colon B \hookrightarrow Y$ of the unit ball $B \subseteq \mathbb{C}^r$ with $\iota(0) = y$ and $\iota(B) \subseteq V$. We shall argue that there is a constant $C \geq 0$ such that $\psi \leq C$ on $U \setminus U \cap Z$.

Fix a point $y \in \overline{U} \setminus Z$. If $\psi(y) = -\infty$, there is nothing to prove, so let us suppose from now on that $\psi(y) \neq -\infty$. By definition of the metric on the dual bundle, we can then find a vector $\alpha \in E_y$ with $|\alpha|_{H,y} = 1$ such that

$$\psi(y) = \log\big|g(\alpha)\big|.$$

Choose an embedding $\iota\colon B \hookrightarrow Y$ such that $\iota(0) = y$ and $\iota(B) \subseteq V$. Using Lemma 23.1, we obtain a holomorphic section $s \in H^0(V, \mathscr{F})$ with $s(0) = \alpha$ and

$$\int_V |s|_H^2 \, d\mu \leq C_0;$$

the integrand is of course only defined on the subset $V \setminus V \cap Z$, but this does not matter because $V \cap Z$ has measure zero. It follows that $\psi(y)$ is equal to the value of $\log|g(s)|$ at the point y, and so the desired upper bound for ψ is a consequence of Lemma 23.4 below. □

LEMMA 23.4. *Fix $K \geq 0$, and consider the set*

$$S_K = \left\{ s \in H^0(V, \mathscr{F}) \,\middle|\, \int_V |s|_H^2 \, d\mu \leq K \right\}.$$

There is a constant $C \geq 0$ such that, for every section $s \in S_K$, the holomorphic function $g(s)$ is uniformly bounded by C on the compact set $\overline{U}$.

PROOF. Since $g(s)$ is holomorphic on V, it is clear that each individual function $g(s)$ is bounded on $\overline{U}$. To get an upper bound that works for every $s \in S_K$ at once, we use a compactness argument. Given a section $s \in H^0(V, \mathscr{F})$, we invert the process from above and define

$$\beta = s \otimes \big(dt_1 \wedge \cdots \wedge dt_r\big) \in H^0(V, \omega_Y \otimes \mathscr{F}) = H^0\big(f^{-1}(V), \omega_X \otimes L \otimes \mathcal{I}(h)\big).$$

If $s \in S_K$, then one has

$$\|\beta\|_h^2 = \int_V |s|_H^2 \, d\mu \leq K.$$

Because $\overline{V}$ is compact and f is proper, we can cover $f^{-1}(V)$ by finitely many open sets W that are biholomorphic to the open unit ball in $\mathbb{C}^n$, and on which L is trivial. Let $z_1, \ldots, z_n$ be a holomorphic coordinate system on W, choose a nowhere vanishing holomorphic section $s_0 \in H^0(W, L)$, and write $|s_0|_h^2 = e^{-\varphi}$, with φ plurisubharmonic on W. Then $\beta|_W = bs_0 \otimes dz_1 \wedge \cdots \wedge dz_n$ for some holomorphic function $b \in H^0(W, \mathscr{O}_W)$, and

$$\int_W |b|^2 e^{-\varphi} (dx_1 \wedge dy_1) \wedge \cdots \wedge (dx_n \wedge dy_n) = \int_W |\beta|_h^2 \leq K.$$

As we are dealing with finitely many open sets, Proposition 12.5 shows that every sequence in S_K has a subsequence that converges uniformly on compact subsets to some $\beta \in H^0\big(f^{-1}(V), \omega_X \otimes L \otimes \mathcal{I}(h)\big)$. This is all that we need.

Indeed, suppose that the assertion was false. Then we could find a sequence $s_0, s_1, s_2, \ldots \in S_K$ such that the maximum value of $|g(s_k)|$ on the compact set $\overline{U}$ was at least k. Let $\beta_0, \beta_1, \beta_2, \ldots$ denote the corresponding sequence of holomorphic sections of $\omega_X \otimes L \otimes \mathcal{I}(h)$ on the open set $f^{-1}(V)$; after passing to a subsequence, the β_k will converge uniformly on compact subsets to $\beta \in H^0\big(f^{-1}(V), \omega_X \otimes L \otimes \mathcal{I}(h)\big)$. Let $s \in H^0(V, \mathscr{F})$ be the unique section of $\mathscr{F}$ such that

$$\beta = s \otimes \big(dt_1 \wedge \cdots \wedge dt_r\big).$$

By Proposition 15.3, the s_k converge to s in the Fréchet space topology on $H^0(V, \mathscr{F})$. Since $g \colon \mathscr{F} \to \mathscr{O}_Y$ is a morphism, the holomorphic functions $g(s_k)$ therefore converge uniformly on compact subsets to $g(s)$. But then $|g(s_k)|$ must be uniformly bounded on $\overline{U}$, contradicting our initial choice. □

The next step is to show that the function $\psi = \log|g|_{H^*}$ is upper semi-continuous on $Y \setminus Z$. The proof is similar to that of Proposition 22.5.

PROPOSITION 23.5. *For every $g \in H^0(Y, \mathscr{F}^*)$, the function $\psi = \log|g|_{H^*}$ is upper semi-continuous on $Y \setminus Z$.*

PROOF. After restricting everything to a suitable open neighborhood of any given point $y \in Y \setminus Z$, we may assume without loss of generality that $Y = B$ and $Z = \emptyset$ and $y = 0$. Then $g \in H^0(B, E^*)$, and it will be enough to show that

$\psi = \log|g|_{H^*}$ is upper semi-continuous at the origin. In other words, we need to argue that

$$\limsup_{k\to+\infty} \psi(y_k) \le \psi(0) \tag{23.6}$$

for every sequence $y_0, y_1, y_2, \ldots \in B$ that converges to the origin. We may assume that $\psi(y_k) \neq -\infty$ for all $k \in \mathbb{N}$, and that the sequence $\psi(y_k)$ actually has a limit.

As we saw before, there is, for each $k \in \mathbb{N}$, a holomorphic section $s_k \in H^0(B, E)$ such that $\psi(y_k)$ equals the value of $\log|g(s_k)|$ at the point y_k; the Ohsawa-Takegoshi theorem allows us to choose these sections in such a way that

$$|s_k(y_k)|_{H,y_k} = 1 \quad \text{and} \quad \int_B |s_k|_H \, d\mu \le K$$

for some constant $K \ge 0$. Passing to a subsequence, if necessary, we can arrange that the s_k converge uniformly on compact subsets to some $s \in H^0(B, E)$. Then the holomorphic functions $g(s_k)$ converge uniformly on compact subsets to $g(s)$, and (23.6) reduces to showing that the value at the origin of $\log|g(s)|$ is less or equal to $\psi(0)$. By definition of the dual metric H^*, we have

$$\psi \ge \log|g(s)| - \log|s|_H,$$

and so this is equivalent to proving that $|s(0)|_{H,0} \le 1$. As in (22.2) and (22.3), each s_k determines a lower semi-continuous function $F_k \colon B \times X_0 \to [0, +\infty]$ with

$$1 = |s_k(y_k)|^2_{H,y_k} = \int_{X_0} F_k(y_k, -) \frac{\omega_0^{n-r}}{(n-r)!}.$$

Likewise, s determines a lower semi-continuous function $F \colon B \times X_0 \to [0, +\infty]$. Since the local weight functions $e^{-\varphi}$ of the pair (L, h) are lower semi-continuous, and since s_k converges uniformly on compact subsets to s, we get

$$F(0, -) \le \liminf_{k\to+\infty} F_k(y_k, -).$$

We can now apply Fatou's lemma and conclude the proof in the same way as in Proposition 22.5. □

24. Proof of the pushforward theorem, Part III. In this section, we complete the proof of Theorem 21.1 by showing that the pair (E, H) has semi-positive curvature, and that H extends to a singular hermitian metric on $\mathscr{F}$ with the minimal extension property. The key point is that we can prove the required mean-value inequalities because the optimal value of the constant in the Ohsawa-Takegoshi theorem is exactly the volume of the unit ball. To illustrate how this works, let us first show that the singular hermitian metric H on $Y \setminus Z$ has the minimal extension property (see §20). For the statement, recall that $r = \dim Y$, and that $B \subseteq \mathbb{C}^r$ is the open unit ball.

PROPOSITION 24.1. *For every embedding $\iota \colon B \hookrightarrow Y$ with $y = \iota(0) \in Y \setminus Z$, and for every $\alpha \in E_y$ with $|\alpha|_{H,y} = 1$, there is a holomorphic section $s \in H^0(B, \iota^*\mathscr{F})$ with*

$$s(0) = \alpha \quad \textit{and} \quad \frac{1}{\mu(B)} \int_B |s|^2_H \, d\mu \le 1.$$

PROOF. The proof is the same as that of Lemma 23.1; we only need to replace the constant C_0 by its optimal value $\mu(B)$. □

Now let us prove that H extends to a singular hermitian metric on $\mathscr{F}$ with semi-positive curvature. Keeping the notation from above, this amounts to proving that the function $\psi\colon Y\setminus Z \to [-\infty, +\infty)$ in (23.2) extends to a plurisubharmonic function on Y. We already know that ψ is upper semi-continuous (by Proposition 23.5) and bounded from above in a neighborhood of every point in Y (by Proposition 23.3). What we need to prove is the mean-value inequality along holomorphic arcs in $Y \setminus Z$. The Ohsawa-Takegoshi theorem with sharp estimates renders the proof of the mean-value inequality almost a triviality.

PROPOSITION 24.2. *For every holomorphic mapping $\gamma\colon \Delta \to Y\setminus Z$, the function $\psi = \log|g|_{H^*}$ satisfies the mean-value inequality*

$$(\psi \circ \gamma)(0) \leq \frac{1}{\pi} \int_\Delta (\psi \circ \gamma)\, d\mu.$$

PROOF. If h is identically equal to $+\infty$ on the preimage of $\gamma(\Delta)$, the inequality is clear, so we may assume that this is not the case. Since $f\colon X \to Y$ is submersive over $Y \setminus Z$, we may then pull everything back to Δ and reduce the problem to the case $Y = \Delta$. If $\psi(0) = -\infty$, then the mean-value inequality holds by default. Assuming from now on that $\psi(0) \neq -\infty$, we choose an element $\alpha \in E_0$ with $|\alpha|_{H,0} = 1$, such that

$$\psi(0) = \log|g|_{H^*,0} = \log|g(\alpha)|.$$

Using the minimal extension property (in Proposition 24.1, with $m = 1$), there is a holomorphic section $s \in H^0(\Delta, E)$ such that

$$s(0) = \alpha \quad \text{and} \quad \frac{1}{\pi}\int_\Delta |s|_H^2\, d\mu \leq 1.$$

The existence of this section is all that we need to prove the mean-value inequality. By definition of the metric H^* on the dual bundle, we have the pointwise inequality

$$|g|_{H^*} \geq \frac{|g(s)|}{|s|_H}$$

and therefore $2\psi \geq \log|g(s)|^2 - \log|s|_H^2$; here $g(s)$ is a holomorphic function on Δ, whose value at the origin equals $g(\alpha)$. Integrating, we get

$$\frac{1}{\pi}\int_\Delta 2\psi\, d\mu \geq \frac{1}{\pi}\int_\Delta \log|g(s)|^2\, d\mu - \frac{1}{\pi}\int_\Delta \log|s|_H^2\, d\mu$$

Now $\log|g(s)|^2$ satisfies the mean-value inequality, and so the first term on the right-hand side is at least $\log|g(\alpha)|^2 = 2\psi(0)$. Since the function $x \mapsto -\log x$ is convex, and since the function $|s|_H^2$ is integrable, the second term can be estimated by Jensen's inequality to be at least

$$-\log\left(\frac{1}{\pi}\int_\Delta |s|_H^2\, d\mu\right) \geq -\log 1 = 0.$$

Putting everything together, we obtain

$$\frac{1}{\pi}\int_\Delta 2\psi\, d\mu \geq 2\psi(0),$$

which is the mean-value inequality (up to a factor of 2). □

We have verified that ψ is plurisubharmonic on $Y \setminus Z$. We already know from Proposition 23.3 that ψ is locally bounded from above in a neighborhood of every point in Y; consequently, it extends uniquely to a plurisubharmonic function on all of Y, using Lemma 12.4. By duality, the singular hermitian metric H is therefore well-defined on the entire open set $Y(\mathscr{F})$ where the sheaf $\mathscr{F} = f_*\big(\omega_X \otimes L \otimes \mathcal{I}(h)\big)$ is locally free. We have already shown that H has the minimal extension property. This finishes the proof of Theorem 21.1. □

PROOF OF COROLLARY 21.4. Suppose that $f\colon X \to Y$ is submersive and that the singular hermitian metric h on the line bundle L is continuous. To prove that H is continuous, it suffices to show that for every locally defined section $s \in H^0(U, \mathscr{F})$, the function $|s|_H^2$ on $U \setminus U \cap Z$ admits a continuous extension to all of U. This is a local problem, and so we may assume that $Y = B$ is the open unit ball in $\mathbb{C}^r$, with coordinates $t_1, \dots, t_r$, and that $s \in H^0(B, \mathscr{F})$. Define

$$\beta = s \wedge (dt_1 \wedge \dots \wedge dt_r) \in H^0(B, \omega_B \otimes \mathscr{F}) = H^0\big(X, \omega_X \otimes L \otimes \mathcal{I}(h)\big).$$

By Ehresmann's fibration theorem, X is diffeomorphic to the product $B \times X_0$, and as in (22.2), we can write

$$|\beta|_h^2 = F \cdot d\mu \wedge \frac{\omega_0^{n-r}}{(n-r)!}$$

with $F\colon B \times X_0 \to [0, +\infty]$ continuous. Now

$$y \mapsto \int_{X_0} F(y, -) \frac{\omega_0^{n-r}}{(n-r)!}$$

defines a continuous function on B that agrees with $|s|_H^2$ on the complement of the bad set Z, due to (22.3). □

25. Positivity of the determinant line bundle. In this section, we show that if a holomorphic vector bundle E has a singular hermitian metric with semi-positive curvature, then the determinant line bundle $\det E$ has the same property. The proof in [**38**, Proposition 1.1] relies on locally approximating a given singular hermitian metric from below by smooth hermitian metrics [**3**, Proposition 3.1].

PROPOSITION 25.1. *If (E, h) has semi-positive curvature, so does* $(\det E, \det h)$.

Let us first analyze what happens over a point. Let V be a complex vector space of dimension r, and $|-|_h$ a positive definite singular hermitian inner product on V; in the notation of §16, we have $V_0 = 0$. Let $\mathbb{P}(V)$ be the projective space parametrizing one-dimensional quotient spaces of V, and denote by $\mathscr{O}(1)$ the universal line bundle on $\mathbb{P}(V)$. We have a surjective morphism $V \otimes \mathscr{O} \to \mathscr{O}(1)$, and so h induces a singular hermitian metric on $\mathscr{O}(1)$, with singularities along the subspace $\mathbb{P}(V/V_{fin}) \subseteq \mathbb{P}(V)$. To see this, choose a basis $e_1, \dots, e_r \in V$ such that $e_1, \dots, e_k$ form an orthonormal basis of V_{fin} with respect to the inner product $\langle -, - \rangle_h$, and denote by $[z_1, \dots, z_r]$ the resulting homogeneous coordinates on $\mathbb{P}(V)$. Then the local weight functions of the metric on $\mathscr{O}(1)$ are given by the formula

$$\log\big(|z_1|^2 + \dots + |z_k|^2\big),$$

with the convention that $z_i = 1$ on the i-th standard affine open subset.

Now the one-dimensional complex vector space $\det V = \bigwedge^r V$ is naturally the space of global sections of an adjoint bundle on $\mathbb{P}(V)$, because

$$\det V \simeq H^0\big(\mathbb{P}(V), \omega_{\mathbb{P}(V)} \otimes \mathscr{O}(r)\big).$$

The isomorphism works as follows. The element $e_1 \wedge \cdots \wedge e_r \in \det V$ determines a holomorphic r-form $dz_1 \wedge \cdots \wedge dz_r$ on the dual vector space V^*; after contraction with the Euler vector field $z_1\partial/\partial z_1 + \cdots + z_r\partial/\partial z_r$, we get a holomorphic $(r-1)$-form

$$\Omega = \sum_{i=1}^{r} (-1)^{i-1} z_i \, dz_1 \wedge \cdots \wedge \widehat{dz_i} \wedge \cdots \wedge dz_r$$

on $\mathbb{P}(V)$ that is homogeneous of degree r, hence a global section of the holomorphic line bundle $\omega_{\mathbb{P}(V)} \otimes \mathscr{O}(r)$. Integration over $\mathbb{P}(V)$ therefore defines a positive definite singular hermitian inner product H on $\det V$. We have

$$|e_1 \wedge \cdots \wedge e_r|_H^2 = \int_{\mathbb{P}(V)} \frac{c_{r-1} \cdot \Omega \wedge \overline{\Omega}}{\big(|z_1|^2 + \cdots + |z_k|^2\big)^r},$$

which simplifies in the affine chart $z_1 = 1$ to

$$|e_1 \wedge \cdots \wedge e_r|_H^2 = \int_{\mathbb{C}^{r-1}} \frac{d\mu}{\big(1 + |z_2|^2 + \cdots + |z_k|^2\big)^r}.$$

Now there are two cases. If $V_{\mathit{fin}} \neq V$, then $k < r$, and the integral is easily seen to be $+\infty$. If $V_{\mathit{fin}} = V$, then $k = r$, and the integral evaluates to $\pi^{r-1}/(r-1)!$, the volume of the open unit ball in $\mathbb{C}^{r-1}$. In conclusion, we always have

$$|e_1 \wedge \cdots \wedge e_r|_H^2 = \frac{\pi^{r-1}}{(r-1)!} \cdot |e_1 \wedge \cdots \wedge e_r|_{\det h}^2.$$

With this result in hand, we can now prove Proposition 25.1.

PROOF. Let $p\colon \mathbb{P}(E) \to X$ denote the associated $\mathbb{P}^{r-1}$-bundle, and let $\mathscr{O}_E(1)$ be the universal line bundle on $\mathbb{P}(E)$. We have a surjective morphism $p^*E \to \mathscr{O}_E(1)$, and by Lemma 18.2, the singular hermitian metric on E induces a singular hermitian metric on the line bundle $\mathscr{O}_E(1)$, still with semi-positive curvature. We have

$$\omega_{\mathbb{P}(E)/X} \simeq p^* \det E \otimes \mathscr{O}_E(-r),$$

and therefore $\det E \simeq p_*\big(\omega_{\mathbb{P}(E)/X} \otimes \mathscr{O}_E(r)\big)$ is the pushforward of an adjoint bundle. The calculation above shows that, up to a factor of $\pi^{r-1}/(r-1)!$, the resulting singular hermitian metric on $\det E$ agrees with $\det h$ pointwise. The assertion about the curvature of $(\det E, \det h)$ is therefore a consequence of Corollary 21.2. □

26. Consequences of the minimal extension property. In this section, we derive a few interesting consequences from the minimal extension property. All of the results below are true for smooth hermitian metrics with Griffiths semi-positive curvature on holomorphic vector bundles; the minimal extension property is what makes them work even in the presence of singularities.

Let $\mathscr{F}$ be a torsion-free coherent sheaf on X, of generic rank $r \geq 1$, and suppose that $\mathscr{F}$ has a singular hermitian metric with semi-positive curvature and the minimal extension property. Let E be the holomorphic vector bundle of rank r obtained by restricting $\mathscr{F}$ to the open subset $X(\mathscr{F})$; by assumption, the pair (E, h) has semi-positive curvature. Proposition 25.1 shows that $(\det E, \det h)$ also has semi-positive curvature. Let $\det \mathscr{F}$ be the holomorphic line bundle obtained as the double dual of $\bigwedge^r \mathscr{F}$; its restriction to $X(\mathscr{F})$ agrees with $\det E$. Since $X \setminus X(\mathscr{F})$ has codimension ≥ 2, the singular hermitian metric on $\det E$ extends uniquely to a singular hermitian metric on $\det \mathscr{F}$. The following result is due to Cao and Păun [**6**, Theorem 5.23], who proved it using results by Raufi [**38**].

THEOREM 26.1. *Suppose that X is compact and that $c_1(\det \mathscr{F}) = 0$ in $H^2(X, \mathbb{R})$. Then $\mathscr{F}$ is locally free, and (E, h) is a hermitian flat bundle on $X = X(\mathscr{F})$.*

PROOF. Since X is compact, the singular hermitian metric on $\det \mathscr{F}$ is smooth and has zero curvature (by Lemma 13.2). Restricting to the open subset $X(\mathscr{F})$, we see that the same is true for $(\det E, \det h)$. Now the idea is to use the minimal extension property to construct, locally on X, a collection of r sections of $\mathscr{F}$ that are orthonormal with respect to h.

We can certainly cover X by open subsets that are isomorphic to the open unit ball $B \subseteq \mathbb{C}^n$ and are centered at points $x \in X \setminus Z$ where the singular hermitian inner product $|-|_{h,x}$ is finite and positive definite. After restricting everything to an open subset of this kind, we may assume that $X = B$, that the point $0 \in B$ lies in the subset $B \setminus Z$, and that $|-|_{h,0}$ is a genuine hermitian inner product on the r-dimensional complex vector space E_0. Choose an orthonormal basis $e_1, \dots, e_r \in E_0$. By the minimal extension property for $\mathscr{F}$, we can find r holomorphic sections $s_1, \dots, s_r \in H^0(B, \mathscr{F})$ such that

$$s_i(0) = e_i \quad \text{and} \quad \frac{1}{\mu(B)} \int_B |s_i|_h^2 \, d\mu \leq 1.$$

Since the logarithm function is strictly concave, Jensen's inequality shows that

$$\frac{1}{\mu(B)} \int_B \log|s_i|_h^2 \, d\mu \leq \log\left(\frac{1}{\mu(B)} \int_B |s_i|_h^2 \, d\mu \right) \leq 0, \tag{26.2}$$

with equality if and only if $|s_i|_h = 1$ almost everywhere.

Now let us analyze the singular hermitian metric on $\det E$. The expression

$$H_{i,j}(x) = \langle s_i(x), s_j(x) \rangle_{h,x}$$

is well-defined outside a set of measure zero, and the resulting function is locally integrable. Denote by $H(x)$ the $r \times r$-matrix with these entries; it is almost everywhere positive definite, and we have

$$\left| s_1 \wedge \cdots \wedge s_r \right|_{\det h}^2 = \det H.$$

Since $\det h$ is actually smooth and flat, we can choose a nowhere vanishing section $\delta \in H^0(B, \det \mathscr{F})$ such that $|\delta|_{\det h} \equiv 1$. We then have $s_1 \wedge \cdots \wedge s_r = g \cdot \delta$ for a holomorphic function $g \in H^0(B, \mathscr{O}_B)$ with $g(0) = 1$, and

$$|g|^2 = \left| s_1 \wedge \cdots \wedge s_r \right|_{\det h}^2 = \det H.$$

From Hadamard's inequality for semi-positive definite matrices, we obtain

$$|g(x)|^2 = \det H(x) \leq \prod_{i=1}^r H_{i,i}(x) = \prod_{i=1}^r \left| s_i(x) \right|_{h,x}^2,$$

with equality if and only if the matrix $H(x)$ is diagonal. Taking logarithms, we get

$$\log|g(x)|^2 \leq \sum_{i=1}^r \log\left| s_i(x) \right|_{h,x}^2.$$

This inequality is valid almost everywhere; integrating, we find that

$$\frac{1}{\mu(B)} \int_B \log|g|^2 \, d\mu \leq \sum_{i=1}^{r} \frac{1}{\mu(B)} \int_B |s_i|_h^2 \, d\mu. \tag{26.3}$$

Now $\log|g|^2$ is plurisubharmonic, and so the mean-value inequality shows that the left-hand side in (26.3) is greater or equal to $\log|g(0)|^2 = 0$. At the same time, the right-hand side is less or equal to 0 by (26.2). The conclusion is that all our inequalities are actually equalities, and so $H(x)$ is almost everywhere equal to the identity matrix of size $r \times r$. In other words, the sections $s_1, \dots, s_r \in H^0(B, \mathscr{F})$ are almost everywhere orthonormal with respect to h.

For any holomorphic section $f \in H^0(B, \mathscr{F}^*)$, we therefore have

$$|f|_{h^*}^2 = \sum_{i=1}^{r} |f \circ s_i|^2$$

almost everywhere on B; because the logarithms of both sides are plurisubharmonic functions on B, the identity actually holds everywhere. The singular hermitian metric h^* is therefore smooth; but then h is also smooth, and the pair (E, h) is a hermitian flat bundle.

To conclude the proof, we need to argue that $\mathscr{F}$ is locally free on all of B. The sections $s_1, \dots, s_r \in H^0(B, \mathscr{F})$ give rise to a morphism of sheaves

$$\sigma\colon \mathscr{O}_B^{\oplus r} \to \mathscr{F}.$$

We already know that σ is an isomorphism on the open subset $B(\mathscr{F})$; by Hartog's theorem, its inverse extends to a morphism of sheaves

$$\tau\colon \mathscr{F} \to \mathscr{O}_B^{\oplus r}$$

with $\tau \circ \sigma = \mathrm{id}$. Because $\mathscr{F}$ is torsion-free, this forces σ to be an isomorphism. □

NOTE. Our proof gives a different interpretation for the fact that $(\det E, \det h)$ has semi-positive curvature. Indeed, without assuming that $\det h$ is smooth and flat, we have $\det H = |g|^2 e^{-\varphi}$, where $\varphi\colon B \to [-\infty, +\infty)$ is locally integrable and $\varphi(0) = 0$. The various inequalities above then combine to give

$$0 \leq \frac{1}{\mu(B)} \int_B \varphi \, d\mu,$$

which is exactly the mean-value inequality for φ.

The next theorem is a new result. It says that when X is compact, all global sections of the dual coherent sheaf $\mathscr{F}^*$ arise from trivial summands in $\mathscr{F}$. Equivalently, every nonzero morphism $\mathscr{F} \to \mathscr{O}_X$ has a section, which means that $\mathscr{F}$ splits off a direct summand isomorphic to $\mathscr{O}_X$.

THEOREM 26.4. *Suppose that X is compact and connected. Then for every nonzero $f \in H^0(X, \mathscr{F}^*)$, there exists a unique global section $s \in H^0(X, \mathscr{F})$ such that $|s|_h$ is a.e. constant and $f \circ s \equiv 1$.*

PROOF. Because the singular hermitian metric on $\mathscr{F}$ has semi-positive curvature, the function $\log|f|_{h^*}$ is plurisubharmonic on X, hence equal to a nonzero constant. After rescaling the metric, we may assume without loss of generality that $|f|_{h^*} \equiv 1$. As in the proof of Theorem 26.1, we cover X by open subsets that are isomorphic to $B \subseteq \mathbb{C}^n$ and are centered at points $x \in X \setminus Z$ where $|-|_{h,x}$ is finite and positive definite. We shall argue that there is a unique section of $\mathscr{F}$ with the desired properties on each open set of this type; by uniqueness, these sections will then glue together to give us the global section $s \in H^0(X, \mathscr{F})$ that we are looking for.

We may therefore assume without loss of generality that $X = B$, that the origin belongs to the subset $B \setminus Z$, and that $|-|_{h,0}$ is a hermitian inner product on the vector space E_0. It is easy to see from

$$\sup \left\{ \frac{|f(v)|}{|v|_{h,0}} \;\middle|\; v \in E_0 \text{ with } |v|_{h,0} \neq 0 \right\} = |f|_{h^*,0} = 1$$

that there exists a vector $v \in E_0$ with $f(v) = 1$ and $|v|_{h,0} = 1$. By the minimal extension property, there is a section $s \in H^0(B, \mathscr{F})$ such that

$$s(0) = v \quad \text{and} \quad \frac{1}{\mu(B)} \int_B |s|_h^2 \, d\mu \leq 1.$$

Now $f \circ s$ is a holomorphic function on B, and by definition of h^*, we have

$$\frac{|f \circ s|}{|s|_h} \leq |f|_{h^*} = 1.$$

Taking logarithms and integrating, we get

$$\frac{1}{\mu(B)} \int_B \log|f \circ s|^2 \, d\mu \leq \frac{1}{\mu(B)} \int_B \log|s|_h^2 \, d\mu \leq \log\left(\frac{1}{\mu(B)} \int_B |s|_h^2 \, d\mu \right) \leq 0,$$

using Jensen's inequality along the way. By the mean-value inequality, the left-hand side is greater or equal to $\log(f \circ s)(0) = 0$, and so once again, all inequalities must be equalities. It follows that $f \circ s \equiv 1$, and that the measurable function $|s|_h$ is equal to 1 almost everywhere.

It remains to prove the uniqueness statement. Suppose that $s' \in H^0(B, \mathscr{F})$ is another holomorphic section with the property that $f \circ s' \equiv 1$ and $|s'|_h = 1$ almost everywhere. Outside a set of measure zero, we have

$$|s' - s|_h^2 + |s' + s|_h^2 = 2|s|_h^2 + 2|s'|_h^2 = 4,$$

and since $f(s' + s) = 2$, we must have $|s' + s|_h^2 \geq 4$. This implies that $|s' - s|_h^2 = 0$ almost everywhere, and hence that $s' = s$. □

E. Pushforwards of relative pluricanonical bundles

27. Introduction. In the previous chapter, we presented a general formalism for constructing singular hermitian metrics with semi-positive curvature on sheaves of the form $f_*(\omega_{X/Y} \otimes L)$. The applications to algebraic geometry come from the fact that the sheaves $f_*\omega_{X/Y}^{\otimes m}$ with $m \geq 2$ naturally fit into this framework. The main result is the following; see [**3**, Corollary 4.2], and also [**43**, Theorem 1.12], [**37**, Theorem 4.2.2].

THEOREM 27.1. *Let $f\colon X \to Y$ be a surjective projective morphism with connected fibers between two complex manifolds. Suppose that $f_*\omega_{X/Y}^{\otimes m} \neq 0$ for some $m \geq 2$.*

(a) *The line bundle $\omega_{X/Y}$ has a canonical singular hermitian metric with semi-positive curvature, called the m-th* Narasimhan-Simha metric. *This metric is continuous on the preimage of the smooth locus of f.*

(b) *If h denotes the induced singular hermitian metric on $L = \omega_{X/Y}^{\otimes(m-1)}$, then*

$$f_*\big(\omega_{X/Y} \otimes L \otimes \mathcal{I}(h)\big) \hookrightarrow f_*\omega_{X/Y}^{\otimes m}$$

is an isomorphism over the smooth locus of f.

We can therefore apply Corollary 21.2 and conclude that for any $m \geq 1$, the torsion-free sheaf $f_*\omega_{X/Y}^{\otimes m}$ has a singular hermitian metric with semi-positive curvature and the minimal extension property. Over the smooth locus of f, this metric is finite and continuous. The minimal extension property has the following remarkable consequences.

COROLLARY 27.2. *Suppose that Y is compact.*

(a) *If $c_1\big(\det f_*\omega_{X/Y}^{\otimes m}\big) = 0$ in $H^2(Y, \mathbb{R})$, then $f_*\omega_{X/Y}^{\otimes m}$ is locally free and the singular hermitian metric on it is smooth and flat.*

(b) *Any nonzero morphism $f_*\omega_{X/Y}^{\otimes m} \to \mathscr{O}_Y$ is split surjective.*

PROOF. This follows from Theorem 26.1 and Theorem 26.4. □

NOTE. There are two or three points in the proof where we need to use invariance of plurigenera. This means that Theorem 27.1 cannot be used to give a new proof for the invariance of plurigenera.

28. The absolute case. Let us start by discussing the absolute case. Take X to be a smooth projective variety of dimension n. Fix an integer $m \geq 1$ for which the vector space

$$V_m = H^0(X, \omega_X^{\otimes m})$$

of all m-canonical forms is nontrivial. Our goal is to construct a singular hermitian metric on the line bundle ω_X, with singularities along the base locus of V_m, such that all elements of V_m have bounded norm. We can measure the *length* of an m-canonical form $v \in V_m$ by a real number $\ell(v) \in [0, +\infty)$, defined by the formula

$$\ell(v) = \left(\int_X (c_n^m v \wedge \overline{v})^{1/m} \right)^{m/2}. \tag{28.1}$$

The constant $c_n = 2^{-n}(-1)^{n^2/2}$ is there to make the expression in parentheses positive. A more concrete definition is as follows. In local coordinates $z_1, \dots, z_n$, we have an expression

$$v = g(z_1, \dots, z_n)(dz_1 \wedge \cdots \wedge dz_n)^{\otimes m},$$

with g holomorphic; the integrand in (28.1) is then locally given by

$$|g|^{2/m} c_n (dz_1 \wedge \cdots \wedge dz_n) \wedge (d\bar{z}_1 \wedge \cdots \wedge d\bar{z}_n). \tag{28.2}$$

For $m \geq 2$, the length function ℓ is not a norm, because the triangle inequality fails to hold. On the other hand, ℓ is continuous on V_m, with $\ell(v) = 0$ iff $v = 0$; we also have $\ell(\lambda v) = |\lambda| \cdot \ell(v)$ for every $\lambda \in \mathbb{C}$.

We can now construct a singular hermitian metric h_m on the line bundle ω_X by using the length function ℓ. Given an element ξ in the fiber of ω_X at a point $x \in X$, we define

$$|\xi|_{h_m,x} = \inf\big\{ \ell(v)^{1/m} \bigm| v \in V_m \text{ satisfies } v(x) = \xi^{\otimes m} \big\} \in [0, +\infty].$$

In other words, we look for the m-canonical form of minimal length whose value at the point x is equal to the m-th power of ξ; if x belongs to the base locus of V_m, then $|\xi|_{h_m,x} = +\infty$ for $\xi \neq 0$. We obtain in this way a singular hermitian metric h_m on the line bundle ω_X, with singularities precisely along the base locus of the linear system V_m. The advantage of this construction is that it is completely canonical: there is no need to choose a basis for V_m.

NOTE. Following Păun and Takayama, we may call h_m the *m-th Narasimhan-Simha metric* on the line bundle ω_X, because Narasimhan and Simha [**29**] used this metric in the special case ω_X ample. A similar construction also appears in Kawamata's proof of Iitaka's conjecture over curves [**22**, §2].

PROPOSITION 28.3. *The Narasimhan-Simha metric h_m on ω_X is continuous, has singularities exactly along the base locus of V_m, and has semi-positive curvature.*

PROOF. We compute the local weights of h_m. Let $z_1, \dots, z_n$ be local holomorphic coordinates on a suitable open subset $U \subseteq X$, and set $s_0 = dz_1 \wedge \dots \wedge dz_n$, which is a nowhere vanishing section of ω_X on the subset U. Consider the function

$$\varphi_m = -\log|s_0|^2_{h_m} : U \to [-\infty, +\infty).$$

The definition of h_m shows that, for every $x \in U$,

$$\varphi_m(x) = \frac{2}{m} \sup\left\{ \log\frac{1}{\ell(v)} \;\middle|\; v \in V_m \text{ satisfies } v(x) = s_0(x)^{\otimes m} \right\}. \tag{28.4}$$

For each $v \in V_m$, there is a holomorphic function $g_v \colon U \to \mathbb{C}$ with $v|_U = g_v \cdot s_0^{\otimes m}$. If $g_v(x) \neq 0$, then the m-canonical form $v/g_v(x)$ contributes to the right-hand side of (28.4), and so we obtain

$$\varphi_m(x) = \frac{2}{m} \sup\Big\{ \log|g_v(x)| \;\Big|\; v \in V_m \text{ satisfies } \ell(v) \leq 1 \Big\}. \tag{28.5}$$

We will see in a moment that the supremum is actually a maximum, because the set of m-canonical forms $v \in V_m$ with $\ell(v) \leq 1$ is compact. Evidently, $\varphi_m(x) = -\infty$ if and only if $x \in U$ belongs to the base locus of V_m.

Now observe that the family of holomorphic functions

$$G_m = \big\{ g_v \in H^0(U, \mathscr{O}_X) \;\big|\; v \in V_m \text{ satisfies } \ell(v) \leq 1 \big\}$$

is uniformly bounded on compact subsets. Indeed, the fact that $\ell(v) \leq 1$ gives us a uniform bound on the $L^{2/m}$-norm of each g_v, and then we can argue as in the proof of Proposition 12.5, using the mean-value inequality. By the n-dimensional version of Montel's theorem, the family G_m is equicontinuous; due to (28.5), our φ_m is therefore continuous, as a function from U into $[-\infty, +\infty)$.

From (28.5), we can also determine the curvature properties of h_m. For each fixed $v \in V_m$, the function $\log|g_v|^{2/m}$ is continuous and plurisubharmonic, and equal to $-\infty$ precisely on the zero locus of g_v. As the upper envelope of an equicontinuous family of plurisubharmonic functions, φ_m is itself plurisubharmonic [**12**, Theorem I.5.7]. This shows that the Narasimhan-Simha metric on ω_X has semi-positive curvature. □

Another good feature of the Narasimhan-Simha metric is that all m-canonical forms are bounded with respect to this metric. Indeed, if we also use h_m to denote the induced singular hermitian metric on $\omega_X^{\otimes m}$, then by construction, we have the pointwise inequality $|v|_{h_m} \leq \ell(v)$ for every $v \in V_m$. In order to fit the Narasimhan-Simha metric into the framework of Chapter D, we write

$$\omega_X^{\otimes m} = \omega_X \otimes \omega_X^{\otimes(m-1)},$$

and endow the line bundle $L = \omega_X^{\otimes(m-1)}$ with the singular hermitian metric h induced by h_m. This metric is continuous and has semi-positive curvature.

LEMMA 28.6. *For every $v \in V_m$, we have $\|v\|_h \leq \ell(v)$.*

PROOF. We keep the notation introduced during the proof of Proposition 28.3. The weight of h with respect to the section $s_0^{\otimes(m-1)}$ of the line bundle $\omega_X^{\otimes(m-1)}$ is

$$e^{-(m-1)\cdot\varphi_m},$$

where φ_m is the function defined in (28.5). Now fix an m-canonical form $v \in V_m$ with $\ell(v) = 1$. On the open set U, the integrand in the definition of $\|v\|_h$ is

$$|g_v|^2 e^{-(m-1)\cdot\varphi_m} \cdot c_n(dz_1 \wedge \cdots \wedge dz_n) \wedge (d\bar{z}_1 \wedge \cdots \wedge d\bar{z}_n).$$

Because of (28.5), we have $\varphi_m \geq \log|g_v|^{2/m}$, and therefore

$$|g_v|^2 e^{-(m-1)\cdot\varphi_m} \leq |g_v|^2 \cdot |g_v|^{-2(m-1)/m} = |g_v|^{2/m}.$$

Looking back at the definition of $\ell(v)$ in (28.2), this shows that $\|v\|_h \leq \ell(v)$. □

Since ℓ is not itself a norm, the inequality will in general be strict. One useful consequence of Lemma 28.6 is the identity

$$H^0\big(X, \omega_X \otimes L \otimes \mathcal{I}(h)\big) = H^0(X, \omega_X \otimes L) = H^0\big(X, \omega_X^{\otimes m}\big). \tag{28.7}$$

Note that the multiplier ideal $\mathcal{I}(h)$ may well be nontrivial; nevertheless, it imposes no extra conditions on global sections of $\omega_X^{\otimes m}$.

29. The Ohsawa-Takegoshi theorem for pluricanonical forms. To analyze how the Narasimhan-Simha metric behaves in families, we will need a version of the Ohsawa-Takegoshi theorem for m-canonical forms. Suppose that $f \colon X \to B$ is a holomorphic mapping to the open unit ball $B \subseteq \mathbb{C}^r$, with f projective and $f(X) = B$, and such that the central fiber $X_0 = f^{-1}(0)$ is nonsingular. To simplify the discussion, let us also assume that f is the restriction of a holomorphic family over a ball of slightly larger radius. As in (28.1), we have length functions ℓ and ℓ_0 on X respectively X_0; because X is not compact, it may happen that $\ell(v) = +\infty$ for certain $v \in H^0(X, \omega_X^{\otimes m})$.

THEOREM 29.1. *For each $u \in H^0\big(X_0, \omega_{X_0}^{\otimes m}\big)$, there is some $v \in H^0\big(X, \omega_X^{\otimes m}\big)$ with*

$$\ell(v) \leq \mu(B)^{m/2} \cdot \ell_0(u),$$

such that the restriction of v to X_0 is equal to $u \wedge (df_1 \wedge \cdots \wedge df_r)^{\otimes m}$.

PROOF. Without loss of generality, we may assume that $\ell_0(u) = 1$. Since X_0 is a projective complex manifold, invariance of plurigenera tells us that the fiber of the coherent sheaf $f_*\omega_X^{\otimes m}$ at the point $0 \in B$ is equal to $H^0\big(X_0, \omega_{X_0}^{\otimes m}\big)$. Because B is a Stein manifold, we can then certainly find a section

$$v \in H^0\big(B, f_*\omega_X^{\otimes m}\big) = H^0\big(X, \omega_X^{\otimes m}\big)$$

with the correct restriction to X_0. By assuming that f comes from a morphism to a ball of slightly larger radius, we can also arrange that the quantity

$$\ell(v) = \left(\int_X (c_n^m v \wedge \overline{v})^{1/m}\right)^{m/2}$$

is finite. Of course, v will not in general satisfy the desired inequality.

The way to deal with this problem is to consider $\omega_X^{\otimes m} = \omega_X \otimes \omega_X^{\otimes(m-1)}$ as an adjoint bundle and to apply the Ohsawa-Takegoshi theorem to get another extension of smaller length. The section $v \in H^0(X, \omega_X^{\otimes m})$ induces a singular hermitian metric on the line bundle $\omega_X^{\otimes m}$, whose curvature is semi-positive. With respect to a local trivialization

$$\varphi \colon \omega_X^{\otimes m}|_U \to \mathscr{O}_U,$$

the weight of this metric is given by $\log|\varphi \circ v|^2$. Endow the line bundle $\omega_X^{\otimes(m-1)}$ with the singular hermitian metric whose local weight is

$$\frac{m-1}{m} \log|\varphi \circ v|^2.$$

It is easy to see that the norm of u with respect to this metric is still equal to $\ell_0(u) = 1$. Theorem 14.4 says that there exists another section $v' \in H^0(X, \omega_X^{\otimes m})$, with the same restriction to X_0, whose norm squared is bounded by $\mu(B)$. To get a useful expression for the norm squared, write

$$v' = Fv,$$

with F meromorphic on X and identically equal to 1 on X_0; then the inequality in the Ohsawa-Takegoshi theorem takes the form

$$\int_X |F|^2 (c_n^m v \wedge \overline{v})^{1/m} \leq \mu(B).$$

We can use this to get an upper bound for the quantity

$$\ell(v') = \left(\int_X |F|^{2/m} (c_n^m v \wedge \overline{v})^{1/m} \right)^{m/2}.$$

To begin with, let us write $(c_n^m v \wedge \overline{v})^{1/m} = L\, d\mu$, where L is a nonnegative real-analytic function on X, and $d\mu$ is some choice of volume form. Using Hölder's inequality with exponents $1/m$ and $(m-1)/m$, we have

$$\ell(v')^{2/m} = \int_X |F|^{2/m} L\, d\mu \leq \left(\int_X |F|^2 L\, d\mu \right)^{1/m} \left(\int_X L\, d\mu \right)^{(m-1)/m},$$

and therefore $\ell(v') \leq \mu(B)^{1/2} \cdot \ell(v)^{(m-1)/m}$, which we may rewrite as

$$\frac{\ell(v')}{\mu(B)^{m/2}} \leq \left(\frac{\ell(v)}{\mu(B)^{m/2}} \right)^{(m-1)/m}.$$

Now we iterate this construction to produce an infinite sequence of m-canonical forms $v_0, v_1, v_2, \ldots \in H^0(X, \omega_X^{\otimes m})$, all with the correct restriction to X_0. The inequality from above shows that one of two things happens: either $\ell(v_k) \leq \mu(B)^{m/2}$ for some $k \geq 0$; or $\ell(v_k) > \mu(B)^{m/2}$ for every $k \in \mathbb{N}$, and

$$\lim_{k \to +\infty} \ell(v_k) = \mu(B)^{m/2}.$$

If the former happens, we are done. If the latter happens, we apply Lemma 29.2: it says that a subsequence converges uniformly on compact subsets to an m-canonical form $v \in H^0(X, \omega_X^{\otimes m})$. Now v satisfies $\ell(v) \leq \mu(B)^{m/2}$ (by Fatou's lemma), and its restriction to X_0 is of course still equal to $u \wedge (df_1 \wedge \cdots \wedge df_r)^{\otimes m}$. □

LEMMA 29.2. *Let X be a complex manifold, and let $v_0, v_1, v_2, \ldots \in H^0(X, \omega_X^{\otimes m})$ be a sequence of m-canonical forms such that $\ell(v_k) \leq C$ for every $k \in \mathbb{N}$. Then a subsequence converges uniformly on compact subsets to a limit $v \in H^0(X, \omega_X^{\otimes m})$.*

PROOF. With respect to a local trivialization of $\omega_X^{\otimes m}$, we have a sequence of holomorphic functions whose $L^{2/m}$-norm is uniformly bounded. Using the mean-value inequality, this implies that the sequence of functions is uniformly bounded on compact subsets; now apply Montel's theorem to get the desired conclusion. □

NOTE. One interesting thing about the proof of Theorem 29.1 is that it looks very similar to Viehweg's covering trick (which we used for example in the proof of Proposition 8.2). The advantage of the metric approach is that one can take a limit to obtain a solution with the same properties as in the case $m = 1$.

30. The relative case. With the help of Theorem 29.1, it is quite easy to analyze the behavior of the Narasimhan-Simha metric in families. Let us first consider the case of a smooth morphism $f\colon X \to Y$; as in the statement of Theorem 27.1, we assume that f is projective with connected fibers, and that $f(X) = Y$. Recall that by invariance of plurigenera, the dimension of the space of m-canonical forms on the fiber $X_y = f^{-1}(y)$ is the same for every $y \in Y$.

The restriction of the relative canonical bundle $\omega_{X/Y}$ to the fiber X_y identifies to the canonical bundle ω_{X_y} of the fiber. We can therefore apply the construction in §28 fiber by fiber to produce a singular hermitian metric h_m on $\omega_{X/Y}$, called the *m-th relative Narasimhan-Simha metric*; we shall give a more careful definition of h_m in a moment. The first result is that h_m is continuous.

PROPOSITION 30.1. *Under the assumptions above, the relative Narasimhan-Simha metric on $\omega_{X/Y}$ is continuous.*

PROOF. Once again, this is an application of the Ohsawa-Takegoshi theorem for pluricanonical forms, which allows us to extend m-canonical forms from the fibers of f, with a uniform upper bound on the length of the extension. After shrinking Y, we can assume that $Y = B$ is the open unit ball in $\mathbb{C}^r$, with coordinates $t_1, \dots, t_r$. We denote by $V_m = H^0(X, \omega_X^{\otimes m})$ the (typically infinite-dimensional) vector space of all m-canonical forms on X. Given $v \in V_m$ and a point $y \in Y$, we have

$$v|_{X_y} = v_y \otimes (dt_1 \wedge \cdots \wedge dt_r)^{\otimes m}$$

for a unique m-canonical form $v_y \in H^0(X_y, \omega_{X_y}^{\otimes m})$. We denote by $\ell(v)$ the length of v on X, and by $\ell_y(v_y)$ the length of v_y on X_y. The Ohsawa-Takegoshi theorem for pluricanonical forms (in Theorem 29.1) implies that, possibly after shrinking Y, there is a constant $C \geq 0$ with the following property:

(30.2) For every $y \in Y$ and every m-canonical form u on X_y of length ≤ 1, there is an m-canonical form $v \in V_m$ such that $v_y = u$ and $\ell(v) \leq C$.

Now let $n = \dim X$. As the morphism f is smooth, every point in X has an open neighborhood U with coordinates $z_1, \dots, z_{n-r}, t_1, \dots, t_r$. Then $s_0 = dz_1 \wedge \cdots \wedge dz_{n-r}$ gives a local trivialization of $\omega_{X/Y}$, and we consider the weight function

$$\varphi_m = -\log|s_0|^2_{h_m} : U \to [-\infty, +\infty)$$

of the relative Narasimhan-Simha metric h_m. On each fiber, φ_m is given by the formula in (28.5); we can use the Ohsawa-Takegoshi theorem to obtain a more uniform description. For each $v \in V_m$, we have

$$v|_U = g_v \cdot (s_0 \wedge dt_1 \wedge \cdots \wedge dt_r)^{\otimes m}$$

for a unique holomorphic function $g_v \colon U \to \mathbb{C}$. By (28.5) and (30.2), we have

$$\varphi_m(x) = \frac{2}{m} \sup\Big\{ \log|g_v(x)| \Bigm| v \in V_m \text{ satisfies } \ell(v) \leq C \text{ and } \ell_y(v_y) \leq 1 \Big\};$$

where $y = f(x)$. We are going to prove that this defines a continuous function on U.

Fix a point $x \in U$, and let $x_0, x_1, x_2, \dots$ be any sequence in U with limit x. Set $y_k = f(x_k)$ and $y = f(x)$. For every $k \in \mathbb{N}$, choose an m-canonical form u_k of length $\ell_{y_k}(u_k) = 1$ on the fiber X_{y_k}, such that u_k computes $\varphi_m(x_k)$. Extend u_k to an m-canonical form v_k of length $\ell(v_k) \leq C$ on X by using (30.2); then

$$\varphi_m(x_k) = \frac{2}{m} \log\big|g_{v_k}(x_k)\big|.$$

After passing to a subsequence, $v_0, v_1, v_2, \dots$ converges uniformly on compact subsets to an m-canonical form $v \in H^0(X, \omega_X^{\otimes m})$. Since $\ell_{y_k}(v_{n,y_k}) = 1$, Fatou's lemma shows that $\ell_y(v_y) \leq 1$. Moreover, the holomorphic functions g_{v_k} converge uniformly on compact subsets to g_v, and therefore

$$\lim_{k \to +\infty} \varphi_m(x_k) = \frac{2}{m} \log\big|g_v(x)\big| \leq \varphi_m(x). \tag{30.3}$$

On the other hand, we can choose an m-canonical form u' of length $\ell_y(u') = 1$ on the fiber X_y, such that u' computes $\varphi_m(x)$. Extend u' to an m-canonical form v' of length $\ell(v') \leq C$ on X by using (30.2); then

$$\varphi_m(x) = \frac{2}{m} \log|g_{v'}(x)|.$$

Now it is easy to see from the definition of the length function that $\ell_{y_k}(v'_{y_k})$ tends to $\ell_y(v'_y)$ as $k \to +\infty$. In particular, the m-canonical form v'_{y_k} on X_{y_k} has nonzero length for $k \gg 0$, which means that

$$\frac{2}{m} \Big(\log|g_{v'}(x_k)| - \log \ell_{y_k}(v'_{y_k}) \Big) \leq \varphi_m(x_k).$$

Since the left-hand side tends to $\varphi_m(x)$, we obtain

$$\varphi_m(x) \leq \liminf_{k \to +\infty} \varphi_m(x_k). \tag{30.4}$$

The two inequalities in (30.3) and (30.4) together say that φ_m is continuous. □

Next, we prove that h_m has semi-positive curvature – just as in the case of adjoint bundles, the proof of this fact is very short, because we know the optimal value of the constant in Theorem 29.1.

PROPOSITION 30.5. *Under the assumptions above, the relative Narasimhan-Simha metric on $\omega_{X/Y}$ has semi-positive curvature.*

PROOF. Keep the notation introduced during the proof of Proposition 30.1. Because the local weight function φ_m is continuous, it suffices to prove that φ_m satisfies the mean-value inequality for mappings from the one-dimensional unit disk Δ into U. If the image of Δ lies in a single fiber, this is okay, because we already know from Proposition 28.3 that φ_m is plurisubharmonic on each fiber. So assume from now on that the mapping from Δ to Y is non-constant. Since the morphism f is smooth, we can then make a base change and reduce the problem to the case where $Y = \Delta$ and where $i \colon \Delta \hookrightarrow X$ is a section of $f \colon X \to \Delta$.

Now let $x_0 = i(0)$ and $X_0 = f^{-1}(0)$, and choose some $u \in H^0(X_0, \omega_{X_0}^{\otimes m})$ with $\ell_0(u) = 1$ that computes $\varphi_m(x_0)$. By Theorem 29.1, there exists an m-canonical form $v \in H^0(X, \omega_X^{\otimes m})$ with $v|_{X_0} = u \wedge df^{\otimes m}$, whose length satisfies the inequality

$$\ell(v) \leq \mu(\Delta)^{m/2} \cdot \ell_0(u) = \pi^{m/2}.$$

In the notation introduced during the proof of Proposition 30.1, we then have

$$\varphi_m(x_0) = \frac{2}{m} \log\lvert g_v(x_0)\rvert.$$

If we define $v_y \in H^0(X_y, \omega_{X_y}^{\otimes m})$ by the formula $v|_{X_y} = v_y \wedge df^{\otimes m}$, then we have

$$\ell(v)^{2/m} = \int_X (c_n^m v \wedge \overline{v})^{1/m} = \int_\Delta \ell_y(v_y)^{2/m} \, d\mu.$$

Now we observe that for almost every $y \in \Delta$, the ratio $v_y/\ell_y(v_y)$ is an m-canonical form on X_y of unit length; by definition of the weight function φ_m, we have

$$\varphi_m(x) \geq \frac{2}{m}\Big(\log\lvert g_v(x)\rvert - \log \ell_y(v_y)\Big) = \frac{2}{m} \log\lvert g_v(x)\rvert - \log \ell_y(v_y)^{2/m}.$$

If we now compute the mean value of $\varphi_m \circ i$ over Δ, we find that

$$\frac{1}{\pi} \int_\Delta \varphi_m(i(y)) \, d\mu \geq \frac{1}{\pi} \int_\Delta \frac{2}{m} \log\lvert g_v(i(y))\rvert \, d\mu - \frac{1}{\pi} \int_\Delta \log \ell_y(v_y)^{2/m} \, d\mu.$$

The first term on the right is greater or equal to $2/m \log\lvert g_v(x_0)\rvert = \varphi_m(x_0)$, because the function $g_v \circ i$ is holomorphic. To estimate the remaining integral, note that

$$\frac{1}{\pi} \int_\Delta \log \ell_y(v_y)^{2/m} \, d\mu \leq \log\left(\frac{1}{\pi} \int_\Delta \ell_y(v_y)^{2/m} \, d\mu\right) = \log\left(\frac{1}{\pi} \cdot \ell(v)^{2/m}\right) \leq 0,$$

by Jensen's inequality and the fact that $\ell(v) \leq \pi^{m/2}$. Consequently, φ_m does satisfy the required mean-value inequality, and h_m has semi-positive curvature. □

NOTE. Compare also Lemma 7 and Lemma 8 in [**22**].

After these preparations, we can now prove Theorem 27.1 in general.

PROOF OF THEOREM 27.1. Suppose that $f \colon X \to Y$ is a projective morphism between two complex manifolds with $f(X) = Y$. Let $Z \subseteq Y$ denote the closed analytic subset where f fails to be submersive. We already know that the restriction of the line bundle $\omega_{X/Y}$ to $f^{-1}(Y \setminus Z)$ has a well-defined singular hermitian metric h_m that is continuous and has semi-positive curvature. To show that h_m extends to a singular hermitian metric with semi-positive curvature on all of X, all we need to prove is that the local weights of h_m remain bounded near $f^{-1}(Z)$; this is justified by Lemma 12.4. Păun and Takayama [**37**, Theorem 4.2.7] observed that this local boundedness again follows very easily from the Ohsawa-Takegoshi theorem for pluricanonical forms.

Fix a point $x_0 \in X$ with $f(x_0) \in Z$. Since the problem is local on Y, we may assume that $Y = B$ is the open unit ball in $\mathbb{C}^r$, with coordinates $t_1, \dots, t_r$, and that $f(x_0) = 0$. On a suitable neighborhood U of the point x_0, we have coordinates $z_1, \dots, z_n$; note that because f is most likely not submersive at x_0, we cannot assert that $t_1, \dots, t_r$ are part of this coordinate system. Let $s_0 \in H^0(U, \omega_{X/Y})$ be a local trivialization of $\omega_{X/Y}$, chosen so that

$$dz_1 \wedge \cdots \wedge dz_n = s_0 \wedge (dt_1 \wedge \cdots \wedge dt_r).$$

Denote by φ_m the weight function of h_m with respect to this local trivialization:

$$\varphi_m(x) = -\log|s_0|^2_{h_m} : U \to [-\infty, +\infty)$$

For $v \in H^0(X, \omega_X^{\otimes m})$, we have $v|_U = g_v \cdot (dz_1 \wedge \cdots \wedge dz_n)^{\otimes m}$ for a holomorphic function $g_v \colon U \to \mathbb{C}$. As explained during the proof of Proposition 30.1, the Ohsawa-Takegoshi theorem for pluricanonical forms implies that there is a constant $C \geq 0$ with the following property: for every $x \in U$, there is some $v \in H^0(X, \omega_X^{\otimes m})$ of length $\ell(v) \leq C$ such that

$$\varphi_m(x) = \frac{2}{m} \log|g_v(x)|.$$

For x sufficiently close to x_0, there is a positive number $R > 0$ such that U contains the closed ball of radius R centered at x. The mean-value inequality and the fact that $\ell(v) \leq C$ now combine to give us an upper bound for $\varphi_m(x)$ that depends only on C and R, but is independent of the point x. In particular, φ_m is uniformly bounded in a neighborhood of the point $x_0 \in f^{-1}(Z)$, and therefore extends uniquely to a plurisubharmonic function on all of U.

The Narasimhan-Simha metric on each fiber X_y with $y \notin Z$ satisfies (28.7); by the Ohsawa-Takegoshi theorem, this means that the inclusion

$$f_*\big(\omega_{X/Y} \otimes L \otimes \mathcal{I}(h)\big) \hookrightarrow f_*\big(\omega_{X/Y} \otimes L\big) = f_*\omega_{X/Y}^{\otimes m}$$

is an isomorphism over $Y \setminus Z$. Due to Corollary 21.4, the singular hermitian metric on $f_*\omega_{X/Y}^{\otimes m}$ is therefore finite and continuous on $Y \setminus Z$. □

References

[1] Bo Berndtsson, *Curvature of vector bundles associated to holomorphic fibrations*, Ann. of Math. (2) **169** (2009), no. 2, 531–560. MR2480611

[2] Bo Berndtsson and László Lempert, *A proof of the Ohsawa-Takegoshi theorem with sharp estimates*, J. Math. Soc. Japan **68** (2016), no. 4, 1461–1472. MR3564439

[3] Bo Berndtsson and Mihai Păun, *Bergman kernels and the pseudoeffectivity of relative canonical bundles*, Duke Math. J. **145** (2008), no. 2, 341–378. MR2449950

[4] Zbigniew Błocki, *On the Ohsawa-Takegoshi extension theorem*, Univ. Iagel. Acta Math. **50** (2013), 53–61. MR3235002

[5] Li Yi, *An Ohsawa-Takegoshi theorem on compact Kähler manifolds*, Sci. China Math. **57** (2014), no. 1, 9–30. MR3146512

[6] Junyan Cao and Mihai Păun, *Kodaira dimension of algebraic fiber spaces over abelian varieties*, Invent. Math. **207** (2017), no. 1, 345–387. MR3592759

[7] Jungkai A. Chen and Christopher D. Hacon, *Characterization of abelian varieties*, Invent. Math. **143** (2001), no. 2, 435–447. MR1835393

[8] Jungkai A. Chen and Christopher D. Hacon, *On algebraic fiber spaces over varieties of maximal Albanese dimension*, Duke Math. J. **111** (2002), no. 1, 159–175. MR1876444

[9] Jungkai A. Chen and Christopher D. Hacon, *On the irregularity of the image of the Iitaka fibration*, Comm. Algebra **32** (2004), no. 1, 203–215. MR2036231

[10] Jungkai A. Chen and Christopher D. Hacon, *Pluricanonical systems on irregular 3-folds of general type*, Math. Z. **255** (2007), no. 2, 343–355. MR2262735

[11] Jungkai Alfred Chen and Christopher D. Hacon, *Kodaira dimension of irregular varieties*, Invent. Math. **186** (2011), no. 3, 481–500. MR2854084

[12] Jean-Pierre Demailly, *Complex analytic and differential geometry*, `https://www-fourier.ujf-grenoble.fr/~demailly/manuscripts/agbook.pdf`, 2012.

[13] Hélène Esnault and Eckart Viehweg, *Lectures on vanishing theorems*, DMV Seminar, vol. 20, Birkhäuser Verlag, Basel, 1992. MR1193913

[14] Mark Green and Robert Lazarsfeld, *Deformation theory, generic vanishing theorems, and some conjectures of Enriques, Catanese and Beauville*, Invent. Math. **90** (1987), no. 2, 389–407. MR910207

[15] Mark Green and Robert Lazarsfeld, *Higher obstructions to deforming cohomology groups of line bundles*, J. Amer. Math. Soc. **4** (1991), no. 1, 87–103. MR1076513

[16] Qi'an Guan and Xiangyu Zhou, *A solution of an L^2 extension problem with an optimal estimate and applications*, Ann. of Math. (2) **181** (2015), no. 3, 1139–1208. MR3296822

[17] Robert C. Gunning and Hugo Rossi, *Analytic functions of several complex variables*, AMS Chelsea Publishing, Providence, RI, 2009. Reprint of the 1965 original. MR2568219

[18] Christopher D. Hacon, *A derived category approach to generic vanishing*, J. Reine Angew. Math. **575** (2004), 173–187. MR2097552

[19] Christopher D. Hacon and Rita Pardini, *On the birational geometry of varieties of maximal Albanese dimension*, J. Reine Angew. Math. **546** (2002), 177–199. MR1900998

[20] Zhi Jiang, *An effective version of a theorem of Kawamata on the Albanese map*, Commun. Contemp. Math. **13** (2011), no. 3, 509–532. MR2813500

[21] Yujiro Kawamata, *Characterization of abelian varieties*, Compositio Math. **43** (1981), no. 2, 253–276. MR622451

[22] Yujiro Kawamata, *Kodaira dimension of algebraic fiber spaces over curves*, Invent. Math. **66** (1982), no. 1, 57–71. MR652646

[23] János Kollár, *Higher direct images of dualizing sheaves. I*, Ann. of Math. (2) **123** (1986), no. 1, 11–42. MR825838

[24] János Kollár, *Shafarevich maps and automorphic forms*, M. B. Porter Lectures, Princeton University Press, Princeton, NJ, 1995. MR1341589

[25] Ching-Jui Lai, *Varieties fibered by good minimal models*, Math. Ann. **350** (2011), no. 3, 533–547. MR2805635

[26] Robert Lazarsfeld, *Positivity in algebraic geometry. II*, Ergebnisse der Mathematik und ihrer Grenzgebiete. 3. Folge. A Series of Modern Surveys in Mathematics [Results in Mathematics and Related Areas. 3rd Series. A Series of Modern Surveys in Mathematics], vol. 49, Springer-Verlag, Berlin, 2004. Positivity for vector bundles, and multiplier ideals. MR2095472

[27] Shigefumi Mori, *Classification of higher-dimensional varieties*, Algebraic geometry, Bowdoin, 1985 (Brunswick, Maine, 1985), Proc. Sympos. Pure Math., vol. 46, Amer. Math. Soc., Providence, RI, 1987, pp. 269–331. MR927961

[28] Shigeru Mukai, *Duality between $D(X)$ and $D(\hat{X})$ with its application to Picard sheaves*, Nagoya Math. J. **81** (1981), 153–175. MR607081

[29] M. S. Narasimhan and R. R. Simha, *Manifolds with ample canonical class*, Invent. Math. **5** (1968), 120–128. MR0236960

[30] Giuseppe Pareschi, *Basic results on irregular varieties via Fourier-Mukai methods*, Current developments in algebraic geometry, Math. Sci. Res. Inst. Publ., vol. 59, Cambridge Univ. Press, Cambridge, 2012, pp. 379–403. MR2931876

[31] Giuseppe Pareschi and Mihnea Popa, *GV-sheaves, Fourier-Mukai transform, and generic vanishing*, Amer. J. Math. **133** (2011), no. 1, 235–271. MR2752940

[32] Giuseppe Pareschi and Mihnea Popa, *Regularity on abelian varieties III: relationship with generic vanishing and applications*, Grassmannians, moduli spaces and vector bundles, Clay Math. Proc., vol. 14, Amer. Math. Soc., Providence, RI, 2011, pp. 141–167. MR2807853

[33] Mihai Păun, *Siu's invariance of plurigenera: a one-tower proof*, J. Differential Geom. **76** (2007), no. 3, 485–493. MR2331528

[34] ———, *Singular Hermitian metrics and positivity of direct images of pluricanonical bundles*, preprint arXiv:1606.00174, 2016.

[35] Mihnea Popa, *Generic vanishing filtrations and perverse objects in derived categories of coherent sheaves*, Derived categories in algebraic geometry, EMS Ser. Congr. Rep., Eur. Math. Soc., Zürich, 2012, pp. 251–278. MR3050706

[36] Mihnea Popa and Christian Schnell, *On direct images of pluricanonical bundles*, Algebra Number Theory **8** (2014), no. 9, 2273–2295. MR3294390

[37] Mihai Păun and Shigeharu Takayama, *Positivity of twisted relative pluricanonical bundles and their direct images*, preprint arXiv:1409.5504, 2014.

[38] Hossein Raufi, *Singular hermitian metrics on holomorphic vector bundles*, Ark. Mat. **53** (2015), no. 2, 359–382. MR3391176

[39] Christian Schnell, *Graduate course on the generic vanishing theorem and its applications*, `http://www.math.stonybrook.edu/~cschnell/pdf/notes/generic-vanishing.pdf`, 2013.

[40] Carlos Simpson, *Subspaces of moduli spaces of rank one local systems*, Ann. Sci. École Norm. Sup. (4) **26** (1993), no. 3, 361–401. MR1222278

[41] Yum-Tong Siu, *Extension of twisted pluricanonical sections with plurisubharmonic weight and invariance of semipositively twisted plurigenera for manifolds not necessarily of general type*, Complex geometry (Göttingen, 2000), Springer, Berlin, 2002, pp. 223–277. MR1922108
[42] Yum-Tong Siu, *Section extension from hyperbolic geometry of punctured disk and holomorphic family of flat bundles*, Sci. China Math. **54** (2011), no. 8, 1767–1802. MR2824973
[43] Hajime Tsuji, *Canonical singular Hermitian metrics on relative canonical bundles*, Amer. J. Math. **133** (2011), no. 6, 1469–1501. MR2863368

Department of Mathematics, University of Utah, Salt Lake City, Utah 84112
Email address: hacon@math.utah.edu

Department of Mathematics, Northwestern University, 2033 Sheridan Road, Evanston, Illinois 60208
Email address: mpopa@math.northwestern.edu

Department of Mathematics, Stony Brook University, Stony Brook, New York 11794
Email address: cschnell@math.sunysb.edu

Contemporary Mathematics
Volume **712**, 2018
http://dx.doi.org/10.1090/conm/712/14347

A strongly geometric general residual intersection

Shihoko Ishii and Wenbo Niu

Dedicated to Professor Lawrence Ein on the occasion of his sixtieth birthday.

ABSTRACT. In this paper, we give formulas for the Grauert-Riemenschneider canonical sheaf and the log canonical threshold for a general residual intersection, and show that minimal log discrepancies are preserved under a general link. We also find evidences suggesting that MJ-singularities are preserved under a general residual intersection.

1. Introduction

The purpose of this paper is to show how general equations can be chosen to produce a residual intersection for a variety and to investigate what properties and invariants of singularities can be preserved under this procedure. The concept of a residual intersection was introduced by Artin-Nagata [**1**] in 1972. Its important case, namely linkage or liaison, was systematically studied by Peskine-Szpiro [**17**] to initiate the modern research of this area. The geometric idea behind the notion of a residual intersection is pretty natural: roughly speaking, any two varieties can be viewed as a residual intersection to each other in their union. When they have the same dimension, they are linked; otherwise, the one of smaller dimension is the residual intersection of another one.

More precisely, let X be a closed subvariety of a nonsingular variety A of codimension c. By choosing t equations from the defining ideal I_X of X, we can define a closed subscheme $M \subset A$ containing X. Then the closure of $M \setminus X$ can be thought of as a t-residual intersection of X. Certainly, different choices of these t equations result in different residual intersections. Among all possible residual intersections, it would be natural to expect that the one given by general equations would have the most typical behavior. This is the idea that motivates our research in this paper.

In the work by Huneke and Ulrich ([**12**], [**8**], [**11**], [**9**], [**10**], etc), the rigorous notions of generic linkage and residual intersection have been established and many fundamental properties have been proved. One of the central problems is to understand how algebraic or geometric properties of a given variety can be preserved under the procedure of generic linkage or residual intersection. As for singularities,

2010 *Mathematics Subject Classification.* Primary 13C40, 14M06.
Key words and phrases. Linkage, log canonical thresholds, Mather-Jacobian singularities, residual intersection.
The first author is partially supported by Grant-In-Aid (c) 1605089 of JSPS.

the behavior of rational singularities under residual intersection has been studied by Chardin and Ulrich [**2**] in connection to questions about Castelnuovo-Mumford regularity. In the second author's work [**15**] and [**16**], it has been shown that singularities such as log-canonical and MJ-singularities are preserved under generic linkage. However, it is not clear if similar results can be established for the more general situation of residual intersections.

In this paper, we focus ourselves on studying singularities and their invariants under a general residual intersection. To put the problem in perspective, we first give a construction of a general residual intersection, where our residual intersection is slightly different from the one defined by Huneke-Ulrich. Then based on techniques from resolutions of singularities, we discuss a couple of problems related to Grauert-Riemenschneider (GR) canonical sheaves, log canonical thresholds and minimal log discrepancies under a general residual intersection. Recall that $X \subset A$ is a closed subvariety. Let Y be a general t-residual intersection (see Definition 2.1) with $t \geq c$. We prove in Theorem 3.5 that the GR canonical sheaf of Y is

$$\omega_Y^{GR} = \mathscr{J}(A, I_X^t) \cdot \mathscr{O}_Y \otimes \omega_A,$$

where $\mathscr{J}(A, I_X^t)$ is the multiplier ideal associated to the pair (A, I_X^t), and the log canonical threshold of Y increases, i.e.,

$$\operatorname{lct}(A, Y) \geq \operatorname{lct}(A, X).$$

If we impose the extra condition that X is locally a complete intersection with rational singularities, the result of Chardin-Ulrich [**2**, Theorem 3.13] implies that Y has rational singularities. In this case, the GR canonical sheaf ω_Y^{RG} is the same as the canonical sheaf ω_Y and therefore we obtain a formula for the canonical sheaf of a general residual intersection (Corollary 3.6).

Turning to the local case, let $x \in X$ be a closed point. In Theorem 3.2, we show that general equations can be chosen to compute the minimal log discrepancies at x. In particular, in Corollary 3.3, we prove that if a complete intersection M is defined by c general equations $a_1, \cdots, a_c$ of I_X, then

$$\operatorname{mld}_{\mathrm{MJ}}(x; X, \mathfrak{a}^{\mathbf{m}}) = \operatorname{mld}_{\mathrm{MJ}}(x; M, \mathfrak{a}^{\mathbf{m}}) = \operatorname{mld}_{\mathrm{MJ}}(x; H, \mathfrak{a}^{\mathbf{m}}),$$

where the hypersurface H is defined by the product $a_1 \cdots a_c$ and $\mathfrak{a}^{\mathbf{m}}$ is a formal product of ideals. Hence the computation of minimal log discrepancies can be reduced to the complete intersection case or, even better, the hypersurface case.

At the end of the paper, we provide evidence (Proposition 3.8) suggesting that MJ-singularities are preserved by general residual intersection (Conjecture 3.9). The notion of MJ-singularities was introduced by Ishii, De Fernex-Docampo, and Ein-Ishii-Mustaţă ([**13**], [**3**], [**5**], and [**4**]). The advantage of this notion is that it can be established for arbitrary varieties without requiring normality and $\mathbb{Q}$-Gorensteiness. We hope that the method in this paper can be useful for this conjecture.

Acknowledgment: We are grateful to Bernd Ulrich for valuable discussions. The second author would like to thank the University of Tokyo for the hospitality of visiting.

2. General residual intersections

Throughout the paper, we work over an algebraically closed field k of characteristic zero. By a variety we mean an equidimensional reduced scheme of finite type over k.

DEFINITION 2.1. Let X be a closed subscheme of an nonsingular affine variety $A = \operatorname{Spec} R$ with ideal sheaf I_X. Let $I_M = (a_1, \cdots, a_t) \subseteq I_X$ such that it defines a closed subscheme M of A. Let Y be the closure of $M \setminus X$ (denoted by $Y = \overline{M \setminus X}$), where the closure of $M \setminus X$ is the closed subscheme defined by the coherent ideal sheaf maximal among ideal sheaves whose restriction on $A \setminus X$ coincide with the ideal sheaf of $M \setminus X$. If $t = \operatorname{codim} Y \geq \operatorname{codim} X$, then Y is called a *t-residual intersection.*

REMARK 2.2. Huneke and Ulrich defined a t-residual intersection in a different way ([**12**]): the closed subscheme defined by $J = (I_M : I_X)$ is called a t-residual intersection if $\operatorname{ht}(J) \geq t \geq \operatorname{ht}(I_X)$. Let t be an integer such that $t \geq \operatorname{codim}(X, A)$, if X satisfies G_t and strongly Cohen-Macaulay, then our residual intersection is Cohen-Macaulay and coincides with the one defined by Huneke-Ulrich. This is proved by applying [**8**, Theorem 3.1].

REMARK 2.3. Our definition of a residual intersection is in some sense "geometric", so it would be nice to call it a geometric t-residual intersection. But the terminology "geometric residual intersection" is already used for a different meaning in [**12**]. A suggestive alternative is to call it a "strongly geometric t-residual intersection" as is written in the title of this paper. But it is too long to be used frequently, so in this paper we call it just "residual intersection".

We are interested in understanding how properties of a variety can be "transfered" to its t-residual intersection. Certainly, in Definition 2.1, a different choice of the generators of I_M will result in a different t-residual intersection. It would be wonderful if the desired properties can be established for all t-residual intersections under every possible choice of I_M. But more realistically, we would like to investigate so called "general" t-residual intersection with respect to a generating set of I_X. We make this point clear in the following construction.

CONSTRUCTION 2.4. Let X be a closed subscheme of a nonsingular affine variety $A = \operatorname{Spec} R$ defined by an ideal I_X. Fix a generating set $\underline{f} = (f_1, \cdots, f_r)$ of I_X. Let $\mathscr{P}$ be a property that we are interested in. For $t \geq \operatorname{codim}_A X$, by saying that a general t-residual intersection Y with respect to $\underline{f}$ has property $\mathscr{P}$ we mean the following: given the ideal

$$I_M := (a_1, \cdots, a_t) = (c_{i,j}) \cdot (f_1, \cdots, f_r)^T,$$

in which

$$a_i := c_{i,1} f_1 + c_{i,2} f_2 + \cdots + c_{i,r} f_r, \qquad \text{for } 1 \leq i \leq t,$$

and $(c_{i,j})$ is a $t \times r$ matrix with elements in k, then there is an open set U in $\mathbb{A}^{t\times r}$ such that for any $(c_{i,j}) \in U$, the corresponding t-residual intersection $Y = \overline{M \setminus X}$ has property $\mathscr{P}$. By saying that a general t-residual intersection Y has property $\mathscr{P}$ without mentioning $\underline{f}$, we mean that it is true for any choice of a generating set $\underline{f}$.

The main framework in this paper is to use resolution of singularities to turn a variety into a divisor and then choose general equations by using Bertini's theorem to produce a t-residual intersection. The crucial point of this approach is that it automatically yields a resolution of singularities of the residual intersection. We show this in the following construction.

CONSTRUCTION 2.5. Let X be a closed subscheme of a nonsingular affine variety $A = \operatorname{Spec} R$ defined by an ideal I_X. Fix a generating set $\underline{f} = (f_1, \cdots, f_r)$ of I_X. So we have a surjective morphism

$$v : \bigoplus^{r} \mathscr{O}_A \xrightarrow{f_1, \cdots, f_r} I_X \longrightarrow 0.$$

Take a log resolution $\varphi : A_1 \to A$ of I_X such that $I_X \cdot \mathscr{O}_{A_1} = \mathscr{O}_{A_1}(-E)$ where E is an effective divisor such that $\operatorname{Exc}(\varphi) \cup E$ is a divisor with simple normal crossing (snc) supports. Pulling back by φ, we obtain a surjective morphism

$$\overline{v} : \bigoplus^{r} \mathscr{O}_{A_1} \xrightarrow{\varphi^* f_1, \cdots, \varphi^* f_r} \mathscr{O}_{A_1}(-E) \longrightarrow 0.$$

So the linear system $\varphi^*|f_1, \cdots, f_r|$ has E as its base locus and therefore it can be decomposed as

$$\varphi^*|f_1, \cdots, f_r| = W + E,$$

in which W is a base point free linear system on A_1. Take t general elements $\alpha_1, \cdots, \alpha_t$ from W and each α_i defines an nonsingular effective divisor F_i on A_1 by Bertini's theorem. So we have

(1) the sections $\alpha_1, \cdots, \alpha_t$ cut out a nonsingular closed subscheme Y_1 (could be reducible or empty) of A_1 of codimension t;
(2) the closed subscheme Y_1 meets E and $\operatorname{exc}(\varphi)$ transversally;
(3) the sections $\alpha_1, \cdots, \alpha_t$ correspond to elements $a_1, \cdots, a_t$ in the vector space $\langle f_1, \cdots, f_r \rangle$.

Set $I_M = (a_1, \cdots, a_t)$ and call the subscheme $Y = \overline{M \setminus X}$ a *general t-residual intersection* (with respect to $\underline{f}$). Note also that

(4) the restriction $\varphi|_{Y_1} : Y_1 \to Y$ is a resolution of singualrities of Y;
(5) $I_M \cdot \mathscr{O}_{A_1} = I_{Y_1} \cdot \mathscr{O}_{A_1}(-E)$.

REMARK 2.6. (1) In Construction 2.5, by Bertini's theorem, we can obtain an open set U such that for any choice of the $t \times r$ matrix $(c_{i,j})$ the corresponding general t-residual intersection Y satisfies the listed properties (1)-(5). Once a property $\mathscr{P}$ is imposed, we need to further shrink this open set U to get the desired one in Construction 2.4. This kind of argument should be clear from the context, so we do not always mention it explicitly.

(2) Roughly speaking, our Construction 2.4 can be viewed as a specialization of the generic residual intersection constructed in [**12**].

PROPOSITION 2.7. *Let X be a closed subscheme of an affine nonsingular variety A. Let Y be a general t-residual intersection of X. Then Y is empty if and only if $\overline{I_M} = \overline{I_X}$, where $\overline{\mathfrak{a}}$ denotes the integral closure of an ideal $\mathfrak{a}$.*

PROOF. We use the notation in Construction 2.5. If Y is empty, then Y_1 must be empty. This in turn implies that $I_M \cdot \mathscr{O}_{A_1} = I_X \cdot \mathscr{O}_{A_1}$. Therefore $\overline{I_M} = \overline{I_X}$ since $\overline{I_M} = \varphi_*(I_M \cdot \mathscr{O}_{A_1})$ and $\overline{I_X} = \varphi_*(I_X \cdot \mathscr{O}_{A_1})$. On the other hand, assume $\overline{I_M} = \overline{I_X}$. Denote by A^+ the normalization of the blowup of A along X.

We have $I_M \cdot \mathcal{O}_{A^+} = I_X \cdot \mathcal{O}_{A^+}$. This implies that on $A \setminus X$, $I_M = \mathcal{O}_A$ so Y is empty. $\square$

PROPOSITION 2.8. *Let X be a closed subscheme of an affine nonsingular variety A defined by an ideal I_X and let $\underline{f} = (f_1, \cdots, f_r)$ be a generating set of I_X. Let Y be a general t-residual intersection with respect to $\underline{f}$ as in Construction 2.5. Consider the regular map*

$$\psi : A \setminus X \longrightarrow \mathbb{A}^r$$

induced by the regular functions $f_1, \cdots, f_r$. Let V be the closure of the image of ψ.

(1) *If V is not a cone in $\mathbb{A}^r$, then Y is nonempty if and only if*

$$\dim k[f_1, \cdots, f_r] \geq t.$$

(2) *If V is a cone in $\mathbb{A}^r$, then Y is nonempty if and only if*

$$\dim k[f_1, \cdots, f_r] > t.$$

PROOF. Let $\pi : \mathbb{A}^r \setminus \{0\} \to \mathbb{P}^{r-1}$ be the canonical projection and let $W \subset \mathbb{P}^{r-1}$ be the image of V by π. Then $\dim W = \dim V + 1$ if V is a cone in $\mathbb{A}^r$, while $\dim W = \dim V$ otherwise. A pull-back of t general hyperplane cuts by $\psi^{-1} \circ \pi^{-1}$ gives $(A \setminus X) \cap M$ for general M. Therefore, Y is non-empty if and only if $\dim W \geq t$. This is equivalent to that $\dim V \geq t+1$ if V is a cone in $\mathbb{A}^r$ and $\dim V \geq t$ if V is not a cone in $\mathbb{A}^r$. The proposition follows from $V = \operatorname{Spec} k[f_1, \cdots, f_r]$. $\square$

EXAMPLE 2.9. For every closed subscheme $X \subset \mathbb{A}^N$, there is a generator system $\underline{g}$ of I_X such that, for any t with $\operatorname{codim}(X, \mathbb{A}^N) \leq t \leq N-1$, a general t-residual intersection of X is nonempty. Indeed, if a generator system $\underline{f} = \{f_1, \ldots, f_r\}$ is given, then we can add more generators of the form $x_1 f_1, x_2 f_1, \ldots, x_N f_1$ to obtain a bigger generator system $\underline{g}$. Then,

$$\dim k[f_1, \ldots, f_r, x_1 f_1, x_2 f_1, \ldots, x_N f_1] = N,$$

where $x_1, x_2, \ldots, x_N$ are the coordinate functions of $\mathbb{A}^N$. By the proposition above, for any t with $\operatorname{codim}(X, \mathbb{A}^N) \leq t \leq N-1$, a general t-residual intersection is not empty.

EXAMPLE 2.10. Let $X \subset \mathbb{A}^4$ be the cone over two skew lines in $\mathbb{P}^3$. Note that it is a two dimensional variety defined by the ideal $I_X = (xz, xw, yz, yw)$. First take the generator system $\underline{f} = \{xz, xw, yz, yw\}$. Then, we observe that the closure V of the image of

$$\psi : \mathbb{A}^4 \setminus X \longrightarrow \mathbb{A}^4$$

is a cone and has dimension 3. Then, a general 3-residual intersection of X with respect to $\underline{f}$ is empty by the Proposition 2.8. We can also calculate this directly by solving the equations.

Next, we take another generator system

$$\underline{g} = \{xz, xw, yz, yw, x(xz), y(xz), z(xz), w(xz)\}$$

as in the previous example. Then, $\dim k[\,\underline{g}\,] = 4$ and a general 3-residual intersection of X with respect to $\underline{g}$ is a curve.

3. Invariants of singularities under a general residual intersection

In this section, we prove our main results about general residual intersections. We start with recalling some basic definitions. Let X and Z be subvarieties of a nonsingular variety A. Take a log resolution $f : A' \to A$ of $I_X \cdot I_Z$ such that $f^{-1}(X) = \sum_{i=1}^{s} a_i E_i$, $f^{-1}(Z) = \sum_{i=1}^{s} z_i E_i$ and the relative canonical divisor $K_{A'/A} = \sum_{i=1}^{s} k_i E_i$. The *log canonical threshold* of (A, X) is defined to be

$$\operatorname{lct}(A, X) = \min_i \left\{ \frac{k_i + 1}{a_i} \right\}.$$

Having fixed $\lambda \in \mathbb{R}_+$, we also define the *generalized log canonical threshold* of $(A, X; \lambda Z)$ by

$$\operatorname{glct}(A, X; \lambda Z) = \min \left\{ \frac{k_i + 1 + \lambda z_i}{a_i} \right\}.$$

For $c \in \mathbb{R}_+$, the *multiplier ideal sheaf* $\mathscr{I}(A, cX)$ associated to the pair (A, cX) is defined by

$$\mathscr{I}(A, cX) = f_* \mathscr{O}_{A'}(K_{A'/A} - \lfloor c \sum a_i E_i \rfloor),$$

where $\lfloor c \sum a_i E_i \rfloor$ is the round down of the $\mathbb{R}$-divisor $c \sum a_i E_i$.

Let W be a proper closed subset of A and let $\mathfrak{a}^{\mathbf{m}} = \mathfrak{a}_1^{m_1} \mathfrak{a}_2^{m_2} \cdots \mathfrak{a}_n^{m_n}$ be a formal product of ideals $\mathfrak{a}_i \subseteq \mathscr{O}_A$ with $\mathbf{m} = (m_1, \cdots, m_n) \in \mathbb{R}^n_{\geq 0}$. We define the *minimal log discrepancy* of $(A, \mathfrak{a}^{\mathbf{m}})$ along W as

$$\operatorname{mld}(W; A, \mathfrak{a}^{\mathbf{m}}) = \inf_{C_A(E) \subseteq W} \{ a(E; A, \mathfrak{a}^{\mathbf{m}}) + 1 \mid E \text{ a prime divisor over } A \}$$

where $a(E; A, \mathfrak{a}^{\mathbf{m}})$ is the discrepancy of E and $C_A(E)$ is the center of E in A. We use the convention that if $\dim A = 1$ and $\operatorname{mld}(W; A, \mathfrak{a}^{\mathbf{m}})$ is negative, then we set it as $-\infty$. For more details on the invariants we defined above, we refer to, for instance, the work [**6**].

We also use the notation of Mather-Jacobian singularities (MJ-singularities for short) which was introduced and studied in [**13**], [**3**], [**5**] and [**4**]. Recall that X is a variety of dimension n and let $f : X' \longrightarrow X$ be a log resolution of the Jacobian ideal Jac_X of X. Then the image of the canonical homomorphism

$$f^*(\wedge^n \Omega_X^1) \longrightarrow \wedge^n \Omega_{X'}^1$$

is an invertible sheaf of the form $\operatorname{Jac}_f \cdot \wedge^n \Omega_{X'}^1$, where Jac_f is the relative Jacobian ideal of f. The ideal Jac_f is invertible and defines an effective divisor $\widehat{K}_{X'/X}$ which is called the *Mather discrepancy divisor* (see also [**5**, Remark 2.3]). For an ideal $\mathfrak{a} \subseteq \mathscr{O}_X$ and $t \in \mathbb{Q}_{\geq 0}$ and for a prime divisor E over X, consider a log resolution $\varphi : X' \longrightarrow X$ of $\operatorname{Jac}_X \cdot \mathfrak{a}$ such that E appears in X' and $\mathfrak{a} \cdot \mathscr{O}_{X'} = \mathscr{O}_{X'}(-Z)$ and $\operatorname{Jac}_X \cdot \mathscr{O}_{X'} = \mathscr{O}_{X'}(-J_{X'/X})$ where Z and $J_{X'/X}$ are effective divisors on X'. We define the *Mather-Jacobian-discrepancy* (*MJ-discrepancy* for short) of E to be

$$a_{\mathrm{MJ}}(E; X, \mathfrak{a}^t) = \operatorname{ord}_E(\widehat{K}_{X'/X} - J_{X'/X} - tZ).$$

The number $a_{\mathrm{MJ}}(E; X, \mathfrak{a}^t) + 1$ is called the *Mather-Jacobian-log discrepancy* (*MJ-log discrepancy* for short). It is independent on the choice of the log resolution φ. Let W be a proper closed subset of X and let η be a point of X such that its closure $\overline{\{\eta\}}$ is a proper closed subset of X. We define the *minimal MJ-log discrepancy* of $(X, \mathfrak{a}^t)$ along W as

$$\operatorname{mld}_{\mathrm{MJ}}(W; X, \mathfrak{a}^t) = \inf_{C_X(E) \subseteq W} \{ a_{\mathrm{MJ}}(E; X, \mathfrak{a}^t) + 1 \mid E \text{ a prime divisor over } X \}$$

and the *minimal MJ-log discrepancy* of $(X, \mathfrak{a}^t)$ at η as

$$\mathrm{mld}_{\mathrm{MJ}}(\eta; X, \mathfrak{a}^t) = \inf_{C_X(E)=\overline{\{\eta\}}} \{ a_{\mathrm{MJ}}(E; X, \mathfrak{a}^t) + 1 \mid E \text{ a prime divisor over } X\}.$$

We say that X is *MJ-canonical* (resp. *MJ-log canonical*) if for every exceptional prime divisor E over X, the MJ-discrepancy $a_{\mathrm{MJ}}(E; X, \mathscr{O}_X) \geq 0$ (resp. ≥ -1) holds. When X is nonsingular, the notion of MJ-singularities is the same as the usual (log) canonical singularities. We shall use the following version of the Inversion of Adjunction. It plays a critical role in transferring singularity information from a variety to its ambient space.

THEOREM 3.1 (Inversion of Adjunction, **[13][3]**). *Let X be a codimension c subvariety of a nonsingular variety A defined by the ideal I_X.*

(1) *Let $W \subset X$ be a proper closed subset of X. Then*

$$\mathrm{mld}_{\mathrm{MJ}}(W; X, \mathscr{O}_X) = \mathrm{mld}(W; A, I_X^c).$$

(2) *Let $\eta \in X$ be a point such that its closure $\overline{\{\eta\}}$ is a proper closed subset of X. Then*

$$\mathrm{mld}_{\mathrm{MJ}}(\eta; X, \mathscr{O}_X) = \mathrm{mld}(\eta; A, I_X^c).$$

Our first main result is to show that minimal log discrepancies can be computed by using a complete intersection or a hypersurface. This result sheds light on how to apply linkage theory to the study of singularities.

THEOREM 3.2. *Let x be a closed point of a nonsingular affine variety A. Let $I \subseteq \mathscr{O}_A$ be an ideal generated by $\{f_1, \ldots, f_r\}$ and let $\tilde{\mathfrak{a}}^{\mathbf{m}} = \tilde{\mathfrak{a}}_1^{m_1}\tilde{\mathfrak{a}}_2^{m_2}\cdots\tilde{\mathfrak{a}}_n^{m_n}$ be a formal product of ideals $\tilde{\mathfrak{a}}_i \subseteq \mathscr{O}_A$ with $\mathbf{m} = (m_1, \cdots, m_n) \in \mathbb{R}^n_{\geq 0}$. Then for $t \geq 1$ general elements $a_1, \cdots, a_t$ in the vector space $\langle f_1, \cdots, f_r\rangle$, the ideals $I_M = (a_1, \cdots, a_t)$ and $I_H = (a_1 a_2 \cdots a_t)$ satisfy the following properties.*

There exist a log resolution $\varphi : A_1 \to A$ of $\tilde{\mathfrak{a}}^{\mathbf{m}} \cdot I \cdot I_H \cdot \mathfrak{m}_x$ and a log resolution $\psi : A_2 \to A$ of $\tilde{\mathfrak{a}}^{\mathbf{m}} \cdot I \cdot I_M \cdot I_H \cdot \mathfrak{m}_x$ such that ψ factors through φ, i.e., $\psi = \varphi \circ \mu$, and μ_* induces a one-to-one correspondence

$$\{\textit{prime divisors in } A_2 \textit{ with center } x\} \xrightarrow{\mu_*} \{\textit{prime divisors in } A_1 \textit{ with center } x\}.$$

Furthermore, for a prime divisor $F \subset A_1$ with center x and a real number $0 \leq \lambda \leq t$,

$$a(F; A, I^\lambda \cdot \tilde{\mathfrak{a}}^{\mathbf{m}}) = a(F; A, I_M^\lambda \cdot \tilde{\mathfrak{a}}^{\mathbf{m}}) = a(F; A, I_H^{\lambda/t} \cdot \tilde{\mathfrak{a}}^{\mathbf{m}}), \textit{ and}$$

$$\mathrm{mld}(x; A, I^\lambda \cdot \tilde{\mathfrak{a}}^{\mathbf{m}}) = \mathrm{mld}(x; A, I_M^\lambda \cdot \tilde{\mathfrak{a}}^{\mathbf{m}}) = \mathrm{mld}(x; A, I_H^{\lambda/t} \cdot \tilde{\mathfrak{a}}^{\mathbf{m}}).$$

PROOF. For simplicity, we prove the case $n = 1$ and $\tilde{\mathfrak{a}}^{\mathbf{m}} = \tilde{\mathfrak{a}}^m$. Take a log resolution $\varphi : A_1 \longrightarrow A$ of $\tilde{\mathfrak{a}} \cdot I \cdot \mathfrak{m}_x$ such that $I \cdot \mathscr{O}_{A_1} = \mathscr{O}_{A_1}(-E)$ and $\tilde{\mathfrak{a}} \cdot \mathscr{O}_{A_1} = \mathscr{O}_{A_1}(-Z)$ where E and Z are effective divisors and $\mathrm{Exc}(\varphi) \cup E \cup Z$ has a snc support. As in Construction 2.5, we have a decomposition

$$\varphi^*|f_1, \cdots, f_r| = W + E$$

where W is free linear system. Choose t general elements $\alpha_1, \cdots, \alpha_t$ in W and denote by F_i the zero locus of α_i. By Bertini's theorem, F_i can be chosen such that $F_1 \cup \cdots \cup F_t \cup \mathrm{Exc}(\varphi) \cup E \cup Z$ has a snc support and therefore the intersection $Y_1 = F_1 \cap \cdots \cap F_t$ is a nonsingular subscheme (not necessarily irreducible) in A_1 of codimension t. Furthermore, the sections $\alpha_1, \cdots, \alpha_t$ correspond to the general elements $a_1, \cdots, a_t$ as desired in the vector space $\langle f_1, \cdots, f_r\rangle$ such that

$$I_M \cdot \mathscr{O}_{A_1} = I_{Y_1} \cdot \mathscr{O}_{A_1}(-E).$$

Now blowup A_1 along Y_1 to obtain $\mu : A_2 = \mathrm{Bl}_{Y_1} A_1 \to A_1$ with an exceptional divisor G. Set $\psi = \varphi \circ \mu : A_2 \to A$. Notice that

$$K_{A_2/A_1} = (t-1)G$$

Hence we obtain

$$K_{A_2/A} - \lambda I \cdot \mathcal{O}_{A_2} - m\tilde{\mathfrak{a}} \cdot \mathcal{O}_{A_2} = \mu^* K_{A_1/A} + (t-1)G - \lambda\mu^* E - m\mu^* Z, \text{ and}$$

$$K_{A_2/A} - \lambda I_M \cdot \mathcal{O}_{A_2} - m\tilde{\mathfrak{a}} \cdot \mathcal{O}_{A_2} = \mu^* K_{A_1/A} - (t-1-\lambda)G - \lambda\mu^* E - m\mu^* Z.$$

We shall use [**6**, Proposition 7.2] to compute the mld at x. First notice that, if a prime divisor with center x appears in A_1 then it must appear in A_2 and vice versa, which gives the one-to-one correspondence in the proposition. Let $F \subset A_1$ be such a prime divisor with center x. Then clearly F is not contained in Y_1 and therefore F is not contained in G. Hence for any real number $0 \leq \lambda \leq t$, we have $a(F; A, I^\lambda \cdot \tilde{\mathfrak{a}}^m) = a(F; A, I_M^\lambda \cdot \tilde{\mathfrak{a}}^m)$.

Next, we consider those prime divisors whose center contains $\{x\}$ as a proper subset. Let F be such a prime divisor. We have two possibilities. First, F is contained in G. In this case, F cannot be in the support of $\mu^* K_{A_1/A} \cup \mu^* E \cup \mu^* Z$. So we have

$$a(F; A, I^\lambda \cdot \tilde{\mathfrak{a}}^m) = \mathrm{ord}_F K_{A_2/A} - \lambda \,\mathrm{ord}_F I - m \,\mathrm{ord}_F \tilde{\mathfrak{a}} + 1 = (t-1) + 1 \geq 0, \text{ and}$$

$$a(F; A, I_M^\lambda \cdot \tilde{\mathfrak{a}}^m) = \mathrm{ord}_F K_{A_2/A} - \lambda \,\mathrm{ord}_F I_M - m \,\mathrm{ord}_F \tilde{\mathfrak{a}} + 1 = (t-1-\lambda) + 1 \geq 0.$$

The second possibility is that F appears in A_2 but is not contained in G. In this case, F must also appear in A_1. This implies $\mathrm{ord}_F I = \mathrm{ord}_F I_M$ and therefore $a(F; A, I^\lambda \cdot \tilde{\mathfrak{a}}^m) = a(F; A, I_M^\lambda \cdot \tilde{\mathfrak{a}}^m)$ (could be negative).

All of the above implies that

$$\mathrm{mld}(x; A, I^\lambda \cdot \tilde{\mathfrak{a}}^m) = \mathrm{mld}(x; A, I_M^\lambda \cdot \tilde{\mathfrak{a}}^m),$$

as desired.

For the case of $I_H = (a_1 \cdots a_t)$, we notice that φ is a log resolution of $I \cdot I_H \cdot \mathfrak{m}_x$ such that

$$I_H \cdot \mathcal{O}_{A_1} = F_1 + \cdots + F_t + tE.$$

Hence

$$K_{A_1/A} - \frac{\lambda}{t} \cdot I_H \cdot \mathcal{O}_{A_1} - m\tilde{\mathfrak{a}} \cdot \mathcal{O}_{A_1} = K_{A_1/A} - \frac{\lambda}{t} F_1 - \cdots - \frac{\lambda}{t} F_t - \lambda E - mZ.$$

Now for a prime divisor F contained in some F_i, the center $C_A(F)$ cannot be x and hence

$$\mathrm{ord}_F K_{A_1/A} - \frac{\lambda}{t} \mathrm{ord}_F \cdot I_H - m \,\mathrm{ord}_F \tilde{\mathfrak{a}} + 1 = -\frac{\lambda}{t} + 1 \geq 0.$$

This proves the last equality. □

As a quick corollary, we see that at a fixed point, minimal log discrepancies are preserved under a general link. For a related result without fixing a point, we refer to [**16**].

COROLLARY 3.3. *If $I = I_X$ is the ideal of a closed subvariety $X \subsetneq A$ of codimension c and $t = \lambda = c$, then the closed subscheme M defined by I_M is a complete intersection containing X and*

$$\mathrm{mld}(x; A, I_X^c \cdot \tilde{\mathfrak{a}}^{\mathbf{m}}) = \mathrm{mld}(x; A, I_M^c \cdot \tilde{\mathfrak{a}}^{\mathbf{m}}) = \mathrm{mld}(x; A, I_H \cdot \tilde{\mathfrak{a}}^{\mathbf{m}}),$$

which yields

$$\mathrm{mld}_{\mathrm{MJ}}(x;X,\mathfrak{a}^{\mathbf{m}})=\mathrm{mld}_{\mathrm{MJ}}(x;M,\mathfrak{a}^{\mathbf{m}})=\mathrm{mld}_{\mathrm{MJ}}(x;H,\mathfrak{a}^{\mathbf{m}}),$$

where $\mathfrak{a}$ *denotes the restrictions of* $\tilde{\mathfrak{a}}$ *on* X, M *and* H. *Also the prime divisors which compute the minimal log discrepancies are common. Moreover, let* $Y=\overline{M\setminus X}$, *then*

$$\mathrm{mld}_{\mathrm{MJ}}(x;Y,\mathfrak{a}^{\mathbf{m}})=\mathrm{mld}(x;A,I_Y^c\cdot\tilde{\mathfrak{a}}^{\mathbf{m}})\geq\mathrm{mld}(x;A,I_X^c\cdot\tilde{\mathfrak{a}}^{\mathbf{m}})=\mathrm{mld}_{\mathrm{MJ}}(x;X,\mathfrak{a}^{\mathbf{m}}).$$

REMARK 3.4. In [**4**, Theorem 5.6, (ii)], it is proved that a two dimensional non-complete intersection MJ-log canonical variety X has a complete intersection model M, i.e., M is a complete intersection containing X and

$$\mathrm{mld}_{\mathrm{MJ}}(x;X)=\mathrm{mld}_{\mathrm{MJ}}(x;M)=0.$$

Therefore, Corollary 3.3 is a generalization of [**4**, Theorem 5.6, (ii)].

In the second main result, we establish a formula for the Grauert-Riemenschneider canonical sheaf of a general residual intersection. One of the central topics in the study of residual intersection is to obtain a formula for canonical sheaves, which is usually difficult. However, the Grauert-Riemenschneider canonical sheaf is always contained in the canonical sheaf and they are the same under certain conditions. We also give a description of log canonical thresholds as well as its generalized form under general residual intersection, which says that after doing residual intersections, singularities improve.

THEOREM 3.5. *Let* X *be a closed subvariety of an affine nonsingular variety* A *and let* Y *be a general* t*-residual intersection of* X. *If* Y *is nonempty, then one has*

(1) $\omega_Y^{GR}=\mathscr{J}(A,I_X^t)\cdot\mathscr{O}_Y\otimes\omega_A$.
(2) $\mathrm{lct}(A,Y)\geq\mathrm{lct}(A,M)=\mathrm{lct}(A,X)$.
(3) $\mathrm{glct}(A,Y;(t-c)X)\geq\mathrm{glct}(A,M;(t-c)X)=\mathrm{lct}(A,X)+(t-c)$.

PROOF. We use the notation in Construction 2.5. Now the sections α_i give a surjective morphism

$$\bigoplus^t\mathscr{O}_{A_1}(E)\xrightarrow{\alpha_1,\cdots,\alpha_t}I_{Y_1}\longrightarrow 0.$$

Since Y_1 is a complete intersection in A_1, restricting the above map onto Y_1 we obtain the conormal bundle of Y_1 inside A_1 as

$$N^*_{Y_1/A_1}=\bigoplus^t\mathscr{O}_{Y_1}(E).$$

Taking the determinant of $N^*_{Y_1/A_1}$ and tensoring with ω_{A_1} yields the dualizing sheaf of Y_1 as

$$\omega_{Y_1}=\omega_{A_1}\otimes\mathscr{O}_{Y_1}(-tE).$$

We also have a short exact sequence

$$0\longrightarrow I_{Y_1}\longrightarrow\mathscr{O}_{A_1}\longrightarrow\mathscr{O}_{Y_1}\longrightarrow 0$$

which induces a short exact sequence

$$(3.1)\ \ 0\longrightarrow I_{Y_1}\otimes\mathscr{O}_{A_1}(K_{A_1/A}-tE)\longrightarrow\mathscr{O}_{A_1}(K_{A_1/A}-tE)\longrightarrow\omega_{Y_1}\otimes\varphi^*\omega_A^{-1}\longrightarrow 0.$$

We further blowup A_1 along Y_1 as $\mu:\tilde{A}=\mathrm{Bl}_{Y_1}A_1\to A$ with an exceptional divisor T on $\tilde{A}$ such that $I_{Y_1}\cdot\mathscr{O}_{\tilde{A}}=\mathscr{O}_{\tilde{A}}(-T)$. Write $\psi=\varphi\circ\mu:\tilde{A}\to A$ and set

$$\mathscr{G}=\mathscr{O}_{\tilde{A}}(-T)\otimes\mu^*(\mathscr{O}_{A_1}(K_{A_1/A}-tE)).$$

Claim. One has $R^1\varphi_*\mathscr{G}=0$.

Proof of Claim: Note that $\mu_*\mathscr{G} = I_{Y_1}\otimes \mathscr{O}_{A_1}(K_{A_1/A}-tE)$ and for $i>0$, $R^i\mu_*\mathscr{G} = R^i\mu_*\mathscr{O}_{\tilde{A}}(-T)\otimes\mathscr{O}_{A_1}(K_{A_1/A}-tE)=0$. On the other hand, notice that $K_{\tilde{A}/A_1}=(t-1)T$ and $K_{\tilde{A}/A}=K_{\tilde{A}/A_1}+\mu^*K_{A_1/A}$ so we have

$$\mathscr{G}=\mathscr{O}_{\tilde{A}}(-T)\otimes\mu^*(\mathscr{O}_{A_1}(K_{A_1/A}-tE))=\mathscr{O}_{\tilde{A}}(K_{\tilde{A}/A}-t(T+\mu^*E)).$$

But by construction, $I_M\cdot\mathscr{O}_{A_1}=I_{Y_1}\cdot\mathscr{O}_{A_1}(-E)$ and therefore $I_M\cdot\mathscr{O}_{\tilde{A}}=\mathscr{O}_{\tilde{A}}(-T-\mu^*E)$. Hence we obtain that $\mathscr{G}=\mathscr{O}_{\tilde{A}}(K_{\tilde{A}/A}-t(I_M\cdot\mathscr{O}_{\tilde{A}}))$. By Local Vanishing Theorem [**14**, 9.4.1], we then obtain that $R^i\psi_*(\mathscr{G})=0$, for $i>0$. Now we use the Leray spectral sequence

$$E_2^{p,q}=R^p\varphi_*R^q\mu_*\mathscr{G}\Rightarrow R^{p+q}(\varphi\circ\mu)_*(\mathscr{G})$$

to get that $R^1\varphi_*\mathscr{G}=0$, completing the proof of claim.

Now we prove (1). For this, we push down the short exact sequence (3.1) and use the claim above. So we obtain an short exact sequence

$$0\longrightarrow\varphi_*(I_{Y_1}\otimes\mathscr{O}_{A_1}(K_{A_1/A}-tE))\longrightarrow\mathscr{J}(A,I_X^t)\longrightarrow\omega_Y^{GR}\otimes\omega_A^{-1}\longrightarrow 0.$$

This implies that $\mathscr{J}(A,I_X^t)\cdot\mathscr{O}_Y=\omega_Y^{GR}\otimes\omega_A^{-1}$, which gives rise to the desired formula by tensoring with ω_A.

For (2), recall that $K_{\tilde{A}/A}=(t-1)T+\mu^*K_{A_1/A}$, $I_M\cdot\mathscr{O}_{\tilde{A}}=\mathscr{O}_{\tilde{A}}(-T-\mu^*E)$, and $I_X\cdot\mathscr{O}_{\tilde{A}}=\mathscr{O}_{\tilde{A}}(-\mu^*E)$. In order to compute $\operatorname{lct}(A,M)$, we consider prime divisors on $\tilde{A}$. For the divisor T, we have

$$\frac{k_T+1}{\operatorname{ord}_T I_M}=\frac{t-1+1}{1}=t.$$

For any prime divisor $F\neq T$ on $\tilde{A}$, we have $\operatorname{ord}_F I_M=\operatorname{ord}_F I_X$ and hence

$$\frac{k_F+1}{\operatorname{ord}_F I_M}=\frac{k_F+1}{\operatorname{ord}_F I_X}.$$

Thus

$$\operatorname{lct}(A,M)=\min_{F\neq T}\{t,\frac{k_F+1}{\operatorname{ord}_F I_M}\}=\min_{F\neq T}\{t,\frac{k_F+1}{\operatorname{ord}_F I_X}\}=\min\{t,\operatorname{lct}(A,X)\}.$$

But since X is generically smooth, we have $\operatorname{lct}(A,X)\leq c\leq t$. So we obtain that $\operatorname{lct}(A,M)=\operatorname{lct}(A,X)$. Finally since $I_M\subseteq I_Y$, we then have $\operatorname{lct}(A,Y)\geq\operatorname{lct}(A,M)=\operatorname{lct}(A,X)$.

For (3), we continue to work on $\tilde{A}$. For the divisor T, we have

$$\frac{k_T+\operatorname{ord}_T I_X^{t-c}+1}{\operatorname{ord}_T I_M}=\frac{t-1+1}{1}=t.$$

For any prime divisor $F\neq T$, we have $\operatorname{ord}_F I_M=\operatorname{ord}_F I_X$ and therefore

$$\frac{k_F+\operatorname{ord}_F I_X^{t-c}+1}{\operatorname{ord}_F I_M}=\frac{k_F+(t-c)\operatorname{ord}_F I_X+1}{\operatorname{ord}_F I_X}=\frac{k_F+1}{\operatorname{ord}_F I_X}+(t-c).$$

Hence we see

$$\operatorname{glct}(A,M;(t-c)I_X)=\min_{F\neq T}\{t,\frac{k_F+1}{\operatorname{ord}_F I_X}+(t-c)\}=\min\{t,\operatorname{lct}(A,X)+(t-c)\}.$$

Since X is a variety of codimension c, we have $\operatorname{lct}(A, X) \leq c$ and therefore $\operatorname{lct}(A, X) + (t - c) \leq t$. Hence $\operatorname{glct}(A, M; (t-c)I_X) = \operatorname{lct}(A, X) + (t - c)$, from which the result follows. □

COROLLARY 3.6. *Let X be a closed subvariety in an affine nonsingular variety A. Assume that X is locally a complete intersection with rational singularities. Then a general t-residual intersection Y of X with $t \geq \operatorname{codim}_A X$ has rational singularities and*

$$\omega_Y = \mathscr{I}(A, I^t) \cdot \mathscr{O}_Y \otimes \omega_A.$$

PROOF. Using [**2**, Theorem 3.13], Y has rational singularities and therefore $\omega_Y^{GR} = \omega_Y$. Now the result follows from Theorem 3.5. □

The following proposition describes the singularities of a general t-residual intersection of a nonsingular variety. The statement (1) is proved in [**2**, Theorem 3.13] and (2) is proved in [**7**, Corollary 8.1.4.]. Here we prove them in different ways; along the line of the construction of general residual intersections.

PROPOSITION 3.7. *Let X be a nonsingular closed subvariety of a nonsingular affine variety A and let Y be a general t-residual intersection of X. If Y is nonempty, then*

(1) *Y has rational singularities.*
(2) $\operatorname{codim}_Y \operatorname{Sing}(Y) \geq t - c + 4$.
(3) $\omega_Y = I_X^{t-c+1} \cdot \mathscr{O}_Y \otimes \omega_A$.

PROOF. We use Construction 2.5. Since X is nonsingular, we can assume X is irreducible. We can take A_1 as the blowup of A along X and therefore the divisor E is irreducible. Since Y_1 is a complete intersection in A_1, we have a Koszul resolution of $\mathscr{O}_{Y_1}$ as follows

$$0 \longrightarrow \mathscr{O}_{A_1}(tE) \longrightarrow \cdots \longrightarrow \bigoplus^{t} \mathscr{O}_{A_1}(E) \longrightarrow \mathscr{O}_{A_1} \longrightarrow \mathscr{O}_{Y_1} \longrightarrow 0.$$

Note that $R^i\varphi_*\mathscr{O}_{A_1}(jE) = 0$ for $i > 0$, $0 \leq j \leq c-1$ and $R^i\varphi_*\mathscr{O}_{A_1}(jE) = 0$ for $i \geq c$ and $j \geq 0$. Hence pushing down the complex above, we deduce that $R^i\varphi_*\mathscr{O}_{Y_1} = 0$ for $i > 0$ and the map $\mathscr{O}_A \to \varphi_*\mathscr{O}_{Y_1}$ is surjective as well. But we have a natural subjection $\mathscr{O}_A \to \mathscr{O}_Y$ through which the map $\mathscr{O}_A \to \varphi_*\mathscr{O}_{Y_1}$ factors, i.e., $\mathscr{O}_A \to \mathscr{O}_Y \to \varphi_*\mathscr{O}_{Y_1}$. Hence we conclude that $\varphi_*\mathscr{O}_{Y_1} = \mathscr{O}_Y$ and therefore Y has rational singularities.

It is clear that $\operatorname{Sing}(Y) \subseteq X \cap Y$. For any $x \in X$, denote by E_x the fiber of the map $\varphi|_E : E \to X$. We have the following three cases for the intersection $Y_1 \cap E_x$.

(a) $Y_1 \cap E_x = \emptyset$, which is equivalent to $x \notin X \cap Y$.
(b) $Y_1 \cap E_x$ is a closed point, which implies that $x \in X \cap Y$ and Y is nonsingular at x.
(c) $Y_1 \cap E_x$ has dimension ≥ 1.

Since $E = \mathbb{P}(I_X/I_X^2)$, the t general sections give t sections in $H^0(\mathscr{O}_E(1))$. Consider the map

$$\psi : \bigoplus^{t} \mathscr{O}_X \longrightarrow I_X/I_X^2$$

induced by the t sections. We see that the locus of points x satisfying case (c) is inside

$$\sigma_{c-2} = \{x \in X \,|\, \operatorname{rank} \psi \otimes k(x) \leq c - 2\}.$$

Hence $\mathrm{Sing}(Y) \subseteq \sigma_{c-2}$. But the codimension of σ_{c-2} in X is $(t-(c-2))(c-(c-2))$. So we obtain $\mathrm{codim}_Y \mathrm{Sing}(Y) \geq t-c+4$.

Now for (3), since X is nonsingular, the multiplier ideal $\mathscr{J}(A, I_X^t) = I_X^{t-c+1}$. It follows from Proposition 3.5 and Y having rational singularities that $\omega_Y = \omega_Y^{GR} = I_X^{t-c+1} \cdot \mathscr{O}_Y \otimes \omega_A$.

□

At the end of this section, we discuss which singularities might be preserved under a general residual intersection. Chardin-Ulrich's result ([**2**]) says that a general residual intersection of a local complete intersection with rational singularities also has rational singularities. A complete answer on rational singularities under a general link has been given in [**15**]. However, the situation is not clear for a general residual intersection. We point out that a local complete intersection with rational singularities has MJ-canonical singularities. In the generic linkage case, it has been proved that MJ-singularities are preserved. So it is reasonable to expect that a similar result can be established for a general residual intersection. Along this line, we provide the following evidence.

PROPOSITION 3.8. *Let X be a closed subvariety of dimension ≤ 3 of a nonsingular affine variety A. Assume that X is locally a complete intersection with MJ-log canonical (resp, MJ-canonical) singularities. Then a general t-residual intersection Y of X with $t \geq \mathrm{codim}_A X$ is also locally a complete intersection with MJ-log canonical (resp. MJ-canonical) singularities.*

PROOF. We keep using the notations in Construction 2.5. If a general t-residual intersection is empty, then there is nothing to prove. So we assume in the sequel that a general t-residual intersection is not empty. Let Y_c be a general link of X. Since $\dim X \leq 3$, by [**9**, Proposition 2.9], Y_c is locally a complete intersection and normal. By shrinking A if necessary, we may assume Y_c is irreducible. Note that

$$\omega_{Y_c} \cong I_X \cdot \mathscr{O}_{Y_c} = \mathscr{O}_{Y_c}(-Z)$$

where $Z = X \cap Y_c$. Hence $\mathscr{O}_{Y_c}(-Z)$ is invertible on Y_c. Let $i : Y_c \to A$ be the inclusion morphism. Restricting the surjective morphism v of Construction 2.5 to Y_c yields a surjective morphism

$$\oplus^r \mathscr{O}_{Y_c} \xrightarrow{i^*f_1, \cdots, i^*f_r} \mathscr{O}_{Y_c}(-Z) \longrightarrow 0.$$

Since $\mathscr{O}_{Y_c}(-Z)$ is invertible, we see that the linear system $i^*|f_1, \cdots, f_r|$ can be decomposed as

$$i^*|f_1, \cdots, f_r| = W_c + Z,$$

where W_c is a free linear system on Y_c. Now a general t-residual intersection Y can be obtained as the locus of $t-c$ general sections of W_c in Y_c.

If X is MJ-log canonical (resp, MJ-canonical), then so is Y_c by the main result of [**16**]. As Y_c is normal and locally a complete intersection, Y_c has log canonical (resp. canonical) singularities [**4**, Remark 2.5]. Now a Bertini type theorem can be established for a general divisor $H \in W_c$. Indeed, take a log resolution $f : \overline{Y}_c \longrightarrow Y_c$ of Y_c. The linear system f^*W_c is base point free on $\overline{Y}_c$. So we take a general member $\overline{H} \in f^*W_c$, which is nonsingular. Accordingly, $\overline{H}$ gives a divisor $H \in W_c$. By the generality of $\overline{H}$, we see that $f^*(H) = \overline{H}$ as Cartier divisors. Now the adjunction formula gives us that

$$K_{\overline{H}/H} = K_{\overline{Y}_c/Y_c}|_{\overline{H}}.$$

Thus H is log canonical (resp. canonical) if so is Y_c. Finally a general t-residual intersection Y is cut by $t-c$ general members of W_c. So by the above Bertini type theorem, we conclude that a general t-residual intersection Y is locally a complete intersection with MJ-log canonical (resp. MJ-canonical) singularities, completing the proof. □

Finally, we propose the following conjecture predicting that MJ-singularities are preserved under a general residual intersection.

CONJECTURE 3.9. Let Y be a general t-residual intersection of X. Then Y is MJ-canonical (resp, MJ-log canonical) if so is X.

References

[1] M. Artin and M. Nagata, *Residual intersections in Cohen-Macaulay rings*, J. Math. Kyoto Univ. **12** (1972), 307–323. MR0301006

[2] Marc Chardin and Bernd Ulrich, *Liaison and Castelnuovo-Mumford regularity*, Amer. J. Math. **124** (2002), no. 6, 1103–1124. MR1939782

[3] Tommaso de Fernex and Roi Docampo, *Jacobian discrepancies and rational singularities*, J. Eur. Math. Soc. (JEMS) **16** (2014), no. 1, 165–199. MR3141731

[4] Lawrence Ein and Shihoko Ishii, *Singularities with respect to Mather-Jacobian discrepancies*, Commutative algebra and noncommutative algebraic geometry. Vol. II, Math. Sci. Res. Inst. Publ., vol. 68, Cambridge Univ. Press, New York, 2015, pp. 125–168. MR3496863

[5] Lawrence Ein, Shihoko Ishii, and Mircea Mustaţă, *Multiplier ideals via Mather discrepancy*, Minimal models and extremal rays (Kyoto, 2011), Adv. Stud. Pure Math., vol. 70, Math. Soc. Japan, [Tokyo], 2016, pp. 9–28. MR3617776

[6] Lawrence Ein and Mircea Mustaţă, *Jet schemes and singularities*, Algebraic geometry—Seattle 2005. Part 2, Proc. Sympos. Pure Math., vol. 80, Amer. Math. Soc., Providence, RI, 2009, pp. 505–546. MR2483946

[7] H. Flenner, L. O'Carroll, and W. Vogel, *Joins and intersections*, Springer Monographs in Mathematics, Springer-Verlag, Berlin, 1999. MR1724388

[8] Craig Huneke, *Strongly Cohen-Macaulay schemes and residual intersections*, Trans. Amer. Math. Soc. **277** (1983), no. 2, 739–763. MR694386

[9] Craig Huneke and Bernd Ulrich, *Divisor class groups and deformations*, Amer. J. Math. **107** (1985), no. 6, 1265–1303 (1986). MR815763

[10] Craig Huneke and Bernd Ulrich, *The structure of linkage*, Ann. of Math. (2) **126** (1987), no. 2, 277–334. MR908149

[11] Craig Huneke and Bernd Ulrich, *Algebraic linkage*, Duke Math. J. **56** (1988), no. 3, 415–429. MR948528

[12] Craig Huneke and Bernd Ulrich, *Residual intersections*, J. Reine Angew. Math. **390** (1988), 1–20. MR953673

[13] Shihoko Ishii, *Mather discrepancy and the arc spaces* (English, with English and French summaries), Ann. Inst. Fourier (Grenoble) **63** (2013), no. 1, 89–111. MR3089196

[14] Robert Lazarsfeld, *Positivity in algebraic geometry. II*, Ergebnisse der Mathematik und ihrer Grenzgebiete. 3. Folge. A Series of Modern Surveys in Mathematics [Results in Mathematics and Related Areas. 3rd Series. A Series of Modern Surveys in Mathematics], vol. 49, Springer-Verlag, Berlin, 2004. Positivity for vector bundles, and multiplier ideals. MR2095472

[15] Wenbo Niu, *Singularities of generic linkage of algebraic varieties*, Amer. J. Math. **136** (2014), no. 6, 1665–1691. MR3282984

[16] ———, *Mather-Jacobian singularities under generic linkage*, to appear in Trans. Amer. Math. Soc.

[17] C. Peskine and L. Szpiro, *Liaison des variétés algébriques. I* (French), Invent. Math. **26** (1974), 271–302. MR0364271

Department of Mathematical Sciences, Tokyo Woman's Christian University, Tokyo, 167-8585, Tokyo

Email address: `shihoko@lab.twcu.ac.jp`

Department of Mathematical Sciences, University of Arkansas, Fayetteville, Arkansas 72701

Email address: `wenboniu@uark.edu`

Contemporary Mathematics
Volume **712**, 2018
http://dx.doi.org/10.1090/conm/712/14348

Quadratic solutions of quadratic forms

János Kollár

ABSTRACT. The aim of this note is to study solutions of a homogeneous quadratic equation $q(x_0,\dots,x_n)=0$, defined over a field k, where the x_i are themselves homogeneous polynomials of some degree d in $r+1$ variables. Equivalently, we are looking at rational maps from projective r-space $\mathbb{P}^r$ to a quadric hypersurface Q, defined over a field k.

The space of maps of $\mathbb{P}^1$ to a quadric Q is stably birational to Q if d is even and to the orthogonal Grassmannian of lines in Q if d is odd. Most of the paper is devoted to obtaining similar descriptions for the spaces parametrizing maps, given by degree 2 polynomials, from $\mathbb{P}^2$ to quadrics. The most interesting case is 4-dimensional quadrics when there are 5 irreducible components. The methods are mostly classical, involving the Veronese surface, its equations and projections.

In the real case, these results provide some of the last steps of a project, started by Kummer and Darboux, to describe all surfaces that contain at least 2 circles through every point.

Given a field k and a quadric hypersurface $Q^n := \bigl(q(x_0,\dots,x_{n+1})=0\bigr)$, we are looking at rational maps, defined over k,

$$\Phi : \mathbb{P}^r \dashrightarrow Q^n := \bigl(q(x_0,\dots,x_{n+1})=0\bigr) \subset \mathbb{P}^{n+1}$$

from projective r-space to Q^n. We denote the set of such maps by $\operatorname{Map}_d(\mathbb{P}^r, Q^n)(k)$. It is easy to see that there is a projective algebraic set $\operatorname{Map}_d(\mathbb{P}^r, Q^n)$, defined over k, whose K points can be naturally identified with $\operatorname{Map}_d(\mathbb{P}^r, Q^n)(K)$ for any field extension $K \supset k$. Thus we aim to describe the algebraic sets $\operatorname{Map}_d(\mathbb{P}^r, Q^n)$ and give concrete parametrization of their k points.

If $d=1$ then the image of Φ is a linear subspace of Q^n, hence the study of linear solutions of quadratic equations is essentially equivalent to the theory of orthogonal Grassmannians. These are homogeneous spaces under the orthogonal group $\operatorname{Aut}(Q^n) = \operatorname{PGO}(q)$ and quite well understood, see Section 8.

Another extreme case is when $r=1$. Solving $q(x_0,\dots,x_{n+1})=0$, where the $x_i = h_i(u,v)$ are homogeneous polynomials in 2 variables u,v is equivalent to studying maps $\mathbb{P}^1 \to Q^n$. In algebraic geometry there has been considerable interest in understanding maps from $\mathbb{P}^1$ to a given variety. Most of these works deal with more general varieties over algebraically closed fields. Quadrics are special cases of homogeneous spaces [**KP01**] and of low degree hypersurfaces [**HRS04**].

2010 *Mathematics Subject Classification.* Primary 14G05, 11E04, 14J26; Secondary 14J70, 14N15, 14P05.

See the introductory notes [**AK03**, **Kol10**] or the more complete treatment given in [**Kol96**] for details. The existence of maps $\mathbb{P}^1 \to X$ over finite fields is studied in [**Kol08**], but not much is known about the structure of these spaces over arbitrary fields. For quadrics, we prove the following in Section 2.

THEOREM 1. *Let Q^n be a smooth quadric of dimension $n \geq 3$. Then*

$$\mathrm{Map}_d(\mathbb{P}^1, Q^n) \overset{bir}{\sim} \begin{cases} Q^n \times \mathbb{P}^{nd} & \text{if } d \text{ is even, and} \\ \mathrm{OG}(\mathbb{P}^1, Q^n) \times \mathbb{P}^{nd-n+3} & \text{if } d \text{ is odd,} \end{cases}$$

where $\overset{bir}{\sim}$ denotes birational equivalence and $\mathrm{OG}(\mathbb{P}^1, Q^n)$ the orthogonal Grassmannian of lines in Q^n.

Instead of parametrizing maps $\mathbb{P}^1 \to Q^n$, it is also of interest to parametrize spaces of rational curves contained in Q^n. For odd degrees, the latter problem is almost equivalent to Theorem 1; see Corollary 31. However, for even degrees parametrizing spaces of rational curves is much harder and we get full answers in only a few cases; see Theorem 32 and Proposition 34.

The rest of the paper is devoted to the next case $r = d = 2$, which is considerably more involved. Thus we aim to understand rational maps

$$\Phi : \mathbb{P}^2 \dashrightarrow Q^n := \big(q(x_0, \dots, x_{n+1}) = 0\big) \subset \mathbb{P}^{n+1}$$

defined by degree 2 homogeneous polynomials $x_i = h_i(u, v, w) \in k[u, v, w]$. As in Theorem 1, our aim is to describe the moduli spaces $\mathrm{Map}_2(\mathbb{P}^2, Q^n)$ up to birational equivalence, using Q^n and its orthogonal Grassmannians. If Q^n is singular then, after a suitable linear coordinate change, we can eliminate some of the variables x_i from q. Thus from now on Q^n denotes a smooth quadric hypersurface of dimension n over a field k.

With the exception of Sections 2–3 we assume from now on that $\mathrm{char}\, k \neq 2$, though this is not always necessary.

In low dimensions we prove in (37.2), (40.2) and (42.3) that

$$\begin{aligned} \mathrm{Map}_2(\mathbb{P}^2, Q^1) &\overset{bir}{\sim} Q^1 \times \mathbb{P}^4, \\ \mathrm{Map}_2(\mathbb{P}^2, Q^2) &\overset{bir}{\sim} Q^2 \times \mathbb{P}^8, \\ \mathrm{Map}_2(\mathbb{P}^2, Q^3) &\overset{bir}{\sim} \big(Q^3 \amalg \mathrm{OG}(\mathbb{P}^1, Q^3)\big) \times \mathbb{P}^{11}. \end{aligned}$$

Let $\mathrm{Map}_2^\circ(\mathbb{P}^2, Q^n) \subset \mathrm{Map}_2(\mathbb{P}^2, Q^n)$ denote the open subset consisting of maps that are everywhere defined and have a 2-dimensional image. If $\Phi \in \mathrm{Map}_2^\circ(\mathbb{P}^2, Q^n)$ and we work over $\mathbb{C}$ then the image of the fundamental class $\Phi_*[\mathbb{P}^2]$ is in $H_4(Q^n, \mathbb{Z})$. If $n = 2$ (resp. 3) then $H_4(Q^n, \mathbb{Z}) \cong \mathbb{Z}$ is generated by Q^2 (resp. $Q^3 \cap H$) where H denotes the hyperplane class. If $n \geq 5$ then $H_4(Q^n, \mathbb{Z}) \cong \mathbb{Z}$ but this time it is generated by the class of any 2-plane $L^2 \subset Q^n$. The most interesting case is $n = 4$. Then $H_4(Q^4, \mathbb{Z}) \cong \mathbb{Z}[A] + \mathbb{Z}[B]$ where $A, B \subset Q^4$ are disjoint planes; see [**HP47**, Book IV, Sec.XIII.4] on linear spaces in quadrics. The same holds over any algebraically closed field if we replace $H_4(Q^n, \mathbb{Z})$ by the group $N_2(Q^n, \mathbb{Z})$ of 2-cycles modulo numerical equivalence. This suggests that the really interesting case is $n = 4$.

The key player in our study is the Veronese surface.

DEFINITION 2. A *Veronese surface* is the image of the map

$$\mathbb{P}^2 \to \mathbb{P}^5 \quad \text{given by} \quad (u : v : w) \mapsto (u^2 : v^2 : w^2 : uv : vw : uw),$$

composed with any automorphism of $\mathbb{P}^5$. They were first studied in [**Ver1884**]. Veronese surfaces have numerous exceptional and extremal properties.

• If a surface contains a conic through any 2 of its points then it is a plane, quadric, cubic scroll or Veronese; this follows from the Kronecker–Castelnuovo theorem; see [**Cas1894**].

• If the secant lines of a surface cover a subvariety of dimension ≤ 4, then it is the Veronese (or it is contained in a linear subspace of dimension ≤ 4). This was proved by Severi [**Sev1901**] (see [**Dal85**] for a short proof) and generalized by Kuiper and Pohl [**Kui62, KP77**] to non-algebraic surfaces.

• Veronese surfaces give rise to remarkable birational maps. As pointed out by F. Russo, this seems to have been known to Bertini (see [**Ber1907**, Secs.XV.9–10] or [**Ber1924**, Secs.XVI.9–10]) and Coble [**Cob22**, Thm.17], but the earliest explicit mention may be in [**SR49**, p.188]. A modern treatment is given by Ein and Shepherd-Barron [**ESB89**]. These connections have been extended by Pirio and Russo to include Jordan algebras [**PR14**]. See the book [**Rus16**] for a detailed treatment of many of these results.

We will use a classification of projections and equations of Veronese surfaces; see (41) and (48).

The geometric version of the main result is the following;

THEOREM 3. *Let Q^4 be a smooth quadric 4-fold defined over an algebraically closed field. Then* $\operatorname{Map}_2(\mathbb{P}^2, Q^4)$ *has 5 irreducible components, each birational to* $\mathbb{P}^{20}$, *and* $\operatorname{Map}_2^\circ(\mathbb{P}^2, Q^4)$ *is open and dense in* $\operatorname{Map}_2(\mathbb{P}^2, Q^4)$. *The 5 components of* $\operatorname{Map}_2^\circ(\mathbb{P}^2, Q^4)$ *can be described as follows*

(1) *(Veronese) In these cases $\Phi(\mathbb{P}^2) \subset Q^4$ is a Veronese surface. There are 2 such components corresponding to $\Phi_*[\mathbb{P}^2] = 3[A] + [B]$ and $\Phi_*[\mathbb{P}^2] = [A] + 3[B]$, where $A, B \subset Q^4$ are disjoint planes.*

(2) *(Projected Veronese) In these cases $\Phi(\mathbb{P}^2) \subset Q^4$ is a singular projection of a Veronese surface. There is 1 such component and $\Phi_*[\mathbb{P}^2] = 2[A] + 2[B]$.*

(3) *(Quadruple plane) In these cases $\Phi(\mathbb{P}^2) \subset Q^4$ is a plane. There are 2 such components corresponding to $\Phi_*[\mathbb{P}^2] = 4[A]$ and $\Phi_*[\mathbb{P}^2] = 4[B]$.*

The corresponding subvarieties of $\operatorname{Map}_2^\circ(\mathbb{P}^2, Q^4)$ will be denoted by

$$\operatorname{Map}_V^\circ(\mathbb{P}^2, Q^4), \quad \operatorname{Map}_{PV}^\circ(\mathbb{P}^2, Q^4) \quad \text{and} \quad \operatorname{Map}_{QP}^\circ(\mathbb{P}^2, Q^4);$$

their closures in $\operatorname{Map}_2(\mathbb{P}^2, Q^4)$ by

$$\operatorname{Map}_V(\mathbb{P}^2, Q^4), \quad \operatorname{Map}_{PV}(\mathbb{P}^2, Q^4) \quad \text{and} \quad \operatorname{Map}_{QP}(\mathbb{P}^2, Q^4).$$

Our main aim is to describe $\operatorname{Map}(\mathbb{P}^2, Q^4)$ over arbitrary fields k, so Theorem 3 is but a special case of the following 3 results when k is algebraically closed. The most interesting is the description of the Veronese locus; it is proved in Section 5. The easier Propositions 5–6 are treated in Section 4. See Notation 20 for the terminology involving quadratic forms.

THEOREM 4 (Veronese). *Let Q^4 be a smooth quadric 4-fold over a field k. Then*

$$\operatorname{Map}_V(\mathbb{P}^2, Q^4) \overset{bir}{\sim} \{\pm\sqrt{-\Delta}\} \times Q^4 \times \mathbb{P}^{16},$$

where Δ is the discriminant of Q^4 and $\{\pm\sqrt{-\Delta}\}$ denotes the 2-point algebraic set defined by the equation $t^2 + \Delta = 0$ over k.

Furthermore, $\mathrm{Map}^\circ_V(\mathbb{P}^2, Q^n)(k)$ *is nonempty iff* $-\Delta$ *is a square in* k *and* $Q^4(k) \neq \emptyset$. *If these hold then* $\mathrm{Map}_V(\mathbb{P}^2, Q^4)$ *is birational to* $\mathbb{P}^{20} \amalg \mathbb{P}^{20}$, *the disjoint union of 2 copies of* $\mathbb{P}^{20}$.

The description of the projected Veronese maps is not hard and the quadruple planes are obvious. The following are special cases of (43.2) and (39.1). (See Notation 20 for the terminology involving quadratic forms.)

PROPOSITION 5 (Projected Veronese). *Let* Q^4 *be a smooth quadric 4-fold. Then*

$$\mathrm{Map}_{PV}(\mathbb{P}^2, Q^4) \overset{bir}{\sim} \mathrm{OG}(\mathbb{P}^1, Q^4) \times \mathbb{P}^{15},$$

where $\mathrm{OG}(\mathbb{P}^1, Q^4)$ *denotes the orthogonal Grassmannian of lines in* Q^4. *Furthermore,* $\mathrm{Map}^\circ_{PV}(\mathbb{P}^2, Q^4)(k)$ *is nonempty iff the anisotropic rank of* Q^4 *is* ≤ 2 *and then* $\mathrm{Map}_{PV}(\mathbb{P}^2, Q^4)$ *is birational to* $\mathbb{P}^{20}$.

PROPOSITION 6 (Quadruple plane). *Let* Q^4 *be a smooth quadric 4-fold. Then*

$$\mathrm{Map}_{QP}(\mathbb{P}^2, Q^4) \overset{bir}{\sim} \mathrm{OG}(\mathbb{P}^2, Q^4) \times \mathbb{P}^{17},$$

where $\mathrm{OG}(\mathbb{P}^2, Q^4)$ *denotes the orthogonal Grassmannian of 2-planes in* Q^4. *Furthermore,* $\mathrm{Map}^\circ_{QP}(\mathbb{P}^2, Q^n)(k) \neq \emptyset$ *iff* Q^4 *is split over* k *and then* $\mathrm{Map}_{QP}(\mathbb{P}^2, Q^4)$ *is birational to* $\mathbb{P}^{20} \amalg \mathbb{P}^{20}$.

Note that if the anisotropic rank of Q^4 is ≤ 2 and $-\Delta$ is a square then Q^4 is split. Thus for non-split forms we have either only 1 irreducible component with smooth k-points (if the anisotropic rank is 2) or 2 irreducible components with smooth k-points (if $-\Delta$ is a square).

The above results count maps from a fixed $\mathbb{P}^2$ to a quadric Q^4. Frequently it is more interesting to count maps up to coordinate changes by $\mathrm{Aut}(\mathbb{P}^2)$ or by $\mathrm{Aut}(Q^4)$.

Taking quotient by $\mathrm{Aut}(\mathbb{P}^2)$ is essentially the same as working with the components of the Chow variety $\mathrm{Chow}(Q^4)$ that parametrize Veronese surfaces (resp. projected Veronese surfaces). Let us denote these by $\mathrm{Chow}_V(Q^4)$ (resp. $\mathrm{Chow}_{PV}(Q^4)$). Let H_5 denote the hyperplane class on $\mathbb{P}^5$ and let $F_k \subset Q^4_k$ be a surface that is a Veronese surface over $\bar{k}$. Then $|2H_{5|F_k} + K_{F_k}|$ is the linear system of lines giving an isomorphism $F_k \cong \mathbb{P}^2_k$. Therefore $\mathrm{Chow}_V(Q^4) \times \mathrm{Aut}(\mathbb{P}^2) \overset{bir}{\sim} \mathrm{Map}_V(\mathbb{P}^2, Q^4)$ and similarly for $\mathrm{Chow}_{PV}(Q^4)$. Hence the previous results determine the stable birational types of $\mathrm{Chow}_V(Q^4)$ and $\mathrm{Chow}_{PV}(Q^4)$. The proofs yield the following slightly stronger forms; see (46.6) and (43.1).

COROLLARY 7 (Chow variety version). *Let* Q^4 *be a smooth quadric 4-fold over a field* k. *Then*

$$\begin{aligned} \mathrm{Chow}_V(Q^4) \times \mathbb{P}^4 &\overset{bir}{\sim} \{\pm\sqrt{-\Delta}\} \times Q^4 \times \mathbb{P}^{12} \quad \textit{and} \\ \mathrm{Chow}_{PV}(Q^4) &\overset{bir}{\sim} \mathrm{OG}(\mathbb{P}^1, Q^4) \times \mathbb{P}^7. \end{aligned}$$

It is quite likely that one can cancel the factor $\mathbb{P}^4$ but my proof does not seem to work without it.

$\mathrm{Chow}_{PV}(Q^4)$ is contained in the irreducible component $\mathrm{Chow}_{CI}(Q^4)$ of the Chow variety $\mathrm{Chow}(Q^4)$ that parametrizes complete intersections of Q^4 with a hyperplane and a quadric. Another interesting irreducible component of $\mathrm{Chow}(Q^4)$ parametrizing degree 4 surfaces is discussed in (47).

We show in Section 6 that working up to coordinate changes by both $\operatorname{Aut}(\mathbb{P}^2)$ and $\operatorname{Aut}(Q^4)$ yields a very nice answer, especially in the Veronese case.

THEOREM 8 (Veronese). *Let k be a field. A smooth quadric hypersurface $Q^4 \subset \mathbb{P}^5$ contains a Veronese surface iff its equation can be written as*

$$Q^4_{a,b} = \left(x_0x_5 = x_1^2 + ax_2^2 + bx_3^2 + abx_4^2\right). \tag{8.1}$$

Up to coordinate changes by $\operatorname{Aut}(\mathbb{P}^2) \times \operatorname{Aut}(Q^4_{a,b})$, the quadric $Q^4_{a,b}$ contains a unique Veronese surface. A representative is given by

$$(u{:}v{:}w) \mapsto \left(u^2 + aw^2 : uv : vw : uw : w^2 : v^2 + bw^2\right). \tag{8.2}$$

COMPLEMENT 9. *Let k be a field. Then $Q^4_{a,b} \leftrightarrow (au^2 + bv^2 + w^2 = 0)$ gives a one-to-one correspondence between*

- *pairs $(V \subset Q^4)$—a Veronese surface contained in a smooth quadric hypersurface defined over k, up to isomorphism—and*
- *smooth plane conics $C \subset \mathbb{P}^2$ defined over k, up to isomorphism.*

A proof of this is given in (48.6). Another way of obtaining a conic out of a pair $(V \subset Q^4)$ is as follows. Pick $p \in V$. Intersecting V with the tangent plane of Q^4 at p gives a hyperplane section of V that is singular at p. Since the hyperplane sections of V are conics, we expect that we get a pair of lines and there is a curve $C_{a,b} \subset V$ where this intersection is a double line.

A direct computation using (8.2) (that I did not find illuminating) shows that $C_{a,b} = (bu^2 + av^2 + abw^2 = 0)$. We can rewrite this as $b(au_1)^2 + a(bv_1)^2 + abw^2 = ab(au_1^2 + bv_1^2 + w^2)$.

REMARK 10. We can fully diagonalize the quadric $Q^4_{a,b}$ by replacing x_0, x_5 by $\frac{1}{2}(x_0 + x_5)$ and $\frac{1}{2}(x_0 - x_5)$. Then we end up with the identity

$$\tfrac{1}{4}\left(u^2+v^2+(a+b)w^2\right)^2 - \tfrac{1}{4}\left(u^2-v^2+(a-b)w^2\right)^2 = (uv)^2 + a(vw)^2 + b(uw)^2 + ab(w^2)^2.$$

Skopenkov pointed out that (if a, b are real and positive) this is equivalent to the quaternionic identity

$$\tfrac{1}{4}(|A|^2 + |B|^2)^2 - \tfrac{1}{4}(|A|^2 - |B|^2)^2 = |AB|^2$$

for the quaternions $A = u + iw\sqrt{a}$ and $B = v + jw\sqrt{b}$.

The following 2 Propositions are also proved in Section 6.

PROPOSITION 11 (Projected Veronese). *Let k be a field. A smooth quadric hypersurface $Q^4 \subset \mathbb{P}^5$ contains a projected Veronese surface iff its equation can be written as*

$$Q^4_a = \left(x_0x_1 + x_2x_3 = x_4^2 + ax_5^2\right).$$

After coordinate changes by $\operatorname{Aut}(\mathbb{P}^2) \times \operatorname{Aut}(Q^4_a)$, projected Veronese surfaces can be brought to the form

$$(u{:}v{:}w) \mapsto \left(0 : q(u{:}v{:}w) : u^2 + av^2 : w^2 : uw : vw\right)$$

where $\deg q = 2$. Restricting q to the line $(w = 0)$ gives an isomorphism between the moduli of projected Veronese surfaces in Q^4_a (up to $\operatorname{Aut}(\mathbb{P}^2) \times \operatorname{Aut}(Q^4_a)$) and $\operatorname{Sym}^2(\mathbb{P}^1)/\mathrm{O}(u^2 + av^2)$.

PROPOSITION 12 (Quadruple plane). *A smooth quadric hypersurface $Q^4 \subset \mathbb{P}^5$ contains a 2-plane iff its equation can be written as*

$$Q^4_{\mathrm{split}} = \big(x_0x_3 + x_1x_4 + x_2x_5 = 0\big).$$

If $\operatorname{char} k \neq 3$ *then, after coordinate changes by* $\operatorname{Aut}\big(Q^4_{\mathrm{split}}\big)$, *quadruple planes can be brought to the form*

$$(u{:}v{:}w) \mapsto \big(\partial C/\partial u : \partial C/\partial v : \partial C/\partial w : 0 : 0 : 0\big),$$

where $C = C(u,v,w)$ is a homogeneous cubic. This gives an isomorphism between the moduli of quadruple planes (up to $\operatorname{Aut}(\mathbb{P}^2) \times \operatorname{Aut}\big(Q^4_{\mathrm{split}}\big)$*) and the moduli of plane cubic curves (up to* $\operatorname{Aut}(\mathbb{P}^2)$*).*

Next we move to quadrics of dimension ≥ 5. Note that $\Phi(\mathbb{P}^2)$ is always contained in a 5-dimensional linear subspace of $\mathbb{P}^{n+1}$, so the cases $n \geq 5$ reduce to the $n \leq 4$ cases is principle but the parametrization of the solutions for $n \geq 5$ proceeds somewhat differently. The following is shown in Section 7.

THEOREM 13. *Assume that $n \geq 5$ and $Q^n(k) \neq \emptyset$. Then*

(13.1) $$\operatorname{Map}_2(\mathbb{P}^2, Q^n) \overset{bir}{\sim} \operatorname{Sym}^2\big(\operatorname{OG}(\mathbb{P}^2, Q^n)\big) \times \mathbb{P}^{14},$$

where Sym^2 *denotes the symmetric square.*

Using (68.2) we can rewrite (13.1) as

(13.2) $$\operatorname{Map}_2(\mathbb{P}^2, Q^n) \overset{bir}{\sim} \operatorname{Sym}^2\big(\operatorname{OG}(\mathbb{P}^1, Q^{n-2}_W)\big) \times \mathbb{P}^{2n+10},$$

where Q^{n-2}_W is the Witt reduction of Q^n (see Notation 20). This variant actually shows the structure of $\operatorname{Map}_2(\mathbb{P}^2, Q^n)$ better than (13.1).

REMARK 14. If $n = 4$ then $\big\{\pm\sqrt{-\Delta}\big\} \times Q^4 \times \mathbb{P}^2$ is birational to an irreducible component of $\operatorname{Sym}^2\big(\operatorname{OG}(\mathbb{P}^2, Q^4)\big)$ (or a union of 2 irreducible components if $-\Delta$ is a square in k) by (69.4). Thus Theorem 13 can be viewed as a generalization of Theorem 4. Note, however that Theorem 13 fails if $Q^n(k) = \emptyset$. Indeed, for $n \geq 5$ consider the diagonal quadric $Q^n := (x_0^2 + \cdots + x_{n+1}^2 = 0)$ over any field $k \subset \mathbb{R}$. Then $\operatorname{Map}_2(\mathbb{P}^2, Q^n)(k) = \emptyset$ but the conjugate pair of linear spaces

$$L^2_\pm := (x_0 \pm ix_1 = x_2 \pm ix_3 = x_4 \pm ix_5 = x_6 = \cdots = x_{n+1} = 0)$$

shows that $\operatorname{Sym}^2\big(\operatorname{OG}(\mathbb{P}^2, Q^n) \neq \emptyset$.

The birational equivalence in Theorem 13 is not completely $\operatorname{Aut}(Q^n)$-equivariant. The reason is that we usually imagine the left hand side as $\operatorname{Chow}_V(Q^n) \times \operatorname{Aut}(\mathbb{P}^2)$. However, the proof works with pointed Veronese surfaces. Thus it views the left had side as $\operatorname{Univ}_V(Q^n) \times \operatorname{Aut}(p \in \mathbb{P}^2)$, where $\operatorname{Univ}_V(Q^n) \to \operatorname{Chow}_V(Q^n)$ denotes the universal family of Veronese surfaces in Q^n and $\operatorname{Aut}(p \in \mathbb{P}^2)$ is the group of automorphisms of $\mathbb{P}^2$ that fix the marked point p. We have $\operatorname{Aut}(Q^n)$-equivariance for the latter presentation.

Theorem 13 implies that $\operatorname{Map}_2(\mathbb{P}^2, Q^n)$ is geometrically irreducible for $n \geq 5$. In particular, projected Veronese surfaces and quadruple planes can be deformed to Veronese surfaces in a 5-dimensional quadric; see Example 55 for explicit descriptions.

I do not know a birational classification of symmetric squares of orthogonal Grassmannians in general. We give an explicit description of $\operatorname{Sym}^2\big(\operatorname{OG}(\mathbb{P}^1, Q^3)\big)$ in (71) and this implies the following.

PROPOSITION 15. *Let* $Q^5 = \big(q(x_0, \ldots, x_4) = x_5x_6\big)$ *be a smooth quadric 5-fold over a field* k. *Then*

$$\mathrm{Map}_2(\mathbb{P}^2, Q^5) \overset{bir}{\sim} \big(\Delta(q)z^2 = q(x_0, \ldots, x_4)\big) \times \mathbb{P}^{22},$$

where $\Delta(q)$ *is the discriminant of* q.

Note that the quadratic form $\Delta(q)z^2 - q(x_0, \ldots, x_4)$ is never positive definite, thus the Hasse-Minkowski theorem (cf. [**Ser73**, p.41]) and Theorem 13 imply the following.

COROLLARY 16. *Let* Q^5 *be a smooth quadric 5-fold over* k *such that* $Q^5(k) \neq \emptyset$. *Assume that* k *is one of the following:*

(1) $\mathbb{R}$ *or any real closed field,*
(2) *a number field,*
(3) *a function field of a curve over a finite field or*
(4) *a function field of a surface over an algebraically closed field.*

Then $\mathrm{Map}_2(\mathbb{P}^2, Q^5) \overset{bir}{\sim} \mathbb{P}^{26}$. □

For other fields the situation is more complicated.

EXAMPLE 17. Let $K := k(t_1, \ldots, t_4)$ where the t_i are algebraically independent. Consider the quadric 5-fold $Q^5 := (x_0^2 + t_1x_1^2 + \cdots + t_4x_4^2 = x_5x_6)$. It is easy to see that

$$x_0^2 + t_1x_1^2 + \cdots + t_4x_4^2 = t_1t_2t_3t_4 \cdot z^2$$

has no solutions in K. Therefore, using Theorem 13 and Proposition 15, we obtain that $\mathrm{Sym}^2\big(\mathrm{OG}(\mathbb{P}^1, Q^3_W)\big)(K) = \emptyset$ and hence $\mathrm{Map}_2^\circ(\mathbb{P}^2, Q^5)(K) = \emptyset$.

Thus the image of any quadratic map $\mathbb{P}^2 \dashrightarrow Q^5$ is a point or a conic. Examples for the latter are $(u{:}v{:}w) \mapsto (0 : \cdots : 0 : uv : u^2 : t_4v^2)$. These correspond to singular K-points of $\mathrm{Map}_2(\mathbb{P}^2, Q^n)$.

The question of uniqueness of Veronese surfaces $V \subset Q^n$ (up to coordinate changes by $\mathrm{Aut}(\mathbb{P}^2) \times \mathrm{Aut}(Q^n)$) is not as simple as in dimension 4.

PROPOSITION 18. *Let* k *be a field. A smooth quadric hypersurface* $Q^5 \subset \mathbb{P}^6$ *contains a Veronese surface iff its equation can be written as*

$$Q^5_{a,b,c} = \big(x_0x_5 = x_1^2 + ax_2^2 + bx_3^2 + abx_4^4 + cx_6^2\big).$$

Up to coordinate changes by $\mathrm{Aut}(\mathbb{P}^2) \times \mathrm{Aut}\big(Q^5_{a,b,c}\big)$, *the quadric* $Q^5_{a,b,c}$ *contains a unique Veronese surface. A representative is given by*

$$(u{:}v{:}w) \mapsto \big(u^2 + aw^2 : uv : vw : uw : w^2 : v^2 + bw^2 : 0\big).$$

The proof is given in (53) and the next examples show that we have non-uniqueness for $n \geq 6$; see also Proposition 24.

EXAMPLE 19. If k is algebraically closed then all smooth subquadrics $Q^4 \subset Q^n$ are isomorphic, thus Theorem 8 implies that there is a unique Veronese surface $V \subset Q^n$, up to coordinate changes by $\mathrm{Aut}(\mathbb{P}^2) \times \mathrm{Aut}(Q^n)$. The same holds over $\mathbb{R}$ if Q^n is the sphere.

However, for most other fields there are quadrics that contain inequivalent Veronese surfaces, up to coordinate changes. Indeed, by Theorem 8, isomorphism classes of quadric 4-folds that contain a Veronese surface are in one-to-one correspondence with isomorphism classes of plane conics. Assume that we have two non-isomorphic ternary forms $x^2 + ay^2 + bz^2$ and $x^2 + ay^2 + cz^2$. Consider the quadric 6-fold

$$Q^6_{a,b,c} = \left(x_0x_7 = x_1^2 + ax_2^2 + bx_3^2 + abx_4^4 + cx_5^2 + acx_6^4\right). \tag{19.1}$$

It contains a Veronese surface V_b that satisfies $(x_5 = x_6 = 0)$ and also a V_c that satisfies $(x_3 = x_4 = 0)$. These are inequivalent since the quadrics they span are not isomorphic.

As another example, let k be a number field and consider

$$Q^7 := \left(x_0^2 + x_1x_2 + x_3x_4 + x_5x_6 + x_7x_8 = 0\right). \tag{19.2}$$

Note that $au^2 - av^2 \sim uv$, so every $Q^4_{a,b}$ has at least one embedding into Q^7. Thus there are infinitely many inequivalent embeddings $V \hookrightarrow Q^7$ defined over k, up to coordinate changes.

NOTATION 20. From now on, $Q^n := \left(q(x_0, \dots, x_{n+1}) = 0\right) \subset \mathbb{P}^{n+1}$ denotes a smooth quadric hypersurface over a field k. Starting with Section 4 we assume that $\operatorname{char} k \neq 2$. The explicit formulas are usually worked out only for diagonal forms.

More generally, $Q^{n,i}$ denotes a quadric hypersurface over a field k whose non-smooth locus has dimension i. We use this mainly for $i = 0$.

A quadratic form q over k—or the corresponding quadric hypersurface Q over k—is called *isotropic* if $Q(k) \neq \emptyset$ and *anisotropic* if $Q(k) = \emptyset$.

The *discriminant* of a symmetric bilinear form $q = \sum_{0 \le i,j \le n+1} a_{ij}x_ix_j$ over k is $\Delta(q) := \det(a_{ij}) \in k/(k^*)^2$. This defines the discriminant of any quadratic form whenever $\operatorname{char} k \neq 2$. (Note that some authors use different signs and 2-powers.) A quadric $Q^n = \left(q(x_0, \dots, x_{n+1}) = 0\right)$ determines the corresponding quadratic form q up to a multiplicative scalar λ. This changes the discriminant by λ^{n+2}. So the *discriminant* of Q^n is only defined for even dimensional quadrics and then it is

$$\Delta(Q^n) := \Delta\big(q(x_0, \dots, x_{n+1})\big) \in k/(k^*)^2.$$

Note that Q^n is smooth iff $\Delta(Q^n) \neq 0$.

For any point $p \in Q^n$, let $Q^{n-1,0}_p \subset Q^n$ denote the intersection of Q^n with its tangent plane at p. Thus $Q^{n-1,0}_p$ is an $(n-1)$-dimensional quadric cone over an $(n-2)$-dimensional smooth quadric—the projectivized tangent cone—denoted by Q^{n-2}_p. Both $Q^{n-1,0}_p$ and Q^{n-2}_p are defined over the field $k(p)$. The Q^{n-2}_p together form the universal projectivized tangent cone $\operatorname{PTC}(Q^n) \to Q^n$.

Assume next that $Q^n(k) \neq \emptyset$. Pick any $p \in Q^n(k)$. In suitable coordinates we may assume that $p = (0, \dots, 0, 1)$ and

$$q(x_0, \dots, x_{n+1}) \sim q_p(x_0, \dots, x_{n-1}) + x_nx_{n+1}.$$

Then $Q^{n-2}_p \cong \left(q_p(x_0, \dots, x_{n-1}) = 0\right) \subset \mathbb{P}^{n-1}$. By Theorem 21, q_p and Q^{n-2}_p are independent of p; we denote this quadric by Q^{n-2}_W and call it the *Witt reduction* of Q^n. We can iterate the Witt reduction until we reach an anisotropic quadric

$$Q_{\mathrm{an}} = Q^{n-2r}_{\mathrm{an}} = \left(q_{\mathrm{an}}(x_0, \dots, x_{n+1-2r}) = 0\right),$$

called the *anisotropic kernel* of Q^n (or of q). The number r is called the *Witt index*. Equivalently, $r - 1$ is the maximum dimension of a linear space that is defined over k and contained in Q^n.

A form is called *split* if $2r = n + 2$, equivalently, if it can be written as

$$x_0x_1 + x_2x_3 + \cdots + x_{2r-2}x_{2r-1}.$$

Let $\pi_p : Q^n \dashrightarrow \mathbb{P}^n$ denote the projection from p. We can factor it as the blow-up of p followed by the contraction of the birational transform $E_p^{n-1,0}$ of $Q_p^{n-1,0}$. Here $E_p^{n-1,0}$ is a $\mathbb{P}^1$-bundle over $Q_p^{n-2} \subset \mathbb{P}^n$. This shows that there is an equivalence

$$\{p \in Q^n\} \leftrightarrow \{Q_W^{n-2} \subset \mathbb{P}^n\} \tag{20.1}$$

between n-dimensional pointed quadrics and $(n-2)$-dimensional quadrics in $\mathbb{P}^n$.

A geometric proof of the special case of Witt's theorem that we used above is the following. Fix a point $p_0 \in Q^n(k)$. For any $p \in Q^n \setminus Q_{p_0}^{n-1,0}$ we get natural isomorphisms (defined over $k(p)$)

$$Q_p^{n-2} \cong Q_p^{n-1,0} \cap Q_{p_0}^{n-1,0} \cong Q_{p_0}^{n-2}. \tag{20.2}$$

This also shows that the universal projectivized tangent cone $\operatorname{PTC}(Q^n) \to Q^n$ is birationally trivial

$$\operatorname{PTC}(Q^n) \overset{bir}{\sim} Q^n \times Q_W^{n-2}. \tag{20.3}$$

21 (Witt's cancellation theorem). Let V be a vector space and q a non-degenerate quadratic form on V. Let $V_1, V_2 \subset V$ be vector subspaces such that $\left(V_i, q|_{V_i}\right)$ are non-degenerate and there is an isometry $\phi : \left(V_1, q|_{V_1}\right) \to \left(V_2, q|_{V_2}\right)$. One form of Witt's cancellation theorem says that ϕ extends to an isometry $\Phi : V \to V$.

This is one example where the correspondence between quadrics and quadratic forms gets cumbersome. For example, consider $(\mathbb{R}^5, q := x_1^2 + x_2^2 + x_3^2 - x_4^2 - x_5^2)$, $V_1 := (x_3 = x_5 = 0)$ and $V_2 := (x_1 = x_3 = 0)$. Note that the quadrics $(q = x_3 = x_5 = 0)$ and $(q = x_1 = x_3 = 0)$ are isomorphic but the quadratic forms $\left(V_1, q|_{V_1}\right)$ and $\left(V_2, q|_{V_2}\right)$ are not. There is no automorphism of the quadric $(q = 0)$ that carries $(q = x_3 = x_5 = 0)$ to $(q = x_1 = x_3 = 0)$.

22 (Effective parametrization). The methods of this paper are effective for 4-dimensional quadrics, that is, the proofs can be converted into explicit formulas giving all elements of $\operatorname{Map}_2(\mathbb{P}^2, Q^4)(k)$ for any field k, provided we know the answer to some basic questions about Q^4.

If there is a k-map $\phi : \mathbb{P}^2 \dashrightarrow Q^n$ then $Q^n(k) \neq \emptyset$. Conversely, if $p \in Q^n(k)$ is any k-point then the constant map $\mathbb{P}^2 \to \{p\} \hookrightarrow Q^n$ is in $\operatorname{Map}_2(\mathbb{P}^2, Q^n)(k)$. Thus $\operatorname{Map}_2(\mathbb{P}^2, Q^n)(k) \neq \emptyset$ iff $Q^n(k) \neq \emptyset$. To the best of my knowledge, there is no algorithm that decides whether a quadric over an arbitrary field has a k-point or not, and we do not claim to say anything about this question.

We show, however, that once we can decide the existence of points, we can also effectively write down all maps in $\operatorname{Map}_2(\mathbb{P}^2, Q^n)(k)$. To be precise, we need to decide the existence of points also for certain subforms and the condition we need is the following. We start with the cases when $n \leq 5$.

Assumption. For any quadratic form $q(x_0, \dots, x_{n+1})$ we are given (or are able to find) a Witt decomposition of q. That is, a linear change of coordinates

$$q(x_0, \dots, x_{n+1}) \sim q_{\text{an}}(y_0, \dots, y_{n-2r+1}) + y_{n-2r+2}y_{n-2r+3} + \dots + y_n y_{n+1}$$

where q_{an} is anisotropic.

Note that such a linear change of coordinates always exists over any field, but in general we do not know how to find it. Given a field k, there is an algorithm to find such a coordinate change for every quadratic form over k iff there is an algorithm that for every quadratic form $q(x_0, \dots, x_{r+1})$ over k

– either proves that q is anisotropic,

– or finds a non-trivial solution $q(a_0, \dots, a_{r+1}) = 0$.

Thus the assumption is always satisfied of k is algebraically closed or real closed. Over any number field, the Hasse–Minkowski theorem (see, for instance [**Ser73**, p.41]) provides a (very inefficient) algorithm.

For $n \geq 6$ we also need to know a parametrization of degree 2 points of the orthogonal Grassmannian $\mathrm{OG}(\mathbb{P}^2, Q^n)$. I do not even know how to do this over nice fields like $\mathbb{Q}$ or $\mathbb{C}(t)$.

1. Surfaces containing many circles

I got interested in these questions after reading the papers [**SK15, Sko15**] which furnish some of the last steps of a project, started by Kummer and Darboux, to describe all surfaces in $\mathbb{R}^n$ that contain at least 2 circles through every point; see also [**Sch01, PSS12**].

The inverse of stereographic projection connects this problem with real algebraic surfaces F on the sphere

$$\mathbb{S}^n := (x_1^2 + \cdots + x_{n+1}^2 = x_0^2) \subset \mathbb{RP}^{n+1}.$$

The formulas for stereographic projection are nicest if we project the sphere

$$\mathbb{S}^n := \left(x_1^2 + \cdots + x_n^2 + z^2 = 1\right)$$

from the south pole $(0, \dots, 0, -1)$. Then

$$\pi(x_1, \dots, x_n, z) = \left(\frac{x_1}{1+z}, \dots, \frac{x_n}{1+z}\right)$$

with inverse

$$\pi^{-1}(x_1, \dots, x_n) = \left(\frac{2x_1}{1+\Sigma}, \dots, \frac{2x_n}{1+\Sigma}, \frac{1-\Sigma}{1+\Sigma}\right),$$

where $\Sigma = \sum_i x_i^2$. A theorem that Ptolemy attributes to Hipparchus ($\sim$ 190-120 BC) says that stereographic projection preserves circles; a modern treatment was given by Halley, [**Hal1695**, Lem.2].

Any conic on a sphere is a circle, thus describing all surfaces in $\mathbb{R}^n$ that contain many circles is equivalent to describing real algebraic surfaces

$$F \subset \mathbb{S}^n := (x_1^2 + \cdots + x_{n+1}^2 = x_0^2) \subset \mathbb{RP}^{n+1}$$

that contain many conics.

Building on works of [**Kac99**] and [**Sch01**], the hardest cases seem to be projections of the Veronese surface or of a degree 8 del Pezzo surface. These are classified in [**Sko15**] for $\mathbb{S}^4$. His idea is to rewrite the equation as

$$x_1^2 + x_2^2 + x_3^2 + x_4^2 = (x_0 - x_5)(x_0 + x_5),$$

think of $x_1^2 + x_2^2 + x_3^2 + x_4^2$ as the norm of the quaternion $\mathbf{X} := x_1 + x_2 i + x_3 j + x_4 k$ and solve the quaternionic equation

$$\mathrm{norm}(\mathbf{X}) = z_0 z_5, \quad \text{where} \quad \mathbf{X} \in \mathbb{H}[u,v],\ z_0, z_5 \in \mathbb{R}[u,v].$$

This is quite nontrivial since $\mathbb{H}[u,v]$ is far from having unique factorization.

In our language, we work with the quadratic form $q := x_1^2 + \cdots + x_5^2 - x_0^2$ over $\mathbb{R}$. Its discriminant is $\Delta = -1$ and the anisotropic kernel is $x_1^2 + x_2^2 + x_3^2 + x_4^2$. Thus we see from Propositions 5–6 and Theorem 4 that the Veronese components

$\mathrm{Map}_V(\mathbb{P}^2, Q^4)(\mathbb{R}) \stackrel{bir}{\sim} \mathbb{RP}^{20} \amalg \mathbb{RP}^{20}$ give all real solutions. By Theorem 8, there is only one solution up to isomorphism. Since a Veronese surface in $\mathbb{P}^{n+1}$ is contained in a linear subspace of dimension ≤ 5, these imply the following.

THEOREM 23. *There are precisely 2 immersed, C^2 surfaces $F \subset \mathbb{S}^n$ (up to the group of Möbius transformations $\mathrm{Aut}(\mathbb{S}^n) = \mathrm{O}(x_1^2 + \cdots + x_{n+1}^2 - x_0^2)$) that contain a circle through any 2 points $p_1, p_2 \in F$:*

(1) *the sphere $\mathbb{S}^2 \subset \mathbb{S}^n$ and*

(2) *the Veronese surface $\mathbb{RP}^2 \hookrightarrow \mathbb{S}^4 \subset \mathbb{S}^n$.* □

There are several ways to describe the real Veronese surface $\mathbb{RP}^2 \hookrightarrow \mathbb{S}^4$. For example, one can think of the Veronese surface V as the set of rank 1 matrices in the projectivization of the space of 3×3 symmetric matrices:

$$(u : v : w) \mapsto \begin{pmatrix} u^2 & uv & uw \\ uv & v^2 & vw \\ uw & vw & w^2 \end{pmatrix}. \tag{23.3}$$

The identity $(uv)^2 + (vw)^2 + (uw)^2 = u^2v^2 + v^2w^2 + u^2w^2$ becomes the quadratic equation

$$x_{01}^2 + x_{12}^2 + x_{02}^2 = x_{00}x_{11} + x_{11}x_{22} + x_{00}x_{22}, \tag{23.4}$$

which defines an ellipsoid that contains V. The right hand side can be diagonalized in a less symmetric form as

$$\tfrac{1}{3}\big(x_{00} + x_{11} + x_{22}\big)^2 - \tfrac{1}{4}\big(x_{00} - x_{11}\big)^2 - \tfrac{1}{12}\big(x_{00} + x_{11} - 2x_{22}\big)^2. \tag{23.5}$$

This gives an embedding of $\mathbb{RP}^2$ into the (affine) sphere $\mathbf{S}^4 = \big(y_1^2 + \cdots + y_5^2 = 1\big)$

$$(u : v : w) \mapsto \tfrac{\sqrt{3}}{u^2+v^2+w^2}\Big(uv : vw : uw : \tfrac{1}{2}(u^2 - v^2) : \tfrac{1}{\sqrt{12}}(u^2 + v^2 - 2w^2)\Big). \tag{23.6}$$

Another form is given in Remark 10:

$$(u : v : w) \mapsto \tfrac{2}{u^2+v^2+2w^2}\Big(uv : vw : uw : w^2 : \tfrac{1}{2}(u^2 - v^2)\Big). \tag{23.7}$$

Another form of the Veronese embedding into $\mathbb{R}^5$ is obtained by looking at the vector space $V^{(3,3)}$ of 3×3 symmetric matrices of trace 0. $\mathrm{SO}(3, \mathbb{R})$ acts on $V^{(3,3)}$ by conjugation and the action preserves the eigenvalues. Furthermore, the orbits are in one-to-one correspondence with the ordered set of eigenvalues $\lambda_1 \leq \lambda_2 \leq \lambda_3$. The generic orbit is large (with stabilizer $(\mathbb{Z}/2)^2$) but if 2 eigenvalues coincide, say for $-1, -1, 2$, then the orbit is isomorphic to the Veronese embedding of $\mathbb{RP}^2$. The $\mathrm{SO}(3, \mathbb{R})$-invariant sphere containing this Veronese surface is given by the equation $\mathrm{tr}(M^2) = 6$.

Note that $M \mapsto M + \mathbf{1}_3$ sends a symmetric matrix with eigenvalues $-1, -1, 2$ to a symmetric matrix of rank 1 and trace 3. This shows that the above affine form is equivalent to the projective one.

Other real quadric hypersurfaces can contain more then one family of Veronese surfaces.

PROPOSITION 24. *Let $Q^n \subset \mathbb{RP}^{n+1}$ be a quadric of dimension $n \geq 4$ and of signature (r, s) where $r \geq s$. Then the number of different Veronese embeddings $\mathbb{RP}^2 \hookrightarrow Q^n$ (up to coordinate changes by $\mathrm{Aut}(\mathbb{RP}^2) \times \mathrm{Aut}(Q^n)$) is*

(1) *zero if $s = 0$ or $r \leq 4, s \leq 2$,*

(2) *one if* $r \geq 5$ *and* $s \in \{1, 2\}$ *or if* $(r, s) = (3, 3), (4, 3), (4, 4)$ *and*
(3) *at most 4 in general.*

Proof. If the linear span of a Veronese surface $V \subset Q^n$ intersects Q^n in a Q^4 then Q^4 has signature $(5, 1), (1, 5)$ or $(3, 3)$. Thus we need to enumerate all solutions to $(r, s) = (r', s') + (r'', s'')$ where (r', s') is one of $(5, 1), (1, 5), (3, 3)$ and $r'', s'' \geq 0$. There are clearly at most 3 solutions but fewer if s is small or for occasional symmetries.

It can also happen that V is contained in a 5-plane $L^5 \subset Q^n$. Then $r, s \geq 6$; usually leading to 1 more family of Veronese embeddings. The exceptional case is $r = s = 6$ when there are 2 families of $L^5 \subset Q^{10}$, but then symmetry gives only 2 families where the linear span of a Veronese surface $V \subset Q^n$ intersects Q^n in a Q^4. □

It would be quite interesting to understand higher dimensional generalizations of our results. The following form seems to be the most promising.

PROBLEM 25. Classify real algebraic varieties $X \subset \mathbb{S}^n$ such that every point pair $x_1, x_2 \in X$ is contained in a circle $C(x_1, x_2) \subset X$.

The paper [**IR10**] classifies complex algebraic varieties $X \subset \mathbb{P}^{n+1}$ that are *conic-connected,* that is, every point pair $x_1, x_2 \in X$ is contained in at least one conic $C(x_1, x_2) \subset X$; see also [**Rus16**, Chap.4]. Thus one needs to decide which conic-connected varieties give real examples in $\mathbb{S}^n$. There are many conic-connected Fano varieties of Picard number 1 and these are not yet fully understood. However, if $\dim X \leq 6$, then these are all so called *Mukai manifolds* and a complete classification of Mukai manifolds is given in [**Muk89, Mel99**]. Thus the list of [**IR10**] is complete for $\dim X \leq 6$, hence it could be feasible to answer the above problem when $\dim X \leq 6$.

An interesting extremal case is the Veronese embedding of $\mathbb{P}^n$ by quadrics. By a theorem of Gallarati (see [**Rus16**, 3.4.4]) if $X^n \subset \mathbb{P}^N$ is a linearly non-degenerate conic-connected variety and $N \geq \frac{1}{2}n(n+3)$ then $N = \frac{1}{2}n(n+3)$ and X^n is the Veronese embedding of $\mathbb{P}^n$ by quadrics. See [**LP71**] for a version for differentiable manifolds. One should study which quadrics contain these higher dimensional Veronese varieties.

2. Maps of $\mathbb{P}^1$ to quadrics

Here we describe the spaces $\mathrm{Map}_d(\mathbb{P}^1, Q^n)$ parametrizing degree d maps from $\mathbb{P}^1$ to a quadric. Equivalently, we describe solutions of a homogeneous quadratic equation $q(x_0, \dots, x_{n+1}) = 0$, defined over a field k, where the $x_i = h_i(u, v)$ are degree d homogeneous polynomials in 2 variables u, v.

The proof of Theorem 1 is a direct consequence of the following 2 propositions.

PROPOSITION 26. *Let* Q^n *be a smooth quadric of dimension* $n \geq 3$. *Then*

$$\mathrm{Map}_d(\mathbb{P}^1, Q^n) \overset{bir}{\sim} \mathrm{Map}_{d-2}(\mathbb{P}^1, Q^n) \times \mathbb{P}^{2n} \quad \textit{for } d \geq 3.$$

PROPOSITION 27. *Let* Q^n *be a smooth quadric of dimension* $n \geq 3$. *Then*

$$\mathrm{Map}_1(\mathbb{P}^1, Q^n) \overset{bir}{\sim} \mathrm{OG}(\mathbb{P}^1, Q^n) \times \mathbb{P}^3 \quad \textit{and} \quad \mathrm{Map}_2(\mathbb{P}^1, Q^n) \overset{bir}{\sim} Q^n \times \mathbb{P}^{2n}.$$

Proof. The image of a degree 1 map is a line, this gives the formula for $\mathrm{Map}_1(\mathbb{P}^1, Q^n)$.

Let $\mathrm{Map}_2^\circ(\mathbb{P}^1, Q^n)$ denote those degree 2 maps that map $\mathbb{P}^1$ isomorphically onto a conic. By [**Kol96**, II.3.14], $\mathrm{Map}_2^\circ(\mathbb{P}^1, Q^n)$ is an open and dense subset of $\mathrm{Map}_2(\mathbb{P}^1, Q^n)$; this can also be seen by a simple dimension count here. A conic is uniquely determined by any 3 of its points, say the images of $0, 1, \infty \in \mathbb{P}^1$. This gives an open embedding (hence a birational equivalence)

$$\mathrm{Map}_2^\circ(\mathbb{P}^1, Q^n) \hookrightarrow Q^n \times Q^n \times Q^n.$$

The birational equivalence $Q^n \times Q^n \times Q^n \overset{bir}{\sim} Q^n \times \mathbb{P}^{2n}$ can be seen as follows. Fix two hyperplanes $H_0, H_1 \subset \mathbb{P}^{n+1}$. Given $p \in Q^n$ and $q_i \in H_i$, there is a unique degree 2 map $\phi : \mathbb{P}^1 \to Q^n$ such that $\phi(\infty) = p$ and, $\phi(i)$ is the residual intersection point of $\langle p, q_i \rangle \cap Q^n$ for $i = 0, 1$. □

Note that $n = 2$ is exceptional for Proposition 27 and

$$\mathrm{Map}_2(\mathbb{P}^1, Q^2) \overset{bir}{\sim} \left(Q^2 \times \mathbb{P}^4\right) \amalg \left(\mathrm{OG}(\mathbb{P}^1, Q^2) \times \mathbb{P}^5\right)$$

where the second component parametrizes double covers of lines.

The plan to prove Proposition 26 is the following. Given a degree d map $\phi : \mathbb{P}^1 \to Q^n$, we aim to write its image as a directrix of a ruled surface S. For suitable choice of S the residual intersection of S and Q^n is the image of a degree $d-2$ map $\psi : \mathbb{P}^1 \to Q^n$. We start with a general discussion on ruled surfaces.

28 (Ruled surfaces). Let C be a smooth projective curve and $\phi_i : C \to \mathbb{P}^n$ two morphisms. We want to understand the ruled surface swept out by the lines $\langle \phi_1(p), \phi_2(p) \rangle$ for $p \in C$.

Consider the abstract ruled surface $S := \mathbb{P}_C\left(\phi_1^*\mathcal{O}_{\mathbb{P}^n}(1) + \phi_2^*\mathcal{O}_{\mathbb{P}^n}(1)\right)$. It has 2 natural disjoint sections $C_1, C_2 \subset S$ and $\mathcal{O}_S(1)|_{C_i} \cong \phi_i^*\mathcal{O}_{\mathbb{P}^n}(1)$. Consider the exact sequence

$$0 \to \mathcal{O}_S(1)(-C_1 - C_2) \to \mathcal{O}_S(1) \to \mathcal{O}_S(1)|_{C_1} + \mathcal{O}_S(1)|_{C_2} \to 0.$$

Since $\mathcal{O}_S(1)(-C_1 - C_2)$ has degree -1 on the rulings, all of its cohomologies are 0. Thus we get that

$$H^0\big(S, \mathcal{O}_S(1)\big) \cong H^0\big(C, \phi_1^*\mathcal{O}_{\mathbb{P}^n}(1)\big) + H^0\big(C, \phi_2^*\mathcal{O}_{\mathbb{P}^n}(1)\big).$$

That is, the pair of morphisms (ϕ_1, ϕ_2) uniquely extends to a morphism $\Phi : S \to \mathbb{P}^n$. The degree of the image, that is, the self-intersection of $\Phi^*\mathcal{O}_{\mathbb{P}^n}(1)$ is $\deg \phi_1 + \deg \phi_2$.

We can do better if there are r points $p_1, \dots, p_r \in C$ such that $\phi_1(p_i) = \phi_2(p_i)$ holds. More generally, let $Z \subset C$ be a 0-dimensional subscheme such that $\phi_1|_Z = \phi_2|_Z$. Let us denote this map by ϕ_Z. We have natural restriction maps $r_i : \phi_i^*\mathcal{O}_{\mathbb{P}^n}(1) \to \phi_Z^*\mathcal{O}_{\mathbb{P}^n}(1)$. We define the sheaf E as the kernel of the map

$$r_1 - r_2 : \phi_1^*\mathcal{O}_{\mathbb{P}^n}(1) \oplus \phi_2^*\mathcal{O}_{\mathbb{P}^n}(1) \to \phi_Z^*\mathcal{O}_{\mathbb{P}^n}(1).$$

Then E is locally free of rank 2 and we have natural surjections $E \twoheadrightarrow \phi_i^*\mathcal{O}_{\mathbb{P}^n}(1)$. Now we set $S := \mathbb{P}_C(E)$. As before it has 2 natural sections $C_1, C_2 \subset S$ such that $\mathcal{O}_S(1)|_{C_i} \cong \phi_i^*\mathcal{O}_{\mathbb{P}^n}(1)$, but now $C_1 \cap C_2 \cong Z$. The pair of morphisms (ϕ_1, ϕ_2) uniquely extends to a morphism $\phi_{12} : C_1 \cup C_2 \to \mathbb{P}^n$. As before, there is a unique extension to $\Phi : S \to \mathbb{P}^n$. The gain is that the self-intersection of $\Phi^*\mathcal{O}_{\mathbb{P}^n}(1)$ is $\deg \phi_1 + \deg \phi_2 - \deg Z$.

Let now $Q^n \subset \mathbb{P}^{n+1}$ be a smooth quadric, C a smooth curve and $\phi : C \to Q^n$ a morphism of degree a. Choose any other morphism $\psi : C \to \mathbb{P}^{n+1}$ of degree b. We get a ruled surface $\Phi : S(\phi, \psi) \to \mathbb{P}^{n+1}$. Assume that its image is not contained in Q^n. Then $\Phi^{-1}(Q^n)$ is in the linear system of sections of $\mathcal{O}_S(2) = \Phi^*\mathcal{O}_{\mathbb{P}^{n+1}}(2)$.

By construction it contains C_1; let $R := R(\phi, \psi)$ denote the residual intersection. Thus R is a union of a section $C_3 \cong C$ of S plus possibly some rulings. These rulings are over those points $p \in C$ for which the line $\langle \phi(p), \psi(p) \rangle$ is contained in Q^n and possibly over those points $p \in C$ for which $\phi(p) = \psi(p)$. Thus we get a new morphism

$$\phi * \psi : \Phi|_{C_3} : C \to Q^n. \tag{28.1}$$

If there is a 0-dimensional subscheme $Z \subset C$ such that $\phi|_Z = \psi|_Z$ then the degree of the residual curve R is $2 \deg S - \deg \phi = \deg \phi + 2 \deg \psi - 2 \deg Z$ hence

$$\deg(\phi * \psi) = \deg \phi + 2 \deg \psi - 2 \deg Z - \#(\text{rulings in } R). \tag{28.2}$$

We get an interesting degenerate case if ψ also maps to Q^n. Then $\Phi^{-1}(Q^n)$ contains both sections C_1, C_2, hence the residual curve $R := R(\phi, \psi)$ is a union of $\deg \phi + \deg \psi$ rulings.

29 (Proof of (26)). We aim to prove the birational equivalence

$$\operatorname{Map}_d(\mathbb{P}^1, Q^n) \overset{bir}{\sim} \operatorname{Map}_{d-2}(\mathbb{P}^1, Q^n) \times \mathbb{P}^{2n} \quad \text{for } d \geq 3. \tag{29.1}$$

We fix the points $0, 1, \infty \in \mathbb{P}^1$, these will play a special role in the construction. To emphasize this, we write $\operatorname{Map}_d(\mathbb{P}^1, 0, 1, \infty; Q^n)$ instead of $\operatorname{Map}_d(\mathbb{P}^1, Q^n)$, though these spaces are isomorphic.

Fix an auxiliary hyperplane $L^n \subset \mathbb{P}^{n+1}$. We construct a morphism

$$\Pi_d : \operatorname{Map}_d(\mathbb{P}^1, 0, 1, \infty; Q^n) \dashrightarrow \operatorname{Map}_{d-2}(\mathbb{P}^1, 0, 1, \infty; Q^n).$$

Let $\phi_d : (\mathbb{P}^1, 0, 1, \infty) \to Q^n$ be a morphism of degree d. Let $\phi_1 : (\mathbb{P}^1, 0, 1, \infty) \to \mathbb{P}^{n+1}$ be a morphism of degree 1 that sends $0 \mapsto \phi_d(0), \infty \mapsto \phi_d(\infty)$ and 1 to the intersection of L^n with the line $\langle \phi_d(0), \phi_d(\infty) \rangle$. As in (28) the pair (ϕ_d, ϕ_1) determines a ruled surface S_{d-1} and by (28.2) $Q^n \cap S_{d-1} = \phi(\mathbb{P}^1) \cup R_{d-2}$ where R_{d-2} is a curve of degree $d-2$. We check below that it is an irreducible rational curve for general ϕ_d. Since both ϕ_d and ψ_{d-2} give sections of S_{d-1}, the 3 marked points on the image of ϕ_d give 3 marked points on the image of ψ_{d-2}. This determines $\psi_{d-2} : (\mathbb{P}^1, 0, 1, \infty) \to Q^n$.

Next we show that the generic fiber of Π_d is birational to $\mathbb{P}^{2n}$. To see this, fix $\psi_{d-2} : (\mathbb{P}^1, 0, 1, \infty) \to Q^n$. Pick points $x_0, x_\infty \in Q^n$. Let $\phi_1 : (\mathbb{P}^1, 0, 1, \infty) \to \mathbb{P}^{n+1}$ be a morphism of degree 1 that sends $0 \mapsto x_0, \infty \mapsto x_\infty$ and 1 to the intersection of L^n with the line connecting x_0, x_∞. As in (28) the pair (ψ_{d-2}, ψ_1) determines a ruled surface S_{d-1}. By construction $Q^n \cap S_{d-1} = \psi(\mathbb{P}^1) \cup R_d$. As before, this determines $\phi_d : (\mathbb{P}^1, 0, 1, \infty) \to Q^n$. We need to check that the curves R_{d-2} and R_d are irreducible for general choices of ϕ and ψ. For this it is enough to find one particular case where the second construction gives an irreducible curve R_d; the same set-up then proves the converse too. We start with the degenerate case where ψ_{d-2} maps onto a line $L \subset Q^n$.

We can now vary Q^n in the linear system $|\mathcal{O}_{\mathbb{P}^{n+1}}(2)|(-L)$, which is base-point free outside L. Bertini theorem gives that for general Q^n the residual intersection $R_d := (S_{d-1} \cap Q^n) \setminus L$ is irreducible. Since all smooth quadrics of dimension n are isomorphic over $\bar{k}$, irreducibility can be achieved by fixing Q^n and changing S_{d-1}.

Note that if $n = 2$ and ϕ_d is a curve of bidegree $(1, d-1)$ on Q^2 then the resulting R_{d-1} is a union of $d-1$ lines. So one really needs to check that the construction gives an irreducible curve for $n \geq 3$.

3. Rational curves on quadrics

Instead of looking at degree d maps from $\mathbb{P}^1$ to Q^n, it is also of interest to study geometrically rational curves contained in Q^n.

DEFINITION 30. Let X be a projective variety and L an ample line bundle on X. For $g, d \geq 0$ let $\mathrm{M}_g^\circ(X, d)$ be the space parametrizing morphisms $\phi : C \to X$ where
- C is a smooth, irreducible curve of genus g,
- ϕ is a morphism such that $C \to \phi(C)$ is birational and
- $\deg_C \phi^* L = d$.

One can think of $\mathrm{M}_g^\circ(X, d)$ as a subset of the Chow variety $\mathrm{Chow}(X)$ and it is frequently denoted by $\mathrm{RatCurve}_d(X)$ [**Kol96**, Sec.II.2].

One can also view $\mathrm{M}_g^\circ(X, d)$ as an open subset of $\bar{M}_g(X, d)$, the moduli space of stable maps of genus g and of degree d [**FP97**]. For most purposes the latter is the best compactification but for the birational properties of $\mathrm{M}_g^\circ(X, d)$ the precise compactification does not matter much.

Note that $\mathrm{Map}_d(\mathbb{P}^1, X)$ is a PGL_2-torsor over $\mathrm{M}_0^\circ(X, d)$ but this torsor is usually not Zariski-locally trivial. However, if $d = 2e + 1$ is odd and $(\phi : C \to X)$ is in $\mathrm{M}_0^\circ(X, d)$ then $\psi^* L \otimes \omega_{C_n}^e$ is a degree 1 line bundle on C. Thus the universal family over $\mathrm{M}_0^\circ(X, d)$ is Zariski-locally trivial and we conclude that

$$\mathrm{Map}_{2e+1}(\mathbb{P}^1, X) \overset{bir}{\sim} \mathrm{M}_0^\circ(X, 2e+1) \times \mathrm{PGL}_2 \overset{bir}{\sim} \mathrm{M}_0^\circ(X, 2e+1) \times \mathbb{P}^3. \tag{30.1}$$

However, for $d = 2e$ even, there can be (and usually there are) degree $2e$ rational curves on X without k-points. Thus we still get a rational map $\mathrm{Map}_{2e}(\mathbb{P}^1, X) \dashrightarrow \mathrm{M}_0^\circ(X, 2e)$ but a typical fiber is a non-trivial principal homogeneous space under PGL_2. It seems much harder to connect the birational properties of $\mathrm{Map}_{2e}(\mathbb{P}^1, X)$ and $\mathrm{M}_0^\circ(X, 2e)$.

We are mostly interested in the spaces $\mathrm{M}_0^\circ(Q^n, d)$. If $n \geq 3$ then a dense open subset of $\mathrm{M}_0^\circ(Q^n, d)$ parametrizes smooth rational curves in Q^n; see [**Kol96**, II.3.14]. For odd values of d, Theorem 1 and (30.1) give the following.

COROLLARY 31. *Let Q^n be a smooth quadric of dimension $n \geq 3$. Then*

$$\mathrm{M}_0^\circ(Q^n, 2e+1) \times \mathbb{P}^3 \overset{bir}{\sim} \mathrm{OG}(\mathbb{P}^1, Q^n) \times \mathbb{P}^{2en+3}. \qquad \square$$

It is quite likely that the $\mathbb{P}^3$ factor can be canceled. This is not hard to see for $d = 3$ but the above argument does not establish it.

We are much less successful for even degrees and for other genera; we get a complete description in 4 cases only.

THEOREM 32. *Let Q^n be a smooth quadric of dimension n. Then*

(1) $\mathrm{M}_0^\circ(Q^n, 2) \overset{bir}{\sim} \mathbb{P}^{3n-3}$ *if* $n \geq 1$,
(2) $\mathrm{M}_1^\circ(Q^n, 4) \overset{bir}{\sim} \mathbb{P}^{4n}$ *if* $n \geq 2$,
(3) $\mathrm{M}_0^\circ(Q^n, 4) \times \mathbb{P}^2 \overset{bir}{\sim} \mathrm{Sym}^2(\mathrm{OG}(\mathbb{P}^1, Q^n)) \times \mathbb{P}^{n+5}$ *if* $n \geq 3$.
(4) $\mathrm{M}_1^\circ(Q^n, 5) \times \mathbb{P}^6 \overset{bir}{\sim} \mathrm{OG}(\mathbb{P}^1, Q^n) \times \mathbb{P}^{5n+3}$ *if* $n \geq 3$.

Proof. The first two claims are clear. A general conic in Q^n is the intersection of Q^n with a 2-plane, so

$$\mathrm{M}_0^\circ(Q^n, 2) \overset{bir}{\sim} \mathrm{Grass}(\mathbb{P}^2, \mathbb{P}^{n+1}) \overset{bir}{\sim} \mathbb{P}^{3n-3}.$$

Similarly, a general degree 4 elliptic curve in Q^n is the intersection of Q^n with a 3-plane and a quadric in it, so

$$\mathrm{M}_1^\circ(Q^n,4) \overset{bir}{\sim} \mathrm{Grass}(\mathbb{P}^3,\mathbb{P}^{n+1}) \times \mathbb{P}^8 \overset{bir}{\sim} \mathbb{P}^{4n}.$$

For (3) and (4) we use degree 4 del Pezzo surfaces and start with the smallest case $n=3$. Pick an auxiliary fixed point $p_0 \in \mathbb{P}^4 \setminus Q^3$.

Let $C_4 \subset Q^3$ be a degree 4 rational normal curve. Pick a point pair $\{p_1,p_2\} \in \mathrm{Sym}^2(C_4) \cong \mathbb{P}^2$ and set $C_2 := Q^3 \cap \langle p_0,p_1,p_2\rangle$. Note that $C_4 \cup C_2$ is a degree 6 curve of arithmetic genus 1. The linear system of quadric sections of Q^3 that contain $C_4 \cup C_2$ has dimension $13-11=2$. We compute in (33.1) that the base locus of such a pencil is $C_4+C_2+\ell_1+\ell_2$ where the ℓ_i are lines. This gives a map

$$\Pi : \mathrm{M}_1^\circ(Q^3,4) \times \mathbb{P}^2 \dashrightarrow \mathrm{Sym}^2\big(\mathrm{OG}(\mathbb{P}^1,Q^3)\big).$$

What are the fibers? Given a pair of lines $\{\ell_1,\ell_2\} \in \mathrm{Sym}^2\big(\mathrm{OG}(\mathbb{P}^1,Q^3)\big)$ pick points $r_1 \in \ell_1, r_2 \in \ell_2$. Set $C_2 := Q^3 \cap \langle p_0,r_1,r_2\rangle$. Note that $B := C_2+\ell_1+\ell_2$ is a degree 4 curve of arithmetic genus 0. Using (33.2) we see that $h^0\big(Q^3,\mathcal{O}_{Q^3}(2)(-B)\big)=5$ and for any 2 sections of $\mathcal{O}_{Q^3}(2)(-B)$, the residual intersection is a C_4. Thus the fiber of Π is given by the choices of r_1,r_2 and the residual C_4. The latter correspond to an open subset of $\mathrm{Grass}(\mathbb{P}^1,\mathbb{P}^4)$ which is birational to $\mathbb{P}^6$. Given a pair of lines $\ell_1,\ell_2 \subset Q^3 \subset \mathbb{P}^4$, we can project them to a $\mathbb{P}^2$ and then the choice of the points $r_i \in \ell_i$ corresponds to the dual $\mathbb{P}^2$. This shows that

$$\mathrm{M}_0^\circ(Q^3,4) \times \mathbb{P}^2 \overset{bir}{\sim} \mathrm{Sym}^2\big(\mathrm{OG}(\mathbb{P}^1,Q^3)\big) \times \mathbb{P}^8.$$

Next we consider higher dimensional quadrics. Any $C_4 \subset Q^n$ spans a $\mathbb{P}^4$. This defines a map

$$\mathrm{M}_0^\circ(Q^n,4) \dashrightarrow \mathrm{Grass}(\mathbb{P}^4,\mathbb{P}^{n+1}),$$

whose fiber over an $L^4 \in \mathrm{Grass}(\mathbb{P}^4,\mathbb{P}^{n+1})$ is $\mathrm{M}_0^\circ(L^4 \cap Q^n,4)$. We wrote down above that this is stably birational to the universal family of pairs of lines. This family parametrizes pairs $\big(\{\ell_1,\ell_2\},L^4\big)$ where ℓ_1,ℓ_2 are lines in $L^4 \cap Q^n$. We can also parametrize this by first choosing $\{\ell_1,\ell_2\} \in \mathrm{Sym}^2\big(\mathrm{OG}(\mathbb{P}^1,Q^n)\big)$ and then an L^4 that contains their span. This corresponds to $\mathrm{Sym}^2\big(\mathrm{OG}(\mathbb{P}^1,Q^n)\big) \times \mathbb{P}^{n-3}$. Putting these together gives that

$$\mathrm{Sym}^2\big(\mathrm{OG}(\mathbb{P}^1,Q^n)\big) \times \mathbb{P}^{n-3} \times \mathbb{P}^8 \overset{bir}{\sim} \mathrm{M}_0^\circ(Q^n,4) \times \mathbb{P}^2.$$

In order to prove (4) we again start with the smallest case $n=3$. The linear system of quadric sections of Q^3 that contain E_5 has dimension $13-9=4$. By (33.2), a general pencil of such quadrics gives a residual intersection which is a degree 3 rational curve. This gives a map

$$\Pi : \mathrm{M}_1^\circ(Q^3,5) \times \mathbb{P}^6 \dashrightarrow \mathrm{M}_0^\circ(Q^3,3).$$

The fiber is given by pencils in $|\mathcal{O}_{Q^3}(2)|(-C_3)$ which is birational to $\mathrm{Grass}(\mathbb{P}^1,\mathbb{P}^7) \overset{bir}{\sim} \mathbb{P}^{12}$. This proves that

$$\Pi : \mathrm{M}_1^\circ(Q^3,5) \times \mathbb{P}^6 \overset{bir}{\sim} \mathrm{M}_0^\circ(Q^3,3) \times \mathbb{P}^{12} \overset{bir}{\sim} \mathrm{OG}(\mathbb{P}^1,Q^3) \times \mathbb{P}^{18},$$

where the second birational equivalence follows from (31). Extending this to higher dimensions works as before since a general $E_5 \subset Q^n$ spans a $\mathbb{P}^4$. This defines a map

$$\mathrm{M}_1^\circ(Q^n,5) \dashrightarrow \mathrm{Grass}(\mathbb{P}^4,\mathbb{P}^{n+1})$$

whose fiber over an $L^4 \in \mathrm{Grass}(\mathbb{P}^4,\mathbb{P}^{n+1})$ is $\mathrm{M}_1^\circ(L^4 \cap Q^n,5)$. □

33 (Curves on degree 4 del Pezzo surfaces). Let $S \subset \mathbb{P}^4$ be a degree 4 del Pezzo surface with its anticanonical embedding. The basic numerical invariants of a curve $C \subset S$ are its degree $d := \deg C = -(C \cdot K_S)$ and its arithmetic genus $p_a := p_a(C) = \big(C \cdot (C + K_S)\big)$. Since

$$h^0\big(S, \mathcal{O}_S(2)\big) = 13 \quad \text{and} \quad h^0\big(C, \mathcal{O}_S(2)|_C\big) \le 2d + 1,$$

we see that if $2d + 1 \le 12$ then we get another curve $C' \subset S$ such that $C + C' \sim \mathcal{O}_S(2)$. The adjunction formula $2p_a(C) - 2 = C(C - H)$ gives that $(C^2) = d + 2p_a - 2$. From this we get that the degree of C' is $d' = 8 - d$ and its arithmetic genus is $p'_a = 4 - d + p_a$. We used some special cases of this.

(33.1) If $d = 6$ and $p_a = 1$ then $d' = 2$ and $p'_a = -1$. Thus C' is the disjoint union of 2 lines. Conversely, if $d = 2$ and $p_a = -1$ then $|C'|$ is a 6-dimensional linear system whose general member is a smooth elliptic curve.

(33.2) If $d = 5$ and $p_a = 1$ then $d' = 3$ and $p'_a = 0$. Thus $|C'|$ is a 4-dimensional linear system whose general member is a smooth rational curve of degree 3. Conversely, if $d = 3$ and $p_a = 0$ then $|C'|$ is a 5-dimensional linear system whose general member is a smooth elliptic curve.

(33.3) If $d = 4$ and $g = 0$ then $d' = 4$ and $g' = 0$. Thus $|C'|$ is a 3-dimensional linear system whose general member is a smooth rational curve.

(33.4) If $d = 6$ and $g = 0$ then we formally get that $d' = 2$ and $g' = -2$ but usually there is no such curve C'. We need to understand this case somewhat differently.

Note that $(C^2) = 4$, thus $|C|$ maps S_4 onto a degree 4 surface in $\mathbb{P}^5$. There are 2 such surfaces over an algebraically closed field.

One is the Veronese, that is, $\mathbb{P}^2$ embedded by $|\mathcal{O}_{\mathbb{P}^2}(2)|$. This represents S_4 as $\mathbb{P}^2$ blown up at 5 points. Note that in this case twice the conic through the 5 points gives the residual curve C'. This is a non-generic situation.

The other surface is $\mathbb{P}^1 \times \mathbb{P}^1$ embedded by $|\mathcal{O}_{\mathbb{P}^1 \times \mathbb{P}^1}(1, 2)|$. This represents S_4 as $\mathbb{P}^1 \times \mathbb{P}^1$ blown up at 4 points and $|C|$ is the birational transform of $|\mathcal{O}_{\mathbb{P}^1 \times \mathbb{P}^1}(1, 2)|$.

Working with degree 4 del Pezzo surfaces as in (32) allows us to describe a few more spaces of maps to Q^3. In all these cases the curves are not linearly normal in $\mathbb{P}^4$, so these results do not yield descriptions for higher dimensional quadrics.

PROPOSITION 34. *Let Q^3 be a smooth quadric. Then*

(1) $\mathrm{M}_0^\circ(Q^3, 6) \times \mathbb{P}^3 \overset{bir}{\sim} \mathbb{P}^{21}$ *and*
(2) $\mathrm{M}_1^\circ(Q^3, 6) \overset{bir}{\sim} \mathrm{Sym}^2\big(\mathrm{OG}(\mathbb{P}^1, Q^3)\big) \times \mathbb{P}^{12}$.

Proof. The elliptic case is easier here. Let $E_6 \subset Q^3$ be a general, smooth curve of genus 1. Computing as in the proof of (32) we see that E_6 is contained in a pencil of quadric sections of Q^3 by (33.1). Intersecting any 2 of them gives a pair of lines as the residual intersection curve. This gives a map $\mathrm{M}_1^\circ(Q^4, 6) \dashrightarrow \mathrm{Sym}^2\big(\mathrm{OG}(\mathbb{P}^1, Q^3)\big)$.

In order to understand the fiber, note that a given pair of disjoint lines $\ell_1 + \ell_2$ is contained in a $\mathbb{P}^7$-of quadrics hence in a $\mathbb{P}^{12}$ of pencils of quadrics. The residual intersections give degree 6 curves of genus 1. This shows (2).

Next let $C_6 \subset Q^3$ be a general, smooth, rational curve. As we checked in (33.4), C_6 lies on a unique degree 4 del Pezzo surface S and on it $|C_6|$ is a linear system mapping onto a surface P that is geometrically isomorphic to $\mathbb{P}^1 \times \mathbb{P}^1$. Note, however, that $|C_6|$ corresponds to $|\mathcal{O}_{\mathbb{P}^1 \times \mathbb{P}^1}(1, 2)|$. Thus, over k, $P \cong \mathbb{P}^1 \times Q^2$

where Q^2 is a conic in some $\mathbb{P}^2$ and $|C_6|$ corresponds to $\pi_1^*\mathcal{O}_{\mathbb{P}^1}(1)\otimes\pi_2^*\mathcal{O}_{\mathbb{P}^2}(1)$ where $\pi_i : P \to \mathbb{P}^i$ are the coordinate projections. Let us now consider

$$|\mathcal{O}_{\mathbb{P}^1}(1)| \times |\mathcal{O}_{\mathbb{P}^1}(2)| \subset |\pi_1^*\mathcal{O}_{\mathbb{P}^1}(1)\otimes\pi_2^*\mathcal{O}_{\mathbb{P}^2}(1)|.$$

Geometrically, a general member of $|\mathcal{O}_{\mathbb{P}^1}(1)| \times |\mathcal{O}_{\mathbb{P}^1}(2)|$ consists of 3 lines. On S_4 these become 3 conics $B_1 + B_2 + B_3$ where B_2 is defined over k and B_1, B_3 are disjoint conjugates. So a general point in $\mathrm{M}_0^\circ(Q^3, 6) \times \mathbb{P}^3$ is represented by $(C_6, B_1 + B_2 + B_3)$.

In order to build this space from the other direction, we first pick B_2, then a pair of points $\{p_1, p_3\}$ on B_2 and after that a pair of conics through p_1, p_3. These choices involve $\mathrm{Grass}(\mathbb{P}^2, \mathbb{P}^4)$, $\mathrm{Sym}^2(Q^2) \cong \mathbb{P}^2$ and then (the Weil restriction of) $\mathrm{Grass}(\mathbb{P}^1, \mathbb{P}^3)$ which parametrizes pairs of 2-planes containing $\{p_1, p_2\}$. Together these give $\mathbb{P}^{16}$. Finally note that $B_1 + B_2 + B_3$ is a curve of degree 6 and genus 0. By (33.3), a generic $B_1 + B_2 + B_3$ lies on a unique degree 4 del Pezzo surface S and on it $|B_1 + B_2 + B_3|$ is a 5-dimensional linear system whose general member is a degree 6 smooth rational curve. □

One can try to use K3 surfaces in a similar way. This leads to some interesting examples but nothing very useful for our present purposes.

35 (Residual intersections on K3 surfaces). It was observed by **[Kov94]** that if a K3 surface contains a smooth rational curve then usually it also contains another one. Computing the class of the other curve gives some interesting examples.

($C_9 \subset Q^3$). $H^0(Q^3, \mathcal{O}_{Q^3}(3))$ has dimension $\binom{7}{4} - 5 = 30$. Thus a general $C_9 \subset Q^3$ is contained in a unique pencil of $|\mathcal{O}_{Q^3}(3)|$. The intersection of any two members gives a curve $C_9 + C_9'$ where C_9' is another degree 9 rational curve meeting C in 29 points. In fact, C_9' is the unique degree 9 rational curve in $\mathbb{P}^4$ meeting C in 29 points. Indeed, any such rational curve C_9^* would meet Q^3 in $29 > 2\cdot 9$ points, thus C_9^* is contained in Q^3. Next, C_9^* would meet any member of the pencil $|\mathcal{O}_{Q^3}(3)|(-C_9)$ in at least $29 > 3 \cdot 9$ points, thus C_9^* is contained in the base locus of $|\mathcal{O}_{Q^3}(3)|(-C_9)$, which is exactly $C_9 \cup C_9'$.

This gives a natural involution on $\mathrm{M}_0^\circ(Q^3, 9)$.

($C_8 \subset Q^3$). Computing as above we see that a general C_8 is contained in a $\mathbb{P}^5$ of K3 surfaces of degree 6. Following the method of **[Kov94]** gives that a general K3 surface contains another smooth rational curve that is linearly equivalent to $39H - 14C_8$. Thus it has degree 200 and meets C_8 in 262 points.

4. Degenerate maps of $\mathbb{P}^2$

Let $\Phi \in \mathrm{Map}_2(\mathbb{P}^2, Q^n)$ be given by its coordinate functions $h_0, \dots, h_{n+1}$. The h_i are in the 6-dimensional vector space V^6 of degree 2 homogeneous polynomials in 3 variables. We say that Φ is *non-degenerate* if the h_i span V^6 and *degenerate* otherwise. Thus the image of a degenerate map lies in a (possibly singular) subquadric $Q^r \subset Q^n$ of dimension ≤ 3. In this section we describe degenerate maps $\mathbb{P}^2 \dashrightarrow Q^n$, based on their image.

36 (Expected dimension of $\mathrm{Map}_2(\mathbb{P}^2, Q^n)$). A map $\Phi : \mathbb{P}^2 \dashrightarrow Q^n$ is a map $\Phi : \mathbb{P}^2 \dashrightarrow \mathbb{P}^{n+1}$ whose image lies in Q^n. Quadratic maps from $\mathbb{P}^2$ to $\mathbb{P}^{n+1}$ are given $n + 2$ quadratic forms

$$h_0(u, v, w), \dots, h_{n+1}(u, v, w) \in H^0(\mathbb{P}^2, \mathcal{O}_{\mathbb{P}^2}(2)). \tag{36.1}$$

We exclude the case when all of the h_i are identically 0 and $(h_0, \dots, h_{n+1})$ is identified with $(\lambda h_0, \dots, \lambda h_{n+1})$ for any nonzero constant λ. Thus these maps naturally form a projective space of dimension $6(n+2)-1 = 6n+11$. The image lies on Q^n iff

$$q\big(h_0(u,v,w), \dots, h_{n+1}(u,v,w)\big) \equiv 0. \tag{36.2}$$

The left hand side is a degree 4 homogeneous polynomial in u, v, w whose coefficients are degree 2 polynomials in the coefficients of the h_i. Thus (36.2) is equivalent to $\binom{6}{2} = 15$ quadratic equations in the coefficients of the h_i. Thus we have a natural realization

$$\operatorname{Map}_2(\mathbb{P}^2, Q^n) \subset \mathbb{P}^{6n+11} \tag{36.3}$$

as the common zero set of 15 quadratic equations. In particular,

$$\dim \operatorname{Map}_2(\mathbb{P}^2, Q^n) \geq 6n + 11 - 15 = 6n - 4. \tag{36.4}$$

More precisely, every irreducible component of $\operatorname{Map}_2(\mathbb{P}^2, Q^n)$ has dimension $\geq 6n - 4$. We will see that equality holds if $n \geq 3$.

37 (Maps with small dimensional image). Maps with 0-dimensional image are parametrized by Q^n.

Maps with 1-dimensional image are obtained as composites $\mathbb{P}^2 \dashrightarrow \mathbb{P}^1 \to Q^n$. The composite can be a quadratic map in two ways. If $\mathbb{P}^1 \to Q^n$ is linear then $\mathbb{P}^2 \dashrightarrow \mathbb{P}^1$ should be a pencil of conics, giving a moduli space that is birational to $\operatorname{Map}_1(\mathbb{P}^1, Q^n) \times \mathbb{P}^8$. If $\mathbb{P}^1 \to Q^n$ is quadratic then $\mathbb{P}^2 \dashrightarrow \mathbb{P}^1$ should be a pencil of lines, giving a moduli space that is birational to $\operatorname{Map}_2(\mathbb{P}^1, Q^n) \times \mathbb{P}^2$. Using Theorem 1, the spaces of maps whose image is a line (resp. conic) are further birational to

$$\begin{array}{lcl} \operatorname{Map}_L(\mathbb{P}^2, Q^n) & \overset{bir}{\sim} & \operatorname{OG}(\mathbb{P}^1, Q^n) \times \mathbb{P}^{11} \quad \text{and} \\ \operatorname{Map}_C(\mathbb{P}^2, Q^n) & \overset{bir}{\sim} & Q^n \times \mathbb{P}^{2n+2}. \end{array} \tag{37.1}$$

In particular, if $n = 1$ and there is a map $\mathbb{P}^2 \dashrightarrow Q^1$ then $Q^1(k) \neq \emptyset$, thus we can write $q \sim x^2 - yz$. The simplest solution is the inverse of the projection from $(0,1,0)$. More generally, we get solutions of the form

$$\begin{array}{lcl} h_x(u,v,w) & = & \ell_1(u,v,w)\ell_2(u,v,w), \\ h_y(u,v,w) & = & \alpha\, \ell_1^2(u,v,w), \\ h_z(u,v,w) & = & \alpha^{-1}\ell_2^2(u,v,w). \end{array} \tag{37.2}$$

38 (Quadric cone image). Assume that the image of $\Phi : \mathbb{P}^2 \to Q^n$ is a singular quadric surface. Let $p \in Q^n$ be the singular point of $\Phi(\mathbb{P}^2)$. Composing with projection from p we get a map $\pi_p \circ \Phi : \mathbb{P}^2 \dashrightarrow Q_p^{n-2}$ whose image is a conic. By (37.1) the later form a space birational to $Q_p^{n-2} \times \mathbb{P}^{2n-2}$. Choosing $p \in Q^n$ and $p' \in Q_p^{n-2}$ is the same as choosing the line $\ell = \langle p, p' \rangle \in \operatorname{OG}(\mathbb{P}^1, Q^n)$ and the point $p \in \ell$. Thus the maps $\pi_p \circ \Phi$ are parametrized by $\operatorname{OG}(\mathbb{P}^1, Q^n) \times \mathbb{P}^{2n-1}$. Lifting a map $\phi^1 : \mathbb{P}^2 \dashrightarrow Q^1 = \big(q(x_0, x_1, x_2) = 0\big) \subset \mathbb{P}^2$ to a map to the corresponding quadric cone $\phi^{2,0} : \mathbb{P}^2 \dashrightarrow Q^{2,0} = \big(q(x_0, x_1, x_2) = 0\big) \subset \mathbb{P}^3$ is equivalent to choosing an arbitrary 3rd component $\phi_3^{2,0}(u,v,w)$. Thus the space of maps with quadric cone image is birational to

$$\operatorname{Map}_{QC}(\mathbb{P}^2, Q^n) \overset{bir}{\sim} \operatorname{OG}(\mathbb{P}^1, Q^n) \times \mathbb{P}^{2n+5}. \tag{38.1}$$

39 (Quadruple planes). If the image of $\Phi(\mathbb{P}^2)$ is a linear subspace $L^2 \subset Q^n$, then in suitable coordinates Φ is given as $(h_0, h_1, h_2, 0, \dots, 0)$. These maps are parametrized by $h_0, h_1, h_2 \in H^0\big(\mathbb{P}^2, \mathcal{O}_{\mathbb{P}^2}(2)\big)$, up to a multiplicative constant. For general choices h_0, h_1, h_2 have no common zero and then Φ is an everywhere defined degree 4 morphism $\mathbb{P}^2 \to L^2$. If $n = 4$ then $\Phi_*[\mathbb{P}^2]$ equals $4[A]$ or $4[B]$ and every deformation again has the same class. If the h_i have common zeros, we get maps of lower degree and even maps whose image is a conic, a line or a point.

All 2-planes in Q^n are parametrized by $\mathrm{OG}(\mathbb{P}^2, Q^n)$. As we noted before Corollary 7, the universal family over $\mathrm{OG}(\mathbb{P}^2, Q^n)$ is birationally trivial. Thus

$$\mathrm{Map}_{QP}(\mathbb{P}^2, Q^n) \overset{bir}{\sim} \mathrm{OG}(\mathbb{P}^2, Q^n) \times \mathbb{P}^{17}. \tag{39.1}$$

If $n = 4$ then there is a 2-plane defined over k iff q is split and in this case $\mathrm{OG}(\mathbb{P}^2, Q^4) \cong \mathbb{P}^3 \amalg \mathbb{P}^3$, hence $\mathrm{Map}_2(\mathbb{P}^2, Q^4)$ is birational to the disjoint union of 2 copies of $\mathbb{P}^{20}$.

40 (Smooth quadric surface image). If Q^2 is split then it is isomorphic to $\mathbb{P}^1 \times \mathbb{P}^1$ and quadratic maps $\mathbb{P}^2 \dashrightarrow \mathbb{P}^1 \times \mathbb{P}^1$ correspond to a pair of linear projections $\mathbb{P}^2 \dashrightarrow \mathbb{P}^1$. If $Q^2 = xy - zt$ then these are given by

$$\begin{array}{rcl} h_x(u,v,w) & = & \ell_1(u,v,w)\ell_3(u,v,w), \\ h_y(u,v,w) & = & \ell_2(u,v,w)\ell_4(u,v,w), \\ h_z(u,v,w) & = & \ell_1(u,v,w)\ell_4(u,v,w), \\ h_t(u,v,w) & = & \ell_2(u,v,w)\ell_3(u,v,w). \end{array} \tag{40.1}$$

Each of the pairs (ℓ_1, ℓ_2) and (ℓ_3, ℓ_4) is determined up to a multiplicative scalar, so $\mathrm{Map}_2(\mathbb{P}^2, Q^2) \overset{bir}{\sim} \mathbb{P}^{10}$.

Geometrically, blow up the 2 points $(\ell_1 = \ell_2)$ and $(\ell_3 = \ell_4)$ and contract the line L connecting them to a point $p \in Q^2$. Thus Φ^{-1} is given by projecting Q^2 from p to $\mathbb{P}^2$ and composing it with an automorphism of $\mathbb{P}^2$. This geometric description holds over any field, even if Q^2 is not split.

Thus if the image of $\phi : \mathbb{P}^2 \dashrightarrow Q^n$ is a smooth quadric surface then ϕ can be uniquely obtained by composing the inverse of a projection $\pi_p : Q^n \dashrightarrow \mathbb{P}^n$ with a linear embedding $\mathbb{P}^2 \dashrightarrow \mathbb{P}^n$. These *unprojection maps* form a variety

$$\mathrm{Map}_{UP}(\mathbb{P}^2, Q^n) \overset{bir}{\sim} Q^n \times \mathbb{P}^{3n+2}. \tag{40.2}$$

In particular, for a quadric surface Q^2, if there is a map $\mathbb{P}^2 \dashrightarrow Q^2$ then $Q^2(k) \neq \emptyset$. This holds iff $q \sim x^2 - ay^2 - zt$. The inverse of the projection from $(0,0,1,0)$ gives the obvious solution $(uw, vw, u^2 - av^2, w^2)$. More generally, if we write $h_x(u,v,w) \pm \sqrt{a}h_y(u,v,w)$ as products of linear factors

$$h_x \pm \sqrt{a}h_y = \big(\ell_1 \pm \sqrt{a}\ell_2\big)\big(\ell_3 \pm \sqrt{a}\ell_4\big),$$

we get the general solutions

$$h_x = \ell_1\ell_3 + a\ell_2\ell_4,\ h_y = \ell_1\ell_4 + \ell_2\ell_3,\ h_z = \ell_1^2 - a\ell_2^2,\ h_t = \ell_3^2 - a\ell_4^2, \tag{40.3}$$

but some non-obvious changes of $\ell_1, \dots, \ell_4$ result in the same map.

Before we go further, we need to recall some classical facts about the Veronese surface; see [**Ber1907**, Cap.XV], [**Ber1924**, Kap.XVI] or [**SR49**, Sec.VII.3] for details.

41 (Veronese surface and its projections). Let $\Phi : \mathbb{P}^2 \dashrightarrow \mathbb{P}^{n+1}$ be a map given by degree 2 homogeneous polynomials. Since the space of degree 2 homogeneous polynomials in 3 variables has dimension 6, $\Phi(\mathbb{P}^2)$ is always contained in a 5-dimensional linear subspace of $\mathbb{P}^{n+1}$. Moreover, up to $\operatorname{Aut}(\mathbb{P}^5)$, there is a unique embedding $\Phi : \mathbb{P}^2 \hookrightarrow \mathbb{P}^5$ given by degree 2 homogeneous polynomials whose image is not contained in any hyperplane. Its image is called a *Veronese surface;* we denote it by $V \subset \mathbb{P}^5$. A typical example is

$$\Phi : (u : v : w) \mapsto (u^2 : v^2 : w^2 : uv : vw : uw).$$

For every other quadratic map $\Phi' : \mathbb{P}^2 \dashrightarrow \mathbb{P}^{n+1}$, we can think of $\Phi'(\mathbb{P}^2)$ as a projection of the Veronese surface.

If we think of the dual $(\mathbb{P}^5)^\vee$ as the space of conics in $\mathbb{P}^2$, then we see that $\operatorname{Aut}(\mathbb{P}^2)$ acts on $\mathbb{P}^5$ with 3 orbits, corresponding to the rank of the conic. The rank 1 orbit is the Veronese surface, the rank 2 orbit is its secant variety and the rank 3 orbit is open.

This makes it easy to determine the projections of V to $\mathbb{P}^4$: we just need to compute one example for each of the 3 orbits.

• If we project from a point p not on the secant variety, the projection is an isomorphism and we get a smooth degree 4 surface V_1. For example, projecting from $(1 : 1 : 1 : 0 : 0 : 0 : 0)$ gives

$$(u : v : w) \mapsto (u^2 - w^2 : v^2 - w^2 : uv : vw : uw).$$

We see that V_1 does not lie on any quadric 3-fold Q^3.

• If we project from a point on the secant variety but not on V, the projection is a degree 4 surface with a double line V_2. In suitable coordinates it is the image of the map

$$(u : v : w) \mapsto (u^2 : v^2 : w^2 : uv : vw).$$

It is contained in the family of quadrics $\lambda(x_0x_1 - x_4^2) + \mu(x_1x_2 - x_5^2)$. All of these are singular, namely at the point $(\mu : 0 : -\lambda : 0 : 0)$.

Thus V_2 is a complete intersection of any 2 quadrics that contain it.

• If we project from a point on V, the projection is a degree 3 surface V_3. In suitable coordinates it is the image of the map $\mathbb{P}^2 \dashrightarrow \mathbb{P}^4$ given by

$$(u : v : w) \mapsto (u^2 : v^2 : uv : vw : uw).$$

The map blows up the point $(0 : 0 : 1)$. The image is a degree 3 ruled surface. V_3 lies on a 3-dimensional family of quadrics best described as the 2×2 subdeterminants of the matrix

$$\begin{pmatrix} x_0 & x_2 & x_4 \\ x_2 & x_1 & x_3 \end{pmatrix}.$$

42 (Quadric 3-folds). We have seen on our list in (41) that a smooth quadric 3-fold can not contain a degree 3 or 4 projection of a Veronese surface. Thus for all non-degenerate maps the image is a quadric surface contained in Q^3.

Maps whose image is a smooth $Q^2 \subset Q^3$ were described in (40.3) and

$$\operatorname{Map}_{UP}(\mathbb{P}^2, Q^3) \overset{bir}{\sim} Q^3 \times \mathbb{P}^{11}. \tag{42.1}$$

This family has members over k whenever $Q^3(k) \neq \emptyset$ and then $\operatorname{Map}_{UP}(\mathbb{P}^2, Q^3) \overset{bir}{\sim} \mathbb{P}^{14}$.

Maps to a quadric cone $Q^{2,0} \subset Q^3$ form another family of dimension $3+11=14$. By (38.1) they form a family

$$\operatorname{Map}_{QC}(\mathbb{P}^2, Q^3) \overset{bir}{\sim} \operatorname{OG}(\mathbb{P}^1, Q^3) \times \mathbb{P}^{11}. \tag{42.2}$$

The second family has members over k whenever Q^3 contains a k-line. This happens only if

$$q(x_0, \dots, x_4) \sim y_0^2 + y_1 y_2 + y_3 y_4.$$

Both of these families have dimension 14, so we conclude that

$$\operatorname{Map}_2(\mathbb{P}^2, Q^3) \overset{bir}{\sim} (Q^3 \times \mathbb{P}^{11}) \amalg (\operatorname{OG}(\mathbb{P}^1, Q^3) \times \mathbb{P}^{11}). \tag{42.3}$$

43 (Projected Veronese surfaces). Let $V_2 \subset Q^n$ be a projected Veronese surface with double line $\ell \subset V_2$.

The linear span $\langle V_2 \rangle$ of V_2 has dimension 4 and V_2 lies in the 3-dimensional quadric $Q^n \cap \langle V_2 \rangle$. As we noted in (42), $Q^n \cap \langle V_2 \rangle$ has a unique singular point p. Thus V_2 determines a singular subquadric $Q^{3,0} \subset Q^n$ plus a line $\ell \subset Q^{3,0}$ through the vertex p of $Q^{3,0}$. We can specify these data by first picking the line ℓ, then the point $p \in \ell$ and finally a 3-dimensional linear subspace $[\ell] \in L^3 \subset \mathbb{P}(T_p Q^n)$.

Fix $Q^{3,0} := (x_1 x_2 = x_3 x_4) \subset \mathbb{P}^4$ and the line $\ell = (x_2 = x_3 = x_4 = 0)$. One easily computes that quadric sections of $Q^{3,0}$ that vanish doubly along ℓ are all of the form $x_2 L(x_0, \dots, x_4) + a_3 x_3^2 + a_4 x_4^2$, thus their linear system has dimension 6.

Putting these together we obtain that the Chow variety of projected Veronese surfaces is

$$\begin{aligned} \operatorname{Chow}_{PV}(Q^n) &\overset{bir}{\sim} \operatorname{OG}(\mathbb{P}^1, Q^n) \times \mathbb{P}^1 \times \mathbb{P}^{3(n-4)} \times \mathbb{P}^6 \\ &\overset{bir}{\sim} \operatorname{OG}(\mathbb{P}^1, Q^n) \times \mathbb{P}^{3n-5}. \end{aligned} \tag{43.1}$$

Since the quadratic maps $\mathbb{P}^2 \to V_2$ are birational, we conclude that

$$\operatorname{Map}_{PV}(\mathbb{P}^2, Q^n) \overset{bir}{\sim} \operatorname{OG}(\mathbb{P}^1, Q^n) \times \mathbb{P}^{3n+3}. \tag{43.2}$$

5. Veronese surfaces in quadric 4-folds

In the previous section we have enumerated all degenerate maps of $\mathbb{P}^2$ to a smooth quadric 4-fold. Among those, $\operatorname{Map}_{PV}(\mathbb{P}^2, Q^4)$ and $\operatorname{Map}_{QP}(\mathbb{P}^2, Q^4)$ have dimension 20 and the others have dimension < 20. Since the expected dimension of $\operatorname{Map}_2(\mathbb{P}^2, Q^4)$ is $6 \cdot 4 - 4 = 20$ by (36), once we prove that $\dim \operatorname{Map}_V(\mathbb{P}^2, Q^4) = 20$, we will know that all the geometric irreducible components of $\operatorname{Map}_2(\mathbb{P}^2, Q^4)$ are given by $\operatorname{Map}_V(\mathbb{P}^2, Q^4), \operatorname{Map}_{PV}(\mathbb{P}^2, Q^4)$ and $\operatorname{Map}_{QP}(\mathbb{P}^2, Q^4)$.

Thus it remains to describe the moduli space of Veronese surfaces in quadric 4-folds.

44 (Homology class of Veronese surfaces). The Chern class of Q^4 is $(1+H_5)^6(1+2H_5)^{-1}$ where H_5 is the hyperplane class on $\mathbb{P}^5$. Restricting to V this gives $(1+2H_2)^6(1+4H_2)^{-1}$ where H_2 is the hyperplane class on $\mathbb{P}^2$ (note that the restriction of H_5 equals $2H_2$). The Chern class of $\mathbb{P}^2$ is $(1+H_2)^3$, hence the Chern class of the normal bundle of $V \subset Q^4$ is

$$c(N_{V,Q}) = (1+2H_2)^6(1+4H_2)^{-1}(1+H_2)^{-3} = 1 + 5H_2 + 10H_2^2.$$

Thus $(V \cdot V) = 10$. Since $(V \cdot H_5^2) = 4$ this leads to the possibilities $[V] = 3[A] + [B]$ or $[V] = [A] + 3[B]$, as computed in [**Ber1907**, Sec.XV.13], [**Ber1924**, Sec.XVI.13]. This suggests, and we will see, that there are 2 distinct Veronese families.

We start with a duality between the 2 families of Veronese surfaces. Then we work with a special case when one side is a very singular specialization of Veronese surfaces to get our moduli description.

45 (Linking the two Veronese families). Let Q^4 be a smooth quadric 4-fold and $V \subset Q^4$ a Veronese surface. V is contained in a 5-dimensional family of quadrics in $\mathbb{P}^5$, thus in a 4-dimensional family of quadric sections of Q^4. Let $\langle Q_1, Q_2\rangle \subset |\mathcal{O}_{Q^4}(2)(-V)|$ be any pencil. Then $Q^4 \cap Q_1 \cap Q_2$ is a degree 8 surface containing V. Thus we get a residual intersection

$$Q^4 \cap Q_1 \cap Q_2 = V \cup V', \tag{45.1}$$

where V' is another degree 4 surface contained in Q^4. If $[V] = 3[A] + [B]$ then $[V'] = [A] + 3[B]$, so the correspondence $V \leftrightarrow V'$ interchanges the 2 families. Most likely, a given V will give all others by repeating such linkages but I have no proof of this.

Let us see some nice examples, using the representation of V in the space of symmetric matrices. Our first pair is

$$V := \begin{pmatrix} u^2 & uv & uw \\ uv & v^2 & vw \\ uw & vw & w^2 \end{pmatrix} \quad \text{and} \quad V' := \begin{pmatrix} u^2 & -uv & -uw \\ -uv & v^2 & -vw \\ -uw & -vw & w^2 \end{pmatrix}. \tag{45.2}$$

Their union is defined by the symmetric subdeterminant equations

$$x_{01}^2 - x_{00}x_{11} = x_{02}^2 - x_{00}x_{22} = x_{12}^2 - x_{11}x_{22} = 0. \tag{45.3}$$

The sum of the 3 equations gives

$$Q^4 := \left(x_{01}^2 + x_{02}^2 + x_{12}^2 - x_{00}x_{11} - x_{00}x_{22} - x_{11}x_{22} = 0\right); \tag{45.4}$$

it is an ellipsoid over $\mathbb{R}$; see (23.4–5).

We can also take the 3 non-symmetric subdeterminant equations

$$x_{01}x_{12} - x_{02}x_{11} = x_{12}x_{02} - x_{01}x_{22} = x_{01}x_{22} - x_{12}x_{02} = 0. \tag{45.5}$$

In this case V'' is the union of four 2-planes, consisting of all matrices of the form

$$\begin{pmatrix} * & 0 & 0 \\ 0 & * & 0 \\ 0 & 0 & * \end{pmatrix}, \begin{pmatrix} * & * & 0 \\ * & * & 0 \\ 0 & 0 & 0 \end{pmatrix}, \begin{pmatrix} 0 & 0 & 0 \\ 0 & * & * \\ 0 & * & * \end{pmatrix} \text{ and } \begin{pmatrix} * & 0 & * \\ 0 & 0 & 0 \\ * & 0 & * \end{pmatrix}. \tag{45.6}$$

Note that the canonical class of the intersection of 3 quadrics in $\mathbb{P}^5$ is trivial, so $V \cup V'$ is a singular K3 surface. Another way to construct these $V \cup V'$ is the following.

A degree 6 elliptic normal curve $\tau : E \hookrightarrow \mathbb{P}^5$ is contained in 4 different Veronese surfaces, an observation going back to Coble [**Cob22**]. Indeed, for a given elliptic curve E its degree 6 embeddings $\tau_6 : E \hookrightarrow \mathbb{P}^5$ correspond to degree 6 line bundles $L_6 := \tau_6^*\mathcal{O}_{\mathbb{P}^5}(1)$ and degree 3 embeddings $\tau_3 : E \hookrightarrow \mathbb{P}^2$ correspond to degree 3 line bundles $L_3 := \tau_3^*\mathcal{O}_{\mathbb{P}^2}(1)$. Combining τ_3 with the Veronese embedding gives a degree 6 elliptic normal curve such that $L_6 = L_3^{\otimes 2}$. Given L_6, there are 4 different L_3 satisfying this equation. These give the 4 different Veronese surfaces containing $\tau_6(E)$. The union of any 2 of them is a singular K3 surface.

Russo pointed out that the various degenerations of Veronese surfaces can best be seen using the dual picture to be discussed in (48).

Next we study a special case when V' is singular, even reducible, but slightly better than the ones exhibited in (45.5–6).

46 (A degenerate version). Let $p \in V \subset Q^4 \subset \mathbb{P}^5$ be a pointed Veronese surface. Instead of working with all quadrics that contain V, we consider only quadric cones with vertex p that contain V. These form a linear system $|2H_5|(-V - 2p)$ of dimension 2, where $|H_5|$ is the linear system of hyperplane sections of Q^4. We claim that if $\langle Q_1, Q_2\rangle \subset |2H_5|(-V - 2p)$ is a general pencil then

$$Q_1 \cap Q_2 = V \cup P_1 \cup P_2 \cup Q^2, \tag{46.1}$$

where the P_i are 2-planes meeting at p and Q^2 is a smooth quadric surface through p. This is easier to see by projecting everything from p. So we have $\pi : \mathbb{P}^5 \dashrightarrow \mathbb{P}^4$ and $\pi(V)$ is a degree 3 surface $F_3 \cong \mathbb{F}_1$, as discussed in (41). Thus $|2H_4|(-F_3)$ has dimension 2 and if $\langle Q_1', Q_2'\rangle \subset |2H_4|(-F_3)$ is a general pencil then

$$Q_1' \cap Q_2' = F_3 \cup L^2, \tag{46.2}$$

where L^2 is a 2-plane that meets F_3 in a conic. This L^2 is the projection of the Q^2.

The planes P_i are less visible after projection; we obtain them as follows. Write $(p \in V)$ as the image of $\phi : (p_0, \mathbb{P}^2) \hookrightarrow \mathbb{P}^5$. Note that $\phi^{-1}\big(V \cap T_pQ^4\big)$ is a degree 2 curve in $\mathbb{P}^2$ that is singular at p_0. Thus $\phi^{-1}\big(V \cap T_pQ^4\big)$ is a pair of lines in $\mathbb{P}^2$ that meet at p_0 and $V \cap T_pQ^4$ is a pair of conics C_i meeting at p. Each C_i spans a 2-plane P_i which is contained every member of $|2H_{\mathbb{P}^5}|(-V - 2p)$. Thus the P_i do not depend on $\langle Q_1, Q_2\rangle$ but Q^2 does.

Note that the π-images of the P_i are lines $L_i \subset F_3$ and L^2 meets both of them. Thus $P_1 \cup P_2 \cup Q^2$ is a surface where all 3 components meet at p, $P_i \cap Q^2$ are lines and $P_i \cap P_2 = \{p\}$. These data specify a unique isomorphism class of surfaces in $\mathbb{P}^5$, up to $\mathrm{Aut}(\mathbb{P}^5)$.

This construction can be reversed. That is, assume that we have $P_1 \cup P_2 \cup Q^2 \subset Q^4$. Let $|2H_{\mathbb{P}^5}|(-Q^2 - P_1 - P_2 - 2p)$ denote that linear system of quadric cones with vertex p that contain $P_1 \cup P_2 \cup Q^2$. We see that it has dimension 2. (This is best seen after projecting from p. Then we have the linear system of quadrics that contain a 2-plane L^2 and two lines L_i that meet it.)

If $\langle Q_1, Q_2\rangle \subset |2H_5|(-Q^2 - P_1 - P_2 - 2p)$ is a general pencil then

$$Q_1 \cap Q_2 = V \cup P_1 \cup P_2 \cup Q^2, \tag{46.3}$$

where V is a Veronese surface passing through p.

Let us next study the induced correspondence

$$\{V\} \leftarrow \{V \cup Q^2 \cup P_1 \cup P_2\} \rightarrow \{P_1 \cup P_2\}. \tag{46.4}$$

We start with $[V] \in \mathrm{Chow}_V(Q^4)$. As we noted before Corollary 7, the universal family over $\mathrm{Chow}_V(Q^4)$ is birationally trivial, thus pointed Veronese surfaces are birationally parametrized by $\mathrm{Chow}_V(Q^4) \times \mathbb{P}^2$. The choice of the pencil $\langle Q_1, Q_2\rangle \subset |2H_{\mathbb{P}^5}|(-V - 2p)$ is another $\mathbb{P}^2$-factor. Thus the left hand side of (46.4) is birational to

$$\mathrm{Chow}_V(Q^4) \leftarrow \mathrm{Chow}_V(Q^4) \times \mathbb{P}^4.$$

On the right hand side of (46.4) we start by choosing $P_1 \cup P_2$. This is the same as picking $P_1, P_2 \subset Q^4$ that are in the same geometric irreducible component of $\mathrm{OG}(\mathbb{P}^2, Q^4)$. This will be denoted by $\mathrm{Sym}^2_a\big(\mathrm{OG}(\mathbb{P}^2, Q^4)\big)$ in (69). Then we choose Q^2. As we noted above, this is equivalent to having 2 lines $L_1, L_2 \subset \mathbb{P}^4$ and choosing a 2-plane L^2 that meet both of them. After fixing an auxiliary $\mathbb{P}^2_2 \subset \mathbb{P}^4$, these are parametrized by $L_1 \times L_2 \times \mathbb{P}^2$. Note, however, that for us the L_1, L_2 are conjugate, thus instead of $L_1 \times L_2$ we pick another auxiliary $\mathbb{P}^2_0 \subset \langle L_1, L_2 \rangle \cong \mathbb{P}^3$ and represent a point pair $\{p_1, p_2 : p_i \in L_i\}$ by the intersection point $\langle p_1, p_2 \rangle \cap \mathbb{P}^2_0$. (This is just a birational construction of the Weil restriction $\Re(L_1)$.) As before, the choice of the pencil $\langle Q_1, Q_2 \rangle \subset |2H_5|(-Q^2 - P_1 - P_2 - 2p)$ is a $\operatorname{Grass}(\mathbb{P}^1, \mathbb{P}^4)$-factor. Thus the diagram (46.4) is birational to

$$(46.5)\qquad \begin{array}{ccc} \operatorname{Chow}_V(Q^4) \times \mathbb{P}^4 & \stackrel{bir}{\sim} & \operatorname{Sym}^2_a\big(\operatorname{OG}(\mathbb{P}^2, Q^4)\big) \times \mathbb{P}^{10} \\ \downarrow & & \downarrow \\ \operatorname{Chow}_V(Q^4) & & \operatorname{Sym}^2_a\big(\operatorname{OG}(\mathbb{P}^2, Q^4)\big). \end{array}$$

We finally use (69.4) to get that

$$(46.6)\qquad \operatorname{Chow}_V(Q^4) \times \mathbb{P}^4 \stackrel{bir}{\sim} (t^2 + \Delta = 0) \times Q^4 \times \mathbb{P}^{12}.$$

REMARK 47. Another family of degree 4 surfaces in Q^4 can be obtained starting with 2 disjoint planes $P_1, P_2 \subset Q^4$. A general member of this family is $\mathbb{P}^1 \times \mathbb{P}^1$ embedded by $\mathcal{O}_{\mathbb{P}^1 \times \mathbb{P}^1}(2,1)$. It is better to view such a surface as the product of a conic with $\mathbb{P}^1$, thus we have

$$C \times \mathbb{P}^1 \subset \mathbb{P}^2 \times \mathbb{P}^1 \stackrel{j}{\hookrightarrow} \mathbb{P}^5,$$

where j is the embedding given by $\mathcal{O}_{\mathbb{P}^2 \times \mathbb{P}^1}(1,1)$. The residual intersection of Q^4 with $j\big(\mathbb{P}^2 \times \mathbb{P}^1\big)$ is a pair of planes $\mathbb{P}^2 \times \{p_1, p_2\}$. Conversely, given a disjoint pair of planes $P_1, P_2 \subset Q^4$ and an isomorphism $\tau : P_1 \cong P_2$, we get an embedding $P_1 \amalg P_2 \subset j\big(\mathbb{P}^2 \times \mathbb{P}^1\big)$. The residual intersection of $j\big(\mathbb{P}^2 \times \mathbb{P}^1\big)$ with Q^4 gives a required degree 4 surface in Q^4. This shows that this irreducible component of $\operatorname{Chow}(Q^4)$ is birational to $\operatorname{Sym}^2_s\big(\operatorname{OG}(\mathbb{P}^2, Q^4)\big) \times \mathbb{P}^8$ where $\operatorname{Sym}^2_s\big(\operatorname{OG}(\mathbb{P}^2, Q^4)\big)$ is the irreducible component of $\operatorname{Sym}^2\big(\operatorname{OG}(\mathbb{P}^2, Q^4)\big)$ parametrizing disjoint pairs of 2-panes; see (69) for details.

6. Quadratic maps up to coordinate changes

Here we study $\operatorname{Map}_2(\mathbb{P}^2, Q^4)$, up to coordinate changes by $\operatorname{Aut}(\mathbb{P}^2) \times \operatorname{Aut}(Q^4)$. The 3 types of irreducible components of $\operatorname{Map}_2(\mathbb{P}^2, Q^4)$ behave differently. There is especially nice geometry behind the Veronese maps. The following classical facts can be found in [**Ber1907**, Cap.XV], [**Ber1924**, Kap.XVI], [**SR49**, p.188] or [**ESB89**].

48 (Equations of Veronese surfaces). The equations of the Veronese surface are especially clear if we think of the ambient $\mathbb{P}^5$ as the projectivization of the space of 3×3 symmetric matrices. Thus we consider $\mathbb{P}^2_{\mathbf{u}}$ with coordinates $\{u_0, u_1, u_2\}$ and $\mathbb{P}^5_{\mathbf{x}}$ with coordinates $\{x_{ij}\}$ for $0 \le i \le j \le 2$. The Veronese surface $V = V_{\mathbf{x}} \subset \mathbb{P}^5_{\mathbf{x}}$ is the image of the map $\phi : \mathbb{P}^2_{\mathbf{u}} \hookrightarrow \mathbb{P}^5_{\mathbf{x}}$ given by

$$(48.1)\qquad x_{ij} = u_i u_j \quad \text{for} \quad 0 \le i \le j \le 2.$$

In order to describe the equations defining V, consider the matrix

$$(48.2)\qquad M := \begin{pmatrix} x_{00} & x_{01} & x_{02} \\ x_{01} & x_{11} & x_{12} \\ x_{02} & x_{12} & x_{22} \end{pmatrix}.$$

V is defined by the equations

$$(48.3)\qquad \operatorname{rank} M \le 1, \quad \text{equivalently,} \quad x_{ij}x_{k\ell} = x_{i\ell}x_{kj},$$

(where we set $x_{ij} = x_{ji}$) and the secant variety of V is given by $\det M = 0$. The quadrics containing V are given by (linear combinations of) the 2×2 subdeterminants of M. These define a birational map

$$\sigma_{\mathbf{x}} : \mathbb{P}^5_{\mathbf{x}} \dashrightarrow \mathbb{P}^5_{\mathbf{y}} \quad \text{given by} \quad y_{ij} = (-1)^{i+j} M_{ji}. \tag{48.4}$$

Geometrically, we obtain $\sigma_{\mathbf{x}}$ by blowing up $V_{\mathbf{x}} \subset \mathbb{P}^5_{\mathbf{x}}$ and then contracting the (birational transform of the) secant variety of $V_{\mathbf{x}} \subset \mathbb{P}^5_{\mathbf{x}}$. Note that $\sigma_{\mathbf{x}} : \mathbb{P}^5_{\mathbf{x}} \dashrightarrow \mathbb{P}^5_{\mathbf{y}}$ and $\sigma_{\mathbf{y}} : \mathbb{P}^5_{\mathbf{y}} \dashrightarrow \mathbb{P}^5_{\mathbf{x}}$ are inverses of each other since, for an $r \times r$ matrix,

$$\operatorname{adj}\big(\operatorname{adj}(M)\big) = (\det M)^{r-2} \cdot M. \tag{48.5}$$

(For this reason, one usually views $\sigma := \sigma_{\mathbf{x}}$ as an involution of $\mathbb{P}^5$, but for us it will be convenient to distinguish the source $\mathbb{P}^5$ from the target $\mathbb{P}^5$.) Note that $\sigma_{\mathbf{x}}$ maps quadrics that contain V to hyperplanes; this correspondence is given as follows. Restricting the above factorization of $\sigma_{\mathbf{x}}$ to a quadric Q^4 we obtain the following. First we blow up $V_{\mathbf{x}} \subset Q^4$ and then contract the (birational transform of the) secant variety of $V_{\mathbf{x}}$ in Q^4; that is, the union of those secant lines of $V_{\mathbf{x}}$ that are contained in Q^4. The image of Q^4 is a hyperplane $L^4 \subset \mathbb{P}^5_{\mathbf{x}}$. For the inverse, we blow up $V_{\mathbf{y}} \cap L^4$ and then contract the (birational transform of the) secant variety of $V_{\mathbf{y}} \cap L^4$. If Q^4 is smooth then the blow-up of $V_{\mathbf{x}}$ is smooth but the blow-up of $V_{\mathbf{y}} \cap L^4$ is smooth iff $V_{\mathbf{y}} \cap L^4$ is smooth. Thus we get the following.

Claim 48.6. $\sigma_{\mathbf{x}}$ establishes a one-to-one correspondence between

(a) pairs $V_{\mathbf{x}} \subset Q^4$, up to isomorphism and
(b) pairs $V_{\mathbf{y}} \cap L^4 \subset L^4$, up to isomorphism.

Furthermore, Q^4 is smooth iff the corresponding $V_{\mathbf{y}} \cap L^4$ is smooth. □

I have not been able to track down the original author of this claim. For the geometric version all the ingredients are in [**Ber1907**, Secs.XV.9–10], [**Cob22**, Thm.17] and it is mentioned as an exercise in [**SR49**, p.188].

49 (Proof of Theorem 8). First observe that $V_{\mathbf{y}} \cap L^4$ in Claim 48.6 is a degree 4 geometrically rational curve in $\mathbb{P}^4$, thus it is obtained as the image of a plane conic C embedded by $|-2K_C|$. In particular, the pair $V_{\mathbf{y}} \cap L^4 \subset L^4$ is uniquely determined by the plane conic C, up to isomorphism. This proves Complement 9 and the uniqueness claim of Theorem 8.

It is clear that the image of the embedding (8.2) lies on the quadric (8.1). Thus it remains to show that all quadrics containing a Veronese surface can be written in the form (8.1).

Any plane conic can be given by a diagonal equation

$$C := \big(av_0^2 + bv_1^2 + cv_2^2 = 0\big) \subset \mathbb{P}^2_{\mathbf{v}}. \tag{49.1}$$

This conic is cut out of the Veronese surface $\mathbb{P}^2_{\mathbf{v}} \hookrightarrow \mathbb{P}^5_{\mathbf{y}}$, by the hyperplane

$$L^4_C := \big(ay_{00} + by_{11} + cy_{22} = 0\big) \subset \mathbb{P}^5_{\mathbf{y}}, \tag{49.2}$$

and taking its preimage by $\sigma_{\mathbf{x}}$ gives the quadric

$$Q^4_C := \big(a(x_{11}x_{22} - x_{12}^2) + b(x_{00}x_{22} - x_{02}^2) + c(x_{00}x_{11} - x_{01}^2) = 0\big) \subset \mathbb{P}^5_{\mathbf{x}}. \tag{49.3}$$

We can assume that $c = 1$ and rearrange the equation into

$$\big(x_{00} + ax_{22}\big)\big(x_{11} + bx_{22}\big) = x_{01}^2 + ax_{12}^2 + bx_{02}^2 + abx_{22}^2. \tag{49.4}$$

This proves (8.1). □

One can construct the inverse correspondence $Q^4 \mapsto C$ explicitly as follows. The Witt reduction of Q^4_C is the quadric surface

$$Q^2_C := \left(x_{01}^2 + ax_{12}^2 + bx_{02}^2 + abx_{22}^2 = 0\right), \tag{49.5}$$

and the latter is isomorphic to $C \times C$ by Paragraph 50. Thus the conic C and the quadric Q^4_C uniquely determine each other, up to isomorphism.

50 (Decomposable quadric surfaces). Let $C := (ax_0^2 + bx_1^2 + cx_2^2 = 0)$. Then

$$\begin{aligned}\psi : &(x_0 : x_1 : x_2) \times (y_0 : y_1 : y_2)\\ &\mapsto \left(x_1y_2 - x_2y_1 : -x_0y_2 + x_2y_0 : x_0y_1 - x_1y_0 : ax_0y_0 + bx_1y_1 + cx_2y_2\right)\end{aligned}$$

is an isomorphism of $C \times C$ and the quadric

$$Q := (bcz_0^2 + acz_1^2 + abz_2^2 + z_3^2 = 0).$$

Note that ψ is given by 4 sections of $\mathcal{O}_{C\times C}(2,2)$ that vanish along the diagonal.

Explanation. The first 3 components of ψ map 2 points on C to the line connecting them. This gives a morphism to (the dual) $\mathbb{P}^2$ and its branch locus is the set of tangent lines of C. Tangent lines have coefficients $(2ax_0, 2bx_1, 2cx_2)$ and these satisfy the equation $bc(2ax_0)^2 + ac(2bx_1)^2 + ab(2cx_2)^2 = 0$. This shows that $bcz_0^2 + acz_1^2 + abz_2^2$ is (a constant times) a square.

The inverse of ψ is given by

$$\begin{aligned}(x_0 : x_1 : x_2) &= \left(z_1z_3 - bz_0z_2 : -z_0z_3 - az_1z_2 : bz_0^2 + az_1^2\right) \quad \text{and}\\ (y_0 : y_1 : y_2) &= \left(z_1z_3 + bz_0z_2 : +z_0z_3 - az_1z_2 : bz_0^2 + az_1^2\right).\end{aligned}$$

Next we prove the other claims about maps up to coordinate changes.

51 (Proof of Proposition 12). If $Q^4 \subset \mathbb{P}^5$ contains a plane, we may assume that it is $(x_3 = x_4 = x_5 = 0)$. Then $\mathbb{P}^2 \to Q^4$ is of the form

$$(u : v : w) \mapsto \left(q_0(u,v,w), q_1(u,v,w), q_2(u,v,w), 0, 0, 0\right).$$

where the q_i are homogeneous of degree 2. The q_i span a base point free net of conics and, if $\operatorname{char} k \neq 3$ then any such net can be written as

$$\left\langle \tfrac{\partial C(u,v,w)}{\partial u}, \tfrac{\partial C(u,v,w)}{\partial v}, \tfrac{\partial C(u,v,w)}{\partial w} \right\rangle,$$

where $C(u,v,w)$ is a homogeneous cubic; see [**SR49**, p.42].

Thus the equation of Q^4 can be written as $x_0\ell_3 + x_1\ell_4 + x_2\ell_5 = 0$ where the ℓ_i are linear. Then Q is smooth iff the 6 linear terms in the equation are linearly independent, hence we can change x_3, x_4, x_5 to $x_i := \ell_i$ for $i = 3, 4, 5$.

This gives an isomorphism between the moduli space of quadruple planes (up to $\operatorname{Aut}(\mathbb{P}^2) \times \operatorname{Aut}(Q^4)$) and the moduli space of plane cubics (up to $\operatorname{Aut}(\mathbb{P}^2)$). This completes the proof of Proposition 12.

Note that we have in fact proved the much more precise statement: there is an isomorphism of stacks

$$\operatorname{Aut}(\mathbb{P}^2)\backslash \operatorname{Map}^\circ_{QP}(\mathbb{P}^2, Q^4)/\operatorname{Aut}(Q^4) \cong \operatorname{Div}^\circ_3(\mathbb{P}^2)/\operatorname{Aut}(\mathbb{P}^2)$$

where $\operatorname{Div}^\circ_3(\mathbb{P}^2)$ parametrizes smooth, degree 3 plane curves. Equivalently, for any field k we have an equality of sets

$$\mathrm{PGL}_3(k)\backslash \operatorname{Map}^\circ_{QP}(\mathbb{P}^2, Q^4)(k)/\mathrm{O}(Q^4, k) \cong \operatorname{Div}^\circ_3(\mathbb{P}^2)(k)/\mathrm{PGL}_3(k).$$

52 (Proof of Proposition 11). Let $S \subset Q^4$ be a projected Veronese surface with singular line $L \subset S$. Then S is contained in a unique $Q^{3,0} \subset Q^4$ and projecting S from the vertex of $Q^{3,0}$ gives a birational map $\pi : S \to Q^2$. The image of the singular line in S is a point $p \in Q^2$.

Conversely, given $p \in Q^2$ we first blow up p to get B_pQ^2 with exceptional curve E and then obtain $\tau : B_pQ^2 \to S$ that is an isomorphism $B_pQ^2 \setminus E \cong S \setminus L$ outside E but whose restriction $\tau_E : E \to L$ has degree 2. We can identify $\tau_E : E \to L$ with a degree 2 homogeneous polynomial on T_pQ^2, up to scalars.

Note that the quadric Q^2 itself gives a quadratic form on T_pQ^2; in suitable coordinates it is $x^2 + \Delta y^2$ where Δ is the discriminant of Q^4. Thus, up to isomorphisms, S is determined by $\operatorname{Sym}^2(\mathbb{P}^1)/\mathrm{O}(x^2 + \Delta y^2)$. □

53 (Proof of Proposition 18). Assume that Q^5 contains a Veronese surface V. The linear span of V intersects Q^5 in a 4-dimensional subquadric Q^4_V. As we noted in Paragraph 41, a Veronese surface is not contained in any 4-dimensional quadric with an isolated singularity, thus Q^4_V is smooth.

Fix now an equation $q(x_0, \dots, x_6) = 0$ for Q^5. Using Proposition 11, after a coordinate change we can write q as

$$A\bigl(x_0x_5 + x_1^2 + ax_2^2 + bx_3^2 + abx_4^2\bigr) + Cx_6^2$$

where $x_6 = 0$ defines Q^4_V. Note that the discriminant of the form in parenthesis is -1, thus $-C/\Delta(q)$ is a square. We can thus choose $C = -\Delta(q)$, independent of the original choice of $V \subset Q^5$. By Witt's cancellation theorem (21) this shows that Q^4_V is uniquely determined by Q^5, up to isomorphism. Thus Proposition 11 shows that the pair $(V \subset Q^5)$ is unique up to isomorphism. □

7. Quadrics of dimension ≥ 5

In Section 4 we proved that every irreducible component of $\operatorname{Map}_2(\mathbb{P}^2, Q^n)$ has dimension $\geq 6n-4$ and we computed the spaces that parametrize degenerate maps. In all cases they have dimension $< 6n - 4$ provided $n \geq 5$. Thus $\operatorname{Map}^\circ_V(\mathbb{P}^2, Q^n)$ is dense in $\operatorname{Map}_2(\mathbb{P}^2, Q^n)$ for every $n \geq 5$. In particular, projected Veronese surfaces and quadruple planes can be deformed to Veronese surfaces in a 5-dimensional quadric. We give some concrete examples of this in (55).

For now we focus on describing the birational type of $\operatorname{Map}^\circ_V(\mathbb{P}^2, Q^n)$.

54 (Proof of Theorem 13). We closely follow the arguments in (46). As in (46.4), the key point is to study the correspondence

$$\{V\} \leftarrow \{V \cup Q^2 \cup P_1 \cup P_2 \subset Q^4\} \to \{P_1 \cup P_2\}. \tag{54.1}$$

Here V is a Veronese surface in Q^n, $Q^4 \subset Q^n$ is a smooth subquadric that contains V and $V \cup Q^2 \cup P_1 \cup P_2$ is a complete intersection of Q^4 with 2 other quadrics and $V \cup Q^2 \cup P_1 \cup P_2 \subset Q^4$ is the same configuration as in (46).

Since a Veronese surface spans a $\mathbb{P}^5$, it uniquely determines the 4-dimensional subquadric $Q^4 \subset Q^n$ that contains it. Thus, as before, the left hand side of (54.1) is birational to

$$\operatorname{Chow}_V(Q^n) \leftarrow \operatorname{Chow}_V(Q^n) \times \mathbb{P}^4. \tag{54.2}$$

On the right hand side of (54.1) we start with the family of pairs of intersecting planes $\{P_1 \cup P_2\}$. Let us denote this by $\operatorname{Sym}^2_i\bigl(\mathrm{OG}(\mathbb{P}^2, Q^n)\bigr)$. (This is a locally closed subset of $\operatorname{Sym}^2\bigl(\mathrm{OG}(\mathbb{P}^2, Q^4)\bigr)$, but, unlike in the $n = 4$ case, it is not an irreducible component. However, it is a dense subset of an irreducible component of $\mathrm{Hilb}_{t^2+3t+1}(Q^n)$, parametrizing subschemes with Hilbert polynomial t^2+3t+1.)

Sending a pair $P_1 \cup P_2$ to the intersection point gives a map

$$\Pi : \mathrm{Sym}^2_i\bigl(\mathrm{OG}(\mathbb{P}^2, Q^n)\bigr) \to Q^n \tag{54.3}$$

whose fiber over $p \in Q^n$ is a dense open subset of $\mathrm{Sym}^2\bigl(\mathrm{OG}(\mathbb{P}^1, Q^{n-2}_p)\bigr)$. In general the fibration given by Π is not birationally trivial, but it is so if $Q^n(k) \neq \emptyset$; this follows from (20.3). We obtain that if $Q^n(k) \neq \emptyset$ then

$$\mathrm{Sym}^2_i\bigl(\mathrm{OG}(\mathbb{P}^2, Q^n)\bigr) \overset{bir}{\sim} Q^n \times \mathrm{Sym}^2\bigl(\mathrm{OG}(\mathbb{P}^1, Q^{n-2}_W)\bigr). \tag{54.4}$$

The rest is very much like before, except that the linear span of $P_1 \cup P_2$ is 4-dimensional, thus the choice of $Q^4 \supset P_1 \cup P_2$ is equivalent to choosing a 5-dimensional linear subspace containing $P_1 \cup P_2$. So we get an extra factor of $\mathbb{P}^{n-4}$. At the end we obtain that the diagram (54.1) is birational to

$$\begin{array}{ccc} \mathrm{Chow}_V(Q^n) \times \mathbb{P}^4 & \overset{bir}{\sim} & \mathrm{Sym}^2_i\bigl(\mathrm{OG}(\mathbb{P}^2, Q^n)\bigr) \times \mathbb{P}^{n-4} \times \mathbb{P}^{10} \\ \downarrow & & \downarrow \\ \mathrm{Chow}_V(Q^n) & & \mathrm{Sym}^2_i\bigl(\mathrm{OG}(\mathbb{P}^2, Q^n)\bigr). \end{array} \tag{54.5}$$

If $Q^n(k) \neq \emptyset$ then (54.4) further gives that

$$\mathrm{Chow}_V(Q^n) \times \mathbb{P}^4 \overset{bir}{\sim} \mathrm{Sym}^2\bigl(\mathrm{OG}(\mathbb{P}^1, Q^{n-2}_W)\bigr) \times \mathbb{P}^{2n+6}. \tag{54.6}$$

Since $\mathrm{Map}^\circ_V(\mathbb{P}^2, Q^n) \overset{bir}{\sim} \mathrm{Chow}_V(Q^4) \times \mathbb{P}^8$, we see that (54.6) implies (13.2). As we already noted, (68.2) shows that (13.1) and (13.2) are equivalent. □

EXAMPLE 55. We give examples of projected Veronese surfaces and quadruple planes deforming to Veronese surfaces in a 5-dimensional quadric.

First consider the quadric $Q^5 := \bigl(x_0x_1 + x_2x_3 = x_4^2 + ax_5^2 + x_6^2\bigr)$. Pick any degree 2 polynomial $q_1(u,v)$ that is relatively prime to $u^2 + av^2$. Pick any degree 2 polynomial $q_6(u,v)$ and write $q_6^2 = q_0q + q_3(u^2 + av^2)$ where q_0, q_3 have degree 2. Consider the family of maps

$$\phi_t : (u{:}v{:}w) \mapsto \bigl(t^2q_0 : q_1 : u^2 + av^2 : w^2 + t^2q_3 : uw : vw : tq_6\bigr)$$

Note that ϕ_0 is a projected Veronese surface but the image of ϕ_t is a Veronese surface for $t \neq 0$.

Second, for the quadric $Q^5 := (x_0x_3 + x_1x_4 + x_2x_5 = x_6^2)$ consider the family of maps $\phi_t : \mathbb{P}^2 \to Q^5$ given by

$$\phi_t : (u{:}v{:}w) \mapsto \bigl(u^2 : v^2 : w^2 : (tv+tw)^2 : (tu+tw)^2 : (tu+tv)^2 : tuv + tuw + tvw\bigr).$$

Note that ϕ_0 is a quadruple plane but the image of ϕ_t is a Veronese surface for $t \neq 0$.

8. Orthogonal Grassmannians

In the next two sections we discuss various facts about orthogonal Grassmannians that we used earlier. For the current applications we mainly need symmetric squares of $\mathrm{OG}(\mathbb{P}^r, Q^n)$ for $r = 1, 2$. We focus on these cases but give more general statements when possible.

DEFINITION 56 (Orthogonal Grassmannians). The scheme parametrizing of all m-dimensional linear spaces contained in a smooth quadric Q^n is called the *orthogonal Grassmannian,* denoted by $\mathrm{OG}(\mathbb{P}^m, Q^n)$.

Over an algebraically closed field the orthogonal Grassmannian is usually denoted by $\mathrm{OG}(m+1, n+2)$; the notation suggests $(m+1)$-dimensional q-isotropic sub-vector spaces of k^{n+2} where q is a non-degenerate quadratic form.

The orthogonal Grassmannian $\mathrm{OG}(m+1, n+2)$ is a smooth projective variety that is homogeneous under the orthogonal group $\mathrm{O}(q)$.

The dimension of orthogonal Grassmannians can be computed several ways. For example, an easy application of (57.2) shows that the universal $\mathbb{P}^m$-bundle over $\mathrm{OG}(m+1, n+2)$ is an $\mathrm{OG}(m, n)$-bundle over Q^n. Thus $\dim \mathrm{OG}(m+1, n+2) = (n-m) + \dim \mathrm{OG}(m, n)$. Repeating this until we reach $\mathrm{OG}(1, n-2m+2) \cong Q^{n-2m}$ we get that

$$\dim \mathrm{OG}(m+1, n+2) = \tfrac{1}{2}(2n-3m)(m+1). \tag{56.1}$$

This also shows that $\mathrm{OG}(m+1, n+2)$ is nonempty iff $2m \leq n$, irreducible for $2m < n$ and has 2 irreducible components for $2m = n$; see (62). (Over an algebraically closed field some authors use $\mathrm{OG}(m+1, 2m+2)$ to denote any one of these components.)

Thus $\mathrm{OG}(\mathbb{P}^m, Q^n)$ is a smooth projective k-variety that is homogeneous under the group $\mathrm{PGO}(Q^n) = \mathrm{Aut}(Q^n)$.

Intersection theory of orthogonal Grassmannians is studied in [**Cos11**, **Cos14**] and their quantum cohomology in [**KT04**, **BKT15**]. Here we aim to study the birational properties of orthogonal Grassmannians and their symmetric powers. Let us start with the existence of k-points.

57 (Basic correspondences). A basic tool comparing different orthogonal Grassmannians is the variety parametrizing pairs of linear spaces $L^r \subset L^m$ contained in Q^n. Let us denote it by $\mathrm{OF}(\mathbb{P}^r, \mathbb{P}^m, Q^n)$ (for orthogonal flags). Forgetting one of the linear spaces gives two morphisms

$$\mathrm{OG}(\mathbb{P}^r, Q^n) \stackrel{\pi_r}{\longleftarrow} \mathrm{OF}(\mathbb{P}^r, \mathbb{P}^m, Q^n) \stackrel{\pi_m}{\longrightarrow} \mathrm{OG}(\mathbb{P}^m, Q^n). \tag{57.1}$$

The fiber of π_m over $L^m \subset Q^n$ is the (ordinary) Grassmannian $\mathrm{Grass}(\mathbb{P}^r, L^m)$.

We claim that the fiber of π_r over $L := L^r \subset Q^n$ is an orthogonal Grassmannian $\mathrm{OG}(\mathbb{P}^{m-r-1}, Q_L^{n-2r-2})$ that can be constructed as follows. First choose coordinates such that

$$q(x_0, \dots, x_{n+1}) \sim y_0 y_1 + \cdots + y_{2r} y_{2r+1} + q_{n-2r}(y_{2r+2}, \dots, y_{n+1})$$

and $L^r = (y_1 = y_3 = \cdots = y_{2r+1} = y_{2r+2} = y_{2r+3} = \cdots = y_{n+1} = 0)$. Then we can identify an L^m satisfying $L^r \subset L^m \subset Q^n$ with a linear space

$$L^{m-r-1} \subset Q_L^{n-2r-2} := \bigl(q_{n-2r}(y_{2r+2}, \dots, y_{n+1}) = 0\bigr).$$

Geometrically one can obtain Q_L^{n-2r-2} as follows. First take

$$Q_L^{n-r-1,r} := \cap_{p \in L} T_p Q^n.$$

It is a quadric of dimension $n-r-1$ that is singular along L. Then $Q_L^{n-r-1,r}$ is the cone over Q_L^{n-2r-2} with vertex L.

Two cases are especially useful. If $r = 0$ then we get

$$Q^n \stackrel{\pi_0}{\longleftarrow} \mathrm{OF}(\mathbb{P}^0, \mathbb{P}^m, Q^n) \stackrel{\pi_m}{\longrightarrow} \mathrm{OG}(\mathbb{P}^m, Q^n), \tag{57.2}$$

where the fiber of π_m over $L^m \subset Q^n$ is L^m while the fiber of π_0 over $p \in Q^n$ is the orthogonal Grassmannian $\mathrm{OG}(\mathbb{P}^{m-1}, Q_p^{n-2})$.

The other interesting case is $r = m-1$. Then we have

$$\text{(57.3)} \qquad \mathrm{OG}(\mathbb{P}^{m-1}, Q^n) \stackrel{\pi_{m-1}}{\longleftarrow} \mathrm{OF}(\mathbb{P}^{m-1}, \mathbb{P}^m, Q^n) \stackrel{\pi_m}{\longrightarrow} \mathrm{OG}(\mathbb{P}^m, Q^n).$$

Here π_m is a $\mathbb{P}^m$-bundle (the dual of the natural $\mathbb{P}^m$-bundle) while the fiber of π_{m-1} over $L^{m-1} \subset Q^n$ is the quadric Q_L^{n-2m}).

Claim 57.4. Using the notation of (57.1),

(1) π_m is a Zariski locally trivial $\mathbb{P}^m$-bundle and
(2) π_r is a Zariski locally trivial $\mathrm{OG}(\mathbb{P}^{m-r-1}, Q_M^{n-2r})$-bundle if there is a linear subspace $M^r \subset Q^n$ defined over k.

Proof. If $r = 0$ then π_m is a $\mathbb{P}^m$-bundle and the pull-back of $\mathcal{O}_{\mathbb{P}^{n+1}}(1)$ gives a relative $\mathcal{O}(1)$. Thus the $\mathbb{P}^m$-bundle is Zariski locally trivial and so are its Grassmann bundles, proving the first part.

By contrast, π_r is usually not Zariski locally trivial. However, assume that we have $M := M^r \subset Q^n$ defined over k. As in (20.2), for any $L := L^r \subset Q^n \setminus Q_M^{n-r-1,r}$ we get natural isomorphisms (defined over $k(L)$)

$$Q_L^{n-2r-2} \cong Q_L^{n-r-1,r} \cap Q_M^{n-r-1,r} \cong Q_M^{n-2r-2}.$$

These trivialize the Q_L^{n-2r-2}-bundle over the open set parametrizing those L that are disjoint from $Q_M^{n-r-1,r}$. Thus the $\mathrm{OG}(\mathbb{P}^{m-r-1}, Q_L^{n-2r-2})$-bundle is also trivialized. □

PROPOSITION 58. *Let Q^n be a smooth quadric such that $Q^n(k) \neq \emptyset$. Then*

$$\mathrm{OG}(\mathbb{P}^m, Q^n) \stackrel{bir}{\sim} \mathrm{OG}(\mathbb{P}^{m-1}, Q_W^{n-2}) \times \mathbb{P}^{n-m}.$$

Proof. Pick a k point $p \in Q^n$ and consider the diagram (57.2)

$$Q^n \stackrel{\pi_0}{\longleftarrow} \mathrm{OF}(\mathbb{P}^0, \mathbb{P}^m, Q^n) \stackrel{\pi_m}{\longrightarrow} \mathrm{OG}(\mathbb{P}^m, Q^n).$$

Let $Q^{n-m} \subset Q^n$ be the intersection of Q^n with a linear subspace of codimension m containing p such that Q^{n-m} is smooth at p. Since the fiber of π_m over $L^m \subset Q^n$ is L^m, we see that the restriction

$$\pi_m : \pi_0^{-1}(Q^{n-m}) \to \mathrm{OG}(\mathbb{P}^m, Q^n)$$

is birational. By (57.4), π_0 is Zariski locally trivial with fiber $\mathrm{OG}(\mathbb{P}^{m-1}, Q_W^{n-2})$. Projecting from p shows that $Q^{n-m} \stackrel{bir}{\sim} \mathbb{P}^{n-m}$. Thus

$$\mathrm{OG}(\mathbb{P}^m, Q^n) \stackrel{bir}{\sim} \mathrm{OG}(\mathbb{P}^{m-1}, Q_W^{n-2}) \times Q^{n-m} \stackrel{bir}{\sim} \mathrm{OG}(\mathbb{P}^{m-1}, Q_W^{n-2}) \times \mathbb{P}^{n-m}. \quad \square$$

COROLLARY 59. *Let Q^n be a quadric with Witt index r and anisotropic reduction Q_{an}^{n-2r}.*

(1) *If $m < r$ then $\mathrm{OG}(\mathbb{P}^m, Q^n)$ is rational.*
(2) *If $m \geq r$ then $\mathrm{OG}(\mathbb{P}^m, Q^n)(k) = \emptyset$ and*

$$\mathrm{OG}(\mathbb{P}^m, Q^n) \stackrel{bir}{\sim} \mathrm{OG}(\mathbb{P}^{m-r}, Q_{\mathrm{an}}^{n-2r}) \times \mathbb{P}^{r(2n-r+1)/2}.$$

Proof. The claims are clear if the Witt index is 0. Otherwise we apply (58) repeatedly.

If $m \geq r$ we stop when we reach the anisotropic kernel Q_{an}^{n-2r} to get (2). If $m < r$ then we stop with $\mathrm{OG}(\mathbb{P}^0, Q_W^{n-2m}) = Q_W^{n-2m}$. Since $m < r$ the quadric

Q_W^{n-2m} still has k-points, hence it is birational to $\mathbb{P}^{n-2m}$. Thus $\mathrm{OG}(\mathbb{P}^m, Q^n)$ is rational. □

We get the following rationality criteria for $\mathrm{OG}(\mathbb{P}^m, Q^n)$.

THEOREM 60. *Let $Q^n = \big(q(x_0, \dots, x_{n+1}) = 0\big)$ be a smooth quadric over a field k. The following are equivalent.*

(1) *$\mathrm{OG}(\mathbb{P}^m, Q^n)$ is rational.*
(2) *$\mathrm{OG}(\mathbb{P}^m, Q^n)$ has a k-point.*
(3) *$\mathrm{OG}(\mathbb{P}^m, Q^n)$ has a k'-point over an odd degree field extension k'/k.*
(4) *Q^n contains a linear subspace of dimension m.*
(5) *The Witt index of q is $\geq m+1$. Equivalently, we can write*

$$q(x_0, \dots, x_{n+1}) \sim y_0y_1 + \cdots + y_{2m}y_{2m+1} + q_W(y_{2m+2}, \dots, y_{n+1}).$$

Proof. The implications (1) ⇒ (2) ⇒ (3) and (2) ⇔ (4) are clear and (4) ⇔ (5) was noted in Paragraph 20. We proved in (59) that (5) ⇒ (1). Thus it remains to show that (3) ⇒ (2).

We use induction on n. If $\mathrm{OG}(\mathbb{P}^m, Q^n)$ has a k'-point then Q^n contains a linear space defined over k', hence also a k'-point. Thus there is a k-point $p \in Q^n$ by Springer's theorem; see [**Spr52**] or (67). By (58)

$$\mathrm{OG}(\mathbb{P}^m, Q^n) \overset{bir}{\sim} \mathrm{OG}(\mathbb{P}^{m-1}, Q_W^{n-2}) \times \mathbb{P}^n, \tag{60.6}$$

hence $\mathrm{OG}(\mathbb{P}^{m-1}, Q_W^{n-2})$ also has a k'-point. By induction $\mathrm{OG}(\mathbb{P}^{m-1}, Q_W^{n-2})$ has a k-point, hence, using (60.6) from right to left shows that $\mathrm{OG}(\mathbb{P}^m, Q^n)$ has a k-point. □

Three of the orthogonal Grassmannians are closely related to Severi-Brauer varieties. The obvious one is $\mathrm{OG}(\mathbb{P}^1, Q^2)$, the more interesting ones are $\mathrm{OG}(\mathbb{P}^1, Q^3)$ and $\mathrm{OG}(\mathbb{P}^2, Q^4)$. For these the symmetric powers can be determined using [**Kol16b**]. (See [**Kol16a**] for a geometric introduction to Severi-Brauer varieties.)

LEMMA 61. *Let $Q^3 = \big(q(x_0, \dots, x_4) = 0\big)$ be a smooth quadric 3-fold over k. Then*

(1) *$\mathrm{OG}(\mathbb{P}^1, Q^3)$ is a 3-dimensional Severi–Brauer variety.*
(2) *$\mathrm{OG}(\mathbb{P}^1, Q^3)$ has period 1 or 2.*
(3) *$\mathrm{OG}(\mathbb{P}^1, Q^3) \cong \mathbb{P}^3 \Leftrightarrow Q^3$ contains a line $\Leftrightarrow q \sim y_0y_1 + y_2y_3 + a_4y_4^2$.*
(4) *$\mathrm{OG}(\mathbb{P}^1, Q^3)$ has index 2 iff $q \sim q_4(y_0, \dots, y_3) + a_4y_4^2$ such that $\mathrm{discr}(q_4)$ is a square. If these hold then $\mathrm{Sym}^2\big(\mathrm{OG}(\mathbb{P}^1, Q^3)\big) \overset{bir}{\sim} \mathbb{P}^6$.*
(5) *$\mathrm{OG}(\mathbb{P}^1, Q^3)$ has index 4 iff $\mathrm{Sym}^2\big(\mathrm{OG}(\mathbb{P}^1, Q^3)\big)$ has no k-points.*
(6) *Let $a_0x_0^2 + \cdots + a_4x_4^2 = 0$ be a diagonal equation of Q^3. Then*

$$\big[\mathrm{OG}(\mathbb{P}^1, Q^3)\big] = \big[(a_0x_0^2 + a_1x_1^2 + a_2x_2^2 = 0)\big] \times \big[(a_0a_1a_2y^2 = a_3x_3^2 + a_4x_4^2)\big],$$

where $\big[*\big]$ denotes the class of an object in the Brauer group.

Proof. Assume first that k is algebraically closed and let $\ell \in Q^3$ be any line. Then $\{\ell' : \ell \cap \ell' \neq \emptyset\}$ defines a divisor $H_\ell \subset \mathrm{OG}(2,5)$. Given three general lines ℓ_1, ℓ_2, ℓ_3, there is a unique line ℓ' that meets all 3. Indeed, the first 2 lines span an $L^3 \subset \mathbb{P}^4$ and $\ell_3 \cap L^3$ is a single point p. Thus the only line ℓ' in $\mathbb{P}^4$ that meets the ℓ_i is the one passing through p and meeting both ℓ_1, ℓ_2. This ℓ' meets Q^3 in 3 points, so is contained in it. This shows that $(H^3) = 1$, so $|H|$ maps $\mathrm{OG}(2,5)$ isomorphically onto $\mathbb{P}^3$, proving (1).

An m-dimensional Severi-Brauer variety is isomorphic to $\mathbb{P}^m$ iff it has a k-point. Thus $\operatorname{OG}(\mathbb{P}^1, Q^3) \cong \mathbb{P}^3$ iff Q^3 contains a line and the rest of (3) is clear.

If we are over a field k, then there may not be lines defined over k but there are conics $C \subset Q^3$ defined over k. Thus $\{\ell' : C \cap \ell' \neq \emptyset\}$ defines a divisor $D_C \subset \operatorname{OG}(\mathbb{P}^1, Q^3)$ that is defined over k. Over $\bar{k}$ we have $D_C \sim 2H$ and $\ell' \mapsto C \cap \ell'$ gives a morphism $D_C \to C$ defined over k. Thus in fact D is a product of 2 conics $D_C \cong C \times C'$. This shows that $\operatorname{OG}(\mathbb{P}^1, Q^3)$ is a Severi-Brauer variety of period 2, proving (2). (By our construction, it contains a quadric that is a product of 2 conics; actually this always holds, by a theorem of Albert, if the period is 2; see [**Kol05**, Thm.18] for a geometric proof.)

By definition, $\operatorname{OG}(\mathbb{P}^1, Q^3)$ has index 2 iff it has a degree 2 point and the latter holds iff Q^3 contains pair of lines ℓ, ℓ' that are either both defined over k or are conjugates over k. These span an $L^3 \subset \mathbb{P}^4$ and $Q^2 := L^3 \cap Q^3$ is a smooth quadric that contains a conjugate pair of lines. By (62) this holds iff its discriminant is a square.

Next we claim that if P is a Severi-Brauer variety that has a point over a degree 2 field extension k'/k then $\operatorname{Sym}^2 P$ is rational. Indeed, let $\pi : U \to G$ denote the universal bundle over the Grassmannian of lines. Then $\operatorname{Sym}^2 P$ is birational to $\mathbb{P}_G(\pi_* \mathcal{O}_U(-K_U))$, hence to $\mathbb{P}^2 \times G$. Let L be a hyperplane in P defined over k'. Then G is birational to the Weil restriction $\Re_{k'/k} L$, hence rational and so is $\operatorname{Sym}^2 P$. This proves (4) and (5) just lists the remaining cases.

For (6) we need to follow the construction of D_C above. So assume that $Q^3 = (a_0 x_0^2 + \cdots + a_4 x_4^2 = 0)$. We can use $C = (x_3 = x_4 = 0)$ as our first conic; it is isomorphic to $(a_0 x_0^2 + a_1 x_1^2 + a_2 x_2^2 = 0) \subset \mathbb{P}^2$. To get our second conic C', we project Q^3 from C to get $\pi : Q^3 \dashrightarrow \mathbb{P}^1_{x_3, x_4}$. The fiber over a point $(s{:}t)$ is

$$\begin{aligned} Q^2(s,t) &= \big(a_0 x_0^2 + \cdots + a_4 x_4^2 = t x_3 - s x_4 = 0\big) \\ &\cong \big(a_0 x_0^2 + a_1 x_1^2 + a_2 x_2^2 + \big(a_3 + (t^2/s^2) a_4\big) x_3^2 = 0\big). \end{aligned}$$

Furthermore, $C' \to \mathbb{P}^1_{x_3, x_4}$ is the double cover whose fiber over $(s{:}t)$ is the 2 families of lines on $Q^2(s,t)$. Thus, by (62), the equation of C' is $w^2 = \operatorname{discr}\big(Q^2(s,t)\big)$. We can divide by $a_0 a_1 a_2$ and set $y := w/(a_0 a_1 a_2)$ to get the form $C' \cong (a_0 a_1 a_2 y^2 = a_3 x_3^2 + a_4 x_4^2)$. □

As we noted, the middle dimensional orthogonal Grassmannians $\operatorname{OG}(\mathbb{P}^r, Q^{2r})$ are geometrically reducible. The discriminant tells us when they are reducible.

LEMMA 62. *Let Q^{2r} be an even dimensional smooth quadric over a field k. Then the scheme parametrizing the 2 geometric components of* $\operatorname{OG}(\mathbb{P}^r, Q^{2r})$ *is* $\big(t^2 + (-1)^r \Delta = 0\big)$, *where Δ is the discriminant of Q^{2r}.*

Thus $\operatorname{OG}(\mathbb{P}^r, Q^{2r})$ *is reducible iff $(-1)^{r+1} \Delta(Q^{2r})$ is a square.*

Proof. Set $m := r + 1$ and consider the universal quadratic form

$$q^{2m}_{\text{univ}}(x_1, \ldots, x_{2m}) := \sum_{1 \le i \le j \le 2m} t_{ij} x_i x_j$$

over the affine space of dimension $m(2m+1)$ over $\operatorname{Spec} \mathbb{Z}[\frac{1}{2}]$. Let Δ be the discriminant and $U := \mathbb{A}^{m(2m+1)} \setminus (\Delta = 0)$ the complement of its zero set. The Stein factorization of the universal orthogonal Grassmannian of m-dimensional isotropic subspaces gives an étale double cover of U. Since $\mathbb{A}^{m(2m+1)}$ is simply connected, this is given by an equation

$$z^2 = (-1)^a \cdot 2^b \cdot \Delta^c$$

for some integers a, b, c which can all be chosen 0 or 1. To pin down these choices, we can check a particular case, for instance

$$q_t := x_1^2 - x_2^2 + \cdots + x_{2m-3}^2 - x_{2m-2}^2 + x_{2m-1}^2 - tx_{2m}^2.$$

We have $\Delta(q_t) = (-1)^m t$ and the 2 families of m-planes correspond to the members

$$(x_1 - x_2 = \cdots = x_{2n-3} - x_{2n-2} = x_{2m-1} \pm \sqrt{t} x_{2m} = 0.$$

Thus the correct equation is $z^2 = (-1)^m \Delta$. □

PROPOSITION 63. *Let Q^{2r} be an even dimensional smooth quadric over a field k and $Q^{2r-1} \subset Q^{2r}$ a smooth quadric hypersurface. Them*

$$\mathrm{OG}(\mathbb{P}^r, Q^{2r}) \cong \mathrm{OG}(\mathbb{P}^{r-1}, Q^{2r-1}) \times \bigl(t^2 + (-1)^r \Delta = 0\bigr).$$

Proof. Intersecting $L^r \subset Q^{2r}$ with Q^{2r-1} gives a morphism $\pi_1 : \mathrm{OG}(\mathbb{P}^r, Q^{2r}) \to \mathrm{OG}(\mathbb{P}^{r-1}, Q^{2r-1})$ which is everywhere defined since a smooth quadric of dimension $2r-1$ can not contain a $\mathbb{P}^r$. We can think of a fiber of π_1 as a fiber of the projection considered in (57.3)

$$\mathrm{OG}(\mathbb{P}^{r-1}, Q^{2r}) \stackrel{\pi_{r-1}}{\longleftarrow} \mathrm{OF}(\mathbb{P}^{r-1}, \mathbb{P}^r, Q^{2r}).$$

We identified this fiber as a 0-dimensional quadric, hence 2 points. Combining with (62) we get a surjection

$$\mathrm{OG}(\mathbb{P}^r, Q^{2r}) \to \mathrm{OG}(\mathbb{P}^{r-1}, Q^{2r-1}) \times \bigl(t^2 + (-1)^r \Delta = 0\bigr),$$

which is then necessarily an isomorphism. □

COROLLARY 64. $\mathrm{OG}(\mathbb{P}^2, Q^4)$ *is a Severi-Brauer scheme over* $\bigl(t^2 + \Delta = 0\bigr)$. □

9. Symmetric squares of orthogonal Grassmannians

In this section we study the rationality of $\mathrm{Sym}^2\bigl(\mathrm{OG}(\mathbb{P}^1, Q^n)\bigr)$. Note that if $n \geq 3$ then $\mathrm{Aut}(Q^n)$ acts with a dense orbit on $\mathrm{Sym}^2\bigl(\mathrm{OG}(\mathbb{P}^1, Q^n)\bigr)$. Thus if the latter has a k-point then it is unirational. (This needs a small argument about singular k-points.)

More generally, one can ask about the structure of $\mathrm{Sym}^d\bigl(\mathrm{OG}(\mathbb{P}^m, Q^n)\bigr)$. Building on the assumption that quadrics are similar to Severi-Brauer varieties, the following questions appear to be quite natural and they should give a general outline of what could hold. The analogous assertions for Severi-Brauer varieties are proved in [**Kol16b**], however, similar problems about other homogeneous spaces turned out to have a negative answer [**Flo04, Par05**]. So there is little reason to believe that the answers should be positive.

QUESTION 65. Let Q^n be a smooth quadric over a field k. Fix $m \geq 0$ and write $d = 2^a d'$ where d' is odd. Which of the following are true?

(1) $\mathrm{Sym}^d\bigl(\mathrm{OG}(\mathbb{P}^m, Q^n)\bigr)$ is rational over k if $2^{m+1} \mid d$.
(2) $\mathrm{Sym}^d\bigl(\mathrm{OG}(\mathbb{P}^m, Q^n)\bigr)$ is rational over k iff it has a k-point.
(3) The stable birational equivalence class of $\mathrm{Sym}^d\bigl(\mathrm{OG}(\mathbb{P}^m, Q^n)\bigr)$ depends only on 2^a.

The simplest case, when $a = 0$, follows from (60).

PROPOSITION 66. *Let Q^n be a smooth quadric over a field k and assume that d is odd. Then*

$$\operatorname{Sym}^d\big(\operatorname{OG}(\mathbb{P}^m, Q^n)\big) \overset{stab}{\sim} \operatorname{OG}(\mathbb{P}^m, Q^n).$$

Proof. We show the stable birational equivalences

$$\operatorname{Sym}^d\big(\operatorname{OG}(\mathbb{P}^m, Q^n)\big) \overset{stab}{\sim} \operatorname{Sym}^d\big(\operatorname{OG}(\mathbb{P}^m, Q^n)\big) \times \operatorname{OG}(\mathbb{P}^m, Q^n) \overset{stab}{\sim} \operatorname{OG}(\mathbb{P}^m, Q^n).$$

First let K be the function field of $\operatorname{OG}(\mathbb{P}^m, Q^n)$. Then $\operatorname{OG}(\mathbb{P}^m, Q^n)(K) \neq \emptyset$ hence $\operatorname{OG}(\mathbb{P}^m, Q^n)$ is rational over K by (60) and so are its symmetric powers. This shows the stable birational equivalence on the right.

Next let L be the function field of $\operatorname{Sym}^d\big(\operatorname{OG}(\mathbb{P}^m, Q^n)\big)$. Then $\operatorname{OG}(\mathbb{P}^m, Q^n)$ has a point in an odd degree field extension L'/L, hence it is rational over L by (60). Thus the stable birational equivalence on the left also holds. □

Another special case of Question 65 is symmetric powers of $\operatorname{OG}(\mathbb{P}^0, Q^n) \cong Q^n$. [**Spr52**] proves that if a quadric has a point over an odd degree field extension of k then it has a k-point. The idea of the proof gives more.

THEOREM 67. *Let Q^n be a smooth quadric of dimension $n \geq 1$. Then*

$$\operatorname{Sym}^d(Q^n) \overset{bir}{\sim} \begin{cases} \mathbb{P}^{nd} & \text{if } d \text{ is even,} \\ Q^n \times \mathbb{P}^{nd-n} & \text{if } d \text{ is odd.} \end{cases}$$

Proof. We prove that, for $d \geq 2$, $\operatorname{Sym}^d(Q^n) \overset{bir}{\sim} \operatorname{Sym}^{d-2}(Q^n) \times \mathbb{P}^{2n}$. In affine coordinates we write $Q^n = \big(q(x_1, \dots, x_n, t) = 0\big)$.

For d general points $p_1, \dots, p_d \in Q^n$ there are unique polynomials $g_1(t), \dots, g_n(t)$ of degree $\leq d-1$ whose graph $\Gamma := \Gamma(p_1, \dots, p_d)$ passes through $p_1, \dots, p_d$. We define $\operatorname{Sym}^d(Q^n) \dashrightarrow \operatorname{Sym}^{d-2}(Q^n) \times \mathbb{A}^n \times \mathbb{A}^n$ by sending $\{p_1, \dots, p_d\}$ to the triple consisting of the $d-2$ other intersection points of Γ with Q^n and the intersection points of Γ with the hyperplanes $(t=0)$ and $(t=1)$. □

68 (Witt reduction and symmetric powers). The rationality problem of symmetric powers can be reduced to the anisotropic case. Indeed, we proved in (58) that if Q^n is a smooth quadric such that $Q^n(k) \neq \emptyset$ then

$$\operatorname{OG}(\mathbb{P}^m, Q^n) \overset{bir}{\sim} \operatorname{OG}(\mathbb{P}^{m-1}, Q_W^{n-2}) \times \mathbb{P}^{n-m}. \tag{68.1}$$

We can next take symmetric powers and, using [**Kol16b**, Lem.7], conclude that

$$\operatorname{Sym}^d\big(\operatorname{OG}(\mathbb{P}^m, Q^n)\big) \overset{bir}{\sim} \operatorname{Sym}^d\big(\operatorname{OG}(\mathbb{P}^{m-1}, Q_W^{n-2})\big) \times \mathbb{P}^{d(n-m)}. \tag{68.2}$$

69 (Symmetric square of $\operatorname{OG}(\mathbb{P}^r, Q^{2r})$). Since $\operatorname{OG}(\mathbb{P}^r, Q^{2r})$ has 2 geometric irreducible components, its symmetric square can be written as

$$\operatorname{Sym}^2\big(\operatorname{OG}(\mathbb{P}^r, Q^{2r})\big) = \operatorname{Sym}^2_s\big(\operatorname{OG}(\mathbb{P}^r, Q^{2r})\big) \amalg \operatorname{Sym}^2_a\big(\operatorname{OG}(\mathbb{P}^r, Q^{2r})\big) \tag{69.1}$$

where the first (symmetric) component parametrizes pairs $L_1^r, L_2^r \subset Q^{2r}$ that are in different geometric irreducible components of $\operatorname{OG}(\mathbb{P}^r, Q^{2r})$ while the second (anti-symmetric) component parametrizes pairs that are in the same geometric irreducible component. Note that $\operatorname{Sym}^2_s\big(\operatorname{OG}(\mathbb{P}^r, Q^{2r})\big)$ is geometrically irreducible while $\operatorname{Sym}^2_a\big(\operatorname{OG}(\mathbb{P}^r, Q^{2r})\big)$ is geometrically reducible.

If $r = 1$ then

$$\begin{aligned} \operatorname{Sym}^2_s\big(\operatorname{OG}(\mathbb{P}^1, Q^2)\big) &\cong Q^2 \quad \text{and} \\ \operatorname{Sym}^2_a\big(\operatorname{OG}(\mathbb{P}^1, Q^2)\big) &\cong (t^2 = \Delta) \times \mathbb{P}^2 \end{aligned} \tag{69.2}$$

Indeed, the first isomorphism is obtained by identifying a pair of intersecting lines on Q^2 with their intersection point. The second isomorphism is obtained by noting that $\mathrm{OG}(\mathbb{P}^1, Q^2)$ is a conic bundle over $(t^2 = \Delta)$ by (62) and the symmetric square of any conic is $\mathbb{P}^2$ by (67).

If $r = 2$ then two general planes of the same family meet at a single point, thus we have a map

$$\operatorname{Sym}^2_a\big(\mathrm{OG}(\mathbb{P}^2, Q^4)\big) \dashrightarrow Q^4. \tag{69.3}$$

The fiber over $p \in Q^4$ can be identified with $\operatorname{Sym}^2_a\big(\mathrm{OG}(\mathbb{P}^1, Q^2_p)\big)$. The $\mathrm{OG}(\mathbb{P}^1, Q^2_p)$-bundle over Q^4 is a conic bundle over an étale double cover of Q^4. Since Q^4 is simply connected, the étale double cover is isomorphic to $Q^4 \times (t^2 + \Delta = 0)$. (Note that Witt reduction from Q^4 to Q^2_p changes the sign of the discriminant.) Thus we have a conic bundle over $Q^4 \times (t^2 + \Delta = 0)$ and we take its relative symmetric square. This is a Zariski locally trivial $\mathbb{P}^2$-bundle over $Q^4 \times (t^2 + \Delta = 0)$, hence

$$\operatorname{Sym}^2_a\big(\mathrm{OG}(\mathbb{P}^2, Q^4)\big) \overset{bir}{\sim} (t^2 + \Delta = 0) \times Q^4 \times \mathbb{P}^2. \tag{69.4}$$

Symmetric squares of $\mathrm{OG}(\mathbb{P}^1, Q^n)$ are related to double covers of Grassmannians.

Proposition 70. *Let Q^n be a smooth quadric of dimension ≥ 3. Then there is a double cover $\pi : G(Q^n) \to \operatorname{Grass}(\mathbb{P}^3, \mathbb{P}^{n+1})$, ramified along a divisor $D \sim 2H$ such that*

$$\operatorname{Sym}^2\big(\mathrm{OG}(\mathbb{P}^1, Q^n)\big) \overset{bir}{\sim} G(Q^n) \times \mathbb{P}^2.$$

Proof. A K-point of $\operatorname{Sym}^2\big(\mathrm{OG}(\mathbb{P}^1, Q^n)\big)$ is the same as a pair of lines $\ell, \ell' \subset Q^n$ that are either both defined over K or are conjugates over K. In general ℓ, ℓ' are disjoint, so their span $\langle \ell, \ell' \rangle$ has dimension 3. Thus we get a map

$$\pi : \operatorname{Sym}^2\big(\mathrm{OG}(\mathbb{P}^1, Q^3)\big) \dashrightarrow \operatorname{Grass}(\mathbb{P}^3, \mathbb{P}^{n+1}). \tag{70.1}$$

Given $L^3 \subset \mathbb{P}^{n+1}$ such that $Q^n \cap L^3$ is a smooth quadric surface Q^2_L, we proved in (69.2) that the fiber of π over $[L^3]$ is

$$\operatorname{Sym}^2_a\big(\mathrm{OG}(\mathbb{P}^1, Q^2)\big) \cong (t^2 = \Delta) \times \mathbb{P}^2.$$

Therefore the Stein factorization of π is

$$\operatorname{Sym}^2\big(\mathrm{OG}(\mathbb{P}^1, Q^3\big) \overset{bir}{\sim} G(Q^n) \times \mathbb{P}^2 \to G(Q^n) \to \operatorname{Grass}(\mathbb{P}^3, \mathbb{P}^{n+1}), \tag{70.2}$$

where $G(Q^n) \to \operatorname{Grass}(\mathbb{P}^3, \mathbb{P}^{n+1})$ is a double cover that ramifies over a point of the Grassmannian $[L^3] \in \operatorname{Grass}(\mathbb{P}^3, \mathbb{P}^{n+1})$ iff $Q^n \cap L^3$ is a singular quadric (or if $L^3 \subset Q^n$).

Let $Q^3 \subset \mathbb{P}^4$ be a smooth quadric 3-fold. A general pencil of hyperplane sections $\langle L^3_\lambda \cap Q^3 \rangle$ of Q^3 has 2 singular members. Since such pencils $\langle L^3_\lambda \rangle \subset L^4 \subset \mathbb{P}^{n+1}$ correspond to lines on $\operatorname{Grass}(\mathbb{P}^3, \mathbb{P}^{n+1})$, we conclude that the branch divisor $D \subset \operatorname{Grass}(\mathbb{P}^3, \mathbb{P}^{n+1})$ is linearly equivalent to $2H$ where H is the hyperplane class on $\operatorname{Grass}(\mathbb{P}^3, \mathbb{P}^{n+1})$. □

Next we show that $G(Q^3)$ is itself a quadric and determine its equation. This will prove Proposition 15.

71 (Proof of Proposition 15). By Theorem 13 and by Proposition 70,

$$\operatorname{Map}_2(\mathbb{P}^2, Q^5) \overset{bir}{\sim} \operatorname{Sym}^2\big(\mathrm{OG}(\mathbb{P}^1, Q^3_W)\big) \times \mathbb{P}^{20} \overset{bir}{\sim} G(Q^3_W) \times \mathbb{P}^{22},$$

where Q^3_W is the Witt reduction of Q^5. Thus we need to understand $\pi : G(Q^3_W) \to \operatorname{Grass}(\mathbb{P}^3, \mathbb{P}^4)$. Note that $\operatorname{Grass}(\mathbb{P}^3, \mathbb{P}^4)$ is the dual of $\mathbb{P}^4$ and π ramifies along the dual quadric $Q^\vee_W$ which is isomorphic to Q^3_W. Thus if $Q^3_W \cong \big(q(x_0, \dots, x_4) = 0\big)$ then

$$G(Q^3_W) \cong \big(c\Delta(q)z^2 = q(x_0, \dots, x_4)\big) \quad \text{for some } c.$$

It is not hard to compute everything explicitly and obtain that $c = 1$. As a possible shortcut, working with the universal quadric over $\mathbb{Z}[\frac{1}{2}]$ as in the proof of (62), we see that c is a unit in $\mathbb{Z}[\frac{1}{2}]$, hence it is enough to compute it in one example. $\square$

ACKNOWLEDGMENTS. I thank I. Coskun, G. Di Cerbo, D. Krashen, R. Kusner, M. Lieblich, N. Lubbes, R. Parimala, B. Poonen, M. Skopenkov and B. Sturmfels for comments, discussions and references. I learned a lot of the early history, especially the role of [**Ber1907**], from F. Russo. Partial financial support was provided by the NSF under grant number DMS-1362960.

References

[AK03] Carolina Araujo and János Kollár, *Rational curves on varieties*, Higher dimensional varieties and rational points (Budapest, 2001), Bolyai Soc. Math. Stud., vol. 12, Springer, Berlin, 2003, pp. 13–68. MR2011743

[Ber1907] Eugenio Bertini, *Introduzione alla geometria proiettiva degli iperspazi con appendice sulle curve algebriche e loro singolaritá*, E. Spoerri, Pisa, 1907.

[Ber1924] ———, *Einführung in die projektive Geometrie mehrdimensionaler Raäume, mit einem Anhang über algebraische Kurven und ihre Singularitäten*, Seidel & Sohn, Wien, 1924.

[BKT15] Anders Skovsted Buch, Andrew Kresch, and Harry Tamvakis, *A Giambelli formula for even orthogonal Grassmannians*, J. Reine Angew. Math. **708** (2015), 17–48. MR3420327

[Cas1894] Guido Castelnuovo, *Sulle superficie algebriche che ammettono un sistema doppiamente infinito di sezioni piane riduttibili*, Rendiconti della E. Acc. dei Lincei **5** (1894), 22–25.

[Cob22] Arthur B. Coble, *Associated sets of points*, Trans. Amer. Math. Soc. **24** (1922), no. 1, 1–20. MR1501210

[Cos11] Izzet Coskun, *Restriction varieties and geometric branching rules*, Adv. Math. **228** (2011), no. 4, 2441–2502. MR2836127

[Cos14] Izzet Coskun, *Rigidity of Schubert classes in orthogonal Grassmannians*, Israel J. Math. **200** (2014), no. 1, 85–126. MR3219572

[Dal85] Magnar Dale, *Severi's theorem on the Veronese-surface*, J. London Math. Soc. (2) **32** (1985), no. 3, 419–425. MR825917

[ESB89] Lawrence Ein and Nicholas Shepherd-Barron, *Some special Cremona transformations*, Amer. J. Math. **111** (1989), no. 5, 783–800. MR1020829

[Flo04] Mathieu Florence, *Zéro-cycles de degré un sur les espaces homogènes* (French), Int. Math. Res. Not. **54** (2004), 2897–2914. MR2097288

[FP97] W. Fulton and R. Pandharipande, *Notes on stable maps and quantum cohomology*, Algebraic geometry—Santa Cruz 1995, Proc. Sympos. Pure Math., vol. 62, Amer. Math. Soc., Providence, RI, 1997, pp. 45–96. MR1492534

[Hal1695] Edmond Halley, *An easie demonstration of the analogy of the logarithmick tangents to the meridian line or sum of the secants: With various methods for computing the same to the utmost exactness*, Phil. Trans. Royal Soc. **19** (1695), 202–214.

[HP47] W. V. D. Hodge and D. Pedoe, *Methods of Algebraic Geometry. Vol. I*, Cambridge, at the University Press; New York, The Macmillan Company, 1947. MR0028055

[HRS04] Joe Harris, Mike Roth, and Jason Starr, *Rational curves on hypersurfaces of low degree*, J. Reine Angew. Math. **571** (2004), 73–106. MR2070144

[IR10] Paltin Ionescu and Francesco Russo, *Conic-connected manifolds*, J. Reine Angew. Math. **644** (2010), 145–157. MR2671777

[Kac99] Yasuyuki Kachi, *Characterization of Veronese varieties and the Hodge index theorem*, (unpublished), 1999.

[Kol96] János Kollár, *Rational curves on algebraic varieties*, Ergebnisse der Mathematik und ihrer Grenzgebiete. 3. Folge. A Series of Modern Surveys in Mathematics [Results in Mathematics and Related Areas. 3rd Series. A Series of Modern Surveys in Mathematics], vol. 32, Springer-Verlag, Berlin, 1996. MR1440180

[Kol05] János Kollár, *Conics in the Grothendieck ring*, Adv. Math. **198** (2005), no. 1, 27–35. MR2183248

[Kol08] János Kollár, *Looking for rational curves on cubic hypersurfaces*, Higher-dimensional geometry over finite fields, NATO Sci. Peace Secur. Ser. D Inf. Commun. Secur., vol. 16, IOS, Amsterdam, 2008, pp. 92–122. Notes by Ulrich Derenthal. MR2484078

[Kol10] János Kollár, *Holomorphic and pseudo-holomorphic curves on rationally connected varieties*, Port. Math. **67** (2010), no. 2, 155–179. MR2662865

[Kol16a] ______, *Severi-Brauer varieties; a geometric treatment*, ArXiv e-prints (2016).

[Kol16b] ______, *Symmetric powers of Severi-Brauer varieties*, to appear in Annales Mathématiques Toulouse.

[Kov94] Sándor J. Kovács, *The cone of curves of a $K3$ surface*, Math. Ann. **300** (1994), no. 4, 681–691. MR1314742

[KP77] Nicolaas H. Kuiper and William F. Pohl, *Tight topological embeddings of the real projective plane in E^5*, Invent. Math. **42** (1977), 177–199. MR0494122

[KP01] B. Kim and R. Pandharipande, *The connectedness of the moduli space of maps to homogeneous spaces*, Symplectic geometry and mirror symmetry (Seoul, 2000), World Sci. Publ., River Edge, NJ, 2001, pp. 187–201. MR1882330

[KT04] Andrew Kresch and Harry Tamvakis, *Quantum cohomology of orthogonal Grassmannians*, Compos. Math. **140** (2004), no. 2, 482–500. MR2027200

[Kui62] Nicolaas H. Kuiper, *On convex maps*, Nieuw Arch. Wisk. (3) **10** (1962), 147–164. MR0145545

[LP71] John A. Little and William F. Pohl, *On tight immersions of maximal codimension*, Invent. Math. **13** (1971), 179–204. MR0293645

[Mel99] Massimiliano Mella, *Existence of good divisors on Mukai varieties*, J. Algebraic Geom. **8** (1999), no. 2, 197–206. MR1675146

[Muk89] Shigeru Mukai, *Biregular classification of Fano 3-folds and Fano manifolds of coindex* 3, Proc. Nat. Acad. Sci. U.S.A. **86** (1989), no. 9, 3000–3002. MR995400

[Par05] R. Parimala, *Homogeneous varieties—zero-cycles of degree one versus rational points*, Asian J. Math. **9** (2005), no. 2, 251–256. MR2176607

[PR14] Luc Pirio and Francesco Russo, *Quadro-quadric Cremona transformations in low dimensions via the JC-correspondence* (English, with English and French summaries), Ann. Inst. Fourier (Grenoble) **64** (2014), no. 1, 71–111. MR3330541

[PSS12] Helmut Pottmann, Ling Shi, and Mikhail Skopenkov, *Darboux cyclides and webs from circles*, Comput. Aided Geom. Design **29** (2012), no. 1, 77–97. MR2871363

[Rus16] Francesco Russo, *On the geometry of some special projective varieties*, Lecture Notes of the Unione Matematica Italiana, vol. 18, Springer, Cham; Unione Matematica Italiana, Bologna, 2016. MR3445582

[Sch01] Josef Schicho, *The multiple conical surfaces*, Beiträge Algebra Geom. **42** (2001), no. 1, 71–87. MR1824751

[Ser73] J.-P. Serre, *A course in arithmetic*, Springer-Verlag, New York-Heidelberg, 1973. Translated from the French; Graduate Texts in Mathematics, No. 7. MR0344216

[Sev1901] Francesco Severi, *Intorno ai punti doppi impropri di una superficie generale dello spazio a quattro dimensioni e ai suoi punti tripli apparenti*, Rend. Circ. Mat. Palermo **15** (1901), 33–51.

[SK15] M. Skopenkov and R. Krasauskas, *Surfaces containing two circles through each point and Pythagorean 6-tuples*, ArXiv e-prints (2015).

[Sko15] M. Skopenkov, *Surfaces containing two circles through each point*, ArXiv e-prints (2015).

[Spr52] Tonny Albert Springer, *Sur les formes quadratiques d'indice zéro* (French), C. R. Acad. Sci. Paris **234** (1952), 1517–1519. MR0047021

[SR49] J. G. Semple and L. Roth, *Introduction to Algebraic Geometry*, Oxford, at the Clarendon Press, 1949. MR0034048

[Ver1884] Giuseppe Veronese, *La superficie omaloide normale a due dimensioni e del quarto ordine dello spazio a cinque dimensioni e le sue projezioni nel piano e nello spazio ordinario*, Memorie della E. Acc. dei Lincei **19** (1884), 344–371.

Princeton University, Princeton, New Jersey 08544-1000
Email address: kollar@math.princeton.edu

Contemporary Mathematics
Volume **712**, 2018
http://dx.doi.org/10.1090/conm/712/14349

Non-Cohen-Macaulay canonical singularities

Sándor J Kovács

Dedicated to Lawrence Ein on the occasion of his 60th birthday

1. Introduction

Since the first counter-example to Kodaira vanishing in positive characteristic was constructed by Raynaud [**Ray78**] many other counter-examples have been found satisfying various prescribed properties [**DI87, Eke88, SB91, Kol96, Lau96, Muk13, DCF15, CT16a, CT16b**]. An elementary counter-example for which the line bundle violating Kodaira vanishing is very ample was constructed by Lauritzen and Rao in [**LR97**]. Let us denote it by X. It is straightforward from the construction that X is a rational variety and for $p = 2$ and $\dim X = 6$ it is Fano. Let Z denote the cone over X using the embedding given by the global sections of the very ample line bundle violating Kodaira vanishing. It is well-known that a cone over a Fano variety has klt singularities if K_Z is $\mathbb{Q}$-Cartier. (See 2.1.) The failure of Kodaira vanishing on X implies that Z will not have Cohen-Macaulay singularities, in particular it does not have rational singularities. As pointed out by Esnault and Kollár, although in this example K_Z is not $\mathbb{Q}$-Cartier, one can easily find a boundary Δ on Z that makes $K_Z + \Delta$ $\mathbb{Q}$-Cartier, and hence the pair (Z, Δ) klt. In other words Lauritzen and Rao's counter-example to Kodaira vanishing produces a klt pair (Z, Δ) such that Z is not Cohen-Macaulay. This provides a counter-example to the positive characteristic analogue of Elkik's theorem [**Elk81**], [**KM98**, 5.22]. Examples of non-Cohen-Macaulay klt singularities were also given by Yasuda in [**Yas14**] and Cascini and Tanaka in [**CT16a**].

We will show that one can use the above X to produce even more interesting singularities. I will demonstrate below that in fact the very ample line bundle ω_X^{-2} also violates Kodaira vanishing and hence leads to a cone, using the polarization given by ω_X^{-1}, whose canonical sheaf is a line bundle, has canonical singularities, and is not Cohen-Macaulay. Of course, then it also does not have rational singularities. In other words, the purpose of this note is to prove the following.

THEOREM 1.1. *Let k be a field of characteristic* 2. *Then there exists a Fano variety X over k such that*

(i) $\dim X = 6$,

Supported in part by NSF Grant DMS-1565352 and the Craig McKibben and Sarah Merner Endowed Professorship in Mathematics at the University of Washington.

(ii) ω_X^{-1} *is very ample, and*

(iii) ω_X^{-2} *violates Kodaira vanishing.*

One might ask if there is a similar example in smaller dimensions. It follows from [**CT16b**, A.1] that there are no Del Pezzo surfaces with this property. Hence a similar example in smaller dimension would be at least 3-dimensional. One might also ask if there is a similar example where ω_X^{-1} violates Kodaira vanishing. The example here is certainly not such and it is well-known that no such example exists for $\dim X = 2, 3$ [**SB97, Sch07, Mad16**]. While this is an interesting question, it is irrelevant for the purposes of the present article. The more interesting question is whether there are similar examples in all positive characteristics.

My main interest in the above result lies in the following application. By taking the cone over X given by the embedding induced by the global sections of ω_X^{-1} we obtain the following.

THEOREM 1.2. *Let k be a field of characteristic 2. Then there exists a variety Z over k with the following properties:*

(a) *Z is of dimension 7, has a single isolated canonical singularity, and admits a resolution of singularities by a smooth variety over k,*

(b) *ω_Z is a line bundle,*

(c) *Z is not Cohen-Macaulay, in particular, Z is not Gorenstein and does not have rational singularities.*

Again, one might ask if there are such singularities in smaller dimensions. Of course, if one finds examples such as in 1.1 in smaller dimensions, that would provide smaller dimensional examples for 1.2 as well. However, as mentioned above there are no examples similar to 1.1 in dimension 2 which makes it an interesting question whether there exist 3-dimensional canonical singularities, perhaps even of index 1, that are not Cohen-Macaulay. And again, the possibly more interesting question is whether there are such singularities in all positive characteristics.

NOTE ADDED IN PROOF. While this paper was under review the above question has been answered. Bernasconi gave examples of non-Cohen-Macaulay klt singularities in characteristic 3 in [**Ber17**] and Totaro and Yasuda gave examples of non-Cohen-Macaulay terminal singularities in all positive characteristics in [**Tot17**] and [**Yas17**].

In the opposite direction Hacon and Witaszek [**HW17**] recently proved that in dimension 3 klt singularities are rational if the characteristic of the base field is sufficiently large.

ACKNOWLEDGMENT. I am grateful to János Kollár, Hiromu Tanaka, Burt Totaro, Takehiko Yasuda, and to the referee for useful comments.

2. Non-Cohen-Macaulay singularities via failure of Kodaira vanishing

DEFINITION 2.1. Let X be a smooth projective variety over k and $\mathscr{L}$ an ample line bundle on X. Then we will say that *$\mathscr{L}$ violates Kodaira vanishing* if there exists an $i < \dim X$ such that $H^i(X, \mathscr{L}^{-1}) \neq 0$. By Serre duality this is equivalent to that $H^{\dim X - i}(X, \mathscr{L} \otimes \omega_X) \neq 0$.

The canonical divisor of a normal variety Z is denoted, as usual, by K_Z and the associated reflexive sheaf of rank 1, the canonical sheaf, is denoted by ω_Z.

I.e., $\omega_Z \simeq \mathscr{O}_Z(K_Z)$. A Weil divisor D on Z is $\mathbb{Q}$-Cartier if there exists a non-zero $m \in \mathbb{N}$ such that mD is Cartier. A normal variety Z is said to have *rational singularities* if for a resolution of singularities $\phi : \widetilde{Z} \to Z$ the following conditions hold:

(i) $\mathcal{R}^i\phi_*\mathscr{O}_{\widetilde{Z}} = 0$ for $i > 0$, and

(ii) $\mathcal{R}^i f_*\omega_{\widetilde{Z}} = 0$ for $i > 0$.

In characteristic 0 (ii) is automatic by the Grauert-Riemenschneider vanishing theorem [**GR70**], [**KM98**, 2.68]. For the definition of *klt* and *canonical* singularities the reader is referred to [**Kol13**, 2.8].

Rational singularities are Cohen-Macaulay by the following well-known lemma. A very short proof is included for the convenience of the reader.

LEMMA 2.2. *Let Z be a scheme with rational singularities. Then Z is Cohen-Macaulay.*

PROOF. Let $d = \dim Z$ and let $\phi : \widetilde{Z} \to Z$ be a resolution of singularities of Z. This implies that $\mathscr{O}_Z \simeq \mathcal{R}\phi_*\mathscr{O}_{\widetilde{Z}}$ and $\omega_Z \simeq \mathcal{R}\phi_*\omega_{\widetilde{Z}}$. Then by Grothendieck duality

$$\omega_Z[d] \simeq \mathcal{R}\phi_*\omega_{\widetilde{Z}}[d] \simeq \mathcal{R}\phi_*\mathcal{R}\mathcal{H}om_{\widetilde{Z}}(\mathscr{O}_{\widetilde{Z}}, \omega^\bullet_{\widetilde{Z}}) \simeq \mathcal{R}\mathcal{H}om_Z(\mathcal{R}\phi_*\mathscr{O}_{\widetilde{Z}}, \omega^\bullet_Z) \simeq \omega^\bullet_Z,$$

and hence Z is Cohen-Macaulay. □

REMARK 2.3. It follows easily that if Z is not Cohen-Macaulay, then for any resolution of singularities $\phi : \widetilde{Z} \to Z$, there exists an $i > 0$ such that either $\mathcal{R}^i\phi_*\mathscr{O}_{\widetilde{Z}} \neq 0$ or $\mathcal{R}^i\phi_*\omega_{\widetilde{Z}} \neq 0$.

Next I will review the more-or-less well-known idea of constructing non-Cohen-Macaulay singularities as cones over varieties violating Kodaira vanishing.

(2.4) Let X be a normal projective variety over a field k of characteristic $p > 0$, $\mathscr{L}$ an ample line bundle on X, and $Z = C_a(X, \mathscr{L}) = \operatorname{Spec}\left(\oplus_{m\geq 0} H^0(X, \mathscr{L}^m)\right)$ the *affine cone* over X with conormal bundle $\mathscr{L}$. (Here we follow the convention of [**Kol13**, 3.8] on cones.)
Then we have the following well-known criterion cf. [**Kol13**, 3.11]:

(2.5) Z is Cohen-Macaulay if and only if $H^i(X, \mathscr{L}^q) = 0$ for all $0 < i < \dim X$ and $q \in \mathbb{Z}$.
This implies for example that cones over varieties whose structure sheaves have non-trivial *middle* cohomology, for instance abelian varieties of dimension at least 2, are not Cohen-Macaulay. It also implies that

(2.6) if a power of $\mathscr{L}$ violates Kodaira vanishing, then Z is not Cohen-Macaulay.
Next recall that the canonical divisor of a canonical singularity is $\mathbb{Q}$-Cartier and observe that in the above construction

(2.7) if $\omega_X^r \simeq \mathscr{L}^q$ for $r, q \in \mathbb{Z}$, $r \neq 0$, then K_Z is $\mathbb{Q}$-Cartier of index at most r.

However, even if (2.7) fails, Z may still provide an example of a klt singularity with an appropriate boundary as we will see in the next statement, which summarizes what we found in this section. Note that this statement is a simple consequence of the combination of [**Kol13**, 3.1, 3.11].

PROPOSITION 2.8. *In addition to the definitions in* (2.4) *assume that* X *is a smooth Fano variety and that some power of* $\mathscr{L}$ *violates Kodaira vanishing. Then there exists a* $\mathbb{Q}$*-divisor* Δ *on* Z *such that*

(i) (Z, Δ) *has klt singularities,*

(ii) Z *is not Cohen-Macaulay, and hence in particular has non-rational singularities, and*

(iii) *if* $\omega_X \simeq \mathscr{L}^q$ *for some* $q \in \mathbb{Z}$*, then* Z *has canonical singularities.*

PROOF. Since ω_X^{-1} is ample, there is an $r \in \mathbb{N}$, $r > 0$, such that $\mathscr{N} = \mathscr{L}^{-1} \otimes \omega_X^{-r}$ is also ample. Let N be a general member of the complete linear system corresponding to $\mathscr{N}^m$ for some $m \gg 0$, $\widehat{N} \subseteq Z$ the cone over N, and $\Delta := \frac{1}{rm}\widehat{N}$. Then $\mathscr{O}_X(rm(K_X + \frac{1}{rm}N)) \simeq \mathscr{L}^{-m}$, so $K_Z + \Delta$ is a Cartier divisor on Z (cf.[**Kol13**, 3.14(4)]), and hence $K_Z + \Delta$ is $\mathbb{Q}$-Cartier. Furthermore, N is smooth and hence $(X, \frac{1}{rm}N)$ is klt, so (i) follows from [**Kol13**, 3.1(3)]. Now, if $\omega_X \simeq \mathscr{L}^q$ for some $q \in \mathbb{Z}$, then ω_Z is a line bundle and hence (iii) follows from (i). Finally, (ii) is simply a restatement of (2.6). □

3. The construction of Lauritzen and Rao

Next, I will recall the construction of Lauritzen and Rao from [**LR97**].

Let V be a vector space of dimension $n+1$ over a field k of characteristic p where $p \geq n-1 \geq 2$, and let $\mathbb{P}(V) \simeq \mathbb{P}^n$ be the associated projective space of dimension n. Let $W := \mathbb{P}(V) \times \mathbb{P}(V^\vee)$ and for $a, b \in \mathbb{Z}$ let $\mathscr{O}_W(a,b)$ denote the line bundle $\mathscr{O}_{\mathbb{P}(V)}(a) \boxtimes \mathscr{O}_{\mathbb{P}(V^\vee)}(b)$ on W. Next let $\mathscr{A}$ be the locally free sheaf defined by the short exact sequence

$$0 \longrightarrow \mathscr{A} \longrightarrow V \otimes \mathscr{O}_{\mathbb{P}(V)} \longrightarrow \mathscr{O}_{\mathbb{P}(V)}(1) \longrightarrow 0, \tag{3.1}$$

and let $\alpha : Y := \mathbb{P}(\mathscr{A}^\vee) \to \mathbb{P}(V)$ be the projective space bundle over $\mathbb{P}(V)$ associated to $\mathscr{A}^\vee$. Let $\mathscr{O}_\alpha(1)$ denote the corresponding tautological line bundle on Y. Then there exists another associated short exact sequence on Y:

$$0 \longrightarrow \mathscr{G} \longrightarrow \alpha^*\mathscr{A}^\vee \longrightarrow \mathscr{O}_\alpha(1) \longrightarrow 0, \tag{3.2}$$

which defines the locally free sheaf $\mathscr{G}$ on Y. It is shown in [**LR97**, p.23] that Y admits a closed embedding into $W \simeq \mathbb{P}^n \times \mathbb{P}^n$ with bihomogenous coordinate ring

$$k[x_0, \dots, x_n; y_0, \dots, y_n]\Big/\left(\sum x_i y_i\right). \tag{3.3}$$

In particular, the ideal sheaf of Y in W is $\mathscr{O}_W(-1,-1)$. Let $\mathscr{O}_Y(a,b) := \mathscr{O}_W(a,b)|_Y$. Then it follows easily that

$$\omega_Y \simeq \mathscr{O}_Y(-n,-n), \quad \alpha^*\mathscr{O}_{\mathbb{P}(V)}(a) \simeq \mathscr{O}_Y(a,0), \quad \text{and} \quad \mathscr{O}_\alpha(b) \simeq \mathscr{O}_Y(0,b). \tag{3.4}$$

Let η be defined as the composition of the natural morphisms induced by the morphisms in 3.1 and 3.2 using the isomorphisms in 3.4:

$$V^\vee \otimes \mathscr{O}_Y \longrightarrow \alpha^*\mathscr{A}^\vee \longrightarrow \mathscr{O}_Y(0,1) \tag{3.5}$$

where the composite arrow $V^\vee \otimes \mathscr{O}_Y \to \mathscr{O}_Y(0,1)$ is labelled η.

Then we have the following commutative diagram, where $\mathscr{B} = \ker \eta$:

$$
\begin{array}{ccccccccc}
 & & 0 & & 0 & & & & \\
 & & \downarrow & & \downarrow & & & & \\
 & & \mathscr{O}_Y(-1,0) & \xrightarrow{=} & \mathscr{O}_Y(-1,0) & & & & \\
 & & \downarrow & & \downarrow & & & & \\
0 & \longrightarrow & \mathscr{B} & \longrightarrow & V^\vee \otimes \mathscr{O}_Y & \xrightarrow{\eta} & \mathscr{O}_Y(0,1) & \longrightarrow & 0 \\
 & & {\scriptstyle\tau}\downarrow & & \downarrow & & \downarrow{\scriptstyle =} & & \\
0 & \longrightarrow & \mathscr{G} & \longrightarrow & \alpha^*\mathscr{A}^\vee & \longrightarrow & \mathscr{O}_Y(0,1) & \longrightarrow & 0 \\
 & & \downarrow & & \downarrow & & & & \\
 & & 0 & & 0 & & & &
\end{array} \qquad (3.6)
$$

Finally, let $\mathscr{G}' = \mathscr{G} \otimes \mathscr{O}_\alpha(1)$ and $\pi : X := \mathbb{P}(F^*\mathscr{G}') \to Y$, where $F : Y \to Y$ is the absolute Frobenius morphism of Y. Note that by construction $\dim X = 3n - 3$ and $\dim Y = 2n - 1$.

(3.7) Again, it is shown in [**LR97**, p.23] that the tautological line bundle of π, denoted by $\mathscr{O}_\pi(1)$, is globally generated and the line bundle $\pi^*\mathscr{O}_Y(1,1) \otimes \mathscr{O}_\pi(1)$ is very ample. It follows that $\pi^*\mathscr{O}_Y(1,1) \otimes \mathscr{O}_\pi(q)$ is also very ample for any $q > 0$.

Using the formula for the canonical bundle of a projective space bundle, one obtains that

$$\omega_X \simeq \pi^*\mathscr{O}_Y(p-n, p(n-2)-n) \otimes \mathscr{O}_\pi(-n+1) \qquad (3.8)$$

As it was pointed out by Hélène Esnault if one chooses the values $p = 2$ and $n = 3$, then X is a Fano variety and hence there exists a klt pair (Z, Δ) where Z is not Cohen-Macaulay, in particular, it does not have rational singularities cf. 2.8.

4. A Fano variety violating Kodaira vanishing

We will use the above construction and prove that if $p = 2$ and $n = 3$, then the very ample line bundle ω_X^{-2} violates Kodaira vanishing. To do this, first we need to compute a few auxiliary cohomology groups. We will keep using the notation introduced in 3.

LEMMA 4.1. *Let $a, b \in \mathbb{Z}$. Then*

$$H^i(Y, \mathscr{O}_Y(a,b)) = 0$$

if either

(i) *a and b are arbitrary and $0 < i < n-1$, or*

(ii) *$a, b > -n$ and $i > 0$, or*

(iii) *at least one of a and b is negative and $i = 0$.*

PROOF. By (3.3) we have the following short exact sequence:

$$0 \longrightarrow \mathscr{O}_W(a-1, b-1) \longrightarrow \mathscr{O}_W(a,b) \longrightarrow \mathscr{O}_Y(a,b) \longrightarrow 0$$

Since $W \simeq \mathbb{P}^n \times \mathbb{P}^n$, using the Künneth formula, the first two (non-zero) sheaves above have no cohomology in the following cases:

(a) for $0 < i < n$ and arbitrary a and b,

(b) for $a, b > -n$ and $i > 0$, and

(c) at least one of a and b is negative and $i = 0$.

Then (a) implies (i), (b) implies (ii), and (a) and (c) together imply (iii). □

COROLLARY 4.2. *Under the same conditions as in 4.1,*

$$H^i(Y, \mathscr{O}_Y(a,b) \otimes F^*(V^\vee \otimes \mathscr{O}_Y)) = 0.$$

PROOF. $F^*(V^\vee \otimes \mathscr{O}_Y)$ is a free $\mathscr{O}_Y$ sheaf, so this is straightforward from 4.1. □

LEMMA 4.3. *Let $a, b \in \mathbb{Z}$. Then*

$$H^1(Y, \mathscr{O}_Y(a,b) \otimes F^*\mathscr{B}) \simeq$$

$$\simeq \operatorname{coker}\left[H^0(Y, \mathscr{O}_Y(a,b) \otimes F^*(V^\vee \otimes \mathscr{O}_Y)) \xrightarrow{\eta_1 := F^*\eta} H^0(Y, \mathscr{O}_Y(a, b+p))\right]$$

where $\eta_1 = F^*\eta$ is induced by the morphism η defined in (3.5). In particular, if either $a < 0$ or $b < -p$, then

$$H^1(Y, \mathscr{O}_Y(a,b) \otimes F^*\mathscr{B}) = 0.$$

PROOF. Consider the Frobenius pull-back of the middle row of the diagram in (3.6) twisted with $\mathscr{O}_Y(a,b)$:

$$0 \longrightarrow \mathscr{O}_Y(a,b) \otimes F^*\mathscr{B} \longrightarrow \mathscr{O}_Y(a,b) \otimes F^*(V^\vee \otimes \mathscr{O}_Y) \longrightarrow \mathscr{O}_Y(a, b+p) \longrightarrow 0.$$

Then, since $n > 2$, both statements follow from 4.2. □

LEMMA 4.4. *Let $a, b \in \mathbb{Z}$. Then the morphism induced by τ in (3.6) is an isomorphism:*

$$H^1(\mathscr{O}_Y(a,b) \otimes F^*\mathscr{B}) \xrightarrow{\simeq} H^1(\mathscr{O}_Y(a,b) \otimes F^*\mathscr{G}).$$

Furthermore, if $a < p$ or $b < -p$, then the natural morphism induced by the same morphism as above is an injection:

$$H^1(\mathscr{O}_Y(a,b) \otimes \operatorname{Sym}^2 F^*\mathscr{B}) \hookrightarrow H^1(\mathscr{O}_Y(a,b) \otimes \operatorname{Sym}^2 F^*\mathscr{G}).$$

PROOF. Consider the Frobenius pull-back of the first vertical short exact sequence from (3.6):

$$0 \longrightarrow \mathscr{O}_Y(-p, 0) \longrightarrow F^*\mathscr{B} \longrightarrow F^*\mathscr{G} \longrightarrow 0$$

Since $n > 2$, this, combined with 4.1, implies the first statement.

Next, observe that this short exact sequence also implies that there exists a filtration

$$\operatorname{Sym}^2 F^*\mathscr{B} \supseteq \mathscr{E} \supseteq \mathscr{O}_Y(-2p, 0)$$

such that (after twisting by $\mathscr{O}_Y(a,b)$) we have the short exact sequences

(4.4.1) $$0 \longrightarrow \mathscr{O}_Y(a-2p, b) \longrightarrow \mathscr{O}_Y(a,b) \otimes \mathscr{E} \longrightarrow \mathscr{O}_Y(a-p, b) \otimes F^*\mathscr{G} \longrightarrow 0$$

and

(4.4.2)

$$0 \longrightarrow \mathscr{O}_Y(a,b) \otimes \mathscr{E} \longrightarrow \mathscr{O}_Y(a,b) \otimes \operatorname{Sym}^2 F^*\mathscr{B} \longrightarrow \mathscr{O}_Y(a,b) \otimes \operatorname{Sym}^2 F^*\mathscr{G} \longrightarrow 0$$

Then by the first statement, 4.3, 4.1, and (4.4.1) it follows that if $a < p$ or $b < -p$, then

$$H^1(Y, \mathscr{O}_Y(a,b) \otimes \mathscr{E}) = 0.$$

By (4.4.2) this implies the second statement. □

LEMMA 4.5. *Let $a, b \in \mathbb{Z}$ such that $a \geq 0$ and $b > -n$. Then*

$$H^1(Y, \mathscr{O}_Y(a,b) \otimes \operatorname{Sym}^2 F^*\mathscr{B}) \neq 0$$

PROOF. Observe that the middle horizontal short exact sequence in the diagram (3.6) implies that there exists a filtration

$$\operatorname{Sym}^2 (F^*(V^\vee \otimes \mathscr{O}_Y)) \supseteq \mathscr{F} \supseteq \operatorname{Sym}^2 F^*\mathscr{B}$$

such that (after twisting by $\mathscr{O}_Y(a,b)$) we have the short exact sequences
(4.5.1)

$$0 \longrightarrow \mathscr{O}_Y(a,b) \otimes \operatorname{Sym}^2 F^*\mathscr{B} \longrightarrow \mathscr{O}_Y(a,b) \otimes \mathscr{F} \longrightarrow \mathscr{O}_Y(a,b+p) \otimes F^*\mathscr{B} \longrightarrow 0$$

and
(4.5.2)

$$0 \longrightarrow \mathscr{O}_Y(a,b) \otimes \mathscr{F} \longrightarrow \mathscr{O}_Y(a,b) \otimes \operatorname{Sym}^2 (F^*(V^\vee \otimes \mathscr{O}_Y)) \longrightarrow \mathscr{O}_Y(a,b+2p) \longrightarrow 0.$$

Since $n > 2$, it follows from (4.5.2) and 4.1 that

$$H^2(Y, \mathscr{O}_Y(a,b) \otimes \mathscr{F}) = 0$$

and that

$$H^1(Y, \mathscr{O}_Y(a,b) \otimes \mathscr{F}) \simeq$$

$$\operatorname{coker}\left[H^0\left(Y, \mathscr{O}_Y(a,b) \otimes \operatorname{Sym}^2 (F^*(V^\vee \otimes \mathscr{O}_Y))\right) \xrightarrow{\eta_2} H^0(Y, \mathscr{O}_Y(a,b+2p))\right].$$

The morphism η_2 here is given by the matrix $[y_i^p y_j^p \mid i,j = 0, \dots, n]$. Furthermore, it follows from 4.3 that

$$H^1(Y, \mathscr{O}_Y(a,b+p) \otimes F^*\mathscr{B}) \simeq$$

$$\simeq \operatorname{coker}\left[H^0(Y, \mathscr{O}_Y(a,b+p) \otimes F^*(V^\vee \otimes \mathscr{O}_Y)) \xrightarrow{\eta_1} H^0(Y, \mathscr{O}_Y(a,b+2p))\right].$$

The morphism η_1 here is given by the matrix $[y_i^p \mid i = 0, \dots, n]$. Note, that by assumption $a \geq 0$ and $b \geq -n+1 \geq -p$, so $H^0(Y, \mathscr{O}_Y(a,b+p) \otimes F^*(V^\vee \otimes \mathscr{O}_Y)) \neq 0$. Then it is easy to see, for example from the description of η_1 and η_2 above, that

$$\operatorname{im} \eta_2 \subsetneq \operatorname{im} \eta_1,$$

and hence combined with (4.5.1) the above imply that

$$H^1(Y, \mathscr{O}_Y(a,b) \otimes \operatorname{Sym}^2 F^*\mathscr{B}) \simeq \operatorname{im} \eta_1 \big/ \operatorname{im} \eta_2 \neq 0. \tag{4.5.3}$$

□

REMARK 4.6. Observe that the previous argument was the place where working in positive characteristic was crucial. The morphisms η_1 and η_2 are given by the pth powers of the global sections of $\mathscr{O}_Y(0,1)$. We obtain the non-trivial cokernels and the "gap" between them from the fact that the global sections of $\mathscr{O}_Y(0,p)$ are not generated by these p^{th} powers.

This argument fails for several reasons in characteristic 0. First of all, p^{th} powers do not define an $\mathscr{O}_Y$-module homomorphism. Of course, they do not define one in any characteristic, which is the reason that we first have to pull-back everything by the Frobenius. However, the pth powers do give an $F^*\mathscr{O}_Y$-module homomorphism. There is of course no Frobenius in characteristic 0, but one might think that then one could use another finite morphism to pull-back these sections and thereby replacing the global sections by an appropriate power. However, in characteristic 0 this would mean switching to an actual cover many of whose properties would change. For instance, very likely that cover would no longer be Fano or even have negative Kodaira dimension and other parts of the proof would break down.

To summarize, the reason this argument works in positive characteristic is that there is a high degree endomorphism which is one-to-one on points. Then again, this is not surprising at all as this is usually the reason when a statement holds in positive characteristic but not in characteristic 0.

THEOREM 4.7. *If $p \leq n = 3$, then $\dim X = 6$, and $H^5(X, \omega_X^2) \neq 0$.*

PROOF. By Serre duality and (3.8) we have that

(4.7.1)
$$\begin{aligned} H^i(X, \omega_X^2)^\vee \simeq H^{6-i}(X, \omega_X^{-1}) &\simeq H^{6-i}(X, \pi^*\mathscr{O}_Y(3-p, 3-p) \otimes \mathscr{O}_\pi(2)) \\ &\simeq H^{6-i}(Y, \mathscr{O}_Y(3-p, 3-p) \otimes \pi_*\mathscr{O}_\pi(2)) \\ &\simeq H^{6-i}(Y, \mathscr{O}_Y(3-p, 3-p) \otimes \operatorname{Sym}^2 F^*\mathscr{G}') \\ &\simeq H^{6-i}(Y, \mathscr{O}_Y(3-p, 3+p) \otimes \operatorname{Sym}^2 F^*\mathscr{G}) \end{aligned}$$

Since $p > a = 3 - p \geq$ and $b = 3 + p > -n = -3$, the statement follows from 4.4 and 4.5. □

This might seem to give a desired example in $p = 3$ as well, but this non-vanishing is only interesting when X is Fano, i.e., when ω_X^{-1} is ample and that only holds when $p = 2$.

COROLLARY 4.8. *If $n = 3$ and $p = 2$, then X is a Fano variety on which ω_X^{-1} is very ample and ω_X^{-2} violates Kodaira vanishing. In particular, 1.1 follows.*

PROOF. If $n = 3$ and $p = 2$, then by (3.8) $\omega_X \simeq \pi^*\mathscr{O}_Y(-1,-1) \otimes \mathscr{O}_\pi(-2)$ and hence ω_X^{-1} is very ample by 3.7. By 4.7, ω_X^{-2} violates Kodaira vanishing. □

COROLLARY 4.9. *1.2 holds.*

PROOF. Let $Z = C_a(X, \omega_X^{-1})$. Then the statement follows from 4.8 and 2.8. □

References

[Ber17] F. Bernasconi, *Kawamata-Viehweg vanishing fails for log del Pezzo surfaces in char 3.* arXiv:1709.09238

[CT16a] P. Cascini and H. Tanaka, *Purely log terminal threefolds with non-normal centres in characteristic two*, 2016. arXiv:1607.08590

[CT16b] P. Cascini and H. Tanaka, *Smooth rational surfaces violating Kawamata-Viehweg vanishing*, 2016. arXiv:1607.08542

[DI87] Pierre Deligne and Luc Illusie, *Relèvements modulo p^2 et décomposition du complexe de de Rham* (French), Invent. Math. **89** (1987), no. 2, 247–270. MR894379

[DCF15] Gabriele Di Cerbo and Andrea Fanelli, *Effective Matsusaka's theorem for surfaces in characteristic p*, Algebra Number Theory **9** (2015), no. 6, 1453–1475. MR3397408

[Eke88] Torsten Ekedahl, *Canonical models of surfaces of general type in positive characteristic*, Inst. Hautes Études Sci. Publ. Math. **67** (1988), 97–144. MR972344

[Elk81] Renée Elkik, *Rationalité des singularités canoniques* (French), Invent. Math. **64** (1981), no. 1, 1–6. MR621766

[GR70] Hans Grauert and Oswald Riemenschneider, *Verschwindungssätze für analytische Kohomologiegruppen auf komplexen Räumen* (German), Invent. Math. **11** (1970), 263–292. MR0302938

[HW17] C. D. Hacon and J. Witaszek, *On the rationality of Kawamata log terminal singularities in positive characteristic.* arXiv:1706.03204

[Kol96] János Kollár, *Rational curves on algebraic varieties*, Ergebnisse der Mathematik und ihrer Grenzgebiete. 3. Folge. A Series of Modern Surveys in Mathematics [Results in Mathematics and Related Areas. 3rd Series. A Series of Modern Surveys in Mathematics], vol. 32, Springer-Verlag, Berlin, 1996. MR1440180

[Kol13] János Kollár, *Singularities of the minimal model program*, Cambridge Tracts in Mathematics, vol. 200, Cambridge University Press, Cambridge, 2013. With a collaboration of Sándor Kovács. MR3057950

[KM98] János Kollár and Shigefumi Mori, *Birational geometry of algebraic varieties*, Cambridge Tracts in Mathematics, vol. 134, Cambridge University Press, Cambridge, 1998. With the collaboration of C. H. Clemens and A. Corti; Translated from the 1998 Japanese original. MR1658959

[LR97] N. Lauritzen and A. P. Rao, *Elementary counterexamples to Kodaira vanishing in prime characteristic*, Proc. Indian Acad. Sci. Math. Sci. **107** (1997), no. 1, 21–25. MR1453823

[Lau96] Niels Lauritzen, *Embeddings of homogeneous spaces in prime characteristics*, Amer. J. Math. **118** (1996), no. 2, 377–387. MR1385284

[Mad16] Zachary Maddock, *Regular del Pezzo surfaces with irregularity*, J. Algebraic Geom. **25** (2016), no. 3, 401–429. MR3493588

[Muk13] Shigeru Mukai, *Counterexamples to Kodaira's vanishing and Yau's inequality in positive characteristics*, Kyoto J. Math. **53** (2013), no. 2, 515–532. MR3079312

[Ray78] M. Raynaud, *Contre-exemple au "vanishing theorem" en caractéristique $p > 0$* (French), C. P. Ramanujam—a tribute, Tata Inst. Fund. Res. Studies in Math., vol. 8, Springer, Berlin-New York, 1978, pp. 273–278. MR541027

[Sch07] Stefan Schröer, *Weak del Pezzo surfaces with irregularity*, Tohoku Math. J. (2) **59** (2007), no. 2, 293–322. MR2347424

[SB91] N. I. Shepherd-Barron, *Unstable vector bundles and linear systems on surfaces in characteristic p*, Invent. Math. **106** (1991), no. 2, 243–262. MR1128214

[SB97] N. I. Shepherd-Barron, *Fano threefolds in positive characteristic*, Compositio Math. **105** (1997), no. 3, 237–265. MR1440723

[Tot17] B. Totaro, *The failure of Kodaira vanishing for Fano varieties, and terminal singularities that are not Cohen-Macaulay.* arXiv:1710.04364

[Yas14] Takehiko Yasuda, *The p-cyclic McKay correspondence via motivic integration*, Compos. Math. **150** (2014), no. 7, 1125–1168. MR3230848

[Yas17] T. Yasuda, *Discrepancies of p-cyclic quotient varieties.* arXiv:1710.06044

University of Washington, Department of Mathematics, Seattle, Washington 98195
Email address: skovacs@uw.edu
URL: http://www.math.washington.edu/~kovacs

Contemporary Mathematics
Volume **712**, 2018
http://dx.doi.org/10.1090/conm/712/14350

Full cones swept out by minimal rational curves on irreducible Hermitian symmetric spaces as examples of varieties underlying geometric substructures

Ngaiming Mok

This paper is dedicated to Professor Lawrence Ein.

ABSTRACT. We explore the application of the general theory of sub-VMRT structures to a special class of singular Schubert cycles on irreducible Hermitian symmetric spaces of the compact type. Specifically we consider germs of complex submanifolds modeled on the germ at a smooth point of the full cone $\mathcal{V}$ of minimal rational curves on an irreducible Hermitian symmetric space X of rank ≥ 2 other than a Lagrangian Grassmannian. Here $\mathcal{V} = \mathcal{V}(x)$ is the union of minimal rational curves emanating from a point $x \in X$. We study sub-VMRT structures $\varpi : \mathscr{C}(S) \to S$ on complex submanifolds $S \subset W$ of connected open subsets $W \subset X$ in the complex topology modeled on the inclusion $\big(\mathscr{C}_y(\mathcal{V}) \subset \mathscr{C}_y(X)\big)$ of VMRTs at a smooth point $y \in \mathcal{V}$, focusing on the question of linear saturation in the current article. On top of providing examples for illustrating the theory of sub-VMRT structures applied to singular uniruled projective subvarieties, $\mathcal{V} \subset X$ is also important for the study of holomorphic isometries of complex unit balls into irreducible bounded symmetric domains in view of the author's earlier result that, for an irreducible bounded symmetric domain $\Omega \subset X$ of rank ≥ 2 embedded into its dual compact manifold X by the Borel embedding, the intersection $\mathcal{V}(x) \cap \Omega$ is the image of a nonstandard holomorphic isometric embedding of the complex unit ball whenever x lies on the smooth locus of $\partial\Omega$.

1. Introduction and motivation

Starting with Hwang-Mok [**HM98**], a program of study on uniruled projective manifolds from a differential-geometric perspective was launched based on the geometry of their varieties of minimal rational tangents (VMRTs). To explain the latter concept, fixing an ample line bundle L on a uniruled projective manifold X we introduced the notion of a minimal rational curve, by which we mean a *free* rational curve on X of minimal degree with respect to L among all free rational curves on X. Denoting by Chow(X) the Chow space of X, a minimal rational component $\mathcal{K}$ on X is the normalization of a dense Zariski open subset $\mathcal{K}'$ of *some*

2010 *Mathematics Subject Classification.* Primary 53B21, 53C35, 53C55.
Key words and phrases. Minimal rational curve, variety of minimal rational tangents, non-degeneracy for substructures, linear saturation, Condition (T), holomorphic isometry.
Research partially supported by the GRF grant 7039/12P of the Hong Kong Research Grants Council.

irreducible subvariety $\mathcal{Q}$ of Chow(X) such that each member of $\mathcal{Q}$ is a rational 1-cycle and such that $\mathcal{K}' \subset \mathcal{Q}$ is the subset consisting of images $f(\mathbb{P}^1)$ of parametrized minimal rational curves $f : \mathbb{P}^1 \to X$. (For details, cf. Mok [**Mo16b**, §1].) Note that the quasi-projective variety $\mathcal{K}$ is nonsingular. We will make no distinction between $\kappa \in \mathcal{K}$ parametrized by $f : \mathbb{P}^1 \to X$ and $\kappa' \in \mathcal{K}'$ corresponding to the image cycle $f(\mathbb{P}^1)$ and refer to both as a minimal rational curve. The VMRT of a uniruled projective manifold $(X, \mathcal{K})$ equipped with a minimal rational component, denoted by $\mathscr{C}_x(X) \subset \mathbb{P}T_x(X)$ at a general point $x \in X$, is the collection of projectivizations of vectors tangent to minimal rational curves passing through x. It is our perspective that, given $(X, \mathcal{K})$, the underlying VMRT structure $\pi : \mathscr{C}(X) \to X$ is a rich geometric object which encodes a lot of the information about X as a projective manifold. (For a reference to the early part of the theory we refer the reader to Hwang-Mok [**HM99**]). Assuming furthermore that X is of Picard number 1 and that the Gauss map of $\mathscr{C}_x(X) \subset \mathbb{P}T_x(X)$ is immersive at a general smooth point, we proved in Hwang-Mok [**HM01**] a general result called the Cartan-Fubini Extension Theorem ascertaining that X is uniquely determined as a projective manifold by $\pi|_U : \mathscr{C}(X)|_U \to U$ for any nonempty connected open subset U of X in the *complex* topology. (Here and in what follows by an open subset of a projective variety (or complex space) we always mean an open subset in the complex topology unless otherwise specified.) In other words, given Fano manifolds $(X, \mathcal{K})$ and $(X', \mathcal{K}')$ of Picard number 1 equipped with minimal rational components, nonempty connected open subsets $U \subset X$ and $U' \subset X'$, and a biholomorphic map $f : U \xrightarrow{\cong} U'$ such that $[df](\mathscr{C}(X)|_U) = \mathscr{C}(X')|_{U'}$, we proved that f extends to a global biholomorphism $F : X \xrightarrow{\cong} X'$.

The study of $\pi : \mathscr{C}(X) \to X$, $\mathscr{C}(X) \subset \mathbb{P}T(X)$, as a fibered subspace of $\mathbb{P}T(X)$ leads to a rich geometric theory, especially when $\mathscr{C}_x(X) \subset \mathbb{P}T_x(X)$ are mutually projectively equivalent to each other as x varies over general points of X. In the latter case Hwang [**Hw10**] [**Hw12**] [**Hw15**] studied the Cartanian geometry of isotrivial families of VMRTs and obtained quite a number of rigidity results notably concerning the flatness of the VMRT structure, where one works primarily with the geometric structure defined on connected open subsets in the complex topology. On the other hand, when we fix a uniruled projective manifold $(X, \mathcal{K})$, we have the class of projective subvarieties of X uniruled by $\mathcal{K}$ and we are interested in the characterization of some of these projective subvarieties $Z \subset X$ by their germs $(Z; z_0)$ of complex-analytic subvarieties at general smooth points $z_0 \in Z$. Restricting $\pi : \mathscr{C}(X) \to X$ to a connected open subset $W \subset X$ and considering a submanifold $S \subset W$ such that the intersection $\mathscr{C}_x(S) := \mathscr{C}_x(X) \cap \mathbb{P}T_x(S)$ for $x \in S$ defines a sub-VMRT structure (cf. two paragraphs below for an example and §2 for the general definition), the study of $S \subset W \subset X$ may be regarded as an analogue of the study of locally closed Riemannian submanifolds of a given Riemannian manifold. Here, a principal problem is to characterize among sub-VMRT structures $\varpi : \mathscr{C}(S) \to S$, $S \subset W \subset X$, those for which the underlying complex manifold S can be analytically continued to a projective subvariety $Z \subset X$. In other words, we are interested in determining whether there exists an irreducible projective subvariety $Z \subset X$ such that $S \subset Z$ and such that $\dim(S) = \dim(Z)$, and we call this the *Algebraicity Problem.* Moreover, when a model uniruled projective subvariety $Z \subset X$ is given, $0 \in Z \subset X$ is a general smooth point, and $\varpi : \mathscr{C}(S) \to S$ is in some precise sense modeled on $\big(\mathscr{C}_0(Z) \subset \mathscr{C}_0(X)\big)$ (cf. two paragraphs below for the example of Grassmann substructures), we are further interested in the rigidity question as to whether any $S \subset W \subset X$ modeled on $\big(\mathscr{C}_0(Z) \subset \mathscr{C}_0(X)\big)$ is necessarily an open subset of a translate γZ of the model projective subvariety $Z \subset X$ by an automorphism $\gamma \in \mathrm{Aut}(X)$. We call this the *Recognition Problem.* When $X = G/P$ is a rational homogeneous space of Picard number 1, for both the Algebraicity Problem and the Recognition Problem we are typically interested in the characterization of certain smooth linear sections $Z \subset X$ (i.e., smooth intersections of X in its minimal canonical embedding into $\mathbb{P}(\Gamma(X, \mathcal{O}(1))^*) =: \mathbb{P}^N$ with projective linear subspaces of $\mathbb{P}^N$), where there is excessive intersection in $\mathscr{C}_0(Z) = \mathscr{C}_0(X) \cap \mathbb{P}T_0(Z)$. More generally, one can raise the same questions for certain possibly singular linear sections $Z \subset X$ where such a phenomenon of excessive intersection occurs at a general smooth point of Z and there exist minimal rational curves of X lying on the smooth part $\mathrm{Reg}(Z) \subset Z$.

When $X = G/P$ is Hermitian symmetric, and $Z \subset X$ is a smooth Schubert cycle, the problem of *recognizing* $Z \subset X$ from geometric properties of its tangent bundle $T(Z)$ is a crucial step in determining whether $Z \subset X$ is Schur rigid in the sense that for any integer $r \geq 1$, any algebraic cycle homologous to rZ is necessarily a sum $\gamma_1 Z + \cdots + \gamma_r Z$, $\gamma_k \in G$ for $1 \leq k \leq r$. When X is Hermitian symmetric, the problem on Schur rigidity for smooth Schubert cycles was solved in special cases by Walters [**Wa97**] and Bryant [**Br01**] and in general by Hong [**Ho07**]. The same problem for singular Schubert cycles in case the ambient space is a Grassmannian was treated by Hong [**Ho05**] and the general case where X is an irreducible Hermitian symmetric space of the compact type and $Z \subset X$ is an arbitrary Schubert cycle was settled by Robles-The [**RT12**]. In the cited works crucial to the proofs in the affirmative cases are algebraic results ascertaining the vanishing of certain cohomology groups defined in terms of Lie algebras.

For illustration consider the question of characterizing Grassmann submanifolds of Grassmann manifolds, in which $Z := G(a,b) \subset G(p,q) =: X$, where $a \leq p, b \leq q, \min(a,b) \geq 2$, and $G(a,b) \subset G(p,q)$ is the obvious inclusion map. Denote by $0 \in Z \subset X$ any base point. Let $r \geq 1$ be an integer and let $Y \subset X$ be an algebraic cycle, $\dim(Y) = ab =: m$, such that $[Y] = r[Z]$ as cohomology classes. Write $n = a + b$. The homology group $H_{2m}(X, \mathbb{Z}) =: \mathbb{Z}^s$ is generated as an Abelian group by Schubert classes, i.e., homology classes of m-dimensional Schubert cycles. Let $\{Z_1 = Z, Z_2, \cdots, Z_s\}$ be a collection of m-dimensional Schubert cycles on X such that the homology classes $[Z_1], \cdots, [Z_s]$ form an integral basis of $H_{2m}(X, \mathbb{Z})$. By Walters [**Wa97**] and Byrant [**Br01**] there are nonnegative $U(n)$-invariant closed (m,m)-forms $\nu_1, \cdots, \nu_s$ on X such that $\int_{Z_i} \nu_j = \delta_{ij}$, where δ_{ij} stands for the Kronecker delta. From $[Y] = r[Z]$ it follows by integrating the closed (m,m)-forms $\nu_2, \cdots, \nu_s$ on the smooth part $\mathrm{Reg}(Y)$ of Y and from the nonnegativity of such forms that for $2 \leq i \leq s$ the restriction of ν_i to $\mathrm{Reg}(Y)$ must vanish identically. This introduces the Schur differential system on the Grassmann bundle $\mathrm{Gr}(m, T(X))$, to be denoted $\mathscr{S} \subset \mathrm{Gr}(m, T(X))$, where for $x \in X$, $\mathscr{S}_x \subset \mathrm{Gr}(m, T_x(X))$ consists of all m-planes $\Pi \subset T_x(X)$ such that the (m,m)-covectors $\nu_i(x)$, $2 \leq i \leq s$, annihilate the real $2m$-dimensional vector space $\mathrm{Re}(\Pi) \subset \mathrm{Re}(T_x(X))$. $X = G(p,q)$ is equipped with universal holomorphic vector bundles U and V of rank p resp. q such that $T(X) \cong U \otimes V$. (In what follows $T(X)$ is identified with $U \otimes V$.) It can be proven that, given a connected open subset $W \subset X$ and a complex submanifold $S \subset W$, the requirement that $T_x(S)$ belongs to $\mathscr{S}_x$ for every $x \in S$ is equivalently the requirement that there exists holomorphic vector subbundles $U' \subset U|_S$ and $V' \subset V|_S$ of rank a resp. b such that $T(S) = U' \otimes V'$. As a consequence, S is an integral submanifold of the Schur differential system if and only if for any point $x \in S$, there exists some $\varphi_x \in \mathrm{Aut}(X)$ such that $x = \varphi_x(0)$ and such that $T_x(S) = d\varphi_x(0)(T_0(Z))$. (In the work of Hong [**Ho07**] analogous statements were established in the case of smooth nonlinear Schubert cycles in an irreducible Hermitian symmetric space of the compact type.) For the Grassmannian $X = G(p,q)$ the VMRT at $x \in X$ is given by $\mathscr{C}_x(X) = \varsigma(\mathbb{P}U_x \times \mathbb{P}V_x)$, where $\varsigma : \mathbb{P}U_x \times \mathbb{P}V_x \to \mathbb{P}(U_x \otimes V_x) = \mathbb{P}T_x(X)$ is the Segre embedding. For the pair $(Z,X) = \big(G(a,b), G(p,q)\big)$ the Schur differential system $\mathscr{S} \subset \mathrm{Gr}(m, T(X))$ is related to VMRTs in that $\Pi \in \mathscr{S}_x$ if and only if $\Pi = \mathscr{C}_x(X) \cap \mathbb{P}T_x(Y)$ for some sub-Grassmannian $Y = \varphi_x Z$ such that $\varphi_x \in \mathrm{Aut}(X)$ and $\varphi_x(0) = x$. Thus $S \subset W \subset X$ is an integral submanifold of $\mathscr{S}$ if and only if, defining $\mathscr{C}_x(S) := \mathscr{C}_x(X) \cap \mathbb{P}T_x(S)$, for any $x \in S$ the triple $\big(\mathscr{C}_x(S) \subset \mathscr{C}_x(X) \subset \mathbb{P}T_x(X)\big)$ is the image of $\big(\mathscr{C}_0(Z) \subset \mathscr{C}_0(X) \subset \mathbb{P}T_0(X)\big)$ under $d\varphi_x(0)$ for some $\varphi_x \in \mathrm{Aut}(X)$ such that $\varphi_x(0) = x$. The latter requirement is in fact equivalent to the *a priori* weaker requirement that $\big(\mathscr{C}_x(S) \subset \mathscr{C}_x(X) \subset \mathbb{P}T_x(X)\big)$ is congruent to $\big(\mathscr{C}_0(Z) \subset \mathscr{C}_0(X) \subset \mathbb{P}T_0(X)\big)$ under some projective linear isomorphism $\mathbb{P}T_x(X) \xrightarrow{\cong} \mathbb{P}T_0(X)$. In the terminology of Mok-Zhang [**MZ17**], writing $\mathscr{C}(S) := \mathscr{C}(X) \cap \mathbb{P}T(S)$ and $\varpi : \mathscr{C}(S) \to S$ for the canonical projection, $\varpi : \mathscr{C}(S) \to S$ defines a sub-VMRT structure modeled on $(Z \subset X)$. We also say that the sub-VMRT structure $\varpi : \mathscr{C}(S) \to S$ is modeled on $\big(\mathscr{C}_0(Z) \subset \mathscr{C}_0(X)\big)$. We may also say in this special case that S inherits a Grassmann substructure of rank ≥ 2 (more precisely a $G(a,b)$-substructure) from $X = G(p,q)$. The Schur rigidity of Z in X was established by showing that connected open subsets of translates of Z under $\mathrm{Aut}(X)$ are the only integral submanifolds of $\mathscr{S}$, which results from the vanishing of certain cohomology groups defined in terms of Lie algebras.

We may say that $S \subset X$ is recognized as an open subset of a Grassmann submanifold γZ from the fact that $\varpi : \mathscr{C}(S) \to S$ is modeled on $\big(\mathscr{C}_0(Z) \subset \mathscr{C}_0(X)\big)$. This gives a prototype for the Recognition Problem for distinguished projective subvarieties of G/P by means of their sub-VMRT structures by the study of integral submanifolds of certain associated differential systems. We note that two differential systems are involved in the example, the Schur differential system $\mathscr{S}$ and the Schubert differential system $\mathscr{S}' \subset \mathscr{S}$ which consists at each point $x \in X$ of the collection of tangent spaces of translates γZ of Z passing through x. When X is a an irreducible Hermitian symmetric spaces of rank ≥ 2 and $Z \subset X$ is a nonlinear smooth Schubert cycle (and hence (Z,X) is of sub-diagram type and X_0 is also an irreducible Hermitian symmetric of rank ≥ 2), both $\mathscr{S}'$ and $\mathscr{S}$ are defined analogously. In [**Ho05**] Hong solved the integrability problem for $\mathscr{S}'$ and showed that $\mathscr{S}' = \mathscr{S}$ always holds, thereby proving Schur rigidity for $Z \subset X$.

From the perspective of the theory of VMRTs it is natural to consider the relevant recognition problem separately as a differential-geometric problem. In a related direction Hong-Mok [**HoM10**] extended the Cartan-Fubini Extension Theorem to the non-equidimensional situation. For its formulation let $(Z, \mathcal{H})$ resp. $(X, \mathcal{K})$ be a uniruled projective manifold of Picard number 1 equipped with a minimal rational component. Let $z_0 \in Z$ resp. $x_0 \in X$ be a base point lying outside the bad locus of $(Z, \mathcal{H})$ resp. $(X, \mathcal{K})$. Let $f : (Z, z_0) \to (X; x_0)$ be a germ of VMRT-respecting holomorphic immersion satisfying the hypotheses that, for z belonging to some neighborhood U of z_0, a general point $[\alpha]$ of $[df](\mathscr{C}_z(Z)) \subset \mathscr{C}_{f(z)}(X)$ is a smooth point and a certain nondegeneracy condition on the second fundamental form of $\mathscr{C}_{f(z)}(X) \subset \mathbb{P}T_{f(z)}(X)$ at $[\alpha]$ applied to the pair $\big([df](\mathscr{C}_z(Z)) \subset \mathscr{C}_{f(z)}(X)\big)$ is satisfied. Here a germ of holomorphic immersion f is said to be VMRT-respecting whenever $[df](\mathscr{C}_z(Z)) = \mathscr{C}_{f(z)}(X) \cap \mathbb{P}T_{f(z)}(X)$ for z belonging to U. In the case where $X = G/P$ is a rational homogeneous space defined by a marked Dynkin diagram $(\mathscr{D}(\mathfrak{g}), \gamma)$ and $Z = G_0/P_0$ is defined by a marked Dynkin sub-diagram $(\mathscr{D}(\mathfrak{g}_0), \gamma_0)$ of $(\mathscr{D}(\mathfrak{g}), \gamma)$, so that Z can be holomorphically and equivariantly embedded into X, denoted by $i : Z \hookrightarrow X$, by means of the inclusion $G_0 \subset G$ of complex Lie groups, applying non-equidimensional Cartan-Fubini extension Hong-Mok [**HoM10**] proved that f extends to a standard embedding in the case where Z is nonlinear and the marking is at a long simple root (cf. also Mok [**Mo08a**] for the case of the Grassmannian). (Here a holomorphic map $F : Z \to X$ is said to be a standard embedding if and only if $F = \varphi \circ i \circ \psi$ where $\varphi \in \mathrm{Aut}(X)$ and $\psi \in \mathrm{Aut}(Z)$.) The same problem for the maximal linear case and for the case of markings at a short simple root were settled by Hong-Park [**HoP11**]. Returning to the special case of the Recognition Problem for Grassmann substructures $\varpi : \mathscr{C}(S) \to S$ of rank ≥ 2, the results of [**HoM10**] only show that S is an open subset of a sub-Grassmannian (i.e., a translate of $G(a, b) \subset G(p, q)$) under the additional hypothesis that the Grassmann structure on S inherited from X is flat, equivalently that a finite number of curvature-like tensors vanish according to Guillemin [**Gui65**]. However, the methodology introduced in [**HoM10**] has now been strengthened to give in particular a new solution to the Recognition Problem for nonlinear smooth Schubert cycles in irreducible Hermitian symmetric spaces as we introduced and considered sub-VMRT structures in Mok-Zhang [**MZ17**] in general rather than only those that arise from images of VMRT-respecting germs of holomorphic immersions. Using the same method we settled in [**MZ17**] the Recognition Problem for a large class of pairs of rational homogeneous spaces of Picard number 1.

In Mok-Zhang [**MZ17**] we defined admissible pairs (X_0, X) of rational homogeneous spaces of Picard number 1 and the notion of rigid pairs (X_0, X) among them, and introduced the general theory of sub-VMRT structures on uniruled projective manifolds. Let $0 \in X_0 \subset X$ be any base point and denote by $\mathscr{C}(X_0) \subset \mathbb{P}T(X_0)$ resp. $\mathscr{C}(X) \subset \mathbb{P}T(X)$ the fibered spaces of VMRTs on X_0 resp. X with respect to the unique minimal rational component of projective lines on X_0 resp. X, where X is identified as a projective submanifold by means of the first canonical embedding $\iota : X \hookrightarrow \mathbb{P}(\Gamma(X, \mathcal{O}(1))^*)$. Letting $W \subset X$ be a connected open subset and $S \subset W$ be a complex submanifold, defining $\mathscr{C}_x(S) := \mathscr{C}_x(X) \cap \mathbb{P}T_x(S)$, and denoting by $\varpi : \mathscr{C}(S) \to S$ the natural projection, we say that $\varpi : \mathscr{C}(S) \to S$ is modeled on $(X_0 \subset X)$, equivalently on $\big(\mathscr{C}_0(X_0) \subset \mathscr{C}_0(X)\big)$, if and only if for every point x on S there exists a projective linear isomorphism $\Phi_x : \mathbb{P}T_0(X) \xrightarrow{\cong} \mathbb{P}T_x(X)$ such that $\Phi_x(\mathscr{C}_0(X)) = \mathscr{C}_x(X)$ and $\Phi_x(\mathscr{C}_0(X_0)) = \mathscr{C}_x(S)$. When $X_0 \subset X$ is nonlinear and (X_0, X) is of sub-diagram type we proved rigidity of the pair (X_0, X). While [**HoM10**] deals with VMRT-respecting germs of holomorphic maps $f : (Z; z_0) \to (X; x_0)$, in [**MZ17**] we consider sub-VMRT structures $\varpi : \mathscr{C}(S) \to S$ on $S \subset W \subset X$, $\mathscr{C}(S) := \mathscr{C}(X) \cap \mathbb{P}T(S)$. In place of the nondegeneracy condition expressed in terms of the projective second fundamental form σ of $\mathscr{C}_x(X) \subset \mathbb{P}T_x(X)$ applied to the pair $(\mathscr{C}_x(S) \subset \mathscr{C}_x(X))$ introduced in [**HoM10**], we defined in [**MZ17**] a new nondegeneracy condition for substructures by introducing a vector-valued quadratic form τ which is roughly speaking derived from σ by quotienting out the tangent subspaces of the complex submanifold $S \subset W$. The proof of the rigidity of pairs (X_0, X) of admissible rational homogeneous spaces of sub-diagram type was achieved in two steps. First, we resolved the Algebraicity Problem in the affirmative by means of the quadratic form τ, showing first of all that for a sub-VMRT structure $\varpi : \mathscr{C}(S) \to S$ on $S \subset W \subset X$ modeled on $(X_0 \subset X)$, S is linearly saturated in the sense that for any projective line ℓ on X passing through $x \in S$ and being tangent to S at x, the germ of holomorphic curve $(\ell; x)$ must lie on $(S; x)$. The second step consists of showing by a process of adjunction of minimal rational curves that S is an open subset of the unique rational homogeneous submanifold $Z \subset X$ such that $Z = \gamma X_0$ for some $\gamma \in \mathrm{Aut}(X)$ and such that S and Z are tangent at x. (Here uniqueness of Z satisfying the latter two conditions is deduced from the fact that (X_0, X) is of sub-diagram type.) The second step is based on the method of parallel transports of VMRTs along minimal rational curves introduced in [**Mo08a**] and [**HoM10**]. In the case of nonlinear admissible pairs (X_0, X) of irreducible Hermitian symmetric spaces of sub-diagram type, rigidity of the pair (X_0, X) is the same as the assertion that the only integral submanifolds of the Schur differential system $\mathscr{S}$ (defined analogously as in the case of the Grassmann in the above) are translates of X_0.

We note that in the case of admissible pairs of nonlinear rational homogeneous spaces (X_0, X) of Picard number 1 of sub-diagram type, Hong-Mok proved in [**HoM13**] a weakened form of Schur rigidity by restricting to reduced cycles, i.e., showing that any projective subvariety $Y \subset X$ homologous to $X_0 \subset X$ must be a translate of X_0. In this case the problem was reduced in [**HoM13**] to proving that small deformations $\{X_t : |t| < 1\}$ of X_0 as a complex submanifold of X are translates of X_0. The proof broke up essentially into two steps. In the first step with minor exceptions it was proven that VMRTs of $X_t, |t| < 1$, agree with those of X_0 modulo the obvious action of $\mathrm{Aut}(X)$, while the issue of linear saturation of $X_t, |t| < 1$, does not arise since small deformations of X_0 are necessarily uniruled by projective lines. The second step, which was a solution of the Recognition Problem, was a precursor to the arguments used in Mok-Zhang [**MZ17**]. Rigidity of the pairs (X_0, X) in [**MZ17**] solves the integrability problem for the Schubert differential system, but apart from the Hermitian symmetric case it is not known whether the problem of Schur rigidity for $X_0 \subset X$ can be reduced to the integrability problem for the Schubert differential system.

In this article, we explore the application of the general theory of sub-VMRT structures of Mok-Zhang [**MZ17**] to a special class of singular Schubert cycles on irreducible Hermitian symmetric spaces of the compact type. Specifically we consider germs of complex submanifolds modeled on the germ at a smooth point of the full cone $\mathcal{V}$ of minimal rational curves on an irreducible Hermitian symmetric space X of rank ≥ 2 other than a Lagrangian Grassmannian. Here by a full cone of minimal rational curves we mean the union $\mathcal{V} = \mathcal{V}(x)$ of minimal rational curves emanating from a point $x \in X$, noting that the associated sub-VMRT structures $\varpi : \mathscr{C}(S) \to S$ have singular and possibly reducible fibers. For these

Schubert cycles $\mathcal{V} \subset X$ we apply the methods and results of [**MZ17**] to study sub-VMRT structures $\varpi : \mathscr{C}(S) \to S$ on complex submanifolds $S \subset W$ of connected open subsets $W \subset X$ in the complex topology modeled on $(\mathcal{V} \subset X)$. (Here and henceforth by the phrase "modeled on $(\mathcal{V} \subset X)$" we will always mean "modeled on the germ of the inclusion $(\mathcal{V} \subset X)$ at a smooth point of $\mathcal{V}$.) Since the problem of Schur rigidity on possibly singular Schubert cycles $Z \subset X$ has been completely settled in Robles-The [**RT12**] in the case where X is an irreducible Hermitian symmetric space of the compact type, our focus is rather on the methodology, proving results by means of checking nondegeneracy conditions arising from the theory of sub-VMRT structures, a method that is potentially applicable to rational homogeneous spaces and horospherical varieties of Picard number 1. By a case-by-case checking it follows from [**RT12**] that with the exception of the case where X is the hyperquadric Q^n, $n \geq 3$, or the rank-2 Grassmann manifold $G(2, q)$, $q \geq 2$, Schur rigidity holds for any full cone $\mathcal{V} = \mathcal{V}(x) \subset X$ of minimal rational curves (including the Lagrangian Grassmannian). We are interested in general in the Recognition Problem for Schubert cycles on rational homogeneous spaces of Picard number 1 and, by way of illustration, we will show that in the special cases considered here, methods of [**MZ17**] apply to settle the Algebraicity Problem (cf. second paragraph of the Introduction) for a sub-VMRT structure $\varpi : \mathscr{C}(S) \to S$ on complex submanifold $S \subset W \subset X$ modeled on $\mathcal{V} \subset X$, and that, under the additional assumption that the sub-VMRT structure on $S \subset W$ is *intrinsically flat* (cf. §5), it remains the case that S is linearly saturated and that $S \subset Z$ for some irreducible projective subvariety $Z \subset X$, $\dim(Z) = \dim(X)$, for the excluded cases of the hyperquadric of dimension $n \geq 4$ and rank-2 Grassmannians. The Recognition Problem for $(\mathcal{V}, X)$ will be dealt with in another article (cf. paragraph below).

On top of providing examples for illustration, the set of full cones of minimal rational curves $\mathcal{V}$ is also important for the study of holomorphic isometries of complex unit balls into irreducible bounded symmetric domains. In fact, taking $\Omega \subset X$ to be the Hermitian symmetric space of the noncompact type dual to X and embedded in X by means of the Borel embedding, and taking $q \in \partial\Omega$ to be a regular boundary point, the author has proven in [**Mo16a**] that $V(q) := \mathcal{V}(q) \cap \Omega$ is the image of a holomorphic isometric embedding of the complex hyperbolic space form into Ω equipped with a canonical Kähler-Einstein metric, and the Recognition Problem for $\mathcal{V} \subset X$ enters into the picture in the uniqueness question for holomorphic isometric embedding of the complex unit ball of maximal admissible dimension. In conjunction with results of the current article, the Recognition Problem for $\mathcal{V} \subset X$ has been settled by differential-geometric means by Mok-Yang [**MY18**] by way of the Thickening Lemma of Mok-Zhang [**MZ17**, Proposition 6.1] and a process of reconstruction analogous to that in Mok [**Mo08a**] and Hong-Mok [**HoM10**] [**HoM13**]. We believe that the theory of sub-VMRT structures on uniruled projective manifolds, beyond its applicability to the study of uniruled projective subvarieties, also provides a useful link for the study of complex-analytic problems such as those on bounded symmetric domains. For the explanation of this perspective we refer the reader to Mok [**Mo16b**] on the theory of geometric structures and sub-structures.

2. Background materials and results

We provide here a number of basic definitions and results taken from Mok-Zhang [**MZ17**] necessary for the current article. Let $(X, \mathcal{K})$ be a uniruled projective manifold equipped with a minimal rational component, $\mathcal{K} \subset \mathcal{Q}$ be the compactification of $\mathcal{K}$ by the normalization of some Chow component of X, $B \subset X$ be the bad locus of $(X, \mathcal{K})$, i.e., the minimal subset of X outside of which every member of $\mathcal{Q}$ passing through x must necessarily belong to $\mathcal{K}$, and $\pi : \mathscr{C}(X) \to X$, $\mathscr{C}(X) \subset \mathbb{P}T(X - B)$, be the VMRT structure of $(X, \mathcal{K})$. We denote by $B' \supset B$ the minimal subset outside of which the tangent map is a birational finite morphism (cf. Mok-Zhang [**MZ17**, §5]) and call B' the enhanced bad locus of $(X, \mathcal{K})$. Let $W \subset X - B'$ be a connected open subset in the complex topology, $S \subset W$ be a complex submanifold, and define $\mathscr{C}(S) := \mathscr{C}(X) \cap \mathbb{P}T(S)$. We have the following definition of sub-VMRT structures given in Mok-Zhang [**MZ17**, Definition 5.1].

Definition 2.1. We say that $\varpi := \pi|_{\mathscr{C}(S)} : \mathscr{C}(S) \to S$ is a sub-VMRT structure on $(X, \mathcal{K})$ if and only if (a) the restriction of ϖ to each irreducible component of $\mathscr{C}(S)$ is surjective, and (b) at a general point $x \in S$ and for any irreducible component Γ_x of $\mathscr{C}_x(S)$, we have $\Gamma_x \not\subset \mathrm{Sing}(\mathscr{C}_x(X))$.

Next we will need to consider pairs consisting of VMRTs and their linear sections. We introduce the notion of proper pairs of projective subvarieties, as follows (cf. Mok-Zhang [**MZ17**, Definition 5.2]).

Definition 2.2. Let V be a Euclidean space and $\mathcal{A} \subset \mathbb{P}(V)$ be an irreducible subvariety and $\mathcal{B} \subset \mathcal{A}$ is a subvariety. We say that $(\mathcal{B}, \mathcal{A})$ is a proper pair of projective subvarieties if and only if $\mathcal{B}$ is a linear section of $\mathcal{A}$, and none of the irreducible components Γ of $\mathcal{B}$ is contained in $\mathrm{Sing}(\mathcal{A})$.

Note that for a uniruled projective manifold X and a complex submanifold $S \subset W \subset X - B'$ inheriting a sub-VMRT structure $\varpi : \mathscr{C}(S) \to S$ as in Definition 2.1, at a general point $x \in S$, $(\mathscr{C}_x(S), \mathscr{C}_x(X))$ is a proper pair of subvarieties. We introduce next two nondegeneracy conditions in terms of second fundamental forms of VMRTs. They concern nondegeneracy for mappings and nondegeneracy for substructures. For the formulation recall that for a finite-dimensional vector space V and for a subset $Z \subset \mathbb{P}(V)$, denoting by $\lambda : V - \{0\} \to \mathbb{P}(V)$ the canonical projection we write $\widetilde{Z} := \lambda^{-1}(Z) \subset V - \{0\}$ for the affinization of Z. The following two definitions are adaptations of Mok-Zhang [**MZ17**, Definition 5.3].

Definition 2.3. Let V be a finite-dimensional vector space, $E \subsetneq V$ be a vector subspace and $(\mathcal{B}, \mathcal{A})$ be a proper pair of projective subvarieties in $\mathbb{P}(V)$, $\mathcal{B} := \mathcal{A} \cap \mathbb{P}(E) \subset \mathcal{A} \subset \mathbb{P}(V)$. Let $\chi \in \widetilde{\mathcal{B}}$ be a smooth point of both $\widetilde{\mathcal{A}}$ and $\widetilde{\mathcal{B}}$, and let $\sigma_\chi : S^2T_\chi(\widetilde{\mathcal{A}}) \to V/T_\chi(\widetilde{\mathcal{A}})$ be the second fundamental form at $\chi \in \widetilde{\mathcal{B}}$ of $\widetilde{\mathcal{A}}$ in V with respect to the Euclidean flat connection on V. We say that $(\mathcal{B}, \mathcal{A})$ is nondegenerate for mappings if and only if for each irreducible component Γ of $\mathcal{B}$ and for a general point $\chi \in \widetilde{\Gamma}$, we have

$$\left\{\eta \in T_\chi(\widetilde{\mathcal{A}}) : \sigma_\chi(\eta, \xi) = 0 \text{ for any } \xi \in T_\chi(\widetilde{\mathcal{B}})\right\} = \mathbb{C}\chi\,.$$

Definition 2.4. In the notation of Definition 2.3 write furthermore $V' \subset V$ for the linear span of $\widetilde{\mathcal{A}}$ and define $E' := E \cap V'$. Let $\nu_\chi : V/T_\chi(\widetilde{\mathcal{A}}) \to V/(T_\chi(\widetilde{\mathcal{A}}) + E')$ be the canonical projection and define $\tau_\chi : S^2T_\chi(\widetilde{\mathcal{A}}) \to V/(T_\chi(\widetilde{\mathcal{A}}) + E')$

by $\tau_\chi := \nu_\chi \circ \sigma_\chi$. We say that $(\mathcal{B}, \mathcal{A}; E)$ is nondegenerate for substructures if and only if for each irreducible component Γ of $\mathcal{B}$ and for a general point $\chi \in \widetilde{\Gamma}$, we have

$$\left\{\eta \in T_\chi(\widetilde{\mathcal{A}}) : \tau_\chi(\eta, \xi) = 0 \text{ for any } \xi \in T_\chi(\widetilde{\mathcal{B}})\right\} = T_\chi(\widetilde{\mathcal{B}}).$$

When $E' = E \cap V'$ is the same as the linear span of $\widetilde{\mathcal{B}}$ we drop the reference to E, with the understanding that the projection map ν is defined by using the linear span of $\widetilde{\mathcal{B}}$ as E'.

We will use interchangeably the second fundamental form of $\widetilde{\mathcal{A}}$ in V, and the projective second fundamental form of $\mathcal{A}$ in $\mathbb{P}V$, denoting both by σ. More precisely, for a smooth point χ of $\mathcal{A}$, we have $T_{[\chi]}(V) \cong V/\mathbb{C}\chi$, and, for $\xi, \eta \in T_\chi(\widetilde{\mathcal{A}})$, $\xi' := \xi + \mathbb{C}\chi \in V/\mathbb{C}\chi, \eta' := \eta + \mathbb{C}\chi \in V/\mathbb{C}\chi$, we have $\sigma_\chi : S^2T_\chi(\widetilde{\mathcal{A}}) \to V/T_\chi(\widetilde{\mathcal{A}})$, $\sigma_{[\chi]} : S^2T_{[\chi]}(\mathcal{A}) \to T_{[\chi]}(\mathbb{P}V)/T_{[\chi]}(\mathcal{A}) = (V/\mathbb{C}\chi)/(T_\chi(\widetilde{\mathcal{A}})/\mathbb{C}\chi) \cong V/T_\chi(\widetilde{\mathcal{A}})$, and, identifying the two normal spaces by means of the latter isomorphism, we have $\sigma_{[\chi]}(\xi', \eta') = \sigma_\chi(\xi, \eta)$. The same consideration applies analogously to the vector-valued quadratic form $\tau_\chi : S^2T_\chi(\widetilde{\mathcal{A}}) \to V/(T_\chi(\widetilde{\mathcal{A}}) + E')$ and its projectivized form $\tau_{[\chi]} : S^2T_\chi(\mathcal{A}) \to T_{[\chi]}(\mathbb{P}V)/(T_{[\chi]}(\mathcal{A}) + T_{[\chi]}(\mathbb{P}(E'))) = (V/\mathbb{C}\chi)/((T_\chi(\widetilde{\mathcal{A}}) + E')/\mathbb{C}\chi) \cong V/(T_\chi(\widetilde{\mathcal{A}}) + E')$, and we have $\tau_{[\chi]}(\xi', \eta') = \tau_\chi(\xi, \eta)$.

For the study of rigidity properties of sub-VMRT structures, on top of non-degeneracy conditions formulated in terms of second fundamental forms, we also need a condition regarding the intersection $\mathscr{C}_x(S) := \mathscr{C}_x(X) \cap \mathbb{P}T_x(S)$, to be called Condition (T), as follows (cf. [**MZ17**, Definition 5.4]).

DEFINITION 2.5. Let $\varpi : \mathscr{C}(S) \to S$, $\mathscr{C}(S) := \mathscr{C}(X) \cap \mathbb{P}T(S)$, be a sub-VMRT structure on $S \subset W \subset X - B'$ as in Definition 2.1. For a point $x \in S$, and $[\alpha] \in \mathrm{Reg}(\mathscr{C}_x(S)) \cap \mathrm{Reg}(\mathscr{C}_x(X))$, we say that $(\mathscr{C}_x(S), [\alpha])$, or equivalently $(\widetilde{\mathscr{C}_x}(S), \alpha)$, satisfies Condition (T) if and only if $T_\alpha(\widetilde{\mathscr{C}_x}(S)) = T_\alpha(\widetilde{\mathscr{C}_x}(X)) \cap T_x(S)$. We say that $\varpi : \mathscr{C}(S) \to S$ satisfies Condition (T) at x if and only if $(\widetilde{\mathscr{C}_x}(S), [\alpha])$ satisfies Condition (T) for a general point $[\alpha]$ of each irreducible component of $\mathrm{Reg}(\mathscr{C}_x(S)) \cap \mathrm{Reg}(\mathscr{C}_x(X))$. We say that $\varpi : \mathscr{C}(S) \to S$ satisfies Condition (T) if and only if it satisfies the condition at a general point $x \in S$.

The following two results are the principal results of Mok-Zhang [**MZ17**] relevant to the current article which are adaptations of [**MZ17**, Theorem 1.4] and [**MZ17**, Main Theorem 2]. In the notation of the preceding paragraphs recall that $\varpi := \pi\big|_{\mathscr{C}(S)}$, and $\varpi : \mathscr{C}(S) \to S$ is a sub-VMRT structure.

THEOREM 2.6. *Suppose the VMRT structure $\varpi : \mathscr{C}(S) \to S$ on $S \subset W \subset X - B'$ satisfies* Condition (T). *Assume furthermore that for a general point x on S and for each of the irreducible components $\Gamma_{k,x}$ of $\mathscr{C}_x(S)$, $1 \le k \le m$, the pair $\big(\Gamma_{k,x}, \mathscr{C}_x(X)\big)$ is nondegenerate for substructures. Then, S is rationally saturated with respect to $(X, \mathcal{K})$.*

By the concluding sentence of Theorem 2.6 we mean that for *any* minimal rational curve ℓ belonging to $\mathcal{K}$ such that ℓ is tangent to S at some point $x \in \ell \cap S$, the germ $(\ell; x)$ of holomorphic curve must lie on $(S; x)$. When the ambient uniruled projective manifold X is of Picard number 1 (and hence Fano) and uniruled by lines, i.e., by minimal rational curves whose homology classes are generators of

$H_2(X, \mathbb{Z}) \cong \mathbb{Z}$, we have in [**MZ17**, Main Theorem 2] the following algebraicity result for germs of sub-VMRT structures on X.

THEOREM 2.7. *In Theorem* 2.6 *suppose furthermore that* $(X, \mathcal{K})$ *is a projective manifold of Picard number 1 uniruled by lines and that the distribution* $\mathscr{D}$ *on* S *defined by* $\mathscr{D}_x := \mathrm{Span}(\widetilde{\mathscr{C}_x}(S))$ *is bracket generating. Then, there exists an irreducible subvariety* $Z \subset X$ *such that* $S \subset Z$ *and such that* $\dim(Z) = \dim(S)$.

Note that the bracket-generating condition on $\mathscr{D}$ is trivially satisfied whenever $\mathscr{C}_x(S) \subset \mathbb{P}T_x(S)$ is linearly nondegenerate at a general point $x \in S$. When $(X, \mathcal{K})$ is a Fano manifold equipped with a uniruling by lines, a rationally saturated sub-VMRT structure S on X is said to be linearly saturated.

3. Sub-VMRTs structures modeled on full cones of minimal rational curves in the Hermitian symmetric case

We discuss here some examples on which Main Theorem 2 of [**MZ17**] and its proof apply to show that sub-VMRT structures modeled on them are algebraic, possibly under additional assumptions. These are subvarieties with isolated singularities of irreducible Hermitian symmetric spaces of the compact type. They are examples of singular Schubert cycles for which methods of VMRT geometry especially sub-VMRT structures apply to study rigidity problems in the spirit of Mok [**Mo08a**], Hong-Mok [**HoM10**] [**HoM13**] and Mok-Zhang [**MZ17**], and they are also particularly interesting in view of their relation to holomorphic isometries in Kähler geometry as in Mok [**Mo16a**]. We will see from these examples that in case nondegeneracy of substructures fails, it may still happen that underlying complex submanifolds $S \subset W$ of sub-VMRT structures arising from *VMRT-respecting maps* remain always linearly saturated and algebraic as germs of manifolds.

Let $(X, \mathcal{K})$ be an irreducible Hermitian symmetric space of the compact type of rank ≥ 2, equipped with the uniruling by projective lines. Write $X = G/P$, where G is the identity component of the group $\mathrm{Aut}(X)$ of biholomorphic automorphisms and $P \subset G$ is a parabolic subgroup. Let $x \in X$ and denote by $\mathcal{V}(x)$ the union of minimal rational curves passing through x. For $\gamma \in G$ we have $\gamma(\mathcal{V}(x)) = \mathcal{V}_{\gamma(x)}$. Since $x \in \mathcal{V}(x)$ is the unique singular point, $\mathcal{V}(x)$ is fixed by $\gamma \in G$ if and only if $\gamma \in P$, from which it follows that $\mathcal{V}(x)$ is a Schubert cycle. At a point $y \in \mathcal{V}(x)$ distinct from x let ℓ be the projective line joining x and y, $T_y(\ell) =: \mathbb{C}\alpha_y$ and consider $\mathscr{C}_y(\mathcal{V}(x)) := \mathscr{C}_y(X) \cap \mathbb{P}T_y(\mathcal{V}(x))$. We may assume that x and y lie on a Harish-Chandra coordinate chart so that T_y and T_x are identified by parallel transport with respect to the Euclidean flat connection. Hence, also $\mathscr{C}_y(X)$ and $\mathscr{C}_x(X)$ are identified with each other. Then, $\xi \in \mathscr{C}_y(\mathcal{V}(x))$ if and only if there exists a projective line ℓ' on $\mathcal{V}(x)$ passing through y such that $T_y(\ell') = \mathbb{C}\xi$. For $[\xi] \neq [\alpha_y]$ this occurs if and only if there exists a projective plane $\Pi \subset \mathcal{V}(x)$ such that $y \in \Pi$ and such that $T_y(\Pi) = \mathbb{C}\alpha_y + \mathbb{C}\xi$. Write $T_X|_\ell \cong \mathcal{O}(2) \oplus \mathcal{O}(1)^p \oplus \mathcal{O}^q$. Then, the positive part $P_\ell := (\mathcal{O}(2) \oplus \mathcal{O}(1)^p) \subset T_X|_\ell$ is uniquely determined independent of the choice of direct summands in the Grothendieck decomposition, and $P_{\ell,y} = T_y(\mathcal{V}(x))$. Defining $P_{\alpha_y} := T_{\alpha_y}(\widetilde{\mathscr{C}_y}(X))$, from the deformation theory of rational curves we have $P_{\alpha_y} = P_{\ell,y} = T_y(\mathcal{V}(x))$. Thus, $\mathscr{C}_y(\mathcal{V}(x)) := \mathbb{P}(P_{\alpha_y}) \cap \mathscr{C}_y(X)$. The pairs $(\mathscr{C}_y(\mathcal{V}(x)), \mathscr{C}_y(X))$ are thus constant along $\ell - \{x\}$ in the Harish-Chandra coordinate chart, hence the pairs $\big(\mathscr{C}_y(\mathcal{V}(x)) \subset \mathscr{C}_y(X)\big)$ are projectively equivalent to each other for $y \in \mathcal{V}(x) - \{x\}$. When $X = G^{III}(n, n)$ is a Lagrangian Grassmannian of rank

$n \geq 2$, $\mathscr{C}_y(\mathcal{V}(x)) = \{[\alpha_y]\}$ is a single point. At $0 \in X$, let $[\alpha] \in \mathscr{C}_0(X)$ and define $\mathscr{S}_{[\alpha]} := \mathscr{C}_0(X) \cap \mathbb{P}(P_\alpha)$. For $X \not\cong G^{III}(n,n)$ we consider now sub-VMRT structures modeled on $(\mathcal{V}(x) \subset X)$, i.e., taking x without loss of generality to be on a line emanating from $0, x \neq 0$, modeled on $\big(\mathscr{S}_{[\alpha]} \subset \mathscr{C}_0(X)\big)$ for any $[\alpha] \in \mathscr{C}_0(X)$, and examine the question whether a sub-VMRT structure $\varpi : \mathscr{C}(S) \to S$ modeled on $\big(\mathscr{S}_{[\alpha]} \subset \mathscr{C}_0(X)\big)$ is necessarily linearly saturated. As a preparation we prove

LEMMA 3.1. *Let X be an irreducible Hermitian symmetric space of rank ≥ 2 not biholomorphic to a Lagrangian Grassmannian, $0 \in X$, and let $[\alpha]$ be any point on $\mathscr{C}_0(X)$. Then, $\mathscr{S}_{[\alpha]} \subset \mathbb{P}(P_\alpha)$ is the cone with vertex $[\alpha]$ over a nonsingular projective subvariety $\mathcal{J} \subset \mathbb{P}T_{[\alpha]}(\mathbb{P}(P_\alpha)) = \mathbb{P}(P_\alpha/\mathbb{C}\alpha) = \mathbb{P}T_{[\alpha]}(\mathscr{C}_0(X))$ which is the VMRT of $\mathscr{C}_0(X)$, i.e., $\mathcal{J} = \mathscr{C}_{[\alpha]}(\mathscr{C}_0(X))$.*

PROOF. Note that also the 3-dimensional hyperquadric Q^3 is excluded as it is biholomorphic to the Lagrangian Grassmannian of rank 2. To prove the lemma observe that $\mathscr{C}_0(X)$ admits at $[\alpha]$ a quadratic expansion (cf. Hwang-Mok [**HM99**, (4.2)]). More precisely, writing $T_0(X) = \mathbb{C}\alpha \oplus \mathcal{H}_\alpha \oplus \mathcal{N}_\alpha$, where $\mathbb{C}\alpha \oplus \mathcal{H}_\alpha = P_\alpha$, there exists a quadratic vector-valued symmetric bilinear form $\mathfrak{q}$ on $T_{[\alpha]}(\mathscr{C}_0(X)) = P_\alpha/\mathbb{C}\alpha \cong \mathcal{H}_\alpha$, $\mathfrak{q} : S^2\mathcal{H}_\alpha \to \mathcal{N}_\alpha$, such that, identifying $\mathbb{P}T_0(X)$ as the Zariski closure of $\mathcal{H}_\alpha \oplus \mathcal{N}_\alpha$, $\mathscr{C}_0(X)$ is the Zariski closure of the graph of $\varphi : \mathcal{H}_\alpha \to \mathcal{N}_\alpha$ given by $\varphi(\xi) = \mathfrak{q}(\xi,\xi)$. Thus, $\mathscr{C}_{[\alpha]}(X) \cap \mathcal{H}_\alpha$ is the union of complex lines $\mathbb{C}\xi$ satisfying $\mathfrak{q}(\xi,\xi) = 0$, and its Zariski closure $\mathscr{S}_{[\alpha]}$ is the union of projective lines Λ on $\mathscr{C}_0(X)$ passing through $[\alpha]$. Hence, $\mathscr{S}_{[\alpha]}$ is the cone with vertex $[\alpha]$ over $\mathcal{J} = \mathscr{C}_{[\alpha]}(\mathscr{C}_0(X)) \subset \mathbb{P}T_{[\alpha]}(\mathscr{C}_0(X)) = \mathbb{P}(P_\alpha/\mathbb{C}\alpha)$, as desired. □

For the parabolic subgroup P at $0 \in X$, let $Q = Q(\alpha) \subset P$ be the subgroup which fixes $[\alpha] \in \mathbb{P}T_0(X)$. Let $J \subset Q$ be a Levi factor, $J \subset K^{\mathbb{C}}$. Since J fixes $[\alpha]$, it acts on $V := T_{[\alpha]}(\mathbb{P}(P_\alpha))$. By examining the VMRTs, which are irreducible Hermitian symmetric spaces of the compact type except in the case of the Grassmannian, it follows that the action of J on V is irreducible excepting Grassmannians $G(p,q)$ with $p, q \geq 2$, where $\mathscr{C}_{[\alpha]}(G(p,q)) = \varsigma(\mathbb{P}^{p-1} \times \mathbb{P}^{q-1})$ for the Segre embedding ς, in which case V splits into the direct sum of two irreducible components. For the latter cases writing $T_0(G(p,q)) = U_0 \otimes V_0$, $\alpha = u \otimes v$, $\mathscr{C}_{[\alpha]}(\mathscr{C}_0(G(p,q)))$ is the disjoint union of a copy of $\mathbb{P}^{p-2}$ and a copy of $\mathbb{P}^{q-2}$, $\mathscr{S}_{[\alpha]} = \mathbb{P}(U_0 \otimes \mathbb{C}v) \cup \mathbb{P}(\mathbb{C}u \otimes V_0)$, the two irreducible components intersecting at one point $[\alpha] = [u \otimes v]$. Thus, $\mathbb{P}(P_\alpha) = \mathbb{P}\big((U_0 \otimes \mathbb{C}v) + (\mathbb{C}u \otimes V_0)\big)$ is the linear span of $\mathbb{P}(U_0 \otimes \mathbb{C}v)$ and $\mathbb{P}(\mathbb{C}u \otimes V_0)$, i.e., of $\mathscr{S}_{[\alpha]}$, hence $\mathscr{S}_{[\alpha]}$ is linearly nondegenerate in $\mathbb{P}(P_\alpha)$. For X being considered other than a Grassmannian of rank ≥ 2, $\mathscr{S}_{[\alpha]} \subset \mathbb{P}(P_\alpha)$ is necessarily linearly nondegenerate by the irreducibility of V under the action of J.

If we further exclude the cases where X is a Grassmannian, then $\mathscr{C}_{[\alpha]}(\mathscr{C}_0(X)) \subset \mathbb{P}T_{[\alpha]}(\mathscr{C}_0(X))$ is nonlinear and homogeneous and the second fundamental form on $\mathscr{C}_{[\alpha]}(\mathscr{C}_0(X))$ has trivial kernels. From the description of $\mathscr{S}_{[\alpha]}$ as a cone of projective lines over $\mathscr{C}_{[\alpha]}(\mathscr{C}_0(X))$ we have readily

COROLLARY 3.2. *Let X be an irreducible Hermitian symmetric space of the compact type of rank ≥ 2 which is neither biholomorphic to a Grassmannian nor to a Lagrangian Grassmannian. For $[\alpha] \in \mathscr{C}_0(X)$ write $\mathscr{S}_{[\alpha]} := \mathscr{C}_0(X) \cap \mathbb{P}(P_\alpha)$. Denote by ζ the second fundamental form of $\mathscr{S}_{[\alpha]} \subset \mathbb{P}(P_\alpha)$ at a smooth point $[\beta] \in \mathscr{S}_{[\alpha]}$. Then, we have $\mathrm{Ker}(\zeta_{[\beta]}(\,\cdot\,, T_{[\beta]}(\mathscr{S}_{[\alpha]}))) = T_{[\beta]}(\Lambda)$, where Λ is the projective line on $\mathscr{S}_{[\alpha]}$ containing $[\alpha]$ and $[\beta]$.*

4. Algebraicity of germs of sub-VMRT structures modeled on certain full cones of minimal rational curves

We consider now sub-VMRT structures $\varpi : \mathscr{C}(S) \to S$ of the VMRT structure $\pi : \mathscr{C}(X) \to X$ on X where $\big(\mathscr{C}_x(S) \subset \mathscr{C}_x(X)\big)$ is projectively equivalent to $\big(\mathscr{S}_{[\alpha]} \subset \mathscr{C}_0(X)\big)$. Recall that X is irreducible and of rank ≥ 2, and $X \not\cong G^{III}(n,n), n \geq 2$. We have

THEOREM 4.1. *Let X be an irreducible Hermitian symmetric space of the compact type of rank ≥ 2 other than a Lagrangian Grassmannian. Then, the proper pair of projective subvarieties $(\mathscr{S}_{[\alpha]}, \mathscr{C}_0(X))$ of $\mathbb{P}T_0(X)$ is nondegenerate for substructures, excepting in the cases of hyperquadrics Q^n, $n \geq 4$, and of Grassmannians $G(2,q)$, $q \geq 2$, where nondegeneracy for substructures fails. Excluding those cases, any locally closed complex submanifold $S \subset X$ inheriting a sub-VMRT structure $\varpi : \mathscr{C}(S) \to S$ with fibers $\big(\mathscr{C}_x(S) \subset \mathscr{C}_x(X)\big)$ projectively equivalent to $\big(\mathscr{S}_{[\alpha]} \subset \mathscr{C}_0(X)\big)$ is linearly saturated, i.e., it is uniruled by open subsets of projective lines. Moreover, there exists a subvariety $Z \subset X$ such that $S \subset Z$ and* $\dim(Z) = \dim(S)$.

In the sequel for brevity we will say that S is algebraic as a germ of submanifold (at any point $x \in S$) whenever there exists an irreducible subvariety $Z \subset X$ such that $S \subset Z$ and $\dim(Z) = \dim(S)$. For the proof of Theorem 4.1 we will study the symmetric bilinear form $\tau_{[\beta]} : S^2T_{[\beta]}(\mathscr{C}_0(X))) \to T_{[\beta]}(\mathbb{P}T_0(X))/(T_{[\beta]}(\mathscr{C}_0(X)) + T_{[\beta]}(\mathbb{P}(P_\alpha)))$ for the pair $\big(\mathscr{S}_{[\alpha]} \subset \mathscr{C}_0(X)\big)$ at a smooth point $[\beta] \in \mathscr{S}_{[\alpha]}$, $\tau_{[\beta]} := \nu_{[\beta]} \circ \sigma_{[\beta]}$, following Definition 2.4. We prove first of all the following result related to Corollary 3.2.

PROPOSITION 4.2. *Let X be any irreducible Hermitian symmetric space of the compact type and of rank ≥ 2 neither biholomorphic to a Grassmannian nor to a Lagrangian Grassmannian. For $[\beta] \in \mathrm{Reg}(\mathscr{S}_{[\alpha]})$ write $\Lambda \subset \mathscr{S}_{[\alpha]}$ for the projective line containing $[\alpha]$ and $[\beta]$, and $\sigma_{[\beta]} : S^2T_{[\beta]}(\mathscr{C}_0(X)) \to T_{[\beta]}(\mathbb{P}T_0(X))/(T_{[\beta]}(\mathscr{C}_0(X))$ for the second fundamental form of $\mathscr{C}_0(X) \subset \mathbb{P}T_0(X)$ at $[\beta]$. Then, we have* $\mathrm{Ker}(\sigma_{[\beta]}(\,\cdot\,, T_{[\beta]}(\mathscr{S}_{[\alpha]}))) = T_{[\beta]}(\Lambda)$.

We will give a differential-geometric proof of Proposition 4.2 which is based on a characterization of projective submanifolds with parallel second fundamental form. On $\mathbb{P}^N$ denote by θ the Fubini-Study metric. We have

THEOREM 4.3. (Nakagawa-Takagi [**NT76**]). *A linearly nondegenerate Kähler projective submanifold $(M, \theta|_M) \hookrightarrow (\mathbb{P}^N, \theta)$ has parallel second fundamental form $\sigma : S^2T_M \to T_{\mathbb{P}^N}|_M/T_M$ if and only if M is biholomorphic to a Hermitian symmetric space (S,g) of the compact type of rank ≤ 2 and $(M, \theta|_M)$ is either the image of S under a holomorphic isometric minimal embedding, or S is a projective space and M is its image under the Veronese embedding.*

Note here that there is the projective second fundamental form σ, which is a holomorphic bundle homomorphism, and there is also the second fundamental form σ' with respect to the Riemannian connection of the Kähler manifold $(\mathbb{P}T_0(X), \theta)$. If we regard σ' as taking values in the *holomorphic* normal bundle $T_{\mathbb{P}^N}|_M/T_M$ (in place of the orthogonal complement of T_M in $T_{\mathbb{P}^N}|_M$), then σ' agrees with σ. Here and in the sequel we use the same symbol σ for the two second fundamental forms, noting that parallelism is always defined in terms of the Riemannian connection.

The relevance of Theorem 4.3 to VMRT geometry lies in the fact that the set $\mathscr{E}$ of linearly nondegenerate projective manifolds with parallel second fundamental form is in one-to-one correspondence with the set of projective submanifolds given by VMRTs of irreducible Hermitian symmetric spaces X of the compact type, as given in Mok [**Mo89**, Appendix III.2]. The proof of the latter result is based on a theorem of Ros [**Ro85**] characterizing members of $\mathscr{E}$ in terms of a pinching condition on holomorphic sectional curvatures.

PROOF OF PROPOSITION 4.2. $\mathscr{C}_0(X) \subset \mathbb{P}T_0(X)$ is the VMRT of an irreducible Hermitian symmetric space (X, g) of the compact type. $\mathscr{C}_0(X)$ is itself a Hermitian symmetric space of rank 2 and $\mathscr{C}_0(X) \subset \mathbb{P}T_0(X)$ is the minimal embedding, which is a holomorphic isometric embedding into $(\mathbb{P}T_0(X), \theta)$ for an appropriate choice of the Fubini-Study metric θ, viz., one that arises from the Hermitian inner product on $T_0(X)$ given by (X, g). $(\mathscr{C}_0(X), \theta|_{\mathscr{C}_0(X)})$ is of nonnegative holomorphic bisectional curvature.

$\mathscr{C}_0(X) \subset \mathbb{P}T_0(X)$ is a homogeneous projective submanifold uniruled by projective lines. For $[\beta] \in \mathrm{Reg}(\mathscr{S}_{[\alpha]})$, $\Lambda := \Lambda(\alpha, \beta) := \mathbb{P}(\mathbb{C}\alpha + \mathbb{C}\beta)$ is a minimal rational curve on $\mathscr{C}_0(X)$, hence a standard rational curve. Let $T(\mathscr{C}_0(X))|_\Lambda \cong \mathcal{O}(2) \oplus (\mathcal{O}(1))^a \oplus \mathcal{O}^b$ be the Grothendieck splitting over Λ, $\dim(\mathscr{S}_{[\alpha]}) = 1 + a$. Denoting by $Q(\Lambda) = \mathcal{O}(2) \oplus (\mathcal{O}(1))^a \subset T(\mathscr{C}_0(X))|_\Lambda$ the positive part, we have $T_{[\beta]}(\mathscr{S}_{[\alpha]}) = Q_{[\beta]}(\Lambda)$, the fiber of $Q(\Lambda)$ at $[\beta]$. Moreover, we have

LEMMA 4.4. *Equipping $\mathscr{C}_0(X) \subset \mathbb{P}T_0(X)$ with $\theta|_{\mathscr{C}_0(X)}$, for the projective line $\Lambda \subset \mathscr{C}_0(X)$, $Q(\Lambda) \subset T(\mathscr{C}_0(X))|_\Lambda$ is a parallel vector subbundle.*

PROOF. $T^*(\mathscr{C}_0(X))|_\Lambda \cong \mathcal{O}(-2) \oplus (\mathcal{O}(-1))^a \oplus \mathcal{O}^b$, and $\mathcal{O}^b \subset T^*(\mathscr{C}_0(X))|_\Lambda$ is a trivial bundle equipped with the restriction of a Hermitian metric h on $T^*(\mathscr{C}_0(X))|_\Lambda$ of nonpositive curvature in the sense of Griffiths. Because of the monotonicity on curvatures of Hermitian holomorphic vector subbundles (cf. Mok [**Mo89**, (3.2)]), the Hermitian metric $h|_{\mathcal{O}^b}$ must be flat and hence $\mathcal{O}^b \subset T^*(\mathscr{C}_0(X))|_\Lambda$ must be parallel. Thus, $Q(\Lambda) \subset T(\mathscr{C}_0(X))|_\Lambda$, being the annihilator of $\mathcal{O}^b \subset T^*(\mathscr{C}_0(X))|_\Lambda$, must also be parallel, as desired. □

PROOF OF PROPOSITION 4.2 CONTINUED. Recalling that $\mathscr{S}_{[\alpha]} = \mathscr{C}_0(X) \cap \mathbb{P}(P_\alpha) \subset \mathscr{C}_0(X)$ is a linear section smooth at $[\beta]$, we have $\mathrm{Ker}(\sigma_{[\beta]}(\cdot, T_{[\beta]}(\mathscr{S}_{[\alpha]}))) \cap T_{[\beta]}(\mathscr{S}_{[\alpha]}) = \mathrm{Ker}(\zeta_{[\beta]}(\cdot, T_{[\beta]}(\mathscr{S}_{[\alpha]}))) = T_{[\beta]}(\Lambda)$, by Corollary 3.2. Suppose $\mathrm{Ker}(\sigma_{[\beta]}(\cdot, T_{[\beta]}(\mathscr{S}_{[\alpha]}))) \neq T_{[\beta]}(\Lambda)$. Since $T_{[\beta]}(\mathscr{S}_{[\alpha]}) = Q_{[\beta]}(\Lambda)$, there exists $\eta \in T_{[\beta]}(\mathscr{C}_0(X)) - Q_{[\beta]}(\Lambda)$ such that $\sigma_{[\beta]}(\eta, Q_{[\beta]}(\Lambda)) = 0$. For $[\gamma] \in \Lambda$ define $U_{[\gamma]} := \mathrm{Ker}(\sigma_{[\gamma]}(\cdot, Q_{[\gamma]}(\Lambda))) \subset T_{[\gamma]}(\mathscr{C}_0(X))$. Since $Q(\Lambda) \subset T(\mathscr{C}_0(X))|_\Lambda$ is a parallel subbundle by Lemma 4.4, and σ is parallel with respect to $(\mathbb{P}T_0(X), \theta)$ by Nakagawa-Takagi [**NT76**] (Theorem 4.3 here), it follows that $U \subset T(\mathscr{C}_0(X))|_\Lambda$ is a parallel subbundle. By Corollary 3.2 we have $U \cap Q(\Lambda) = T(\Lambda)$, hence $U = T(\Lambda) \oplus V$ for some parallel subbundle $V \subset T(\mathscr{C}_0(X))|_\Lambda$ transversal to $Q(\Lambda)$, so that $V \cong \mathcal{O}^c$ for some integer c, $1 \le c \le b$. Hence $\sigma_{[\beta]}$ induces a *parallel* bundle homomorphism $\varphi : V \otimes (T(\mathscr{C}_0(X))|_\Lambda / Q(\Lambda)) \to (T(\mathbb{P}T_0(X))/T(\mathscr{C}_0(X)))|_\Lambda := N$. Now from Grothendieck splitting $T(\mathscr{C}_0(X))|_\Lambda / Q(\Lambda) \cong \mathcal{O}^b$, while the normal bundle N, being a quotient bundle of $T(\mathbb{P}T_0(X))|_\Lambda \cong \mathcal{O}(2) \oplus \mathcal{O}(1)^{n-2}$, is necessarily a direct sum of positive line bundles over Λ. By comparing degrees the parallel bundle homomorphism $\varphi : \mathcal{O}^c \otimes \mathcal{O}^b \to N$ must necessarily be 0. In particular, $\sigma_{[\beta]}(\eta, \xi) = 0$ for any $\xi \in T_{[\beta]}(\mathscr{C}_0(X))$, contradicting the standard fact that $\mathrm{Ker}(\sigma_{[\beta]}(\cdot, T_{[\beta]}(\mathscr{C}_0(X)))) = 0$

on the homogeneous nonlinear submanifold $\mathscr{C}_0(X) \subset \mathbb{P}T_0(X)$, proving Proposition 4.2. $\square$

REMARK 4.5. By the process of reconstructing S as an open subset of some projective subvariety $Z \subset X$ by adjunction of minimal rational curves following Mok [**Mo08a**] and Hong-Mok [**HoM10**] [**HoM13**], it can be established that $Z = \gamma(\mathcal{V})$ for some $\gamma \in \mathrm{Aut}(X)$. The arguments will be given in Mok-Yang [**MY18**] in the proof of uniqueness results of holomorphic isometric embeddings of the complex unit ball of maximal admissible dimension into irreducible bounded symmetric domains of rank ≥ 2.

In order to apply Theorem 2.6 and Theorem 2.7, we need to check that the pair $(\mathscr{S}_{[\alpha]}, \mathscr{C}_0(X))$ satisfies Condition (T) as defined in Mok-Zhang [**MZ17**, Definition 5.4] and recalled here in Definition 2.5. In what follows the symbol $\mathcal{V}$ stands for $\mathcal{V}(x)$ for any point $x \neq 0$ lying on some minimal rational curve ℓ on X passing through 0. We have more precisely the following formulation of Condition (T) for the pair $(\mathscr{S}_{[\alpha]}, \mathscr{C}_0(X))$.

PROPOSITION 4.6. *Let X be an irreducible Hermitian symmetric space of the compact type and of rank ≥ 2 not biholomorphic to a Lagrangian Grassmannian. Then $(\mathscr{S}_{[\alpha]}, \mathscr{C}_0(X))$ satisfies* Condition (T). *More precisely, for any nonzero vector $\beta \in \widetilde{\mathscr{S}_{[\alpha]}}$, $[\beta] \neq [\alpha]$, $T_\beta(\widetilde{\mathscr{S}_{[\alpha]}}) = T_\beta(\widetilde{\mathscr{C}_0}(X)) \cap T_0(\mathcal{V}) = P_\beta \cap P_\alpha$.*

For the proof of Proposition 4.6 we will make use of the following lemma (which replaces a more involved argument of the author's, cf. Remark after the proof of Proposition 4.6 below). Both the lemma and its proof given here are provided by the referee.

LEMMA 4.7. *Let $Z \subset \mathbb{P}^k$ be a nonsingular projective subvariety whose homogeneous ideal I_Z is generated by quadratic polynomials. Let $\Pi \subset \mathbb{P}^k$ be a linear subspace which is tangent to Z at two distinct points $a, b \in \Pi \cap Z$. Then Π is tangent to Z along the line joining a and b.*

PROOF. For each quadratic polynomial $q \in I_Z$ which does not vanish identically on Π, let $q' \subset \Pi$ be the quadric hypersurface on Π defined by $q = 0$. It suffices to prove that q' is singular along the line joining a and b for any $q \in I_Z$ which does not vanish identically on Π. But our assumption says that q' is singular at a and b. Since the singular locus of a quadric hypersurface is a linear subspace, this implies that q' is singular along the line joining a and b, as desired. $\square$

PROOF OF PROPOSITION 4.6. Under the hypothesis of the proposition we are going to show equivalently that $T_{[\beta]}(\mathscr{S}_{[\alpha]}) = T_{[\beta]}(\mathscr{C}_0(X)) \cap T_{[\beta]}(\mathbb{P}T_0(\mathcal{V}))$, which is the same as $(P_\beta \cap P_\alpha)/\mathbb{C}\beta$. Clearly, $[\beta]$ is a smooth point of $\mathscr{S}_{[\alpha]}$ and $T_{[\beta]}(\mathscr{S}_{[\alpha]}) = T_{[\beta]}(\mathscr{C}_0(X) \cap \mathbb{P}T_0(\mathcal{V})) \subset T_{[\beta]}(\mathscr{C}_0(X)) \cap T_{[\beta]}(\mathbb{P}T_0(\mathcal{V})) = (P_\beta \cap P_\alpha)/\mathbb{C}\beta$. Recall that $\Lambda \subset \mathbb{P}T_0(X)$ denotes the projective line containing both $[\alpha]$ and $[\beta]$. Let $\gamma \in T_0(X)$ be a nonzero vector such that $(\gamma + \mathbb{C}\beta)/\mathbb{C}\beta \in (P_\beta \cap P_\alpha)/\mathbb{C}\beta$. We have to prove that $(\gamma + \mathbb{C}\beta)/\mathbb{C}\beta$ is tangent to $\mathscr{S}_{[\alpha]}$. When $\gamma \in \mathbb{C}\alpha + \mathbb{C}\beta$, $(\gamma + \mathbb{C}\beta)/\mathbb{C}\beta$ is tangent to Λ at $[\beta]$ and *a fortiori* tangent to $\mathscr{S}_{[\alpha]}$, and it remains to consider the case where α, β and γ are linearly independent. Let Π be the projective 2-plane on $\mathbb{P}T_0(X)$ spanned by $[\alpha], [\beta]$ and $[\gamma]$. We have $\Lambda \subset \Pi$ and Π is tangent to $\mathscr{C}_0(X)$ both at $[\alpha]$ and at $[\beta]$, and our task is to prove that Π is tangent to $\mathscr{S}_{[\alpha]}$ at $[\beta]$. Since by Lemma 3.1 $\mathscr{S}_{[\alpha]} \subset \mathbb{P}(P_\alpha)$ is the cone with vertex $[\alpha]$ over a nonsingular projective subvariety $\mathscr{J} \subset \mathbb{P}T_{[\alpha]}(\mathscr{C}_0(X))$, equivalently it remains to prove that Π is tangent to $\mathscr{C}_0(X)$ along Λ.

For any irreducible Hermitian symmetric space X of the compact type, $0 \in X$, the VMRT $\mathscr{C}_0(X) \subset \mathbb{P}T_0(X)$ is a Hermitian symmetric space of the compact type equivariantly embedded in the projective space $\mathbb{P}T_0(X)$. As is well-known, for any equivariantly embedded rational homogeneous manifold $M \subset \mathbb{P}^\ell$ the homogeneous ideal I_M is generated by quadratic polynomials (cf. Lichtenstein [**Li82**]). Hence, Lemma 4.7 applies to $Z := \mathscr{C}_0(X) \subset \mathbb{P}T_0(X)$, and we conclude that Π is tangent to $\mathscr{C}_0(X)$ along the line Λ joining $[\alpha]$ and $[\beta]$, as desired. □

REMARK 4.8. Originally the author gave a proof of Proposition 4.6 relying on the parallelism of the Riemannian second fundamental form as given in Nakagawa-Takagi [**NT76**] (Theorem 4.3 here), in which the case of Grassmannians was treated separately. The proof given here using Lemma 4.7 as suggested by the referee is both simpler and uniform.

We are now ready to prove Theorem 4.1.

PROOF OF THEOREM 4.1. In the notation of Definition 2.4 (cf. also the paragraph after it), for a smooth point $[\beta]$ on $\mathscr{S}_{[\alpha]}$, we have $\tau_{[\beta]} = \nu_{[\beta]} \circ \sigma_{[\beta]}$, where $\nu_{[\beta]} : T_0(\mathbb{P}T_0(X))/T_{[\beta]}(\mathscr{C}_0(X)) \to T_0(\mathbb{P}T_0(X))/(T_{[\beta]}(\mathscr{C}_0(X)) + T_{[\beta]}(\mathscr{S}_{[\alpha]}))$ is the canonical projection, $T_{[\beta]}(\mathbb{P}(P_\alpha)) \cong P_\alpha/\mathbb{C}\beta$. It remains to check whether $\mathrm{Ker}(\tau_{[\beta]}(\,\cdot\,, T_{[\beta]}(\mathbb{P}(P_\alpha))) \subset T_{[\beta]}(\mathscr{S}_{[\alpha]})$ holds. Using notation as in §3 and noting that in the Hermitian symmetric case we have $\mathfrak{g} = \mathfrak{g}_{-1} \oplus \mathfrak{g}_0 \oplus \mathfrak{g}_1$, for a root $\rho \in \Phi^+$ we define $\Psi_\rho := \{\lambda \in \Phi^+ : \rho - \lambda \in \Phi\}$. We may take $\alpha = E_\mu$, where $\mu \in \Phi^+$ is the highest root. Recall that $P \subset G$ is the parabolic subgroup at $0 \in X$ so that $X = G/P$, and $Q(\alpha) \subset P$ is the subgroup which fixes $\mathbb{C}\alpha$. Since X is not biholomorphic to a Lagrangian Grassmannian, there exists $\nu \in \Psi_\mu$ such that the projective line $\Lambda := \mathbb{P}(\mathbb{C}E_\mu + \mathbb{C}E_\nu) \subset \mathbb{P}T_0(X)$ lies on $\mathscr{C}_0(X)$, which is the case if and only if $\nu \in \Psi_\mu$ is a long root. If X is not biholomorphic to a Grassmannian, then the space of lines $\Lambda \subset \mathscr{C}_0(X)$ passing through $[\alpha]$ is given by $\mathscr{C}_{[\alpha]}(\mathscr{C}_0(X))$, which is itself an irreducible Hermitian symmetric space of the compact type on which $Q(\alpha)$ acts transitively. We may therefore take $\beta = E_\nu$, so that $\mu, \nu \in \Phi^+$ are long roots, $\nu \in \Psi_\mu$, $\mu \in \Psi_\nu$. In the case where X is the Grassmannian $G(p,q)$, $p, q \geq 2$, taking $\alpha = E_\mu$, $\mathscr{S}_{[\alpha]}$ is the union of two projective subspaces of dimension $p-1$ resp. $q-1$ intersecting at $[\alpha]$, and we have to consider $[\beta] \neq [\alpha]$ belonging to either of them. In both cases, modulo the action of $Q(\alpha)$ clearly we may take $\beta = E_\nu$ for some root $\nu \in \Phi^+$.

We take now $\alpha = E_\mu$ and $\beta = E_\nu$, $[\beta] \in \mathscr{S}_{[\alpha]}$ being a smooth point. We have $T_{[\beta]}(\mathscr{C}_0(X)) = P_\beta/\mathbb{C}\beta$. Identifying $P_\beta/\mathbb{C}\beta$ as the orthogonal complement of β in P_β, $T_{[\beta]}(\mathscr{C}_0(X)) = \mathrm{Span}\{E_\rho : \rho \in \Psi_\nu\}$, while by Proposition 4.6 we have $T_{[\beta]}(\mathscr{S}_{[\alpha]}) = (P_\alpha \cap P_\beta)/\mathbb{C}\beta = \mathrm{Span}\{E_\rho : \rho \in (\Psi_\nu \cap \Psi_\mu) \cup \{\mu\}\}$. As in the proof of Mok-Zhang [**MZ17**, Lemma 3.4], to show that $(\mathscr{C}_{[\beta]}(\mathscr{S}_{[\alpha]}), \mathscr{C}_{[\beta]}(\mathscr{C}_0(X)))$ is nondegenerate for substructures, it suffices to check that for any root vector E_π, $\pi \in \Psi_\nu - ((\Psi_\nu \cap \Psi_\mu) \cup \{\mu\}) = \Psi_\nu - (\Psi_\mu \cup \{\mu\})$, there must exist $\omega \in (\Psi_\nu \cap \Psi_\mu) \cup \{\mu\}$ such that $\tau_{[\beta]}(E_\pi, E_\omega) \neq 0$. By Proposition 4.2, there exists $\omega \in (\Psi_\nu \cap \Psi_\mu) \cup \{\mu\}$ such that $\sigma_{[\beta]}(E_\pi, E_\omega) \neq 0$, i.e., $\lambda := \pi + \omega - \nu \in \Phi^+$. We have $\tau_{[\beta]}(E_\pi, E_\omega) \neq 0$ if and only if E_λ mod $\mathbb{C}\beta \notin T_{[\beta]}(\mathbb{P}(P_\alpha)) \cong P_\alpha/\mathbb{C}\beta$, i.e., $\lambda \notin \Psi_\mu \cup \{\mu\}$. Now $\lambda(H_\mu) = \pi(H_\mu) + \omega(H_\mu) - \nu(H_\mu)$. From Grothendieck splitting over minimal rational curves ℓ

on X we see that $\mu(H_\mu)=2$, $\rho(H_\mu)=1$ for $\rho\in\Psi_\mu$ and $\rho(H_\mu)=0$ for $\rho\notin\Psi_\mu\cup\{\mu\}$. (Thus, $\pi(H_\mu)=0$.) We know that $\omega\in(\Psi_\nu\cap\Psi_\mu)\cup\{\mu\}$. In case $\omega=\mu$ we have $\lambda(H_\mu)=1$ so that $\tau_{[\beta]}(E_\pi,E_\omega)=0$. On the other hand, when $\omega,\nu\in\Psi_\mu$ we have $\lambda(H_\mu)=0$, in which case $\lambda\notin\Psi_\mu\cup\{\mu\}$, so that $\tau_{[\beta]}(E_\pi,E_\omega)\neq 0$. We conclude therefore that $E_\pi\notin\mathrm{Ker}(\tau_{[\beta]}(\,\cdot\,,T_{[\beta]}(\mathscr{S}_{[\alpha]})))$ whenever $\dim(\sigma_{[\beta]}(E_\pi,T_{[\beta]}(\mathscr{S}_{[\alpha]})))\geq 2$.

Denote by $s(X)$ the minimum of $\dim(\sigma_{[\beta]}(E_\pi,T_{[\beta]}(\mathscr{C}_0(X))))$ as π ranges over Ψ_ν, which is independent of the choice of $[\beta]\in\mathrm{Reg}(\mathscr{S}_{[\alpha]})$ by homogeneity. Note that $\sigma_{[\beta]}(E_{\pi_1},E_{\pi_2})=0$ whenever $\pi_1,\pi_2\in\Psi_\nu-(\Psi_\mu\cup\{\mu\})$ as $(\pi_1+\pi_2-\nu)(H_\mu)=-1<0$, so that $\dim(\sigma_{[\beta]}(E_\pi,T_{[\beta]}(\mathscr{S}_{[\alpha]})))=\dim(\sigma_{[\beta]}(E_\pi,T_{[\beta]}(\mathscr{C}_0(X))))=s(X)$. Writing $X(E_6)$ for the 16-dimensional Hermitian symmetric space of the compact type of type E_6, and $X(E_7)$ for the 27-dimensional one of type E_7, a straightforward checking gives $s(G(p,q))=\min(p-1,q-1)$ $(p,q\geq 2)$, $s(G^{II}(n,n))=n-3$ $(n\geq 4$, noting that $G^{II}(4,4)\cong Q^6)$, $s(Q^n)=1$ $(n\geq 3)$, $s(X(E_6))=3$, $s(X(E_7))=5$, showing that $\tau_{[\beta}(E_\pi,T_{[\beta]}(\mathscr{S}_{[\alpha]}))\neq 0$ for any $\pi\in\Psi_\nu$ with the exception of $X=Q^n$, $n\geq 3$, and $X=G(2,q)$, $q\geq 2$. Thus, excluding the latter cases $(\mathscr{S}_{[\alpha]},\mathscr{C}_0(X))$ is nondegenerate for substructures.

When $X=Q^n, n\geq 3$, we have $\nu_{[\beta]}\equiv 0$ since $P_\beta+T_{[\beta]}(\mathscr{S}_{[\alpha]})=T_{[\beta]}(\mathbb{P}T_0(X))$. When X is $G(2,q)$, $q\geq 2$, where $s(G(2,q))=1$, with respect to the preceding arguments it does occur that $\sigma_{[\beta]}(E_\pi,E_\mu)\neq 0$ while $\tau_{[\beta]}(E_\pi,E_\mu)=0$ for $\pi\in\Psi_\nu-(\Psi_\mu\cup\{\mu\})$, so that $\mathrm{Ker}(\tau_{[\beta]}(\,\cdot\,,T_{[\beta]}(\mathscr{S}_{[\alpha]})))\supsetneq T_{[\beta]}(\mathscr{S}_{[\alpha]})$. Thus, in both cases $(\mathscr{S}_{[\alpha]},\mathscr{C}_0(X))$ fails to be nondegenerate for substructures. □

REMARK 4.9.

(1) By computing holomorphic bisectional curvatures on $(\mathscr{C}_0(X),\theta|_{\mathscr{C}_0(X)})$, conceptually $s(X)$ is the minimum of the number of flat direct summands in the Grothendieck splitting of $T(\mathscr{C}_0(X))$ over a projective line $\Lambda\subset\mathscr{C}_0(X)$. (When $X\ncong G(p,q)$, there is up to the action of the parabolic subgroup P only one isomorphism class of Λ. Hence, there is no need to take the minimum excepting for $G(p,q)$ of rank ≥ 2, in which case there are two isomorphism classes of Λ.)

(2) For the proof of Theorem 4.1 instead of the parallelism of σ one can also use the combinatorial argument as above. We gave the proof exploiting splitting types as the latter is more geometric and of independent interest.

(3) Robles-The [**RT12**] and Robles [**Ro13**] have completely determined the set of Schubert cycles on irreducible Hermitian symmetric spaces of the compact type which are Schur rigid. In [**RT12**] the authors used cohomological methods due to Kostant [**Ko63**]. In the cases where the desired vanishing of cohomological groups fails, it was established in [**Ro13**] that the underlying Schubert cycle is flexible. More precisely, it was established in [**Ro13**, Theorem 4.1] that there are irreducible (projective algebraic) integral varieties of the associated Schubert system which are not translates of the given Schubert cycle. An integral variety of the associated Schubert system is precisely a subvariety where the isomorphism classes of tangent spaces at smooth points are equivalent to those at smooth points of the corresponding Schubert cycle under the action of automorphisms of the ambient Hermitian symmetric space. In particular, [**Ro13**] applies to the cases $\mathcal{V}=\mathcal{V}_{\mathrm{exc}}$ of the full cones of minimal rational tangents of hyperquadrics and rank-2 Grassmannians, which are the *exceptional* cases

excluded in the statement of Theorem 4.1, to show that there exist projective algebraic subvarieties of X which are integral subvarieties S of the Schubert differential systems associated to the Schubert cycle $\mathcal{V}_{\rm exc}$, which implies, writing $\mathscr{C}_x(S) := \mathscr{C}_x(X) \cap \mathbb{P}T_x(S)$, that $\big(\mathscr{C}_x(S) \subset \mathscr{C}_x(X)\big)$ is projectively equivalent to $\big(\mathscr{S}_{[\alpha]} \subset \mathscr{C}_0(X)\big)$ at a general smooth point $x \in S$. It is however not clear how the construction there leads to *transcendental* sub-VMRT structures modeled on $\big(\mathscr{S}_{[\alpha]} \subset \mathscr{C}_0(X)\big)$ (equivalently modeled on $\big(\mathcal{V} \subset X\big)$). In an indirect way, by the method of reconstruction of $\mathcal{V}$ by parallel transport along minimal rational curves (Mok-Yang [**MY18**]), linear saturation of the smooth locus $\mathrm{Reg}(S)$ of S would imply that S is $\gamma(\mathcal{V})$ for some $\gamma \in \mathrm{Aut}(X)$, hence the relevant examples of [**Ro13**] for $\mathcal{V} = \mathcal{V}_{\rm exc}$ must fail to be linearly saturated.

In §6 we will construct examples of *transcendental* integral varieties of the Schubert differential system associated to $\mathcal{V}_{\rm exc}$ in the cases where the ambient space X is the hyperquadric Q^n of dimension $n \geq 4$.

5. Intrinsically flat sub-VMRT structures modeled on certain full cones of minimal rational curves

In Theorem 4.1, $(\mathscr{S}_\alpha, \mathscr{C}_0(X))$ fails to be nondegenerate for substructures in the case where X is either a hyperquadric $Q^n, n \geq 4$, or a rank-2 Grassmannian $G(2,q)$, $q \geq 2$. We will show that for certain sub-VMRT structures $\varpi : \mathscr{C}(S) \to S$ modeled on $\big(\mathscr{S}_\alpha \subset \mathscr{C}_0(X)\big)$, S remains linearly saturated and algebraic as a germ of submanifold at any point $x \in S$. Let $(Z, \mathcal{H})$ and $(X, \mathcal{K})$ be uniruled projective manifolds equipped with minimal rational components. Denote by $\mathscr{C}(Z) \subset \mathbb{P}T(Z)$ resp. $\mathscr{C}(X) \subset \mathbb{P}T(X)$ the VMRT structures of $(Z, \mathcal{H})$ resp. $(X, \mathcal{K})$. Let $A \subset Z$ resp. $B \subset X$ be the bad locus of $(Z, \mathcal{H})$ resp. $(X, \mathcal{K})$, $W \subset X - B$ be an open subset, $S \subset W$ be a complex submanifold such that $S = f(U)$ for some holomorphic embedding $f : U \to X - B$ from a connected open subset $U \subset Z - A$ which respects VMRTs at a general point of U. Defining $\mathscr{C}(S) = \mathscr{C}(X) \cap \mathbb{P}T(S)$, assume that the canonical projection $\varpi : \mathscr{C}(S) \to S$ is a sub-VMRT structure in the sense of Definition 2.1. By the proofs of Hong-Mok [**HoM10**, Proposition 2.1], Hong-Park [**HoP11**, Proposition 2.3] and Mok-Zhang [**MZ17**, Proposition 5.2], it remains the case that S is linearly saturated with respect to $(X, \mathcal{K})$ under the weaker assumption (†) $\mathrm{Ker}(\sigma_{[\chi]}(\,\cdot\,, T_{[\chi]}(\mathscr{C}(S)) \subset T_{[\chi]}(\mathscr{C}(S)))$ at a general point $[\chi]$ of any irreducible component of $\mathscr{C}_x(S)$ for a general point $x \in S$, where the second fundamental form $\sigma_{[\chi]}$ is used in place of $\tau_{[\chi]} = \nu_{[\chi]} \circ \sigma_{[\chi]}$. Following [**HoP11**] and relating to the notion of substructures in this article we will say that $\varpi : \mathscr{C}(S) \to S$ is *weakly nondegenerate for substructures* whenever (†) holds.

Note that the condition (†) depends only on S and $(Z, \mathcal{H})$ disappears from the definition of weak nondegeneracy for substructures. The assumption that S arises from a map allows one to make use of Hessians $\nabla^2 f$ in place of $\tau = \nu \circ \sigma$, i.e., without taking quotients modulo $T(S)$, and this explains why a weaker nondegeneracy condition is sufficient for proving linear saturation. In place of a uniruled projective manifold $(X, \mathcal{K})$ the notions and proofs go through even in the case when Z has singularities, provided that we consider a "minimal rational component" $\mathcal{H}$ on Z such that a general member of $\mathcal{H}$ is a standard minimal rational curve on Z lying on $\mathrm{Reg}(Z)$. This is the case when $(X, \mathcal{K})$ stands for an irreducible

Hermitian symmetric space of rank ≥ 2 other than a Lagrangian Grassmannian, $Z = \mathcal{V}$ stands for a full cone of minimal rational curves on X, and $\mathcal{H}$ resp. $\mathcal{K}$ stands for the space of projective lines lying on $\mathcal{V}$ resp X. We will refer to $\pi_Z : \mathscr{C}(Z) \to Z$ thus obtained as a generalized VMRT-structure. In what follows we will show that the generalized VMRT-structure is *flat* in the sense that there exist at each smooth point of $\mathcal{V}$ local holomorphic coordinates $(U; z_i)$ with respect to which the generalized VMRTs $\{\mathscr{C}_x(Z) : z \in U\}$ form a constant family. We have

LEMMA 5.1. *For the cone $\mathcal{V} := \mathcal{V}(x) \subset X$ of minimal rational curves at some $x \in X$, the generalized VMRT-structure $\pi_{\mathcal{V}} : \mathscr{C}(\mathcal{V}) \to \mathcal{V}$ is flat.*

PROOF. Let $\mathbb{C}^n \subset X$ be a Harish-Chandra coordinate chart. Let $\alpha \in \widetilde{\mathscr{C}_0}(X)$. Parametrize a neighborhood of $[\alpha] \in \mathscr{C}_0(X)$ by an open holomorphic embedding $\varphi : U \to \mathscr{C}_0(X)$ from a neighborhood U of 0 in $T_{[\alpha]}(\mathscr{C}_0(X))$ by $\varphi(\xi) = \alpha + \xi + O(\|\xi\|^2)$, where $T_{[\alpha]}(\mathscr{C}_0(X)) \cong P_\alpha/\mathbb{C}\alpha$ is identified with a complementary linear subspace $H_\alpha \subset P_\alpha$ of $\mathbb{C}\alpha$. Consider $F : \mathbb{C} \times U \to X$ defined by $F(s,\xi) = s\varphi(\xi) - \alpha = s(\alpha + \xi + O(\|\xi\|^2)) - \alpha$. Thus $F(0,0) = -\alpha$ and $F(1,0) = 0$, so that F maps a neighborhood of $(1,0)$ in $\mathbb{C} \times H_\alpha$ biholomorphically onto a neighborhood of 0 in $\mathcal{V}(-\alpha)$, $dF(1,0)\big(a\frac{\partial}{\partial s} + \xi\big) = a\alpha + \xi$. Noting that for any $\eta \in \mathbb{C}^n$ the Euclidean translation $T_\eta(z) := z + \eta$ on $\mathbb{C}^n$ extends to $\Phi_\eta \in \mathrm{Aut}(X)$, for $t \geq 1$ we define $F_t : \mathbb{C} \times tU \to X$ by

$$F_t(s,\xi) := t(1 + \frac{s-1}{t})(\varphi(\frac{\xi}{t})) - t\alpha = (s-1+t)\Big(\alpha + \frac{\xi}{t} + O(\|\frac{\xi}{t}\|^2)\Big) - t\alpha\,,$$

so that $F_1 \equiv F$, $F_t(1-t,0) = -t\alpha$, $F_t(1,0) = 0$ and $dF_t(1,0)\big(a\frac{\partial}{\partial s} + \xi\big) = a\alpha + \xi$. F_t maps $\Delta(1;t) \times tU$ biholomorphically onto a neighborhood of 0 in $\mathcal{V}(-t\alpha)$,

$$F_t(s,\xi) = (s-1)\alpha + \frac{s-1+t}{t}\,\xi + \frac{s-1+t}{t^2}\,O(\|\xi\|^2)\,.$$

As $t \to \infty$ the holomorphic mappings F_t converge uniformly on compact subsets of $\mathbb{C} \times H_\alpha$ to $G : \mathbb{C} \times H_\alpha \to \mathbb{C}^n$ given by $G(s,\xi) = (s-1)\alpha + \xi$. On the other hand, on X the cones of minimal rational curves $\mathcal{V}(-t\alpha) \subset X$ converge as subvarieties to $\mathcal{V}(\infty_{\alpha 0})$, where $\infty_{\alpha 0}$ is the point at infinity of the projective line $\ell(\alpha, 0) \subset X$ containing the points α and 0 on $\mathbb{C}^n \subset X$. Thus $\mathcal{V}(\infty_{\alpha 0}) \cap \mathbb{C}^n \subset \mathbb{C}^n$ is the linear subspace $P_\alpha = \mathbb{C}\alpha \oplus H_\alpha$. Moreover for any $x \in P_\alpha$ we have $\mathscr{C}_x(\mathcal{V}(\infty_{\alpha 0})) = \mathbb{P}(P_\alpha) \cap \mathscr{C}_x(X)$, which forms a constant family in the Harish-Chandra coordinates, proving Lemma 5.1. □ □

If a sub-VMRT structure $\varpi : \mathscr{C}(S) \to S$ on X modeled on $(\mathcal{V} \subset X)$ arises as the image of a flat generalized VMRT structure under a VMRT-respecting map, we call it an *intrinsically flat* sub-VMRT structure. In the cases excluded from Theorem 4.1, we still have an algebraicity result under the assumption that the underlying sub-VMRT structure $\varpi : \mathscr{C}(S) \to S$ on $S \subset W \subset X$ is intrinsically flat. By the proof of Theorem 2.7 (which is Mok-Zhang [**MZ17**, Main Theorem 2]) given in [**MZ17**], to find $Z \supset S$ projective, $\dim(Z) = \dim(S)$, it suffices to show that S is linearly saturated. In the case of sub-VMRT structures arising from the image of VMRT-respecting holomorphic maps we have the following result of Hong-Park [**HoP11**, Proposition 2.3].

PROPOSITION 5.2. *Let $(Z, \mathcal{H})$ and $(X, \mathcal{K})$ be two uniruled projective manifolds each equipped with a minimal rational component. Assume that $\mathscr{C}_z(Z)$ is irreducible and is positive-dimensional for a general point $z \in Z$. Let $f : U \to X$ be a* *holomorphic embedding defined on a connected open subset $U \subset Z$. If f respects varieties of minimal rational tangents and is weakly nondegenerate for substructures with respect to $(\mathcal{K}, \mathcal{H})$, then $f(U)$ is rationally saturated, i.e., for any $z \in U$ and for a general member $[\alpha] \in \mathscr{C}_z(Z)$ a germ of minimal rational curve in X tangent to $df(\alpha)$ at $f(z)$ is necessarily contained in $f(U)$.*

In [**HoP11**, Proposition 2.3] the authors used the term "weakly nondegenerate" which corresponds to the term "weakly nondegenerate for substructures" in the current article. Proposition 5.2 follows from a modification of [**HoM10**, Proposition 2.1] in which, under the stronger assumption that f respects varieties of minimal rational tangents and is nondegenerate for mappings (in the sense of Definition 2.3 here) with respect to $(\mathcal{K}, \mathcal{H})$, the image of a germ $(\ell; z)$ of a minimal rational curve at $z \in Z$ was shown to be a germ of minimal rational curve $(\ell', f(z))$ on X. Here $T_{f(z)}(\ell') = \mathbb{C}df(\alpha)$. In the statement of Proposition 5.2, the assumption that $\mathscr{C}_z(Z)$ is irreducible for a general point $z \in Z$ is unimportant as can be seen from the proofs of [**HoM10**, Proposition 2.1] and [**HoP11**, Proposition 2.3], provided that the nondegeneracy condition is satisfied at a general point of each of the irreducible components of the VMRT. Moreover the same arguments apply to the generalized VMRT structures in this article in which $(Z, \mathcal{H})$ is given by $(\mathcal{V}, \mathcal{H})$, where $\mathcal{H} \subset \mathcal{V}$ consists of projective lines lying on $\mathcal{V}$ (when the latter is considered as a uniruled projective subvariety of the ambient Hermitian symmetric space $X = G/P$ of rank ≥ 2 not biholomorphic to a Lagrangian Grassmannian).

Finally, we formulate the algebraicity result in the cases excluded from Theorem 4.1.

THEOREM 5.3. *In the exceptional cases in Theorem 4.1 of hyperquadrics $X = Q^n$, $n \geq 4$, and of Grassmannians $X = G(2,q)$, $q \geq 2$, where $(\mathscr{S}_{[\alpha]}, \mathscr{C}_0(X))$ fails to be nondegenerate for substructures, $(\mathscr{S}_{[\alpha]}, \mathscr{C}_0(X))$ remains weakly nondegenerate for substructures. In particular, if $S \subset X$ inherits an intrinsically flat sub-VMRT structure modeled on $\big(\mathscr{S}_{[\alpha]} \subset \mathscr{C}_0(X)\big)$, then S is linearly saturated and there exists a subvariety $Z \subset X$ such that $S \subset Z$ and* $\dim(Z) = \dim(S)$.

PROOF. Theorem 5.3 for the case of the hyperquadric Q_n, $n \geq 4$, follows readily from Proposition 4.2 and Proposition 5.2. In the case of the rank-2 Grassmannian $G(2,q), q \geq 2$, it suffices to check directly that $\mathrm{Ker}(\sigma_{[\beta]}(\,\cdot\,, T_{[\beta]}(\mathscr{S}_{[\alpha]}))) \subset T_{[\beta]}(\mathscr{S}_{[\alpha]})$. Here $\mathscr{C}_0(G(2,q)) \subset \mathbb{P}T_0(G(2,q))$ is given by the image of the Segre embedding $\varsigma : \mathbb{P}^1 \times \mathbb{P}^{q-1} \to \mathbb{P}^{2q-1}$. Writing $T_0(G(2,q)) = U_0 \otimes V_0$, $\dim(U_0) = 2$, $\dim(V_0) = q$, for $\alpha = u \otimes v$, $\mathscr{S}_{[\alpha]} \subset \mathscr{C}_0(G(2,q)) = \varsigma(\mathbb{P}^1 \times \mathbb{P}^{q-1})$ is the union $\mathbb{P}(U_0 \otimes \mathbb{C}v) \cup \mathbb{P}(\mathbb{C}u \otimes V_0)$ of a projective line and a projective $(q-1)$-plane passing through $[\alpha] = [u \otimes v]$. Recall that $T_{u\otimes v}(\widetilde{\mathscr{C}_0}(G(2,q)) = (\mathbb{C}u \otimes V_0) + (U_0 \otimes \mathbb{C}v)$, $(\mathbb{C}u \otimes V_0) \cap (U_0 \otimes \mathbb{C}v) = \mathbb{C}(u \otimes v)$. A smooth point $[\beta]$ of $\mathscr{S}_{[\alpha]}$ is either given by $\beta = u' \otimes v$ where $u' \notin \mathbb{C}u$, or $\beta = u \otimes v'$ where $v' \notin \mathbb{C}v$. In the first case from the quadratic expansion of the affinized VMRT at β we have $\sigma_\beta(u' \otimes \eta, \xi \otimes v) = \xi \otimes \eta \mod ((\mathbb{C}u' \otimes V_0) + (U_0 \otimes \mathbb{C}v))$, while $\sigma_\beta(u_1 \otimes v, u_2 \otimes v) = 0$, and it follows readily that $\mathrm{Ker}(\sigma_{[\beta]}(\,\cdot\,, T_{[\beta]}(\mathscr{S}_{[\alpha]}))) = (U_0 \otimes \mathbb{C}v)/\mathbb{C}(u' \otimes v) = T_{[\beta]}(\mathscr{S}_{[\alpha]})$. In the second case interchanging the roles of U_0 and V_0 exactly the same argument yields again $\mathrm{Ker}(\sigma_{[\beta]}(\,\cdot\,, T_{[\beta]}(\mathscr{S}_{[\alpha]}))) = T_{[\beta]}(\mathscr{S}_{[\alpha]})$. The proof of Theorem 5.3 is complete. □

REMARK 5.4. As for Theorem 4.1, by the process of reconstructing S by adjunction of minimal rational curves it can be established that $Z = \gamma(\mathcal{V})$ for some $\gamma \in \mathrm{Aut}(X)$.

6. Examples of non-standard sub-VMRT structures modeled on full cones of rational curves

In this section we give an elementary construction of examples of *transcendental* sub-VMRT structures modeled on the full cone $\mathcal{V}$ of minimal rational curves in the case where the ambient manifold X is a hyperquadric Q^n for dimension ≥ 4. Fix $n \geq 4$. Let $\mathbb{C}^n \subset Q^n$ be a Harish-Chandra coordinate chart on which the holomorphic conformal structure on $\mathbb{C}^n$ with Euclidean coordinates $(z_1, \cdots, z_n)$ is given by the class of holomorphic nondegenerate quadratic forms $\lambda(z)((dz^1 \otimes dz^1) + \cdots + (dz^n \otimes dz^n))$, where λ is any nowhere zero holomorphic function on $\mathbb{C}^n$. We are going to write down explicit examples of transcendental hypersurfaces $S \subset \mathbb{C}^n \subset Q^n$ such that S inherits a sub-VMRT structure modeled on $\big(\mathscr{S}_{[\alpha]}(Q^n) \subset \mathscr{C}_0(Q^n)\big)$. Here $[\alpha] \in \mathscr{C}_0(Q^n) \cong Q^{n-2} \subset \mathbb{P}T_0(Q^n) \cong \mathbb{P}^{n-1}$, and $\mathscr{S}_{[\alpha]}(Q^n) = \mathscr{C}_0(Q^n) \cap \mathbb{P}(P_\alpha)$ is the singular hyperplane section of $\mathbb{P}(P_\alpha)$ with the isolated singularity at $[\alpha]$. We have

PROPOSITION 6.1. *Let n, m be positive integers such that $n \geq 4$ and $2 \leq m \leq n-2$. Let $A(z_1, \cdots, z_m) = a_1 z_1 + \cdots + a_m z_m$ be a linear function in $(z_1, \cdots, z_m)$ such that $(a_1, \cdots, a_m) \neq 0$ and $a_1^2 + \cdots + a_m^2 = 0$. Let $(b_{m+1}, \cdots, b_n) \neq 0$ be such that $b_{m+1}^2 + \cdots + b_n^2 = 0$. Let $f : \mathbb{C}^n \to \mathbb{C}$ be defined by $f(z_1, \cdots, z_n) = e^{A(z_1, \cdots, z_m)} + b_{m+1} z_{m+1} + \cdots + b_n z_n$. Then, for $c \in \mathbb{C}$ the level set $S = \{f(z) = c\}$ is a transcendental smooth hypersurface on $\mathbb{C}^n \subset Q^n$ such that, defining $\mathscr{C}(S) := \mathbb{P}T(S) \cap \mathscr{C}(Q^n)$ and writing $\varpi : \mathscr{C}(S) \to S$ for the canonical projection, the latter is a sub-VMRT structure modeled on $\big(\mathscr{S}_{[\alpha]}(Q^n) \subset \mathscr{C}_0(Q^n)\big)$ which is neither linearly saturated nor intrinsically flat.*

PROOF. We may assume that $b_n = -1$, hence the level set $S = \{f(z) = c\}$ is given by $z_n = h(z_1, \cdots, z_{n-1}) = e^{A(z_1, \cdots, z_m)} + b_{m+1} z_{m+1} + \cdots + b_{n-1} z_{n-1} - c$ such that $S \subset \mathbb{C}^n$ is a transcendental smooth hypersurface. Write $q = (dz^1 \otimes dz^1) + \cdots + (dz^n \otimes dz^n)$. It induces an isomorphism $L : T^*(\mathbb{C}^n) \xrightarrow{\cong} T(\mathbb{C}^n)$ such that for any $z \in \mathbb{C}^n$, $\omega \in T_z^*(\mathbb{C}^n), \xi \in T_z(\mathbb{C}^n)$, we have $q(L(\omega), \xi) = \omega(\xi)$. Write $p = \frac{\partial}{\partial z_1} \otimes \frac{\partial}{\partial z_1} + \cdots + \frac{\partial}{\partial z_n} \otimes \frac{\partial}{\partial z_n}$. We have $df = e^{A(z_1, \cdots, z_m)}(a_1 dz^1 + \cdots + a_m dz^m) + (b_{m+1} dz^{m+1} + \cdots + b_n dz^n)$, so that $p(df(z), df(z)) = e^{A(z_1, \cdots, z_m)}(a_1^2 + \cdots + a_m^2) + (b_{m+1}^2 + \cdots + b_n^2) = 0$. Now fix c and let $x \in S = \{f(z) = c\}$. We have $T_x(S) = \mathrm{Ker}(df(x)) \subset T_x(\mathbb{C}^n)$. Thus, for $\eta := L(df(x))$ and $\xi \in T_x(\mathbb{C}^n)$ we have $q(\eta, \xi) = q(L(df(x)), \xi) = df(x)(\xi)$, and $\xi \in T_x(S)$ if and only if $q(\eta, \xi) = 0$. On the other hand, $q(\eta, \eta) = p(df(x), df(x)) = 0$, so that $\eta \in T_x(S)$, and $q|_{T_x(S)}$ is degenerate, with kernel spanned by $\eta = e^{A(z_1, \cdots, z_m)}(a_1 \frac{\partial}{\partial z_1} + \cdots + a_m \frac{\partial}{\partial z_m}) + (b_{m+1} \frac{\partial}{\partial z_{m+1}} + \cdots + b_n \frac{\partial}{\partial z_n})$. Hence, $\mathscr{C}_x(S) = \mathbb{P}T_x(S) \cap \mathscr{C}_x(X)$ is the singular hyperplane section of $\mathscr{C}_x(X)$ with the isolated singularity at $[\eta]$, and $\varpi : \mathscr{C}(S) \to S$ inherits a sub-VMRT structure modeled on $\big(\mathscr{S}_{[\alpha]}(Q^n) \subset \mathscr{C}_0(Q^n)\big)$. Consider the projective line ℓ passing through $x = (z_1, \cdots, z_{n-1}, h(z_1, \cdots, z_{n-1}))$ such that $T_x(\ell) = \mathbb{C}\eta$. From the explicit description of S one checks readily that in general $\ell \not\subset S$, so that $S \subset Q^n$ is not linearly saturated. Alternatively, linear saturation of S would imply by the proof of Theorem 2.7 (i.e., proof of Main Theorem 2 of [**MZ17**]) that $\overline{S} \subset Q^n$ is projective, contradicting with the transcendence of S. Finally, again from the transcendence of $S \subset \mathbb{C}^n$ it follows by Theorem 5.3 that $\varpi : \mathscr{C}(S) \to S$ is not intrinsically flat, completing the proof of Proposition 5.2. □

REMARK 6.2. Replacing $f(z_1, \cdots, z_n) = e^{A(z_1,\cdots,z_m)} + b_{m+1}z_{m+1} + \cdots + b_n z_n$ by $g(z_1, \cdots, z_n) = A(z_1, \cdots, z_m)^q + b_{m+1}z_{m+1} + \cdots + b_n z_n$ for any integer $q \geq 2$, by the same arguments as above we obtain examples of non-standard sub-VMRT structures modeled on $\big(\mathscr{S}_{[\alpha]}(Q^n) \subset \mathscr{C}_0(Q^n)\big)$ on nonsingular level sets $S = \{g(z) = c\} \subset \mathbb{C}^n$ which are affine-algebraic.

7. Concluding Remarks

Cohomological methods play an important role in earlier known approaches on various problems of rigidity concerning Schubert cycles on Hermitian symmetric spaces of the compact type. By contrast, in the approach stemming from the geometric theory of uniruled projective manifolds our perspective is to treat rigidity concerning special subvarieties as a problem in differential geometry revolving around varieties of minimal rational tangents. We note however that there are links between the cohomological and the differential-geometric methods worthy of further exploration.

As an example, in the work of Hong [**Ho05**], in which Schur rigidity is established for nonsingular Schubert cycles of irreducible Hermitian symmetric spaces $X = G/P$ of the compact type, the author made use of the result of Goncharov [**Gon87**] from the theory of generalized conformal structures on integral varieties of F-structures, where the cohomology groups concerned are defined on orbits under the action (of the semisimple part) of P on the Grassmann of k-planes of $T_0(X)$. From our perspective much information is already stored in the VMRT, which is the highest weight orbit of the semisimple part of P under the isotropy action on $\mathbb{P}T_0(X)$. In a certain sense, in place of requiring vanishing results on cohomological spaces defined on the orbits, which are rationally homogeneous spaces, we replace them by nondegeneracy conditions defined from projective geometry, introduced by Hong-Mok [**HoM10**] for mappings and by Mok-Zhang [**MZ17**] for substructures. One may say that our approach is microlocal in nature, imposing conditions at a general point of the pair consisting of a VMRT, which in the case of irreducible Hermitian symmetric spaces of the compact type is an orbit in the projectivized tangent space under the isotropy representation, and a sub-VMRT (which is a linear section of the VMRT) rather than global conditions on certain projective varieties which are orbits in Grassmannians under actions derived from the isotropy representation. In this way we relax the requirement from global vanishing results on projective manifolds to microlocal vanishing results (on kernels arising from certain quadratic forms).

A general form of the problem for characterizing special subvarieties of uniruled projective manifolds was formulated as the Recognition Problem in Mok [**Mo16b**, Problem 4.5.1]. In a nutshell the theory of sub-VMRT structure gives an approach for resolving the Recognition Problem for special subvarieties Σ of Hermitian symmetric spaces of the compact type and more generally those of rational homogeneous spaces whenever there exists a projective line lying on the smooth locus $\mathrm{Reg}(\Sigma)$ of Σ. In cohomological approaches the Recognition Problem is a problem of integrability of certain geometric substructures. Our approach breaks the problem down into two steps, the first step being the verification of a microlocal condition implying partial integrability, more precisely the property that the support of the sub-VMRT structure is linearly saturated. The second step is that of reconstruction by a finite process of adjunction of minimal rational curves, i.e., projective lines. The current article is an an illustration of the first step of our scheme beyond smooth Schubert cycles by an examination of a very special class of singular Schubert varieties (which are nonetheless of special interest to Kähler geometry). The issue of reconstruction by an improvement of the method of adjunction of minimal rational curves will be taken up elsewhere. One advantage of our scheme is that, where applicable, it may lead to results ascertaining linear saturation when additional intrinsic conditions are imposed on the underlying complex submanifolds S of sub-VMRT structures, which is illustrated by Theorem 5.3, where linear saturation and algebraicity are proven in the cases excluded by Theorem 4.1 when S is further assumed to be intrinsically flat.

In another direction, we can study the Recognition Problem modeled on a family of sub-VMRTs which come from linear sections of Schubert cycles by the introduction of a quantitative measure of nondegeneracy for substructures, called p-nondegeneracy, $p \geq 1$, where the case $p = 1$ corresponds to the usual notion of nondegeneracy for substructures as given in Definition 2.4. In the case at hand, for sub-VMRT structures modeled on $\big(\mathscr{S}_{[\alpha]} \subset \mathscr{C}_0(X)\big)$ in the notation of Theorem 4.1, the optimal value of p is $s(X)-1$, where $s(X)$ was listed in the second last paragraph of the proof of Theorem 4.1. Taking $[\beta] \in \mathscr{S}_{[\alpha]}$ distinct from $[\alpha]$, $s(X)$ is the minimal value of $\dim\big(\sigma_\beta(\mathbb{C}\eta \otimes T_\beta(\widetilde{\mathscr{S}_{[\alpha]}}))\big)$ as η varies over nonzero vectors in $T_\beta(\widetilde{\mathscr{C}_0}(X)) - T_\beta(\widetilde{\mathscr{S}_{[\alpha]}})$, while the minimal value of $\dim\big(\tau_\beta(\mathbb{C}\eta \otimes T_\beta(\widetilde{\mathscr{S}_{[\alpha]}}))\big)$ is equal to $s(X) - 1$. Let k be a positive integer so that $\dim(\mathscr{S}_{[\alpha]}) - k > 0$. Consider the Grassmannian $\mathrm{Gr}(n-k, T_0(X))$ of vector subspaces $V \subset T_0(X)$ of codimension k in $T_0(X), n := \dim(X)$. There is a dense Zariski open subset $\Phi_k \subset \mathrm{Gr}(n-k, T_0(X))$ such that for every linear subspace $V \subset T_0(X)$ of codimension k belonging to Φ_k, we have (a) $\mathbb{P}(V)$ intersects $\mathscr{S}_{[\alpha]}$ in pure codimension k, (b) the intersection is transversal at a general point of each irreducible component of $\mathscr{S}_{[\alpha]} \cap \mathbb{P}(V)$ and (c) $\mathscr{S}_{[\alpha]} \cap \mathbb{P}(V)$ is linearly nondegenerate in $\mathbb{P}(V)$. Note that from (b) it follows that the pair $(\mathscr{S}_{[\alpha]} \cap \mathbb{P}(V), \mathscr{C}_0(X) \cap \mathbb{P}(V))$ satisfies Condition (T) in the sense of Definition 2.5. A member of Φ_k will be called a Φ_k-general vector subspace of $T_0(X)$ of codimension k. In analogy to Mok-Zhang [**MZ17**, Theorem 9.1] we have the following result Theorem 7.1 on linear saturation and algebraicity for sub-VMRT structures on X modeled on the family of pairs $(\mathscr{S}_{[\alpha]} \cap \mathbb{P}(V), \mathscr{C}_0(X) \cap \mathbb{P}(V))$ for Φ_k-general vector subspaces $V \subset T_0(X)$ of codimension k.

For the formulation of the result let X be an irreducible Hermitian symmetric space of the compact type of rank ≥ 2 not biholomorphic to a Lagrangian Grassmannian. Let $0 \in X$ be a reference point and $[\alpha] \in \mathscr{C}_0(X)$ be an arbitrary point. Let k be a positive integer satisfying $k \leq s(X) - 2$. Identify X as a projective submanifold by means of the first canonical embedding $\iota : X \hookrightarrow \mathbb{P}^N$. Let $\Pi \subset \mathbb{P}^N$ be a projective linear subspace of codimension k such that $Z = X \cap \Pi$ is a smooth linear section of codimension k in X. Since $Z \subset X \subset \mathbb{P}^N$ is a smooth linear section of codimension k, each irreducible component of the set of projective lines on Z passing through a general point $x \in Z$ must be of dimension at least equal to $\dim(\mathscr{C}_0(X)) - k > s(X) - k > 2$, so that in particular Z is uniruled by projective lines. Let $E \subset Z$ be the bad locus (which is here the same as the enhanced bad locus) of Z as a projective submanifold uniruled by lines. From the deformation theory of rational curves the VMRT $\mathscr{C}_x(Z)$ of Z as a projective submanifold uniruled by lines is of dimension exactly equal to $\dim(\mathscr{C}_0(X)) - k$ for $x \in Z - E$. We have

THEOREM 7.1. *Suppose $W \subset Z - E$ is a nonempty open subset in the complex topology, and $S \subset W$ is a complex submanifold of dimension $\dim(\mathscr{C}_0(X)) + 1 - k$ such that, writing $\mathscr{C}(Z) := \mathscr{C}(X) \cap \mathbb{P}T(\Pi) \cap \mathbb{P}T(Z-E)$ and $\mathscr{C}(S) := \mathscr{C}(Z) \cap \mathbb{P}T(S)$, the canonical projection $\varpi : \mathscr{C}(S) \to S$ defines a sub-VMRT structure on S such that for a general point $x \in S$, $\big(\mathscr{C}_x(S) \subset \mathscr{C}_x(Z)\big)$ is projectively equivalent to $\big(\mathscr{S}_{[\alpha]} \cap \mathbb{P}(V_x) \subset \mathscr{C}_0(X) \cap \mathbb{P}(V_x)\big)$ for some Φ_k-general vector subspace $V_x \subset T_0(X)$ of codimension k. Then, S is linearly saturated. Moreover, there exists a subvariety $Z \subset X$ such that $S \subset Z$ and $\dim(Z) = \dim(S)$.*

PROOF. By the proof of Theorem 4.1, $(\mathscr{S}_{[\alpha]}, \mathscr{C}_0(X))$ is $(s(X) - 1)$-nondegenerate for substructures according to the definition of p-nondegeneracy for substructures in the preceding paragraphs. The key to the proof of Theorem 7.1 is the observation that whenever $k \leq s(X) - 2$, the pair $(\mathscr{S}_{[\alpha]} \cap \mathbb{P}(V_x), \mathscr{C}_0(X) \cap \mathbb{P}(V_x))$ remains r-nondegenerate for $r := s(X) - 1 - k \geq 1$ for a Φ_k-general vector subspace $V_x \subset T_0(X)$ of codimension k, which follows from Mok-Zhang [**MZ17**, Proposition 9.1]. By Theorem 2.6, S is linearly saturated. The last statement follows from Theorem 2.7, completing the proof of Theorem 7.1. □

In order for Theorem 7.1 to be applicable we need to have $s(X) \geq 3$, which rules out the cases of Grassmannians of rank ≤ 3, $G^{II}(5,5)$ and hyperquadrics. It is for instance applicable to the Grassmannian $X = G(n,n)$, $n \geq 4$, to prove results of linear saturation and algebraicity for germs of complex submanifolds on smooth codimension-k linear sections admitting sub-VMRT structures modeled on Φ_k-general linear sections of $\mathscr{S}_{[\alpha]}$ for $0 < k \leq n - 3$, and to $X(E_7)$ for $0 < k \leq 3$.

ACKNOWLEDGMENT. This current article is dedicated to Professor Lawrence Ein and written for the special volume on the Conference "Local and global methods in Algebraic Geometry" held at University of Illinois, Chicago in May 2016 in honor of Professor Ein on the occasion of his sixtieth birthday. We would like to thank the organizers of the conference for their kind invitation and their hospitality. The author would also like to thank the referees for their careful reading of the article and for their many suggestions which have led to improvements on the article both in the contents and in the exposition. Especially, he would like to thank one of the referees for suggesting Lemma 4.7 (and providing a proof of the lemma) to give a uniform and simpler proof of Proposition 4.6.

References

[Br01] Bryant, R.: Rigidity and quasi-rigidity of extremal cycles in compact Hermitian symmetric spaces, arXiv:math/0006186, to appear in Annals of Mathematics Studies, Princeton University Press.

[Gui65] Victor Guillemin, *The integrability problem for G-structures*, Trans. Amer. Math. Soc. **116** (1965), 544–560. MR0203626

[Gon87] A. B. Goncharov, *Generalized conformal structures on manifolds*, Selecta Math. Soviet. **6** (1987), no. 4, 307–340. Selected translations. MR925263

[Ho05] Jaehyun Hong, *Rigidity of singular Schubert varieties in* $\mathrm{Gr}(m,n)$, J. Differential Geom. **71** (2005), no. 1, 1–22. MR2191767

[Ho07] Jaehyun Hong, *Rigidity of smooth Schubert varieties in Hermitian symmetric spaces*, Trans. Amer. Math. Soc. **359** (2007), no. 5, 2361–2381. MR2276624

[HoM10] Jaehyun Hong and Ngaiming Mok, *Analytic continuation of holomorphic maps respecting varieties of minimal rational tangents and applications to rational homogeneous manifolds*, J. Differential Geom. **86** (2010), no. 3, 539–567. MR2785842

[HoM13] Jaehyun Hong and Ngaiming Mok, *Characterization of smooth Schubert varieties in rational homogeneous manifolds of Picard number 1*, J. Algebraic Geom. **22** (2013), no. 2, 333–362. MR3019452

[HoP11] Jaehyun Hong and Kyeong-Dong Park, *Characterization of standard embeddings between rational homogeneous manifolds of Picard number 1*, Int. Math. Res. Not. IMRN **10** (2011), 2351–2373. MR2806567

[Hw10] Jun-Muk Hwang, *Equivalence problem for minimal rational curves with isotrivial varieties of minimal rational tangents* (English, with English and French summaries), Ann. Sci. Éc. Norm. Supér. (4) **43** (2010), no. 4, 607–620. MR2722510

[Hw12] Jun-Muk Hwang, *Geometry of varieties of minimal rational tangents*, Current developments in algebraic geometry, Math. Sci. Res. Inst. Publ., vol. 59, Cambridge Univ. Press, Cambridge, 2012, pp. 197–226. MR2931871

[Hw15] Jun-Muk Hwang, *Geometry of varieties of minimal rational tangents*, Current developments in algebraic geometry, Math. Sci. Res. Inst. Publ., vol. 59, Cambridge Univ. Press, Cambridge, 2012, pp. 197–226. MR2931871

[HM98] Jun-Muk Hwang and Ngaiming Mok, *Rigidity of irreducible Hermitian symmetric spaces of the compact type under Kähler deformation*, Invent. Math. **131** (1998), no. 2, 393–418. MR1608587

[HM99] Jun-Muk Hwang and Ngaiming Mok, *Varieties of minimal rational tangents on uniruled projective manifolds*, Several complex variables (Berkeley, CA, 1995), Math. Sci. Res. Inst. Publ., vol. 37, Cambridge Univ. Press, Cambridge, 1999, pp. 351–389. MR1748609

[HM01] Jun-Muk Hwang and Ngaiming Mok, *Cartan-Fubini type extension of holomorphic maps for Fano manifolds of Picard number 1*, J. Math. Pures Appl. (9) **80** (2001), no. 6, 563–575. MR1842290

[Ko63] Bertram Kostant, *Lie algebra cohomology and generalized Schubert cells*, Ann. of Math. (2) **77** (1963), 72–144. MR0142697

[Li82] Woody Lichtenstein, *A system of quadrics describing the orbit of the highest weight vector*, Proc. Amer. Math. Soc. **84** (1982), no. 4, 605–608. MR643758

[Mo89] Ngaiming Mok, *Metric rigidity theorems on Hermitian locally symmetric manifolds*, Series in Pure Mathematics, vol. 6, World Scientific Publishing Co., Inc., Teaneck, NJ, 1989. MR1081948

[Mo08a] Ngaiming Mok, *Characterization of standard embeddings between complex Grassmannians by means of varieties of minimal rational tangents*, Sci. China Ser. A **51** (2008), no. 4, 660–684. MR2395412

[Mo08b] Ngaiming Mok, *Geometric structures on uniruled projective manifolds defined by their varieties of minimal rational tangents* (English, with English and French summaries), Astérisque **322** (2008), 151–205. Géométrie différentielle, physique mathématique, mathématiques et société. II. MR2521656

[Mo16a] Ngaiming Mok, *Holomorphic isometries of the complex unit ball into irreducible bounded symmetric domains*, Proc. Amer. Math. Soc. **144** (2016), no. 10, 4515–4525. MR3531198

[Mo16b] Ngaiming Mok, *Geometric structures and substructures on uniruled projective manifolds*, Foliation theory in algebraic geometry, Simons Symp., Springer, Cham, 2016, pp. 103–148. MR3644245

[MY18] Mok, N., and Yang, X.: In preparation.

[MZ17] Mok, N. and Zhang Y.: Rigidity of pairs of rational homogeneous spaces of Picard number 1 and analytic continuation of geometric substructures on uniruled projective manifolds, to appear in *J. Differential Geom.* *url*: http://hkumath.hku.hk/∼imr/IMRPreprintSeries/2015/IMR2015- 8_v2(2016).pdf

[NT76] Hisao Nakagawa and Ryoichi Takagi, *On locally symmetric Kaehler submanifolds in a complex projective space*, J. Math. Soc. Japan **28** (1976), no. 4, 638–667. MR0417463

[Ro13] C. Robles, *Schur flexibility of cominuscule Schubert varieties*, Comm. Anal. Geom. **21** (2013), no. 5, 979–1013. MR3152970

[RT12] C. Robles and D. The, *Rigid Schubert varieties in compact Hermitian symmetric spaces*, Selecta Math. (N.S.) **18** (2012), no. 3, 717–777. MR2960030

[Ro85] Ros, A.: A characterization of seven compact Kähler manifolds in a complex projective space, *Ann. Math.* **121** (1985), 377–382.

[Wa97] Walters, M.: *Geometry and uniqueness of some extremal subvarieties in complex Grassmannians*, Ph.D. thesis, U. Michigan, 1997.

DEPARTMENT OF MATHEMATICS, THE UNIVERSITY OF HONG KONG, POKFULAM, HONG KONG
Email address: nmok@hku.hk

Contemporary Mathematics
Volume **712**, 2018
http://dx.doi.org/10.1090/conm/712/14351

A boundedness conjecture for minimal log discrepancies on a fixed germ

Mircea Mustaţă and Yusuke Nakamura

Dedicated to Lawrence Ein, on the occasion of his sixtieth birthday

ABSTRACT. We consider the following conjecture: on a klt germ (X, x), for every finite set I there is a positive integer ℓ with the property that for every $\mathbf{R}$-ideal $\mathfrak{a}$ on X with exponents in I, there is a divisor E over X that computes the minimal log discrepancy $\mathrm{mld}_x(X, \mathfrak{a})$ and such that its discrepancy k_E is bounded above by ℓ. We show that this implies Shokurov's ACC conjecture for minimal log discrepancies on a fixed klt germ and give some partial results towards the conjecture.

1. Introduction

One of the outstanding open problems in birational geometry is the Termination of Flips conjecture, which predicts that there are no infinite chains of certain birational transformations (*flips*). It is an insight due to Shokurov that this global problem can be reduced to conjectural properties of invariants of singularities. A typical such property is the Ascending Chain Condition (ACC, for short) which predicts that in a fixed dimension, and with suitable restrictions on the coefficients of the divisors involved, there are no infinite strictly increasing sequences of such invariants. There are two types of invariants that are important in this setting: the log canonical thresholds and the minimal log discrepancies. As a rule, log canonical thresholds are easier to study and they are related to many other points of view on singularities. In particular, Shokurov's ACC conjecture for log canonical thresholds has been proved (see [**dFEM10**] for the smooth case, [**dFEM11**] for the case of varieties with bounded singularities, and [**HMX14**] for the general case). However, while the ACC property in this setting implies the termination of certain families of flips in an inductive setting (see [**Bir07**] for the precise statement), it does not allow proving any termination result in arbitrary dimension. It turns out that in order to do this one has to work with minimal log discrepancies (mlds, for short). In fact, Shokurov showed in [**Sho04**] that two conjectural properties of mlds (the Semicontinuity conjecture and the ACC conjecture) imply termination of flips. The Semicontinuity conjecture is believed to be the easier of the two problems.

2010 *Mathematics Subject Classification.* Primary 14B05; Secondary 14J17, 14E15.
Key words and phrases. Minimal log discrepancy, ascending chain condition.
The first author was partially supported by NSF grants DMS-1265256 and DMS-1401227.
The second author was partially supported by the Grant-in-Aid for Scientific Research (KAKENHI No. 16H06710).

In fact, this is known in some cases (see [**EMY03**] for the case of smooth varieties and [**Nak16a**] for the case of varieties with quotient singularities). In this paper we propose an approach towards Shokurov's ACC conjecture when we only consider mlds on a fixed germ of variety (X, x). In particular, this would cover the case of smooth ambient varieties.

Before stating our main results, let us introduce some notation. We always assume that we work over an algebraically closed field of characteristic 0. Let X be a variety and $x \in X$ a (closed) point. We work with $\mathbf{R}$-ideals $\mathfrak{a}$, that is, formal products $\mathfrak{a} = \prod_{j=1}^{r} \mathfrak{a}_j^{\lambda_j}$, where the λ_j are nonnegative real numbers and the $\mathfrak{a}_j$ are nonzero coherent ideals in $\mathcal{O}_X$. We say that $\mathfrak{a}$ *has exponents in* a set $I \subseteq \mathbf{R}_{\geq 0}$ if $\lambda_j \in I$ for all j. We assume that X is $\mathbf{Q}$-Gorenstein and denote by $\mathrm{mld}_x(X, \mathfrak{a})$ the minimal log discrepancy of $(X, \mathfrak{a})$ at x. This is a nonnegative real number if and only if $(X, \mathfrak{a})$ is log canonical in some neighborhood of x; otherwise, if $\dim(X) \geq 2$, then $\mathrm{mld}_x(X, \mathfrak{a}) = -\infty$. For a divisor E over X, we denote by k_E the coefficient of E in the discrepancy divisor; given an $\mathbf{R}$-ideal $\mathfrak{a}$ on X, we put $a_E(X, \mathfrak{a}) = k_E + 1 - \mathrm{ord}_E(\mathfrak{a})$, where ord_E is the valuation corresponding to E. By definition, $\mathrm{mld}_x(X, \mathfrak{a})$ is the infimum of the set of real numbers $a_E(X, \mathfrak{a})$, where E varies over the divisors with image x in X.

In this paper we consider the following boundedness conjecture for mlds on a fixed germ.

CONJECTURE 1.1. *Let X be a klt variety and let $x \in X$. Given a finite subset $I \subset \mathbf{R}_{\geq 0}$, there is a positive integer ℓ (depending on (X, x) and I) such that for every $\mathbf{R}$-ideal $\mathfrak{a}$ on X with exponents in I, there is a divisor E that computes $\mathrm{mld}_x(X, \mathfrak{a})$ and such that $k_E \leq \ell$.*

We use the theory of generic limits of ideals developed in [**dFM09**], [**Kol08**], and [**dFEM11**] to show the weaker statement in which we bound the order along E of the ideal defining the point $x \in X$ (we expect this result to be useful for attacking the above conjecture). More precisely, we show the following:

THEOREM 1.2. *Let X be a klt variety and $x \in X$ a point defined by the ideal $\mathfrak{m}_x$. For every finite subset $I \subset \mathbf{R}_{\geq 0}$, there is a positive integer ℓ (depending on (X, x) and I) such that the following conditions hold:*

i) *For every $\mathbf{R}$-ideal $\mathfrak{a}$ with exponents in I such that $\mathrm{mld}_x(X, \mathfrak{a}) > 0$ and* **every** *divisor E over X that computes $\mathrm{mld}_x(X, \mathfrak{a})$, we have $\mathrm{ord}_E(\mathfrak{m}_x) \leq \ell$.*
ii) *For every $\mathbf{R}$-ideal $\mathfrak{a}$ with exponents in I such that $\mathrm{mld}_x(X, \mathfrak{a}) \leq 0$, there is* **some** *divisor E over X that computes $\mathrm{mld}_x(X, \mathfrak{a})$ and such that*

$$\mathrm{ord}_E(\mathfrak{m}_x) \leq \ell.$$

In a related direction, we also show that if I is a finite set and (X, x) is fixed, then there is a positive integer ℓ such that for every $\mathbf{R}$-ideal $\mathfrak{a}$ on X with exponents in I, in order to check that $(X, \mathfrak{a})$ is log canonical at x it is enough to check that $a_E(X, \mathfrak{a}) \geq 0$ for all divisors E with center x and with $k_E \leq \ell$ (see Proposition 3.3). This result admits a nice consequence concerning the characterization of log canonical pairs in terms of jet schemes (see Proposition 3.4).

As farther evidence for the conjecture, we handle the two-dimensional case and the case of monomial ideals.

THEOREM 1.3. *Conjecture* 1.1 *holds if* $\dim(X) = 2$.

THEOREM 1.4. *Conjecture* 1.1 *holds if* $(X, x) = (\mathbf{A}^n, 0)$ *and* $\mathfrak{a}$ *is a monomial* $\mathbf{R}$*-ideal.*

Our interest in the above conjecture is motivated by the following connection with Shokurov's ACC conjecture for minimal log discrepancies. Recall that a subset $I \subseteq \mathbf{R}$ *satisfies ACC* (respectively, *DCC*) if it contains no infinite strictly increasing (respectively, decreasing) sequences.

THEOREM 1.5. *Let* X *be a klt variety and* $x \in X$ *be a point such that the assertion in Conjecture* 1.1 *holds for* (X, x) *and for every finite subset* $I \subset \mathbf{R}_{\geq 0}$. *For every fixed DCC set* $J \subset \mathbf{R}_{\geq 0}$*, the set*

$$\{\mathrm{mld}_x(X, \mathfrak{a}) \mid \mathfrak{a} \textit{ is an } \mathbf{R}\textit{-ideal on } X \textit{ with exponents in } J, (X, \mathfrak{a}) \textit{ log canonical at } x\}$$

satisfies ACC.

We show that Conjecture 1.1 is equivalent to two other conjectures on minimal log discrepancies. One of these is (a uniform version of) the Ideal-adic Semicontinuity conjecture for mlds (see Conjecture 7.3 for the precise formulation). This has been studied by Kawakita and various partial answers have been obtained in [**Kaw13a**], [**Kaw13b**], and [**Kaw15**]. The other conjecture is the Generic Limit conjecture on minimal log discrepancies, also studied by Kawakita in [**Kaw14**] (see Conjecture 7.1).

THEOREM 1.6. *Conjectures* 1.1, 7.1, *and* 7.3 *are equivalent.*

The paper is organized as follows. In §2 we recall the definition and some basic facts related to minimal log discrepancies. The following section is devoted to a review of generic limits and to the proof of Theorem 1.2. In §4 and §5 we prove Theorems 1.3 and 1.4, respectively. In §6 we prove Theorems 1.5 and in §7 we prove Theorem 1.6.

Acknowledgments. We would like to thank Dale Cutkosky, Atsushi Ito, Mattias Jonsson, Masayuki Kawakita, Pierre Milman, and Michael Temkin for some useful discussions in connection with this work. We are especially indebted to Masayuki Kawakita for pointing out an error in an earlier version of this paper.

It is a pleasure to dedicate this paper to Lawrence Ein, on the occasion of his sixtieth birthday. Lawrence's work has had a profound influence on the understanding of singularities of algebraic varieties and their role in geometry. The first author, in particular, was introduced to this area through their conversations and collaboration. He would like to express his thanks and admiration.

2. Minimal log discrepancies: definition and basic facts

In this section we review the definition of minimal log discrepancies and set up the notation that we will use later in the paper. For more details and for the proofs of some of the facts that we state, we refer to [**Amb99**].

We work over an algebraically closed ground field, of characteristic 0. Let X be a variety (always assumed to be reduced and irreducible). A *divisor over* X is a prime divisor E on some normal variety Y, proper and birational over X. Such a divisor defines a discrete valuation ord_E of the function field of X and we identify two divisors if they give the same valuation. The image of E on X is the *center* of E on X and it is denoted by $c_X(E)$. For a nonzero coherent ideal sheaf $\mathfrak{a}$ on X, one defines $\mathrm{ord}_E(\mathfrak{a})$ as follows. If E is a prime divisor on Y and t is a uniformizer of the DVR $\mathcal{O}_{Y,E}$, then we can write $\mathfrak{a}\cdot\mathcal{O}_{Y,E}=(t^e)$ for some nonnegative integer e and $\mathrm{ord}_E(\mathfrak{a}):=e$. Note that $\mathrm{ord}_E(\mathfrak{a})>0$ if and only if $c_X(E)\subseteq\mathrm{Cosupp}(\mathfrak{a})$, where $\mathrm{Cosupp}(\mathfrak{a})$ is the support of $\mathcal{O}_X/\mathfrak{a}$.

Let X be a normal variety. One says that X is $\mathbf{Q}$-Gorenstein if the canonical divisor K_X is $\mathbf{Q}$-Cartier. In this case, for every proper, birational morphism $f\colon Y\to X$, with Y normal, we consider the discrepancy divisor $K_{Y/X}$. If E is a divisor over X that appears as a prime divisor on Y, then we denote by k_E the coefficient of E in $K_{Y/X}$ (this is independent of the choice of model Y).

Recall that an $\mathbf{R}$-ideal on X is a formal product $\mathfrak{a}=\prod_{j=1}^r\mathfrak{a}_j^{\lambda_j}$, where each $\mathfrak{a}_j$ is a nonzero coherent ideal sheaf on X and each λ_j is a nonnegative real number. Given such $\mathfrak{a}$ and a divisor E over X, we put

$$\mathrm{ord}_E(\mathfrak{a}):=\sum_{j=1}^r\lambda_j\cdot\mathrm{ord}_E(\mathfrak{a}_j).$$

If $\mathfrak{a}=\prod_{j=1}^r\mathfrak{a}_j^{\lambda_j}$ and $\mathfrak{b}=\prod_{i=1}^s\mathfrak{b}_i^{\mu_i}$ are two $\mathbf{R}$-ideals and δ is a positive real number, then we define the ideals

$$\mathfrak{a}\cdot\mathfrak{b}:=\prod_{j=1}^r\mathfrak{a}_j^{\lambda_j}\cdot\prod_{i=1}^s\mathfrak{b}_i^{\mu_i}$$

and

$$\mathfrak{a}^\delta:=\prod_{j=1}^r\mathfrak{a}_j^{\delta\lambda_j}.$$

It is clear that in this case, if E is a divisor over X, then $\mathrm{ord}_E(\mathfrak{a}\cdot\mathfrak{b})=\mathrm{ord}_E(\mathfrak{a})+\mathrm{ord}_E(\mathfrak{b})$ and $\mathrm{ord}_E(\mathfrak{a}^\delta)=\delta\cdot\mathrm{ord}_E(\mathfrak{a})$.

Suppose now that X is normal and $\mathbf{Q}$-Gorenstein and $\mathfrak{a}$ is an $\mathbf{R}$-ideal on X. For every divisor E over X, the *log discrepancy* of E with respect to $(X,\mathfrak{a})$ is

$$a_E(X,\mathfrak{a}):=k_E+1-\mathrm{ord}_E(\mathfrak{a}).$$

The pair $(X,\mathfrak{a})$ is *log canonical* (respectively, *klt*) if and only if $a_E(X,\mathfrak{a})\geq 0$ (respectively, >0) for every divisor E over X. When $\mathfrak{a}=\mathcal{O}_X$, one simply says that X is log canonical (respectively, klt).

Consider a pair $(X,\mathfrak{a})$, with X a normal, $\mathbf{Q}$-Gorenstein variety and $\mathfrak{a}$ an $\mathbf{R}$-ideal on X. For every (closed) point $x\in X$, the *minimal log discrepancy* of $(X,\mathfrak{a})$ is given by

$$\mathrm{mld}_x(X,\mathfrak{a}):=\inf\{a_E(X,\mathfrak{a})\mid E\text{ is a divisor over }X\text{ with }c_X(E)=x\}.$$

It is a basic fact that $\mathrm{mld}_x(X,\mathfrak{a})\geq 0$ if and only if $(X,\mathfrak{a})$ is log canonical in a neighborhood of x. Moreover, if $\mathrm{mld}_x(X,\mathfrak{a})<0$ and $\dim(X)\geq 2$, then $\mathrm{mld}_x(X,\mathfrak{a})=-\infty$. One can also show that if $\mathrm{mld}_x(X,\mathfrak{a})\geq 0$, then the infimum in the definition is in fact a minimum. Under this assumption, we say that a divisor E over X *computes* $\mathrm{mld}_x(X,\mathfrak{a})$ if $c_X(E)=x$ and $a_E(X,\mathfrak{a})=\mathrm{mld}_x(X,\mathfrak{a})$. When $\mathrm{mld}_x(X,\mathfrak{a})<0$, we will say that E computes $\mathrm{mld}_x(X,\mathfrak{a})$ if $c_X(E)=x$ and $a_E(X,\mathfrak{a})<0$.

Recall that if $\mathfrak{a}$ is a nonzero ideal on X, then a *log resolution* of $(X,\mathfrak{a})$ is a proper, birational morphism $\pi\colon Y\to X$ such that Y is a smooth variety, the exceptional locus $\mathrm{Exc}(\pi)$ is a divisor, $\mathfrak{a}\cdot\mathcal{O}_Y=\mathcal{O}_Y(-F)$ for some effective divisor F on Y, and $F+\mathrm{Exc}(\pi)$ has simple normal crossings. Since we are in characteristic 0, log resolutions exist by Hironaka's theorem. It is a basic result that if X is a normal, $\mathbf{Q}$-Gorenstein variety, $x\in X$, and $\mathfrak{a}=\prod_{j=1}^r\mathfrak{a}_j^{\lambda_j}$ is an $\mathbf{R}$-ideal on X, then for every log resolution $\pi\colon Y \to X$ of $(X, \mathfrak{m}_x \cdot \prod_{j=1}^r \mathfrak{a}_j)$, there is a divisor E on Y which computes $\mathrm{mld}_x(X, \mathfrak{a})$.

PROPOSITION 2.1. *Let X be a normal, $\mathbf{Q}$-Gorenstein variety, $\mathfrak{a}$ an $\mathbf{R}$-ideal on X, and $x \in X$ a point defined by $\mathfrak{m}_x$. If $\mathrm{mld}_x(X, \mathfrak{a}) > 0$, then there is $\delta > 0$ such that we have* $\mathrm{mld}_x(X, \mathfrak{a} \cdot \mathfrak{m}_x^\delta) = 0$.

PROOF. Let $\pi\colon Y \to X$ be a log resolution of $(X, \mathfrak{m}_x \cdot \prod_{j=1}^r \mathfrak{a}_j)$, where $\mathfrak{a} = \prod_{j=1}^r \mathfrak{a}_j^{\lambda_j}$. We see that we may take

$$\delta = \min\left\{ \frac{a_E(X, \mathfrak{a})}{\mathrm{ord}_E(\mathfrak{m}_x)} \;\middle|\; E \text{ divisor on } Y \text{ with } c_X(E) = x \right\}.$$

□

In what follows we will also make use of the notion of log canonical threshold. Suppose that X is a log canonical variety and $x \in X$. If $\mathfrak{a}$ is an $\mathbf{R}$-ideal on X, then the *log canonical threshold* of $(X, \mathfrak{a})$ at x is given by

$$\mathrm{lct}_x(X, \mathfrak{a}) := \inf\left\{ \frac{k_E + 1}{\mathrm{ord}_E(\mathfrak{a})} \;\middle|\; E \text{ divisor over } X \text{ with } x \in c_X(E) \right\}.$$

In fact, if $\mathfrak{a} = \prod_{j=1}^r \mathfrak{a}_j^{\lambda_j}$ and $\pi\colon Y \to X$ is a log resolution of $(X, \prod_{j=1}^r \mathfrak{a}_j)$, then there is a divisor E on Y that computes $\mathrm{lct}_x(X, \mathfrak{a})$, that is, $\mathrm{lct}_x(X, \mathfrak{a}) = (k_E + 1)/\mathrm{ord}_E(\mathfrak{a})$ and $x \in c_X(E)$. Note that we have $\mathrm{mld}_x(X, \mathfrak{a}) \geq 0$ if and only if $\mathrm{lct}_x(X, \mathfrak{a}) \geq 1$.

We collect in the next proposition a few well-known properties of minimal log discrepancies and log canonical thresholds. The proof is straightforward and we omit it.

PROPOSITION 2.2. *Let X be a log canonical variety and let $x \in X$ be defined by $\mathfrak{m}_x$. If $\mathfrak{a}_1, \ldots, \mathfrak{a}_r, \mathfrak{b}_1, \ldots, \mathfrak{b}_r$ are nonzero ideals on X and $\lambda_1, \ldots, \lambda_r, \mu_1, \ldots, \mu_r$ are nonnegative real numbers, then the following hold:*

i) *If $\mathfrak{a}_j \subseteq \mathfrak{b}_j$ for every j, then*

$$\mathrm{mld}_x(X, \mathfrak{a}_1^{\lambda_1} \cdots \mathfrak{a}_r^{\lambda_r}) \leq \mathrm{mld}_x(X, \mathfrak{b}_1^{\lambda_1} \cdots \mathfrak{b}_r^{\lambda_r}) \quad \textit{and}$$
$$\mathrm{lct}_x(X, \mathfrak{a}_1^{\lambda_1} \cdots \mathfrak{a}_r^{\lambda_r}) \leq \mathrm{lct}_x(X, \mathfrak{b}_1^{\lambda_1} \cdots \mathfrak{b}_r^{\lambda_r}).$$

ii) *If $\lambda_j \leq \mu_j$ for every j, then*

$$\mathrm{mld}_x(X, \mathfrak{a}_1^{\lambda_1} \cdots \mathfrak{a}_r^{\lambda_r}) \geq \mathrm{mld}_x(X, \mathfrak{a}_1^{\mu_1} \cdots \mathfrak{a}_r^{\mu_r}) \quad \textit{and}$$
$$\mathrm{lct}_x(X, \mathfrak{a}_1^{\lambda_1} \cdots \mathfrak{a}_r^{\lambda_r}) \geq \mathrm{lct}_x(X, \mathfrak{a}_1^{\mu_1} \cdots \mathfrak{a}_r^{\mu_r}).$$

iii) *For every $\delta > 0$, we have*

$$\mathrm{lct}_x(X, \mathfrak{a}_1^{\delta\lambda_1} \cdots \mathfrak{a}_r^{\delta\lambda_r}) = \delta^{-1} \cdot \mathrm{lct}_x(X, \mathfrak{a}_1^{\lambda_1} \cdots \mathfrak{a}_r^{\lambda_r}).$$

iv) *If E is a divisor over X with $c_X(E) = x$ and E computes $\mathrm{mld}_x(X, \mathfrak{a}_1^{\lambda_1} \cdots \mathfrak{a}_r^{\lambda_r})$ (respectively, $\mathrm{lct}_x(X, \mathfrak{a}_1^{\lambda_1} \cdots \mathfrak{a}_r^{\lambda_r})$) and if d is a positive integer such that $d \cdot \mathrm{ord}_E(\mathfrak{m}_x) \geq \mathrm{ord}_E(\mathfrak{a}_j)$ for all j, then E computes*

$$\mathrm{mld}_x\left(X, \prod_{j=1}^r (\mathfrak{a}_j + \mathfrak{m}_x^d)^{\lambda_j}\right)$$

and this is equal to $\mathrm{mld}_x\left(X, \prod_{j=1}^r \mathfrak{a}_j^{\lambda_j}\right)$ *(respectively, E computes*

$$\mathrm{lct}_x\left(X, \prod_{j=1}^r (\mathfrak{a}_j + \mathfrak{m}_x^d)^{\lambda_j}\right)$$

and this is equal to $\mathrm{lct}_x\left(X, \prod_{j=1}^r \mathfrak{a}_j^{\lambda_j}\right)$*).*

In the next section we will need to work in a more general setting than the one described above, in which X is allowed to be a normal, excellent, $\mathbf{Q}$-Gorenstein scheme of characteristic 0 (that is, all the residue fields of X have characteristic 0). All the above definitions extend to this setting. For details, in particular for the precise definitions of K_X and $K_{Y/X}$ in this framework, we refer to [**dFEM11**, Appendix A].

3. Generic limits: bounding the order of the ideal of the point

Our goal in this section is to prove Theorem 1.2. The proof uses generic limits of sequences of ideals. Such a construction based on nonstandard methods was given in [**dFM09**] and a different one, with the same properties but based on sequences of generic points was later given in [**Kol08**]. In what follows we simply recall the basic properties of such a construction, following [**dFEM11**].

Let X be a klt variety over k and $x \in X$ a closed point. Given a positive integer r and r sequences of coherent sheaves of ideals $(\mathfrak{a}_j^{(i)})_{i\geq 1}$ on X for $1 \leq j \leq r$, we get an affine klt scheme $\widetilde{X}$, a closed point $\widetilde{x} \in \widetilde{X}$, and r ideals $\widetilde{\mathfrak{a}}_1, \ldots, \widetilde{\mathfrak{a}}_r$ on $\widetilde{X}$ (note that it can happen for some $\widetilde{\mathfrak{a}}_j$ to be zero). In [**dFEM11**] one allows the variety X to vary as well; since we assume that this is not the case, it is easy to describe $\widetilde{X}$. If some affine neighborhood of x in X is defined in some $\mathbf{A}_k^N$ by $h_1, \ldots, h_s$, then $\widetilde{X} = \mathrm{Spec}(K[\![x_1, \ldots, x_N]\!]/(h_1, \ldots, h_s))$ for some algebraically closed field extension K of k, and $\widetilde{x}$ is the unique closed point of $\widetilde{X}$. If for some j we have $\mathfrak{a}_j^{(i)} = \mathfrak{m}_x$ for all $i \gg 0$, then $\widetilde{\mathfrak{a}}_j$ is the ideal $\mathfrak{m}_{\widetilde{x}}$ defining $\widetilde{x}$. We collect in the next proposition some basic properties of this construction.

PROPOSITION 3.1. *With the above notation, the following hold:*

i) *If* $\widetilde{\mathfrak{a}}_j = 0$*, then for every* q*, there are infinitely many* i *such that* $\mathfrak{a}_j^{(i)} \subseteq \mathfrak{m}_x^q$*.*

ii) *For every* d*, there is an infinite subset* $\Lambda = \Lambda_d \subset \mathbf{Z}_{>0}$ *such that for every* $i \in \Lambda$ *and for every* $\lambda_1, \ldots, \lambda_r \in \mathbf{R}_{\geq 0}$*, we have*

$$\mathrm{lct}_{\widetilde{x}}\left(\widetilde{X}, \prod_{j=1}^r (\widetilde{\mathfrak{a}}_j + \mathfrak{m}_{\widetilde{x}}^d)^{\lambda_j}\right) = \mathrm{lct}_x\left(X, \prod_{j=1}^r (\mathfrak{a}_j^{(i)} + \mathfrak{m}_x^d)^{\lambda_j}\right).$$

iii) *For every* $\lambda_1, \ldots, \lambda_r \in \mathbf{R}_{>0}$*, if we consider the* $\mathbf{R}$*-ideals* $\mathfrak{a}^{(i)} = \prod_{j=1}^r (\mathfrak{a}_j^{(i)})^{\lambda_j}$ *and* $\widetilde{\mathfrak{a}} = \prod_{j=1}^r \widetilde{\mathfrak{a}}_j^{\lambda_j}$*, then* $\mathrm{lct}_{\widetilde{x}}(\widetilde{X}, \widetilde{\mathfrak{a}})$ *is a limit point of the set*

$$\{\mathrm{lct}_x(X, \mathfrak{a}^{(i)}) \mid i \geq 1\}$$

(with the convention that if some $\widetilde{\mathfrak{a}}_j = 0$*, then* $\mathrm{lct}_{\widetilde{x}}(\widetilde{X}, \widetilde{\mathfrak{a}}) = 0$*).*

iv) *Suppose that* $\widetilde{\mathfrak{a}}_j \neq 0$ *for all* j*. If* E *is a divisor over* $\widetilde{X}$ *with* $c_{\widetilde{X}}(E) = \widetilde{x}$ *and such that* E *computes* $\mathrm{lct}_{\widetilde{x}}(\widetilde{X}, \widetilde{\mathfrak{a}})$*, then for every* $d \gg 0$ *there is an infinite subset* $\Lambda' = \Lambda'_d(E, \lambda_1, \ldots, \lambda_r) \subset \mathbf{Z}_{>0}$ *with the following property: for every* $i \in \Lambda'$ *there is a divisor* E_i *over* X *that computes*

$\mathrm{lct}_x\left(X, \prod_{j=1}^r(\mathfrak{a}_j^{(i)}+\mathfrak{m}_x^d)^{\lambda_j}\right)$, *which is equal to* $\mathrm{lct}_{\widetilde{x}}\left(\widetilde{X}, \prod_{j=1}^r(\widetilde{\mathfrak{a}}_j+\mathfrak{m}_{\widetilde{x}}^d)^{\lambda_j}\right)$ *and we have* $\mathrm{ord}_E(\mathfrak{m}_{\widetilde{x}})=\mathrm{ord}_{E_i}(\mathfrak{m}_x)$ *(in particular, we have* $c_X(E_i)=x$*),* $k_{E_i}=k_E$*, and* $\mathrm{ord}_E(\widetilde{\mathfrak{a}}_j+\mathfrak{m}_{\widetilde{x}}^d)=\mathrm{ord}_{E_i}(\mathfrak{a}_j^{(i)}+\mathfrak{m}_x^d)$ *for* $1\le j\le r$.

PROOF. For the assertion in i), see [**dFEM11**, Lemma 3.1]. The statements in ii), iii), and iv) follow from [**dFEM11**, Proposition 3.3 and Corollary 3.4]. The only assertion that is not explicitly mentioned in *loc. cit.* is the one in iv) saying that $k_E=k_{E_i}$. However, by taking d such that $d\ge\mathrm{ord}_E(\widetilde{\mathfrak{a}}_j)$ for every j, we may assume that with $\widetilde{\mathfrak{b}}=\prod_{j=1}^r(\widetilde{\mathfrak{a}}_j+\mathfrak{m}_{\widetilde{x}}^d)^{\lambda_j}$, we have $\mathrm{ord}_E(\widetilde{\mathfrak{a}})=\mathrm{ord}_E(\widetilde{\mathfrak{b}})$ and $\mathrm{lct}_{\widetilde{x}}(\widetilde{X},\widetilde{\mathfrak{b}})=\mathrm{lct}_{\widetilde{x}}(\widetilde{X},\widetilde{\mathfrak{a}})$ (see Proposition 2.2). We now conclude that $k_E=k_{E_i}$ from the other assertions. □

REMARK 3.2. With the notation in the above proposition, we also have the following variant of the assertion in Proposition 3.1: for every d, there is an infinite subset $\Lambda=\Lambda_d\subset\mathbf{Z}_{>0}$ such that for every $i\in\Lambda$ and for every $\lambda_1,\ldots,\lambda_r\in\mathbf{R}_{\ge0}$, we have

$$\mathrm{mld}_{\widetilde{x}}\left(\widetilde{X}, \prod_{j=1}^r(\widetilde{\mathfrak{a}}_j+\mathfrak{m}_{\widetilde{x}}^d)^{\lambda_j}\right)=\mathrm{mld}_x\left(X, \prod_{j=1}^r(\mathfrak{a}_j^{(i)}+\mathfrak{m}_x^d)^{\lambda_j}\right).$$

The proof is the same as in the case of log canonical thresholds (see [**dFEM11**, Proposition 3.3]), the key point being that minimal log discrepancies are constant generically in a family. More precisely, suppose that $x\in X$ is fixed, T is an arbitrary variety, and $\mathfrak{b}_1,\ldots,\mathfrak{b}_r$ are ideals on $X\times T$ such that each $\mathfrak{b}_{j,t}=\mathfrak{b}_j\cdot\mathcal{O}_{X\times\{t\}}$, with $1\le j\le r$ and $t\in T$, is nonzero. In this case, there is an open subset U of T such that for each $\lambda_1,\ldots,\lambda_r\in\mathbf{R}_{\ge0}$, the minimal log discrepancy

$$\mathrm{mld}_x\left(X, \prod_{j=1}^r\mathfrak{b}_{j,t}^{\lambda_j}\right)$$

is constant for $t\in U$. Moreover, the set Λ can be chosen such that the ideals $\widetilde{\mathfrak{a}}_1,\ldots,\widetilde{\mathfrak{a}}_r$ are again generic limits of the sequences $(\mathfrak{a}_1^{(i)})_{i\in\Lambda},\ldots,(\mathfrak{a}_r^{(i)})_{i\in\Lambda}$.

We can now prove the main result of this section.

PROOF OF THEOREM 1.2. We argue by contradiction. If the conclusion of the theorem fails, then we can find a sequence of $\mathbf{R}$-ideals $(\mathfrak{a}^{(i)})_{i\ge1}$ with exponents in I such that one of the following things happens:
Case 1. We have $\mathrm{mld}_x(X,\mathfrak{a}^{(i)})>0$ for all i and for every i there is a divisor E_i over X that computes $\mathrm{mld}_x(X,\mathfrak{a}^{(i)})$ and such that $\lim_{i\to\infty}\mathrm{ord}_{E_i}(\mathfrak{m}_x)=\infty$.
Case 2. We have $\mathrm{mld}_x(X,\mathfrak{a}^{(i)})=0$ for all i and for every choice of divisors E_i over X such that E_i computes $\mathrm{mld}_x(X,\mathfrak{a}^{(i)})$, we have $\lim_{i\to\infty}\mathrm{ord}_{E_i}(\mathfrak{m}_x)=\infty$.
Case 3. We have $\mathrm{mld}_x(X,\mathfrak{a}^{(i)})<0$ for all i and for every choice of divisors E_i over X such that E_i computes $\mathrm{mld}_x(X,\mathfrak{a}^{(i)})$, we have $\lim_{i\to\infty}\mathrm{ord}_{E_i}(\mathfrak{m}_x)=\infty$.

Suppose that $\lambda_1,\ldots,\lambda_r$ are the nonzero elements of I. We may assume that for every i we can write $\mathfrak{a}^{(i)}=\prod_{j=1}^r(\mathfrak{a}_j^{(i)})^{\lambda_j}$. We use the generic limit construction to construct $\widetilde{x}\in\widetilde{X}$ and an ideal $\widetilde{\mathfrak{a}}_j$ on $\widetilde{X}$ corresponding to the sequence $(\mathfrak{a}_j^{(i)})_{i\ge1}$ for $1\le j\le r$. Let $\widetilde{\mathfrak{a}}$ be the $\mathbf{R}$-ideal on $\widetilde{X}$ given by $\widetilde{\mathfrak{a}}=\prod_{j=1}^r\widetilde{\mathfrak{a}}_j^{\lambda_j}$. When some $\widetilde{\mathfrak{a}}_j$ is zero, we make the convention that $\widetilde{\mathfrak{a}}=0$ and $\mathrm{lct}_{\widetilde{x}}(\widetilde{X},\widetilde{\mathfrak{a}})=0$.

Suppose first that we are either in Case 1 or in Case 2. Note that since $\mathrm{lct}_x(X,\mathfrak{a}^{(i)})\ge1$ for every i, it follows from Proposition 3.1 that $\mathrm{lct}_{\widetilde{x}}(\widetilde{X},\widetilde{\mathfrak{a}})\ge1$.

In particular, each $\widetilde{\mathfrak{a}}_j$ is nonzero and we have $\mathrm{mld}_{\widetilde{x}}(\widetilde{X},\widetilde{\mathfrak{a}}) \geq 0$. Let us consider first the case when $\mathrm{mld}_{\widetilde{x}}(\widetilde{X},\widetilde{\mathfrak{a}}) > 0$. It follows from Proposition 2.1 that there is $\delta > 0$ such that $\mathrm{lct}_{\widetilde{x}}(\widetilde{X},\widetilde{\mathfrak{a}}\cdot\mathfrak{m}_{\widetilde{x}}^{\delta}) = 1$. In this case there are infinitely many i such that $\mathrm{lct}_x(X,\mathfrak{a}^{(i)}\cdot\mathfrak{m}_x^{\delta}) \geq 1$. Indeed, if this is not the case, then $\mathrm{lct}_x(X,\mathfrak{a}^{(i)}\cdot\mathfrak{m}_x^{\delta}) < 1$ for all $i \gg 0$. On the other hand, it follows from Proposition 3.1 that $\mathrm{lct}_{\widetilde{x}}(\widetilde{X},\widetilde{\mathfrak{a}}\cdot\mathfrak{m}_{\widetilde{x}}^{\delta}) = 1$ is a limit point of the set $\{\mathrm{lct}_x(X,\mathfrak{a}^{(i)}\cdot\mathfrak{m}_x^{\delta}) \mid i \geq 1\}$. This contradicts the fact that the set $\{\mathrm{lct}_x(X,\mathfrak{a}^{(i)}\cdot\mathfrak{m}_x^{\delta}) \mid i \geq 1\}$ satisfies ACC (see [**dFEM11**, Theorem 4.2]).

For every i such that $\mathrm{lct}_x(X,\mathfrak{a}^{(i)}\cdot\mathfrak{m}_x^{\delta}) \geq 1$ and for every divisor E_i that computes $\mathrm{mld}_x(X,\mathfrak{a}^{(i)})$, we obtain

$$\mathrm{mld}_x(X,\mathfrak{a}^{(i)}) = k_{E_i} + 1 - \mathrm{ord}_{E_i}(\mathfrak{a}^{(i)}) \geq \delta\cdot\mathrm{ord}_{E_i}(\mathfrak{m}_x).$$

Therefore

$$\mathrm{ord}_{E_i}(\mathfrak{m}_x) \leq \frac{\mathrm{mld}_x(X,\mathfrak{a}^{(i)})}{\delta} \leq \frac{\mathrm{mld}_x(X)}{\delta}$$

for infinitely many i, contradicting the fact that, by assumption, we can choose such divisors E_i with $\lim_{i\to\infty}\mathrm{ord}_{E_i}(\mathfrak{m}_x) = \infty$.

We now consider the case when $\mathrm{mld}_{\widetilde{x}}(\widetilde{X},\widetilde{\mathfrak{a}}) = 0$ (still assuming that we are either in Case 1 or in Case 2). If F is a divisor over $\widetilde{X}$ that computes $\mathrm{mld}_{\widetilde{x}}(\widetilde{X},\widetilde{\mathfrak{a}})$, then it follows from Proposition 3.1 that for $d \gg 0$, we can find an infinite subset $\Gamma' = \Gamma'_d(F,\lambda_1,\ldots,\lambda_r) \subset \mathbf{Z}_{>0}$ such that the following holds. For every $i \in \Gamma'$ we have a divisor F_i over X with $k_F = k_{F_i}$, $\mathrm{ord}_F(\mathfrak{m}_{\widetilde{x}}) = \mathrm{ord}_{F_i}(\mathfrak{m}_x)$ (in particular, $c_X(F_i) = x$), and such that if we put $\mathfrak{b}^{(i)} = \prod_{j=1}^r(\mathfrak{a}_j^{(i)} + \mathfrak{m}_x^d)^{\lambda_j}$ and $\widetilde{\mathfrak{b}} = \prod_{j=1}^r(\widetilde{\mathfrak{a}}_j + \mathfrak{m}_{\widetilde{x}}^d)^{\lambda_j}$, then $\mathrm{ord}_F(\widetilde{\mathfrak{b}}) = \mathrm{ord}_{F_i}(\mathfrak{b}^{(i)})$. By taking $d \geq \mathrm{ord}_F(\widetilde{\mathfrak{a}}_j)$ for every j, we may assume that $\mathrm{ord}_F(\widetilde{\mathfrak{a}}) = \mathrm{ord}_F(\widetilde{\mathfrak{b}})$. We conclude that

$$\begin{aligned}0 = a_F(\widetilde{X},\widetilde{\mathfrak{a}}) &= k_F + 1 - \mathrm{ord}_F(\widetilde{\mathfrak{a}}) = k_{F_i} + 1 - \mathrm{ord}_{F_i}(\mathfrak{b}^{(i)})\\ &= a_{F_i}(X,\mathfrak{b}^{(i)}) \geq a_{F_i}(X,\mathfrak{a}^{(i)}) \geq 0\end{aligned}$$

for every $i \in \Gamma'$. In Case 1, this already gives a contradiction, since the last inequality is strict. If we are in Case 2, we conclude that the divisor F_i computes $\mathrm{mld}_x(X,\mathfrak{a}^{(i)})$. By assumption, we must have $\mathrm{ord}_{F_i}(\mathfrak{m}_x) \to \infty$, contradicting the fact that $\mathrm{ord}_{F_i}(\mathfrak{m}_x)$ is constant for $i \in \Gamma'$.

Finally, suppose that we are in Case 3. Let us assume first that every $\widetilde{\mathfrak{a}}_j$ is nonzero. Since $\mathrm{lct}_x(X,\mathfrak{a}^{(i)}) < 1$ for every i and the set $\{\mathrm{lct}_x(X,\mathfrak{a}^{(i)}) \mid i \geq 1\}$ has $\mathrm{lct}_{\widetilde{x}}(\widetilde{X},\widetilde{\mathfrak{a}})$ as a limit point by Proposition 3.1, it follows that $\mathrm{lct}_{\widetilde{x}}(\widetilde{X},\widetilde{\mathfrak{a}}) < 1$ (recall that the set $\{\mathrm{lct}_x(X,\mathfrak{a}^{(i)}) \mid i \geq 1\}$ satisfies ACC by [**dFEM11**, Theorem 4.2]). Therefore $\mathrm{mld}_{\widetilde{x}}(\widetilde{X},\widetilde{\mathfrak{a}}) < 0$ and consider a divisor G over $\widetilde{X}$, with $c_{\widetilde{X}}(G) = \widetilde{x}$ and with $a_G(\widetilde{X},\widetilde{\mathfrak{a}}) < 0$. We now argue as above: we can find an infinite set $\Gamma'' \subset \mathbf{Z}_{>0}$ such that the following holds. For every $i \in \Gamma''$ we have a divisor G_i over X with $k_G = k_{G_i}$, $\mathrm{ord}_G(\mathfrak{m}_{\widetilde{x}}) = \mathrm{ord}_{G_i}(\mathfrak{m}_x)$ (in particular, $c_X(G_i) = x$), and such that $\mathrm{ord}_G(\widetilde{\mathfrak{b}}) = \mathrm{ord}_{G_i}(\mathfrak{b}^{(i)})$, where $\widetilde{\mathfrak{b}}$ and $\mathfrak{b}^{(i)}$ are defined as above. Furthermore, we may assume that $\mathrm{ord}_G(\widetilde{\mathfrak{a}}) = \mathrm{ord}_G(\widetilde{\mathfrak{b}})$ and we conclude that

$$0 > a_G(\widetilde{X},\widetilde{\mathfrak{a}}) = k_G+1-\mathrm{ord}_G(\widetilde{\mathfrak{a}}) = k_{G_i}+1-\mathrm{ord}_{G_i}(\mathfrak{b}^{(i)}) = a_{G_i}(X,\mathfrak{b}^{(i)}) \geq a_{G_i}(X,\mathfrak{a}^{(i)})$$

for every $i \in \Gamma''$. Since $\mathrm{ord}_{G_i}(\mathfrak{m}_x)$ is constant for all $i \in \Gamma''$, this gives a contradiction.

Let us consider now the case when some $\widetilde{\mathfrak{a}}_j$ is zero. Let T be a fixed divisor over X with $c_X(T) = x$ and let q be a positive integer with $q > \frac{k_T+1}{\lambda_j\cdot\mathrm{ord}_T(\mathfrak{m}_x)}$. Since $\widetilde{\mathfrak{a}}_j$ is zero, it follows from Proposition 3.1 that there are infinitely many i with $\mathfrak{a}_j^{(i)} \subseteq \mathfrak{m}_x^q$. In this case we have

$$a_T(X, \mathfrak{a}^{(i)}) \leq a_T(X, \mathfrak{m}_x^{\lambda_j q}) = k_T + 1 - \lambda_j q \cdot \mathrm{ord}_T(\mathfrak{m}_x) < 0.$$

Therefore T computes $\mathrm{mld}_x(X, \mathfrak{a}^{(i)})$ for infinitely many i, a contradiction. This completes the proof of the theorem. □

While by using generic limits we cannot get a proof for the full statement in Conjecture 1.1, we also obtain the following related statement.

PROPOSITION 3.3. *Let X be a klt variety and $x \in X$ a closed point. If $I \subset \mathbf{R}_{\geq 0}$ is a finite set, then there is a positive integer ℓ such that for every $\mathbf{R}$-ideal with exponents in I, if $a_E(X, \mathfrak{a}) \geq 0$ for all divisors E over X with $c_X(E) = x$ and $k_E \leq \ell$, then $(X, \mathfrak{a})$ is log canonical at x.*

PROOF. Suppose that the conclusion of the proposition fails. In this case we can find a sequence of $\mathbf{R}$-ideals $\mathfrak{a}^{(i)}$ on X, with exponents in I, such that each $(X, \mathfrak{a}^{(i)})$ is not log canonical at x, but $a_E(X, \mathfrak{a}^{(i)}) \geq 0$ for all divisors E over X with $c_X(E) = x$ and $k_E \leq i$. Let $\lambda_1, \ldots, \lambda_r$ be the nonzero elements in I and let us write

$$\mathfrak{a}^{(i)} = \prod_{j=1}^{r} (\mathfrak{a}_j^{(i)})^{\lambda_j}.$$

We use the generic limit construction to produce $\widetilde{x} \in \widetilde{X}$ and ideals $\widetilde{\mathfrak{a}}_j$ on $\widetilde{X}$ corresponding to the sequences $(\mathfrak{a}_j^{(i)})_{i \geq 1}$ for $1 \leq j \leq r$. Let $\widetilde{\mathfrak{a}}$ be the $\mathbf{R}$-ideal on $\widetilde{X}$ given by

$$\widetilde{\mathfrak{a}} = \prod_{j=1}^{r} \widetilde{\mathfrak{a}}_j^{\lambda_j}.$$

When some $\widetilde{\mathfrak{a}}_j$ is zero, we make the convention that $\widetilde{\mathfrak{a}} = 0$.

Our assumption implies $\mathrm{lct}_x(X, \mathfrak{a}^{(i)}) < 1$ for every i. Recall that $\mathrm{lct}_{\widetilde{x}}(\widetilde{X}, \widetilde{\mathfrak{a}})$ is a limit point of the sequence $\left(\mathrm{lct}_x(X, \mathfrak{a}^{(i)})\right)_{i \geq 1}$ by Proposition 3.1 iii). On the other hand, this sequence contains no strictly increasing subsequences by [**dFEM11**, Theorem 4.2]. Therefore $\mathrm{lct}_{\widetilde{x}}(\widetilde{X}, \widetilde{\mathfrak{a}}) < 1$ and the pair $(\widetilde{X}, \widetilde{\mathfrak{a}})$ is not log canonical at $\widetilde{x}$. Let E be a divisor over $\widetilde{X}$ with center $\widetilde{x}$ and such that $a_E(\widetilde{X}, \widetilde{\mathfrak{a}}) < 0$. If $d \in \mathbf{Z}_{>0}$ is large enough, but fixed, then we clearly have

$$a_E\big(\widetilde{X}, \prod_j (\widetilde{\mathfrak{a}}_j + \mathfrak{m}_{\widetilde{x}}^d)^{\lambda_j}\big) = a_E(\widetilde{X}, \widetilde{\mathfrak{a}}) < 0.$$

On the other hand, it follows from Proposition 3.1 iv) that there are infinitely many i for which we can find divisors E_i over X with center x, such that $k_{E_i} = k_E$ and

$$a_{E_i}\big(X, \prod_j (\mathfrak{a}_j^{(i)} + \mathfrak{m}_x^d)^{\lambda_j}\big) = a_E\big(\widetilde{X}, \prod_j (\widetilde{\mathfrak{a}}_j + \mathfrak{m}_{\widetilde{x}}^d)^{\lambda_j}\big) < 0.$$

Since

$$a_{E_i}\big(X, \prod_j (\mathfrak{a}_j^{(i)})^{\lambda_j}\big) \leq a_{E_i}\big(X, \prod_j (\mathfrak{a}_j^{(i)} + \mathfrak{m}_x^d)^{\lambda_j}\big) < 0$$

and $k_{E_i} = k_E$ for infinitely many i, we contradict our assumption. This completes the proof of the proposition. □

The assertion in Proposition 3.3 has an interesting consequence in connection with the description of log canonical pairs in terms of jet schemes, when the ambient variety is smooth. This will not play any role in the following sections, so the reader not interested in jet schemes could skip this part.

Recall that if X is a smooth variety, Y is a closed subscheme of X defined by the nonzero ideal $\mathfrak{a}$, and $q \in \mathbf{R}_{\geq 0}$, then the pair $(X, \mathfrak{a}^q)$ is log canonical if and only if

$$\dim(Y_m) \leq (m+1)(\dim(X) - q) \quad \text{for all} \quad m \geq 0,$$

where Y_m is the mth jet scheme of Y (see [**ELM04**, Corollary 3.2]). For the definition and basic properties of jet schemes and contact loci, we refer to [**ELM04**]. Proposition 3.3 implies that if the dimension of X and $q \in \mathbf{R}_{\geq 0}$ are fixed, then it is enough to check the dimensions of only a prescribed number of jet schemes.

PROPOSITION 3.4. *Given $n \geq 1$ and $q \in \mathbf{R}_{\geq 0}$, there is a positive integer N that satisfies the following property. For every smooth n-dimensional variety X and for every closed subscheme Y of X defined by a nonzero ideal $\mathfrak{a}$, the pair $(X, \mathfrak{a}^q)$ is log canonical if and only if*

$$\dim(Y_m) \leq (m+1)(n-q) \quad \textit{for all} \quad m \leq N.$$

PROOF. The case $q = 0$ is trivial (the pair is always log canonical in this case, hence any N will work), hence we assume from now on $q > 0$. We first consider the case when $X = \mathbf{A}^n$ and choose ℓ given by Proposition 3.3, such that for every nonzero ideal $\mathfrak{a}$ in $\mathbf{A}^n$, if $a_E(\mathbf{A}^n, \mathfrak{a}^q) \geq 0$ for all divisors E over $\mathbf{A}^n$ with center at the origin and $k_E \leq \ell$, then $(\mathbf{A}^n, \mathfrak{a}^q)$ is log canonical at 0. Let $N = \lfloor \frac{\ell+1}{q} \rfloor$, where $\lfloor u \rfloor$ denotes the largest integer $\leq u$. We show that if $\mathfrak{a}$ is a nonzero ideal defining the subscheme Y of $\mathbf{A}^n$ such that $\dim(Y_m) \leq (m+1)(n-q)$ for all $m \leq N$, then $(\mathbf{A}^n, \mathfrak{a}^q)$ is log canonical at 0.

Indeed, if $(\mathbf{A}^n, \mathfrak{a})$ is not log canonical at 0, then it follows by assumption that there is a divisor E over $\mathbf{A}^n$ with center 0 such that $k_E \leq \ell$ and $k_E + 1 < q \cdot \alpha_E$, where $\alpha_E = \operatorname{ord}_E(\mathfrak{a})$. Since α_E is an integer, it follows that $\alpha_E \geq m+1$, where $m = \lfloor \frac{k_E+1}{q} \rfloor$. Let $f \colon Y \to \mathbf{A}^n$ be a log resolution of $(\mathbf{A}^n, \mathfrak{a})$ such that E appears as a divisor on Y. It follows from [**ELM04**, Theorem 2.1] that if $C = \overline{f_\infty(\operatorname{Cont}^{\geq 1}(E))}$, then

$$C \subseteq \operatorname{Cont}^{\geq \alpha_E}(\mathfrak{a}) \subseteq \operatorname{Cont}^{\geq (m+1)}(\mathfrak{a}) \quad \text{and} \quad \operatorname{codim}(C) = k_E + 1.$$

We thus conclude that

$$\dim(Y_m) = (m+1)n - \operatorname{codim}(\operatorname{Cont}^{\geq (m+1)}(\mathfrak{a})) \geq (m+1)n - \operatorname{codim}(C) > (m+1)(n-q).$$

Since $m \leq N$, this proves our assertion.

Suppose now that X is an arbitrary smooth n-dimensional variety and $\mathfrak{a}$ is a nonzero ideal, defining the closed subscheme Y of X, such that

$$\dim(Y_m) \leq (m+1)(n-q) \quad \text{for all} \quad m \leq N.$$

We show that for every $x \in X$, the pair $(X, \mathfrak{a}^q)$ is log canonical at x. Since X is smooth, after possibly replacing X by an open neighborhood of X, we may assume that we have an étale morphism $g \colon X \to \mathbf{A}^n$, with $g(x) = 0$. Let $\mathfrak{m}_x$ denote the ideal defining x and for every $d \geq 1$, let $\mathfrak{a}_d = \mathfrak{a} + \mathfrak{m}_x^d$, defining the subscheme $V(\mathfrak{a}_d)$ of X. For every such d, there is an ideal $\mathfrak{b}_d$ on $\mathbf{A}^n$ defining a subscheme $V(\mathfrak{b}_d)$ supported at 0 and such that $\mathfrak{b}_d \cdot \mathcal{O}_X = \mathfrak{a}_d$. Note that for every d and m, we have

$$V(\mathfrak{b}_d)_m \simeq V(\mathfrak{a}_d)_m \hookrightarrow Y_m,$$

hence by assumption

$$\dim\left(V(\mathfrak{b}_d)_m\right) \leq (m+1)(n-q) \quad \text{for all} \quad m \leq N.$$

As we have seen, this implies that $(\mathbf{A}^n, \mathfrak{b}_d^q)$ is log canonical. Since g is étale, we have $\mathrm{lct}_x(X, \mathfrak{a}_d) = \mathrm{lct}_0(\mathbf{A}^n, \mathfrak{b}_d) \geq q$ for every d, while

$$\mathrm{lct}_x(X, \mathfrak{a}) = \lim_{d \to \infty} \mathrm{lct}_x(X, \mathfrak{a}_d)$$

(see, for example, [**dFEM11**, Proposition 2.15]). We conclude that $\mathrm{lct}_x(X, \mathfrak{a}) \geq q$, that is, the pair $(X, \mathfrak{a}^q)$ is log canonical at x. This completes the proof of the proposition. □

4. A proof of the conjecture in dimension 2

We begin with the following convexity property of log discrepancies from [**Kol13**, Proposition 2.37].

PROPOSITION 4.1. *Let X be a surface and $\mathfrak{a}$ an $\mathbf{R}$-ideal on X such that $(X, \mathfrak{a})$ is log canonical, and $f\colon Y \to X$ a birational morphism from a smooth surface Y. Assume that $a_E(X, \mathfrak{a}) \leq 1$ for every f-exceptional divisor E. If E_1, E_2, and E_3 are f-exceptional prime divisors that satisfy the following conditions:*

(1) *E_1 meets both E_2 and E_3, and*
(2) *E_1 has the self-intersection number $E_1^2 \leq -2$,*

then $a_1 \leq \frac{1}{2}(a_2 + a_3)$, where $a_i = a_{E_i}(X, \mathfrak{a})$.

PROOF. Since the statement is local, we may assume that X is affine. We may write $\mathfrak{a} = \prod \mathfrak{a}_i^{\lambda_i}$ for nonzero ideal sheaves $\mathfrak{a}_i$ and $\lambda_i \in \mathbf{R}_{>0}$. We fix a positive integer c which satisfies $c \geq \lambda_i$ for every i. Take general elements $f_{i1}, \ldots, f_{ic} \in \mathfrak{a}_i$, and let $D_{i1}, \ldots, D_{ic}$ be the corresponding effective Cartier divisors. If $\Delta = \frac{1}{c}\sum_{i,j} \lambda_i D_{ij}$, then (X, Δ) is log canonical and $a_E(X, \Delta) = a_E(X, \mathfrak{a})$ for every f-exceptional divisor E (see [**Nak16b**, Lemma 4.2]).

Let $\{E_i\}$ be the set of all f-exceptional divisors. We write

$$f^*(K_X + \Delta) = K_Y + \widetilde{\Delta} + \sum_i (1 - a_i) E_i,$$

where $\widetilde{\Delta}$ is the strict transform of Δ and $a_i = a_{E_i}(X, \Delta)$. Note that $1 - a_i \geq 0$ for every i, by assumption. We have

$$\begin{aligned} 0 = f^*(K_X + \Delta) \cdot E_1 &= (K_Y + E_1) \cdot E_1 + \widetilde{\Delta} \cdot E_1 - a_1 E_1^2 \\ &\quad + (1 - a_2) E_1 \cdot E_2 + (1 - a_3) E_1 \cdot E_3 + \sum_{i \neq 1,2,3} (1 - a_i) E_1 \cdot E_i. \end{aligned}$$

It is clear that we have

$$(K_Y + E_1) \cdot E_1 \geq -2, \quad \widetilde{\Delta} \cdot E_1 \geq 0, \quad \sum_{i \neq 1,2,3} (1 - a_i) E_1 \cdot E_i \geq 0,$$

and the assumptions (1) and (2) give

$$-a_1 E_1^2 \geq 2a_1, \quad (1 - a_2) E_1 \cdot E_2 \geq 1 - a_2, \quad (1 - a_3) E_1 \cdot E_3 \geq 1 - a_3.$$

By combining all these, we obtain the desired inequality $2a_1 - a_2 - a_3 \leq 0$. □

Proof of Theorem 1.3. Let X be a klt surface, $x \in X$ a point and $I \subset \mathbf{R}_{\geq 0}$ a finite set. The non-log-canonical case follows from Proposition 3.3, hence we only consider the log canonical case.

Let $\mathfrak{a}$ be an $\mathbf{R}$-ideal on X with exponents in I such that $(X, \mathfrak{a})$ is log canonical around x. Let $X_0 \to X$ be the minimal resolution of X. Suppose that $\mathrm{mld}_x(X, \mathfrak{a})$ is not computed by any $(X_0 \to X)$-exceptional divisor. Then, there is a sequence of blow-ups

$$X_n \to X_{n-1} \to \cdots \to X_1 \to X_0 \to X,$$

with the following properties:

(1) For every i with $0 \leq i \leq n-1$, the map $X_{i+1} \to X_i$ is the blow-up of X_i at a point $p_i \in X_i$ with exceptional divisor $E_i \subset X_{i+1}$.
(2) p_0 maps to x by the map $X_0 \to X$.
(3) p_{i+1} maps to p_i by the map $X_{i+1} \to X_i$ for every i with $0 \leq i \leq n-2$ (equivalently, $p_{i+1} \in E_i$).
(4) $a_{E_i}(X, \mathfrak{a}) > \mathrm{mld}_x(X, \mathfrak{a})$ for i with $0 \leq i \leq n-2$ and $a_{E_{n-1}}(X, \mathfrak{a}) = \mathrm{mld}_x(X, \mathfrak{a})$.

The next lemma gives a bound for $k_{E_{n-1}}$ in terms of n.

Lemma 4.2. *With the above notation, we have* $k_{E_{n-1}} \leq 2^{n-1}$.

Proof. We first show that $\mathrm{ord}_{E_{n-1}} F \leq 2^{n-1-i}$ for every prime divisor F on X_i which is exceptional over X_0 with $0 \leq i \leq n-1$. We argue by descending induction on i. The case $i = n-1$ is trivial since each exceptional prime divisor over X_0 is smooth. If $i < n-1$, then the pull-back of F to X_{i+1} is either equal to the strict transform F' of F on X_{i+1} or it is equal to $F' + E_i$. By induction, we conclude that

$$\mathrm{ord}_{E_{n-1}}(F) \leq \mathrm{ord}_{E_{n-1}}(F' + E_i) \leq 2 \cdot 2^{n-2-i} = 2^{n-1-i}.$$

Note now that we have

$$k_{E_{n-1}} = \mathrm{ord}_{E_{n-1}}(K_{X_n/X}) = \mathrm{ord}_{E_{n-1}}(K_{X_0/X}) + \sum_{i=1}^{n} \mathrm{ord}_{E_{n-1}}(K_{X_i/X_{i-1}}).$$

On the other hand, since X_0 is the minimal resolution of X, we have $K_{X_0/X} \leq 0$, hence $\mathrm{ord}_{E_{n-1}}(K_{X_0/X}) \leq 0$. Using the assertion at the beginning of the proof, we conclude

$$k_{E_{n-1}} \leq \sum_{i=1}^{n} \mathrm{ord}_{E_{n-1}}(K_{X_i/X_{i-1}}) = \sum_{i=1}^{n} \mathrm{ord}_{E_{n-1}}(E_{i-1}) \leq 1 + \sum_{i=1}^{n-1} 2^{n-1-i} = 2^{n-1},$$

which gives the desired inequality. □

Returning to the proof of Theorem 1.3, it follows from Lemma 4.2 that in order to conclude the proof of the theorem it is enough to prove the following lemma, giving a bound on the number n of blow-ups of X_0. □

Lemma 4.3. *There exists a positive integer $\ell(I)$ depending on the finite set I that satisfies the following condition: for every $\mathbf{R}$-ideal $\mathfrak{a}$ on X with exponents in I, if $(X, \mathfrak{a})$ is log canonical and $\mathrm{mld}_x(X, \mathfrak{a})$ is not computed by any $(X_0 \to X)$-exceptional divisor, then for every sequence of blow-ups satisfying the condition (1)-(4) above, we have $n \leq \ell(I)$.*

PROOF. If $\mathrm{mld}_x(X, \mathfrak{a}) > 1$, then it is known that X is smooth at x (hence $X_0 = X$) and $n = 1$ (see [**KM98**, Theorem 4.5] and its proof). From now on we suppose $\mathrm{mld}_x(X, \mathfrak{a}) \leq 1$.

We begin by proving the following assertion, which we will need in order to apply Proposition 4.1:

$$a_E(X, \mathfrak{a}) \leq 1 \quad \text{for every} \quad (X_n \to X)\text{-exceptional divisor } E. \tag{4.1}$$

Since $X_0 \to X$ is the minimal resolution, we have $K_{X_0/X} \leq 0$, hence

$$a_E(X, \mathfrak{a}) \leq a_E(X) \leq 1$$

for every $(X_0 \to X)$-exceptional divisor E. Suppose that j is the smallest index with $a_{E_j}(X, \mathfrak{a}) > 1$. We define an $\mathbf{R}$-ideal $\mathfrak{a}_j$ on X_j as follows: if $\mathfrak{b}$ is the $\mathbf{R}$-ideal on X_j such that

$$\mathfrak{a} \cdot \mathcal{O}_{X_j} = \mathfrak{b} \cdot \prod_E \mathcal{O}_X(-E)^{\mathrm{ord}_E(\mathfrak{a})},$$

where the product is over the $(X_j \to X)$-exceptional divisors, then

$$\mathfrak{a}_j = \mathfrak{b} \cdot \prod_E \mathcal{O}_X(-E)^{\mathrm{ord}_E(\mathfrak{a}) - k_E}$$

(note that this is well-defined since $\mathrm{ord}_E(\mathfrak{a}) - k_E = 1 - a_E(X, \mathfrak{a}) \geq 0$ for every such E). It follows from definition that

$$a_E(X, \mathfrak{a}) = a_E(X_j, \mathfrak{a}_j) \quad \text{for every divisor } E \text{ over } X.$$

Since $a_{E_j}(X_j, \mathfrak{a}_j) = a_{E_j}(X, \mathfrak{a}) > 1$, we have $\mathrm{mult}_{p_j} \mathfrak{a}_j < 1$. By [**KM98**, Theorem 4.5], it follows that $\mathrm{mld}_{p_j}(X_j, \mathfrak{a}_j) > 1$. However, this contradicts

$$\mathrm{mld}_{p_j}(X_j, \mathfrak{a}_j) \leq a_{E_{n-1}}(X_j, \mathfrak{a}_j) = a_{E_{n-1}}(X, \mathfrak{a}) = \mathrm{mld}_x(X, \mathfrak{a}) \leq 1.$$

This completes the proof of (4.1).

Suppose now that $F_0, F_1, \ldots, F_c$ are $(X_n \to X_0)$-exceptional divisors that satisfy the following conditions:

(α) $F_0 = E_{n-1}$ and $F_i \neq E_{n-1}$ for $1 \leq i \leq c$, and
(β) F_i meets F_{i+1} for $0 \leq i \leq c-1$.

In this case we have the following sequence of inequalities:

$$a_{E_{n-1}} = a_{F_0} < a_{F_1} < \cdots < a_{F_c}, \tag{4.2}$$

where we set $a_{F_i} = a_{F_i}(X, \Delta)$. In order to see this, note first that by the assumption on the sequence of blow-ups, we have $a_{E_{n-1}} < a_F$ for every $(X_n \to X_0)$-exceptional divisor F except for $F = E_{n-1}$. This gives the first inequality $a_{F_0} < a_{F_1}$. We next use the fact that $F^2 \leq -2$ for every $(X_n \to X_0)$-exceptional divisor F, except for $F = E_{n-1}$; in particular, we have $F_1^2 \leq -2$. It follows from Proposition 4.1 that

$$a_{F_1} \leq \frac{1}{2}(a_{F_0} + a_{F_2}) < \frac{1}{2}(a_{F_1} + a_{F_2}).$$

Therefore $a_{F_1} < a_{F_2}$. We deduce in this way (4.2) by repeatedly applying Proposition 4.1.

By the discreteness of log discrepancies proved by Kawakita [**Kaw14**], there exists a finite subset $U(I) \subset [0,1]$ depending only on I satisfying the following condition:

- For every $\mathbf{R}$-ideal $\mathfrak{a}$ with exponents in I such that $(X, \mathfrak{a})$ is log canonical, if $a_F(X, \mathfrak{a}) \in [0,1]$, then $a_F(X, \mathfrak{a}) \in U(I)$.

Set $\ell_1(I) := \#U(I)$. By the choice of $\ell_1(I)$ and the bound (4.1), if we can find a sequence $F_0, \ldots, F_c$ of exceptional divisors that satisfies the conditions (α) and (β) above, with $c \geq \ell_1(I)$, we contradict the sequence of inequalities (4.2).

The graph-theoretic Lemma 4.4 below thus implies $n < \frac{1}{2}(3^{\ell_1(I)} - 1)$. Indeed, we apply the lemma for the dual graph Γ of $(X_n \to X_0)$-exceptional divisors (the vertices of this graph are given by these exceptional divisors and two vertices are connected by an edge if and only if the divisors intersect on X_n); note that Γ has n vertices and each vertex has degree at most three. This completes the proof of Lemma 4.3. □

LEMMA 4.4. *Let ℓ be a positive integer and G be a connected graph of order $n \geq \frac{1}{2}(3^\ell - 1)$. If every vertex of G has degree ≤ 3, then for any vertex v of G, the graph G contains a chain of length ℓ containing v with degree 1.*

PROOF. We argue by induction on ℓ, the case $\ell = 1$ being trivial. Consider the graph G' obtained by removing the vertex v and the edges containing v. Since G is connected and $\deg(v) \leq 3$, the number of the connected components of G' is at most three. Let G'' be a connected component of G' of order at least $\frac{1}{3}\big(\frac{1}{2}(3^\ell - 1) - 1\big) = \frac{1}{2}(3^{\ell-1} - 1)$. Let v' be a vertex in G'' which is connected to v by an edge in G. By induction, G'' contains a chain of length $\ell - 1$ containing v' with degree 1. By adding v to this chain, we obtain a chain in G which contains v with degree 1. □

5. A proof of the conjecture in the monomial case

In this section we give a proof of Theorem 1.4. More precisely, we prove the following result. A *monomial* $\mathbf{R}$-ideal on $\mathbf{A}^n$ is an $\mathbf{R}$-ideal of the form $\mathfrak{a} = \prod_{j=1}^r \mathfrak{a}_j^{\lambda_j}$, where each ideal $\mathfrak{a}_j$ is generated by monomials.

THEOREM 5.1. *Given a positive integer n and a finite subset $I \subset \mathbf{R}_{\geq 0}$, there is a positive integer ℓ (depending on n and I) such that for every monomial $\mathbf{R}$-ideal $\mathfrak{a}$ on $\mathbf{A}^n$ with exponents in I, there is a divisor E that computes $\mathrm{mld}_0(X, \mathfrak{a})$ and such that $k_E \leq \ell$.*

We will use the following result of Maclagan [**Mac01**, Theorem 1.1]: given an infinite set $\mathcal{U}$ of monomial ideals in $k[x_1, \ldots, x_n]$, then there are two ideals $I, J \in \mathcal{U}$ such that $I \subseteq J$. This implies that given any sequence $(I_m)_{m \geq 1}$ of monomial ideals in $k[x_1, \ldots, x_n]$, there is a subsequence $(I_{j_m})_{m \geq 1}$ such that $I_{j_m} \supseteq I_{j_{m+1}}$ for all m. Indeed, note first that we may assume that each ideal I is equal to I_m for only finitely values of m, since otherwise our assertion is trivial. Since $k[x_1, \ldots, x_n]$ is Noetherian, we can find ideals in $\{I_m \mid m \geq 1\}$ that are maximal with respect to inclusion. By Maclagan's result, there are only finitely many such ideals and by our assumption there are only finitely many m with the property that I_m is such a maximal ideal. Therefore we can find $m_1 \geq 1$ such that $I_{m_1} \supseteq I_m$ for infinitely many values of m. By repeating now the argument for the ideals I_m, with $m > m_1$ and $I_m \subseteq I_{m_1}$, we obtain our assertion.

PROOF OF THEOREM 5.1. If the conclusion of the theorem fails, then there is a sequence $(\mathfrak{a}_m)_{m \geq 1}$ of monomial $\mathbf{R}$-ideals on $\mathbf{A}^n$ and a sequence $(\ell_m)_{m \geq 1}$ with $\lim_{m \to \infty} \ell_m = \infty$ such that for every divisor E over $\mathbf{A}^n$ that computes $\mathrm{mld}_0(\mathbf{A}^n, \mathfrak{a}_m)$, we have $k_E \geq \ell_m$. We will show that this leads to a contradiction.

Let $\lambda_1, \ldots, \lambda_r$ be the elements of I. By assumption, we can write each $\mathfrak{a}_m$ as

$$\mathfrak{a}_m = \prod_{j=1}^{r} \mathfrak{a}_{m,j}^{\lambda_j},$$

where all $\mathfrak{a}_{m,j}$ are monomial ideals. As we have seen, it follows from Maclagan's result that after passing to a subsequence, we may assume that $\mathfrak{a}_{m,1} \supseteq \mathfrak{a}_{m+1,1}$ for all $m \geq 1$. Repeating this for the $\mathfrak{a}_{m,2}, \ldots, \mathfrak{a}_{m,r}$, it follows that after r such steps, we may assume that $\mathfrak{a}_{m,j} \supseteq \mathfrak{a}_{m+1,j}$ for all $m \geq 1$ and all j, with $1 \leq j \leq r$.

In particular, it follows from Proposition 2.2 that $(\mathrm{mld}_0(\mathbf{A}^n, \mathfrak{a}_m))_{m\geq 1}$ is a weakly decreasing sequence. On the other hand, a result of Kawakita [**Kaw14**, Theorem 1.2] says that the set of mld's on a fixed klt germ, for $\mathbf{R}$-ideals with exponents in the finite set I, is finite. We thus conclude that after passing one more time to a subsequence, we may assume that all $\mathrm{mld}_0(\mathbf{A}^n, \mathfrak{a}_m)$ take the same value (possibly infinite).

Let E be a divisor over $\mathbf{A}^n$ that computes $\mathrm{mld}_0(\mathbf{A}^n, \mathfrak{a}_1)$. Given $m \geq 1$, since $\mathfrak{a}_{1,j} \supseteq \mathfrak{a}_{m,j}$ for all j, it follows that

$$\mathrm{mld}_0(\mathbf{A}^n, \mathfrak{a}_m) \leq k_E + 1 - \mathrm{ord}_E(\mathfrak{a}_m) \leq k_E + 1 - \mathrm{ord}_E(\mathfrak{a}_1) = \mathrm{mld}_0(\mathbf{A}^n, \mathfrak{a}_1).$$

Therefore all the above inequalities are equalities. In particular, E computes $\mathrm{mld}_0(\mathbf{A}^n, \mathfrak{a}_m)$ for all $m \geq 1$, a contradiction. This completes the proof of the theorem. □

6. Connection with ACC

Our goal in this section is to prove Theorem 1.5, relating Conjecture 1.1 to the ACC conjecture.

Proof of Theorem 1.5. Suppose that we have a sequence $(\mathfrak{a}_i)_{i\geq 1}$ of $\mathbf{R}$-ideals on X with exponents in J such that each $(X, \mathfrak{a}_i)$ is log canonical around x and with $q_i = \mathrm{mld}_x(X, \mathfrak{a}_i)$, the sequence $(q_i)_{i\geq 1}$ is strictly increasing. Since $q_i \leq \mathrm{mld}_x(X)$ for every i, it follows that $q := \lim_{i\to\infty} q_i < \infty$.

We may write $\mathfrak{a}_i = \prod_{j=1}^{r_i} \mathfrak{a}_{i,j}^{\lambda_{i,j}}$, where each $\mathfrak{a}_{i,j}$ is a nonzero ideal on X with $x \in \mathrm{Cosupp}(\mathfrak{a}_{i,j})$ and each $\lambda_{i,j}$ is a nonzero element of I. Since J is a DCC set, it follows that there is $\varepsilon > 0$ such that $\lambda_{i,j} \geq \varepsilon$ for all i and all j with $1 \leq j \leq r_i$. Let F be a fixed divisor over X with $c_X(F) = x$. For every $i \geq 1$, it follows from the fact that $(X, \mathfrak{a}_i)$ is log canonical around x that

$$r_i\varepsilon \leq \sum_{j=1}^{r_i} \lambda_{i,j} \leq \sum_{j=1}^{r_i} \lambda_{i,j} \cdot \mathrm{ord}_F(\mathfrak{a}_{i,j}) \leq k_F + 1.$$

First, this implies that the r_i are bounded. Second, it implies that the $\lambda_{i,j}$ are bounded. After possibly passing to a subsequence, we may assume that $r_i = r$ for all $i \geq 1$. Furthermore, since J is a DCC set, it follows that after possibly passing again to a subsequence, we may assume that each sequence $(\lambda_{i,j})_{i\geq 1}$ is nondecreasing. Since we have seen that the sequence is bounded, it follows that $\lambda_j := \lim_{i\to\infty} \lambda_{i,j} < \infty$.

We consider new $\mathbf{R}$-ideals $\mathfrak{a}'_i = \prod_{j=1}^{r} \mathfrak{a}_{i,j}^{\lambda_j}$ for $i \geq 1$. We now show that $(X, \mathfrak{a}'_i)$ is log canonical around x for $i \gg 0$. Note that also the set $J' = J \cup \{\lambda_1, \ldots, \lambda_r\}$ satisfies DCC, hence

$$\mathcal{A} := \{\mathrm{lct}_x(X, \mathfrak{b}) \mid \mathfrak{b} \text{ is an } \mathbf{R}\text{-ideal on } X \text{ with exponents in } J'\}$$

satisfies ACC (since we work on a fixed variety, this follows from [**dFEM11**, Theorem 4.2]; for the general statement, see [**HMX14**, Theorem 1.1]). In particular, there is M such that $\mathrm{lct}_x(X, \mathfrak{b}) \leq M$ for every $\mathbf{R}$-ideal $\mathfrak{b}$ on X with exponents in J. Note that we have

$$\lim_{i\to\infty} (\mathrm{lct}_x(X, \mathfrak{a}'_i) - \mathrm{lct}_x(X, \mathfrak{a}_i)) = 0. \tag{6.1}$$

Indeed, it follows from Proposition 2.2 that for every $\delta > 0$ and for every i such that $\lambda_{i,j} \geq (1+\delta)^{-1}\lambda_j$ for all j, we have

$$\frac{1}{\delta+1} \cdot \mathrm{lct}_x(X, \mathfrak{a}_i) \leq \mathrm{lct}_x(X, \mathfrak{a}'_i) \leq \mathrm{lct}_x(X, \mathfrak{a}_i),$$

hence

$$0 \leq \mathrm{lct}_x(X, \mathfrak{a}_i) - \mathrm{lct}_x(X, \mathfrak{a}'_i) \leq \frac{\delta}{\delta+1} \cdot \mathrm{lct}_x(X, \mathfrak{a}_i) \leq \frac{M\delta}{\delta+1}.$$

This gives (6.1). On the other hand, we have by assumption $\mathrm{lct}_x(X, \mathfrak{a}_i) \geq 1$ for all $i \geq 1$. Since the set $\mathcal{A}$ satisfies ACC, we conclude from (6.1) that $\mathrm{lct}_x(X, \mathfrak{a}'_i) \geq 1$ (hence $(X, \mathfrak{a}'_i)$ is log canonical around x) for all $i \gg 0$. After possibly ignoring the first few terms, we may assume that $(X, \mathfrak{a}'_i)$ is log canonical around x for every $i \geq 1$.

We now choose for every i a divisor E_i over X which computes $\mathrm{mld}_x(X, \mathfrak{a}'_i)$. Since we assume that X satisfies the assertion in Conjecture 1.1 for $I = \{\lambda_1, \ldots, \lambda_r\}$, we may and will assume that the set $\{k_{E_i} \mid i \geq 1\}$ is bounded above. Since we have

$$0 \leq \mathrm{mld}_x(X, \mathfrak{a}'_i) = k_{E_i} + 1 - \sum_{j=1}^{r} \lambda_j \cdot \mathrm{ord}_{E_i}(\mathfrak{a}_{i,j}),$$

it follows that there is $B > 0$ such that $\mathrm{ord}_{E_i}(\mathfrak{a}_{i,j}) \leq B$ for all i and j. On the other hand, since $\lambda_{i,j} \leq \lambda_j$ for all i and j, we have by Proposition 2.2

$$a_{E_i}(X, \mathfrak{a}'_i) = \mathrm{mld}_x(X, \mathfrak{a}'_i) \leq \mathrm{mld}_x(X, \mathfrak{a}_i) \leq a_{E_i}(X, \mathfrak{a}_i) \tag{6.2}$$

while

$$a_{E_i}(X, \mathfrak{a}_i) = a_{E_i}(X, \mathfrak{a}'_i) + \sum_{j=1}^{r} (\lambda_j - \lambda_{i,j}) \cdot \mathrm{ord}_{E_i}(\mathfrak{a}_{i,j}).$$

Since the $\mathbf{R}$-ideals $\mathfrak{a}'_i$ have exponents in the finite set $\{\lambda_1, \ldots, \lambda_r\}$, it follows from a result of Kawakita [**Kaw14**, Theorem 1.2] that the set

$$\{\mathrm{mld}_x(X, \mathfrak{a}'_i) \mid i \geq 1\}$$

is finite. After possibly passing to a subsequence, we may thus assume that $\mathrm{mld}_x(X, \mathfrak{a}'_i) = A$ for every $i \geq 1$. We then conclude from (6.2) that

$$A \leq q_i \leq A + B \cdot \sum_{j=1}^{r} (\lambda_j - \lambda_{i,j}). \tag{6.3}$$

Since $\lim_{i\to\infty} \lambda_{i,j} = \lambda_j$ for all j, it follows from (6.3) by passing to limit that $q = A$. Using one more time (6.3), we obtain $A \leq q_i \leq q = A$ for every i, hence the sequence $(q_i)_{i\geq 1}$ is constant, a contradiction. $\square$

7. Three equivalent conjectures

We begin by stating the Generic Limit conjecture and the Ideal-adic Semicontinuity conjecture for minimal log discrepancies.

Let X be a klt variety over k and $x \in X$ a closed point. Given a positive integer r and r sequences of nonzero coherent sheaves of ideals $(\mathfrak{a}_j^{(i)})_{i\geq 1}$ on X, for $1 \leq j \leq r$, the generic limit construction (see §3) gives an affine klt scheme $\widetilde{X}$, a closed point $\widetilde{x} \in \widetilde{X}$, and r ideals $\widetilde{\mathfrak{a}}_1, \ldots, \widetilde{\mathfrak{a}}_r$ on $\widetilde{X}$.

CONJECTURE 7.1 (Generic Limit conjecture, [**Kaw14**, Conjecture 4.5]). *For positive real numbers $\lambda_1, \ldots, \lambda_r$, there exists an infinite subset $S \subseteq \mathbb{Z}_{>0}$ such that the following hold:*

- *The ideals $\widetilde{\mathfrak{a}}_1, \ldots, \widetilde{\mathfrak{a}}_r$ are again generic limits of the sequences of ideals $(\mathfrak{a}_1^{(i)})_{i\in S}, \ldots, (\mathfrak{a}_r^{(i)})_{i\in S}$, and*
- *For every $i \in S$, we have*

$$\mathrm{mld}_{\widetilde{x}}(\widetilde{X}, \prod_{j=1}^{r} \widetilde{\mathfrak{a}}_j^{\lambda_j}) = \mathrm{mld}_x(X, \prod_{j=1}^{r} (\mathfrak{a}_j^{(i)})^{\lambda_j}).$$

REMARK 7.2. Note that in the setting of the above conjecture, the inequality

$$\mathrm{mld}_{\widetilde{x}}(\widetilde{X}, \prod_{j=1}^{r} \widetilde{\mathfrak{a}}_j^{\lambda_j}) \geq \mathrm{mld}_x(X, \prod_{j=1}^{r} (\mathfrak{a}_j^{(i)})^{\lambda_j})$$

can easily be guaranteed. Indeed, let E be a divisor computing $\mathrm{mld}_{\widetilde{x}}(\widetilde{X}, \prod_{j=1}^{r} \widetilde{\mathfrak{a}}_j^{\lambda_j})$. Take a positive integer ℓ such that $\ell \cdot \mathrm{ord}_E(\mathfrak{m}_{\widetilde{x}}) > \mathrm{ord}_E \widetilde{\mathfrak{a}}_j$ holds for each j. Then we have

$$\mathrm{mld}_{\widetilde{x}}(\widetilde{X}, \prod_{j=1}^{r} \widetilde{\mathfrak{a}}_j^{\lambda_j}) = \mathrm{mld}_{\widetilde{x}}(\widetilde{X}, \prod_{j=1}^{r} (\widetilde{\mathfrak{a}}_j + \mathfrak{m}_{\widetilde{x}}^{\ell})^{\lambda_j}).$$

By Remark 3.2, there exists an infinite subset $S \subseteq \mathbf{Z}_{>0}$ such that the first condition in the conjecture holds and

$$\mathrm{mld}_{\widetilde{x}}(\widetilde{X}, \prod_{j=1}^{r} (\widetilde{\mathfrak{a}}_j + \mathfrak{m}_{\widetilde{x}}^{\ell})^{\lambda_j}) = \mathrm{mld}_x(X, \prod_{j=1}^{r} (\mathfrak{a}_j^{(i)} + \mathfrak{m}_x^{\ell})^{\lambda_j})$$

for every $i \in S$. Since

$$\mathrm{mld}_x(X, \prod_{j=1}^{r} (\mathfrak{a}_j^{(i)} + \mathfrak{m}_x^{\ell})^{\lambda_j}) \geq \mathrm{mld}_x(X, \prod_{j=1}^{r} (\mathfrak{a}_j^{(i)})^{\lambda_j}),$$

we obtain the claimed inequality.

We now turn to the (uniform version of) Ideal-adic Semicontinuity conjecture for minimal log discrepancies.

CONJECTURE 7.3. *Let X be a klt variety and let $x \in X$ be a point defined by the ideal $\mathfrak{m}_x$. Given a finite set $I \subseteq \mathbf{R}_{\geq 0}$, there is a positive integer s (depending on (X, x) and I) such that the following holds: for every $\mathbf{R}$-ideals $\mathfrak{a} = \prod_{j=1}^{r} \mathfrak{a}_j^{\lambda_j}$ and $\mathfrak{b} = \prod_{j=1}^{r} \mathfrak{b}_j^{\lambda_j}$, with $\lambda_j \in I$ for all j, if $\mathfrak{a}_j + \mathfrak{m}_x^s = \mathfrak{b}_j + \mathfrak{m}_x^s$ for all j, then $\mathrm{mld}_x(X, \mathfrak{a}) \geq 0$ if and only if $\mathrm{mld}_x(X, \mathfrak{b}) \geq 0$, and if this is the case*[1], *then $\mathrm{mld}_x(X, \mathfrak{a}) = \mathrm{mld}_x(X, \mathfrak{b})$.*

[1]If this is not the case and $\dim(X) \geq 2$, then the two mlds are equal since they are both $-\infty$.

We now prove the result stated in the Introduction, saying that Conjectures 1.1, 7.1, and 7.3 are equivalent.

PROOF OF THEOREM 1.6. We first show that Conjecture 1.1 implies Conjecture 7.3. Suppose that Conjecture 1.1 holds for (X,x) and every finite set I. Let I be such a set. By assumption, there is a positive integer ℓ such that for every $\mathbf{R}$-ideal $\mathfrak{a}=\prod_{j=1}^{r}\mathfrak{a}_j^{\lambda_j}$ on X, with $\lambda_j\in I$ for all j, there is a divisor E computing $\mathrm{mld}_x(X,\mathfrak{a})$ with $k_E\leq\ell$. Let ε be the smallest nonzero element of I and let s be a positive integer that satisfies $s>\frac{\ell+1}{\varepsilon}$. Suppose that $\mathfrak{a}$ and $\mathfrak{b}$ are as in Conjecture 7.3, with $\mathrm{mld}_x(X,\mathfrak{a})\geq 0$. We may and will assume that $\lambda_j>0$ for all j.

Let $\mathfrak{a}'_s=\prod_{j=1}^{r}(\mathfrak{a}_j+\mathfrak{m}_x^s)^{\lambda_j}$ and $\mathfrak{b}'_s=\prod_{j=1}^{r}(\mathfrak{b}_j+\mathfrak{m}_x^s)^{\lambda_j}$. We assume that $\mathfrak{a}_j+\mathfrak{m}_x^s=\mathfrak{b}_j+\mathfrak{m}_x^s$ for all j, hence $\mathfrak{a}'_s=\mathfrak{b}'_s$.

Let E be a divisor over X which computes $\mathrm{mld}_x(X,\mathfrak{a})$ such that $k_E\leq\ell$. In this case we have $0\leq\mathrm{mld}_x(X,\mathfrak{a})=k_E+1-\sum_{j=1}^{r}\lambda_j\cdot\mathrm{ord}_E(\mathfrak{a}_j)$, hence

$$\sum_{j=1}^{r}\lambda_j\cdot\mathrm{ord}_E(\mathfrak{a}_j)\leq\ell+1.$$

It follows from the choice of ε and s that

$$s\cdot\mathrm{ord}_E(\mathfrak{m}_x)\geq s>\frac{\ell+1}{\lambda_j}\geq\mathrm{ord}_E(\mathfrak{a}_j)$$

for every j. Using Proposition 2.2, we obtain $\mathrm{mld}_x(X,\mathfrak{a})=\mathrm{mld}_x(X,\mathfrak{a}'_s)$. Since $\mathfrak{a}'_s=\mathfrak{b}'_s$, we have $\mathrm{mld}_x(X,\mathfrak{b}'_s)=\mathrm{mld}_x(X,\mathfrak{a}'_s)=\mathrm{mld}_x(X,\mathfrak{a})$ and since $\mathrm{mld}_x(X,\mathfrak{b})\leq\mathrm{mld}_x(X,\mathfrak{b}'_s)$, we conclude that $\mathrm{mld}_x(X,\mathfrak{b})\leq\mathrm{mld}_x(X,\mathfrak{a})$.

On the other hand, we have $\mathrm{mld}_x(X,\mathfrak{b})\geq 0$. Indeed, if this is not the case, then by assumption we can find a divisor F that computes $\mathrm{mld}_x(X,\mathfrak{b})$, with $k_F\leq\ell$. Therefore we have $k_F+1<\mathrm{ord}_F(\mathfrak{b})$. We now use the fact that $\mathrm{mld}_x(X,\mathfrak{b}'_s)\geq 0$. First, this implies that $\mathrm{ord}_F(\mathfrak{b}'_s)<\mathrm{ord}_F(\mathfrak{b})$, and since we can write

$$\mathrm{ord}_F(\mathfrak{b}'_s)=\sum_{j=1}^{r}\lambda_j\cdot\min\{s\cdot\mathrm{ord}_F(\mathfrak{m}_x),\mathrm{ord}_F(\mathfrak{b}_j)\},$$

we conclude that there is j such that $s\cdot\mathrm{ord}_F(\mathfrak{m}_x)<\mathrm{ord}_F(\mathfrak{b}_j)$. Second, it gives

$$\ell+1\geq k_F+1\geq\mathrm{ord}_F(\mathfrak{b}'_s)\geq\lambda_j s\cdot\mathrm{ord}_F(\mathfrak{m}_x)\geq\varepsilon s>\ell+1,$$

a contradiction. We thus conclude that $\mathrm{mld}_x(X,\mathfrak{b})\geq 0$.

We can now run the same argument with the roles of $\mathfrak{a}$ and $\mathfrak{b}$ reversed, to conclude that $\mathrm{mld}_x(X,\mathfrak{b})\geq\mathrm{mld}_x(X,\mathfrak{a})$. Therefore $\mathrm{mld}_x(X,\mathfrak{b})=\mathrm{mld}_x(X,\mathfrak{a})$, This completes the proof of the fact that Conjecture 1.1 implies Conjecture 7.3.

We now show that Conjecture 7.3 implies Conjecture 7.1. Suppose that we are in the setting of Conjecture 7.1 and let s be the positive integer provided by Conjecture 7.3 for the set $I=\{\lambda_1,\ldots,\lambda_r\}$. By assumption, we have

$$\mathrm{mld}_x(X,\prod_{j=1}^{r}(\mathfrak{a}_j^{(i)}+\mathfrak{m}_x^{\ell})^{\lambda_j})=\mathrm{mld}_x(X,\prod_{j=1}^{r}(\mathfrak{a}_j^{(i)})^{\lambda_j})$$

for every $\ell\geq s$ and every i. The argument in Remark 7.2 then implies that there is an infinite subset $S\subseteq\mathbf{Z}_{>0}$ that satisfies the first condition in Conjecture 7.1 and such that

$$\mathrm{mld}_x(X, \prod_{j=1}^{r}(\mathfrak{a}_j^{(i)} + \mathfrak{m}_x^{\ell})^{\lambda_j}) = \mathrm{mld}_{\widetilde{x}}(\widetilde{X}, \prod_{j=1}^{r} \widetilde{\mathfrak{a}}_j^{\lambda_j}),$$

for every $i \in S$. We thus have the conclusion in Conjecture 7.1.

Finally, we show that Conjecture 7.1 implies Conjecture 1.1. Let $\lambda_1, \ldots, \lambda_r$ be the nonzero elements of the finite set I. If the assertion in Conjecture 1.1 is not true, then for each positive integer i there exist coherent ideals $\mathfrak{a}_1^{(i)}, \ldots, \mathfrak{a}_r^{(i)}$ with the following property:

- $k_{E_i} \geq i$ holds for every divisor E_i that computes $\mathrm{mld}_x(X, \prod_{j=1}^{r}(\mathfrak{a}_j^{(i)})^{\lambda_j})$.

We use the generic limit construction for $(\mathfrak{a}_j^{(i)})_{\geq 1}$ to obtain coherent ideal sheaves $\widetilde{\mathfrak{a}}_1, \ldots, \widetilde{\mathfrak{a}}_r$ on $\widetilde{X}$. By applying successively[2] Conjecture 7.1 and Remark 7.2, we get an infinite subset $S \subseteq \mathbf{Z}_{>0}$ such that

$$\mathrm{mld}_x(X, \prod_{j=1}^{r}(\mathfrak{a}_j^{(i)} + \mathfrak{m}_x^{\ell})^{\lambda_j}) = \mathrm{mld}_x(X, \prod_{j=1}^{r}(\mathfrak{a}_j^{(i)})^{\lambda_j}) \tag{7.1}$$

for every $i \in S$. Let ℓ' be the ℓ provided by Theorem 1.2. It follows that for every $i \in S$, there is a divisor E_i that computes $\mathrm{mld}_x(X, \prod_{j=1}^{r}(\mathfrak{a}_j^{(i)} + \mathfrak{m}_x^{\ell})^{\lambda_j})$ such that $\mathrm{ord}_{E_i}(\mathfrak{m}_x) \leq \ell'$. The equality (7.1) implies that E_i also computes $\mathrm{mld}_x(X, \prod_{j=1}^{r}(\mathfrak{a}_j^{(i)})^{\lambda_j})$. Therefore we have

$$\mathrm{ord}_{E_i}(\mathfrak{a}_j^{(i)} + \mathfrak{m}_x^{\ell}) = \mathrm{ord}_{E_i}(\mathfrak{a}_j^{(i)})$$

for every j, hence

$$\mathrm{ord}_{E_i}(\mathfrak{a}_j^{(i)}) \leq \mathrm{ord}_{E_i}(\mathfrak{m}_x^{\ell}) \leq \ell\ell'.$$

If $i \in S$ satisfies $i > \mathrm{mld}_x(X) - 1 + \ell\ell' \sum_{j=1}^{r} \lambda_j$, then we have

$$\begin{aligned} a_{E_i}(X, \prod_{j=1}^{r}(\mathfrak{a}_j^{(i)})^{\lambda_j}) &= k_{E_i} + 1 - \sum_{j=1}^{r} \lambda_j \cdot \mathrm{ord}_{E_i}(\mathfrak{a}_j^{(i)}) \\ &\geq i + 1 - \ell\ell' \sum_{j=1}^{r} \lambda_j > \mathrm{mld}_x(X). \end{aligned}$$

This contradicts the fact that

$$a_{E_i}(X, \prod_{j=1}^{r}(\mathfrak{a}_j^{(i)})^{\lambda_j}) = \mathrm{mld}_x(X, \prod_{j=1}^{r}(\mathfrak{a}_j^{(i)})^{\lambda_j}) \leq \mathrm{mld}_x(X).$$

We thus showed that Conjecture 7.1 implies Conjecture 1.1, completing the proof of the theorem. □

REMARK 7.4. By the equivalence of Conjectures 1.1 and 7.1, Theorem 1.3 and Theorem 1.5 also follow from results of Kawakita, see [**Kaw15**, Proposition 4.8, Theorem 5.3].

---

[2] We need the first condition in Conjecture 7.1 in order to be able to apply Remark 7.2 to the resulting subsequences of ideals.

References

[Amb99] Florin Ambro, *On minimal log discrepancies*, Math. Res. Lett. **6** (1999), no. 5-6, 573–580. MR1739216

[Bir07] Caucher Birkar, *Ascending chain condition for log canonical thresholds and termination of log flips*, Duke Math. J. **136** (2007), no. 1, 173–180. MR2271298

[dFEM10] Tommaso de Fernex, Lawrence Ein, and Mircea Mustaţă, *Shokurov's ACC conjecture for log canonical thresholds on smooth varieties*, Duke Math. J. **152** (2010), no. 1, 93–114. MR2643057

[dFEM11] Tommaso de Fernex, Lawrence Ein, and Mircea Mustaţă, *Log canonical thresholds on varieties with bounded singularities*, Classification of algebraic varieties, EMS Ser. Congr. Rep., Eur. Math. Soc., Zürich, 2011, pp. 221–257. MR2779474

[dFM09] Tommaso de Fernex and Mircea Mustaţă, *Limits of log canonical thresholds* (English, with English and French summaries), Ann. Sci. Éc. Norm. Supér. (4) **42** (2009), no. 3, 491–515. MR2543330

[ELM04] Lawrence Ein, Robert Lazarsfeld, and Mircea Mustaţă, *Contact loci in arc spaces*, Compos. Math. **140** (2004), no. 5, 1229–1244. MR2081163

[EMY03] Lawrence Ein, Mircea Mustaţă, and Takehiko Yasuda, *Jet schemes, log discrepancies and inversion of adjunction*, Invent. Math. **153** (2003), no. 3, 519–535. MR2000468

[HMX14] Christopher D. Hacon, James McKernan, and Chenyang Xu, *ACC for log canonical thresholds*, Ann. of Math. (2) **180** (2014), no. 2, 523–571. MR3224718

[Kaw13a] Masayuki Kawakita, *Ideal-adic semi-continuity problem for minimal log discrepancies*, Math. Ann. **356** (2013), no. 4, 1359–1377. MR3072804

[Kaw13b] Masayuki Kawakita, *Ideal-adic semi-continuity of minimal log discrepancies on surfaces*, Michigan Math. J. **62** (2013), no. 2, 443–447. MR3079272

[Kaw14] Masayuki Kawakita, *Discreteness of log discrepancies over log canonical triples on a fixed pair*, J. Algebraic Geom. **23** (2014), no. 4, 765–774. MR3263668

[Kaw15] Masayuki Kawakita, *A connectedness theorem over the spectrum of a formal power series ring*, Internat. J. Math. **26** (2015), no. 11, 1550088, 27. MR3413983

[Kol08] J. Kollár, *Which powers of holomorphic functions are integrable?*, available at *arXiv:0805.0756v1*.

[Kol13] János Kollár, *Singularities of the minimal model program*, Cambridge Tracts in Mathematics, vol. 200, Cambridge University Press, Cambridge, 2013. With a collaboration of Sándor Kovács. MR3057950

[KM98] János Kollár and Shigefumi Mori, *Birational geometry of algebraic varieties*, Cambridge Tracts in Mathematics, vol. 134, Cambridge University Press, Cambridge, 1998. With the collaboration of C. H. Clemens and A. Corti; Translated from the 1998 Japanese original. MR1658959

[Mac01] Diane Maclagan, *Antichains of monomial ideals are finite*, Proc. Amer. Math. Soc. **129** (2001), no. 6, 1609–1615. MR1814087

[Nak16a] Yusuke Nakamura, *On semi-continuity problems for minimal log discrepancies*, J. Reine Angew. Math. **711** (2016), 167–187. MR3456762

[Nak16b] Yusuke Nakamura, *On minimal log discrepancies on varieties with fixed Gorenstein index*, Michigan Math. J. **65** (2016), no. 1, 165–187. MR3466821

[Sho04] V. V. Shokurov, *Letters of a bi-rationalist. V. Minimal log discrepancies and termination of log flips* (Russian, with Russian summary), Tr. Mat. Inst. Steklova **246** (2004), no. Algebr. Geom. Metody, Svyazi i Prilozh., 328–351; English transl., Proc. Steklov Inst. Math. **3(246)** (2004), 315–336. MR2101303

Department of Mathematics, University of Michigan, 530 Church Street, Ann Arbor, Michigan 48109

Email address: `mmustata@umich.edu`

Graduate School of Mathematical Sciences, University of Tokyo, 3-8-1 Komaba, Meguro-Ku, Tokyo, 153-8914, Japan

Email address: `nakamura@ms.u-tokyo.ac.jp`

Contemporary Mathematics
Volume **712**, 2018
http://dx.doi.org/10.1090/conm/712/14352

The Wahl map of one-nodal curves on K3 surfaces

Edoardo Sernesi

This paper is dedicated to L. Ein on the occasion of his 60th birthday.

ABSTRACT. We consider a general primitively polarized K3 surface (S, H) of genus $g+1$ and a 1-nodal curve $\widetilde{C} \in |H|$. We prove that the normalization C of $\widetilde{C}$ has surjective Wahl map provided $g = 40, 42$ or ≥ 44.

Introduction

In this Note we consider 1-nodal curves lying on a K3 surface and we study the gaussian map, or Wahl map, on their normalization. If we consider a primitive linear system $|H|$ on a K3 surface S, then it is well known that every nonsingular $\widetilde{C} \in |H|$ has a non-surjective Wahl map

$$\Phi_K : \bigwedge^2 H^0(\widetilde{C}, \omega_{\widetilde{C}}) \longrightarrow H^0(\widetilde{C}, \omega^3_{\widetilde{C}})$$

(see §1 for the definition) and that, if moreover $\widetilde{C} \in |H|$ is general then it is Brill-Noether-Petri general [**W1, BM, L**]. It is of some interest to decide whether the same properties hold for the normalization C of a 1-nodal $\widetilde{C} \in |H|$, and more generally for the normalization of a singular $\widetilde{C}$ in $|H|$. The Brill-Noether theory of singular curves on a K3 surface has received quite a lot of attention in recent times, see e.g. [**Go, FKP, BFT, CK, Ke**]. On the other hand to our knowledge very little is known on their Wahl map. In [**Hal2, Ke**] the authors consider a modified version of the Wahl map, which does not seem to have a direct and simple relation with the ordinary Wahl map Φ_K; in particular their results point towards the non-surjectivity of such modified map. A different point of view is taken in [**BF**], where the authors give necessary conditions for a singular curve to be hyperplane section of a smooth surface, again in terms of non-surjectivity of certain maps. On the other hand in [**FKPS**] it is proved that the normalization C (of genus 10) of a general 1-nodal curve $\widetilde{C} \in |H|$ on a general polarized (S, H) of genus 11 has general moduli; then the main result of [**CHM**] implies that Φ_K is surjective for such a curve. This surjectivity result is extended in the present paper in the

2010 *Mathematics Subject Classification.* Primary 14J28, 14H10; Secondary 14H51.
Key words and phrases. K3 surface, Wahl map.
We thank M. Halic and M. Kemeny for helpful e-mail correspondence, A. L. Knutsen and A. Bruno for very useful comments on a preliminary version of this work, and the referee for his careful report.

following form:

THEOREM 0.1. *Let (S, H) be a general primitively polarized K3 surface of genus $g+1$. Assume that $g = 40, 42$ or ≥ 44. Let $\widetilde{C} \in |H|$ be a 1-nodal curve and C its normalization. Then the Wahl map*

$$\Phi_K : \bigwedge^2 H^0(C, \omega_C) \longrightarrow H^0(C, \omega_C^3)$$

is surjective.

This of course gives another proof of the main result of [**CHM**] for the values of g as in the statement, since 1-nodal curves are known to exist in $|H|$ for a general primitively polarized (S, H) of any genus $g + 1 \geq 2$ [**MM, Ch**].

Now a few words about the method of proof. Letting $P \in S$ be the unique singular point of $\widetilde{C}$ we consider the blow-up $\sigma : X := \mathrm{Bl}_P S \longrightarrow S$ at P and we let $E \subset X$ be the exceptional curve. Then the normalization of $\widetilde{C}$ is the strict transform $C = \sigma^*\widetilde{C} - 2E \subset X$. The Wahl map Φ_K on C can be decomposed as

$$\Phi_K = H^0(\rho) \cdot \Phi_{K_X+C}$$

where Φ_{K_X+C} is a gaussian map on X and $H^0(\rho)$ is induced in cohomology by a restriction homomorphism:

$$\rho : \Omega^1_X(2K_X + 2C) \longrightarrow \omega_C^3$$

on X. We study these two maps and prove their surjectivity separately. This method of proof is analogous to the one adopted in the work of several authors before, notably [**BM, CLM1, CLM2, DM, W2**]. The restriction on the genus depends on the proof: one would expect the result to hold for $g = 10$ (as it does indeed, as already remarked) and for $g \geq 12$. In fact the surjectivity of $H^0(\rho)$ holds for $g = 10$ or $g \geq 12$ (Lemma 2.1). On the other hand the proof of the surjectivity of Φ_{K_X+C}, which consists in adapting an analogous proof given in [**CLM2**] for plane curves, leads to the restrictions on g in Theorem 0.1: indeed this proof requires that we decompose a certain divisor on X as the sum of three very ample ones and this decomposition forces the genus to increase.

Recent work by M. Kemeny [**Ke**] implies that the curves C considered here, i.e. normalizations of 1-nodal curves on a general primitive K3 surface, are generically Brill-Noether-Petri general and fill a locus in $\mathcal{M}_g$, the coarse moduli space of curves of genus g, whose closure has dimension $19 + g$. Theorem 0.1 and [**W1**] imply that this naturally defined locus is not contained in the closure of the so-called K3-locus (i.e. the locus of smooth curves that can be embedded in a K3 surface).

One can ask whether a result analogous to Theorem 0.1 can be proved for the normalization of curves on K3 surfaces having a more complicated singular point. We did not consider this case. Note though that, to our knowledge, such curves are known to exist only in the case of A_k-singularities or ordinary triple points (see [**GK, Ga**]).

The paper is organized as follows. In §1 we introduce the gaussian maps and explain the strategy of proof of the surjectivity of the Wahl map of a curve lying in a regular surface. In §2 we prove the surjectivity of $H^0(\rho)$ and in §3 we prove the surjectivity of Φ_{K_X+C}. We work over $\mathbb{C}$.

1. Generalities on Gaussian maps

In this section we recall a few definitions and basic facts concerning gaussian maps. Given line bundles L, M on a nonsingular projective variety Y we consider:

$$\mathcal{R}(L, M) = \ker[H^0(Y, L) \otimes H^0(Y, M) \to H^0(Y, L \otimes M)]$$

Then we have a canonical map:

$$\Phi_{L,M} : \mathcal{R}(L, M) \longrightarrow H^0(Y, \Omega^1_Y \otimes L \otimes M)$$

called the *gaussian map*, or *Wahl map*, of L, M, which is defined as follows. Let $\Delta \subset Y \times Y$ be the diagonal and $p_1, p_2 : Y \times Y \to Y$ the projections. Then

$$\mathcal{R}(L, M) = H^0(Y \times Y, p_1^* L \otimes p_2^* M \otimes \mathcal{I}_\Delta)$$

Since $\mathcal{I}_\Delta \otimes \mathcal{O}_\Delta = \Omega^1_Y$, the restriction to Δ:

$$p_1^* L \otimes p_2^* M \otimes \mathcal{I}_\Delta \longrightarrow p_1^* L \otimes p_2^* M \otimes \mathcal{I}_\Delta \otimes \mathcal{O}_\Delta$$

induces $\Phi_{L,M}$ on global sections. The exact sequence:

$$0 \longrightarrow p_1^* L \otimes p_2^* M \otimes \mathcal{I}_\Delta^2 \longrightarrow p_1^* L \otimes p_2^* M \otimes \mathcal{I}_\Delta \longrightarrow p_1^* L \otimes p_2^* M \otimes \mathcal{I}_\Delta \otimes \mathcal{O}_\Delta \longrightarrow 0$$

shows that the vanishing:

$$H^1(Y \times Y, p_1^* L \otimes p_2^* M \otimes \mathcal{I}_\Delta^2) = 0$$

is a sufficient condition for the surjectivity of $\Phi_{L,M}$.

In case $L = M$ we have $\mathcal{R}(L, L) = I_2(Y) \oplus \bigwedge^2 H^0(Y, L)$, where

$$I_2(Y) = \ker[S^2 H^0(Y, L) \to H^0(Y, L^2)]$$

and $\Phi_{L,L}$ is zero on $I_2(Y)$. Therefore $\Phi_{L,L}$ is equivalent to its restriction to $\bigwedge^2 H^0(Y, L)$, which is denoted by

$$\Phi_L : \bigwedge^2 H^0(Y, L) \longrightarrow H^0(Y, \Omega^1_Y \otimes L^2)$$

In particular, for a non-hyperelliptic curve C we are interested in $\Phi_{K,K}$ or rather in

$$\Phi_K : \bigwedge^2 H^0(C, \omega_C) \longrightarrow H^0(C, \omega_C^3)$$

where $\mathcal{O}_C(K) = \omega_C$ is the canonical invertible sheaf.

Suppose that $C \subset X$ where X is a nonsingular regular surface. Then the exact sequence:

$$0 \to \mathcal{O}_X(K_X) \longrightarrow \mathcal{O}_X(K_X + C) \longrightarrow \omega_C \to 0$$

shows that $H^0(X, K_X + C) \to H^0(C, \omega_C)$ is surjective. Moreover it is easy to show that Φ_K fits in the commutative diagram:

$$\begin{array}{ccc} \bigwedge^2 H^0(X, K_X + C) & \xrightarrow{\Phi_{K_X + C}} & H^0(X, \Omega^1_X(2K_X + 2C)) \\ \downarrow & & \downarrow {\scriptstyle H^0(\rho)} \\ \bigwedge^2 H^0(C, \omega_C) & \xrightarrow{\Phi_K} & H^0(\omega_C^3) \end{array}$$

where

$$\rho : \Omega^1_X(2K_X + 2C) \to \omega_C^3$$

is the restriction map. Since the left vertical map is surjective we have:

LEMMA 1.1. *In the above situation*

$$\mathrm{Im}(\Phi_K) \subset \mathrm{Im}(H^0(\rho))$$

and equality holds if Φ_{K_X+C} *is surjective. In particular, if both* Φ_{K_X+C} *and* $H^0(\rho)$ *are surjective, so is* Φ_K.

2. The surjectivity of $H^0(\rho)$

As in the Introduction, we let (S,H) be a K3 surface with a polarization of genus $g+1 \geq 3$ and let $\widetilde{C} \in |H|$ be a curve with one node, i.e. an ordinary double point, at $P \in S$ and no other singularities. Consider the blow-up $\sigma : X := \mathrm{Bl}_P S \longrightarrow S$ of S at P, let $E \subset X$ be the exceptional curve and $C = \sigma^*\widetilde{C} - 2E \subset X$ the strict transform of $\widetilde{C}$. We have an exact sequence on X:

$$0 \longrightarrow \Omega^1_X(\log C)(-C) \longrightarrow \Omega^1_X \longrightarrow \omega_C \longrightarrow 0$$

where $\Omega^1_X(\log C)$ is the sheaf of 1-forms with logarithmic poles along C [**EV**]. Tensoring with $\mathcal{O}_X(2K_X + 2C)$ we obtain:

$$0 \longrightarrow \Omega^1_X(\log C)(2K_X + C) \longrightarrow \Omega^1_X(2K_X + 2C) \xrightarrow{\rho} \omega_C^3 \longrightarrow 0 \tag{2.1}$$

LEMMA 2.1. *Suppose that* (S,H) *is a general primitively polarized K3 surface of genus* $g+1$, *with* $g = 10$ *or* $g \geq 12$. *Then, with the same notations as above, we have:*

$$h^1(X, \Omega^1_X(\log C)(2K_X + C)) = 0$$

In particular $H^0(\rho)$ *is surjective.*

PROOF. The last assertion follows from the exact sequence (2.1). Since $K_X = E$ we have $\mathcal{O}_X(2K_X + C) = \sigma^*H$. Consider the relative cotangent sequence of σ:

$$0 \longrightarrow \sigma^*\Omega^1_S \longrightarrow \Omega^1_X \xrightarrow{\eta} \omega_E \longrightarrow 0$$

and tensor it by σ^*H:

$$0 \longrightarrow \sigma^*\Omega^1_S(H) \longrightarrow \Omega^1_X(\sigma^*H) \xrightarrow{\eta_H} \omega_E \longrightarrow 0$$

We have:

$$H^2(X, \sigma^*\Omega^1_S(H)) = H^2(S, \Omega^1_S(H)) = H^0(S, T_S(-H))^\vee = 0$$

From the assumption about the genus and from the generality of (S,H) it follows that we also have:

$$H^1(X, \sigma^*\Omega^1_S(H)) = H^1(S, \Omega^1_S(H)) = 0$$

(see [**B**], 5.2). Therefore η_H induces an isomorphism:

$$H^1(\eta_H) : H^1(X, \Omega^1_X(\sigma^*H)) \cong H^1(E, \omega_E) \cong \mathbb{C}$$

Then in order to prove the lemma it suffices to show that in the exact sequence:

$$0 \longrightarrow \Omega^1_X(\sigma^*H) \longrightarrow \Omega^1_X(\log C) \otimes \sigma^*H \longrightarrow \sigma^*H \otimes \mathcal{O}_C \longrightarrow 0$$

$$\sigma^*H \otimes \mathcal{O}_C = \omega_C(E)$$

the coboundary map:

$$\partial_H : H^0(C, \omega_C(E)) \longrightarrow H^1(X, \Omega^1_X(\sigma^* H)) \tag{2.2}$$

is non-zero. The above sequence is part of the following exact and commutative diagram:

$$\begin{array}{ccccccccc}
 & & \sigma^*\Omega^1_S(H) & = & \sigma^*\Omega^1_S(H) & & & & \\
 & & \downarrow & & \downarrow & & & & \\
0 & \longrightarrow & \Omega^1_X(\sigma^* H) & \longrightarrow & \Omega^1_X(\log C) \otimes \sigma^* H & \longrightarrow & \omega_C(E) & \longrightarrow & 0 \\
 & & \downarrow{\scriptstyle \eta_H} & & \downarrow & & \| & & \\
0 & \longrightarrow & \omega_E & \longrightarrow & \mathcal{L} & \longrightarrow & \omega_C(E) & \longrightarrow & 0 \\
 & & \downarrow & & \downarrow & & \downarrow & & \\
 & & 0 & & 0 & & 0 & &
\end{array}$$

where $\mathcal{L}$ is an invertible sheaf on $C+E$. Since $H^1(\eta_H)$ is an isomorphism it suffices to show that the coboundary map $H^0(\omega_C(E)) \to H^1(\omega_E)$ of the last row is non-zero or, equivalently, that $H^1(\mathcal{L}) = 0$. Observe that $\mathcal{L} = \mathcal{M} \otimes \sigma^* H$, where $\mathcal{M}$ is defined by the following diagram:

$$\begin{array}{ccccccccc}
 & & \sigma^*\Omega^1_S & = & \sigma^*\Omega^1_S & & & & \\
 & & \downarrow & & \downarrow & & & & \\
0 & \longrightarrow & \Omega^1_X & \longrightarrow & \Omega^1_X(\log C) & \longrightarrow & \mathcal{O}_C & \longrightarrow & 0 \\
 & & \downarrow{\scriptstyle \eta} & & \downarrow & & \| & & \\
0 & \longrightarrow & \omega_E & \longrightarrow & \mathcal{M} & \longrightarrow & \mathcal{O}_C & \longrightarrow & 0 \\
 & & \downarrow & & \downarrow & & \downarrow & & \\
 & & 0 & & 0 & & 0 & &
\end{array}$$

It suffices to show that $\mathcal{M} \ncong \mathcal{O}_{C+E}$ because this will imply that $\mathcal{L} \ncong \omega_{C+E}$, and in turn that $H^1(\mathcal{L}) = 0$. The coboundary map of the middle row

$$\partial : H^0(C, \mathcal{O}_C) \longrightarrow H^1(X, \Omega^1_X)$$

associates to $1 \in H^0(C, \mathcal{O}_C)$ the Atiyah-Chern class of $\mathcal{O}_X(C)$ and $H^1(\eta)(\partial(1))$ is its restriction to E. Moreover

$$\ker(H^1(\eta)) \cong H^1(X, \sigma^*\Omega^1_S)$$

is generated by the Atiyah-Chern classes of the total trasforms under σ of curves in S, which are trivial when restricted to E. Since $E \cdot C = 2$ we see that $\partial(1) \notin \ker(H^1(\eta))$. It follows that the coboundary of the last row:

$$H^0(C, \mathcal{O}_C) \longrightarrow H^1(E, \omega_E)$$

is non-zero. Hence $\mathcal{M} \ncong \mathcal{O}_{C+E}$. □

3. The gaussian map on X

We keep the notations of §2. We will prove the following:

PROPOSITION 3.1. *Suppose that (S,H) is a general primitively polarized K3 surface of genus $g+1$, with $g = 40, 42$ or ≥ 44; let $\widetilde{C} \in |H|$ be a 1-nodal curve and C its normalization. Let $X = \mathrm{Bl}_P(S)$ where $P \in \widetilde{C}$ is the node. Then the gaussian map*

$$\Phi_{K_X+C} : \bigwedge^2 H^0(X, \mathcal{O}_X(K_X + C)) \longrightarrow H^0(X, \Omega^1_X(2K_X + 2C))$$

is surjective.

We will need the following result and its corollary:

PROPOSITION 3.2. *a) For every $d \geq 10$ there is a K3 surface S containing two very ample nonsingular curves A, B such that $\mathrm{Pic}(S) = \mathbb{Z}[A] \oplus \mathbb{Z}[B]$, with intersection matrix:*

$$\begin{pmatrix} A^2 & A \cdot B \\ B \cdot A & B^2 \end{pmatrix} = \begin{pmatrix} 8 & d \\ d & 8 \end{pmatrix}$$

and A and B non-trigonal.

b) For every $d \geq 12$ there is a K3 surface S containing two very ample nonsingular curves A, B such that $\mathrm{Pic}(S) = \mathbb{Z}[A] \oplus \mathbb{Z}[B]$, with intersection matrix:

$$\begin{pmatrix} A^2 & A \cdot B \\ B \cdot A & B^2 \end{pmatrix} = \begin{pmatrix} 10 & d \\ d & 8 \end{pmatrix}$$

and A and B non-trigonal.

In both cases (a) and (b) the surface S does not contain rational nonsingular curves R such that $A \cdot R = 1$ or $B \cdot R = 1$.

PROOF. The Proposition is a special case of [**Kn**], Theorem 4.6. We obtain the Proposition by taking (with the notations used there) $(n, g-1) = (4,4)$ and $(n, g-1) = (5,4)$ respectively. The restriction on d is forced by the requirement that the hypotheses of the theorem apply symmetrically w.r. to A an B so that both are very ample. The non-trigonality follows from the fact that S is embedded by both $|A|$ and $|B|$ so to be an intersection of quadrics. The last assertion is proved as done in loc. cit. for (-2) curves, by comparing discriminants. □

COROLLARY 3.3. *If S is as in Prop. 3.2(a) then $H = A+2B$ defines a primitive very ample divisor class of genus $g+1 = 21+2d$ for every $d \geq 10$. If S is as in Prop. 3.2(b) then $H = A+2B$ defines a primitive very ample divisor class of genus $g+1 = 22+2d$ for every $d \geq 12$.*

PROOF. Clearly H is very ample and it is primitive because the generator A of $\mathrm{Pic}(S)$ appears with coefficient 1. The genus in either case is readily computed using the intersection matrix. □

PROOF OF PROPOSITION 3.1. By semicontinuity it suffices to prove the Proposition for just one primitively polarized K3 surface for each value of g as in the statement. We take (S,H) as in Corollary 3.3, distinguishing cases (a) and (b) according to the parity of g. Letting $\widetilde{C}, C, P$ and X as in the statement, consider the product $X \times X$ and the blow-up $\pi : Y = \mathrm{Bl}_\Delta(X \times X) \to X \times X$ along the diagonal Δ. Let $\Lambda \subset Y$ be the exceptional divisor. For any coherent sheaf $\mathcal{F}$ on X we define $\mathcal{F}_i = (p_i \cdot \pi)^* \mathcal{F}$, $i = 1, 2$. It suffices to prove that

$$H^1(X \times X, p_1^* \mathcal{O}_X(K_X + C) \otimes p_2^* \mathcal{O}_X(K_X + C) \otimes \mathcal{I}_\Delta^2) = 0$$

which is equivalent to:

$$H^1(Y, (K_X + C)_1 + (K_X + C)_2 - 2\Lambda) = 0$$

Note that we have:

$$(K_X + C)_1 + (K_X + C)_2 - 2\Lambda = K_Y + C_1 + C_2 - 3\Lambda$$

and therefore we want to prove that:

(3.1) $$H^1(Y, \mathcal{L}) = 0$$

where we set $\mathcal{L} = \mathcal{O}_Y(K_Y + C_1 + C_2 - 3\Lambda)$. The proof is an adaptation of the proofs of Lemmas (3.1) and (3.10) of [**CLM2**]. One uses the following:

LEMMA 3.4. *Assume that D is a very ample divisor on X. Then $D_1 + D_2 - \Lambda$ is big and nef on Y.*

PROOF. See [**BEL**], Claim 3.3. □

Let $M := \sigma^* H(-3E)$ on X. We have:

$$M = \sigma^* A(-E) + 2(\sigma^* B(-E))$$

Since A and B are non-trigonal S has no trisecant lines through P and does not contain a line through P whether it is embedded by $|A|$ or by $|B|$ (Proposition 3.2). Therefore every curvilinear subscheme of S of length 3 containing P imposes independent conditions to both $|A|$ and $|B|$. Then it follows from [**Co**], Prop. 1.3.4, that both $\sigma^* A(-E)$ and $\sigma^* B(-E)$ are very ample on X. Therefore by the Lemma we have that both $\sigma^* A(-E)_1 + \sigma^* A(-E)_2 - \Lambda$ and $\sigma^* B(-E)_1 + \sigma^* B(-E)_2 - \Lambda$ are big and nef. It follows that the divisor $M_1 + M_2 - 3\Lambda$ is big and nef, being the sum of three big and nef divisors.

Since $C \sim M + E$ we have $C_1 + C_2 \sim M_1 + M_2 + E_1 + E_2$ and therefore we have an exact sequence on Y:

$$0 \to \mathcal{O}_Y(K_Y + M_1 + M_2 - 3\Lambda) \to \mathcal{L} \to \mathcal{O}_{E_1 + E_2}(\mathcal{L}) \to 0$$

By Kawamata-Vieweg we have $H^1(Y, K_Y + M_1 + M_2 - 3\Lambda) = 0$: therefore in order to prove (3.1) it suffices to show that

(3.2) $$H^1(E_1 + E_2, \mathcal{O}_{E_1 + E_2}(\mathcal{L})) = 0$$

Letting $W := E_1 \cap E_2$ we have an exact sequence:

$$0 \to \mathcal{O}_{E_1}(\mathcal{L} - W) \to \mathcal{O}_{E_1 + E_2}(\mathcal{L}) \to \mathcal{O}_{E_2}(\mathcal{L}) \to 0$$

and, by symmetry, it suffices to prove that:

(3.3) $$H^1(E_1, \mathcal{O}_{E_1}(\mathcal{L} - W)) = 0$$

and

$$H^1(E_1, \mathcal{O}_{E_1}(\mathcal{L})) = 0$$

We can then consider the exact sequence on E_1:

$$0 \to \mathcal{O}_{E_1}(\mathcal{L} - W) \to \mathcal{O}_{E_1}(\mathcal{L}) \to \mathcal{O}_W(\mathcal{L}) \to 0$$

and finally we are reduced to prove (3.3) and

$$H^1(W, \mathcal{O}_W(\mathcal{L})) = 0 \tag{3.4}$$

Let $U \cong E \times E$ be the proper transform of $E \times E$ in Y. Then in E_1 we have $W = U + \Lambda_{|E_1}$. As in [**CLM2**], proof of Lemma (3.1), one shows that $\Lambda_{|E_1} \cong \mathbb{P}\mathcal{E}$, where $\mathcal{E} = \mathcal{O}_{\mathbb{P}^1}(1) \oplus \mathcal{O}_{\mathbb{P}^1}(-2)$, and $\mathcal{L}_{|\mathbb{P}\mathcal{E}} = \mathcal{O}_{\mathbb{P}\mathcal{E}}(2C_0 + 2f)$ where $C_0 \in |\mathcal{O}_{\mathbb{P}\mathcal{E}}(1)|$ and f is a fibre of $\mathbb{P}\mathcal{E} \to \mathbb{P}^1$. Now the proof proceeds as in [**CLM2**], Lemma (3.1), after having proved that $\mathcal{O}_{E_1}(\mathcal{L} - W) = \mathcal{O}_{E_1}(K_Y + M_1 + M_2 - 3\Lambda)$ is big and nef. This last fact is obtained exactly as in the proof of Lemma (3.10) of [**CLM2**], using the fact that both $\sigma^*A(-E)$ and $\sigma^*B(-E)$ are very ample on X. □

PROOF OF THEOREM 0.1. Recalling Lemma 1.1, the theorem follows immediately from Lemma 2.1 and from Proposition 3.1. □

References

[BF] Edoardo Ballico and Claudio Fontanari, *Gaussian maps, the Zak map and projective extensions of singular varieties*, Results Math. **44** (2003), no. 1-2, 29–34. MR2011902

[BFT] E. Ballico, C. Fontanari, and L. Tasin, *Singular curves on K3 surfaces*, Sarajevo J. Math. **6(19)** (2010), no. 2, 165–168. MR2757611

[B] Arnaud Beauville, *Fano threefolds and K3 surfaces*, The Fano Conference, Univ. Torino, Turin, 2004, pp. 175–184. MR2112574

[BM] A. Beauville and J.-Y. Mérindol, *Sections hyperplanes des surfaces K3* (French), Duke Math. J. **55** (1987), no. 4, 873–878. MR916124

[BEL] Aaron Bertram, Lawrence Ein, and Robert Lazarsfeld, *Vanishing theorems, a theorem of Severi, and the equations defining projective varieties*, J. Amer. Math. Soc. **4** (1991), no. 3, 587–602. MR1092845

[Ch] Chen X.: Nodal curves on K3 surfaces. arXiv:1611.07423.

[CHM] Ciro Ciliberto, Joe Harris, and Rick Miranda, *On the surjectivity of the Wahl map*, Duke Math. J. **57** (1988), no. 3, 829–858. MR975124

[CK] Ciro Ciliberto and Andreas Leopold Knutsen, *On k-gonal loci in Severi varieties on general K3 surfaces and rational curves on hyperkähler manifolds*, J. Math. Pures Appl. (9) **101** (2014), no. 4, 473–494. MR3179751

[CLM1] Ciro Ciliberto, Angelo Felice Lopez, and Rick Miranda, *On the corank of Gaussian maps for general embedded K3 surfaces*, Proceedings of the Hirzebruch 65 Conference on Algebraic Geometry (Ramat Gan, 1993), Israel Math. Conf. Proc., vol. 9, Bar-Ilan Univ., Ramat Gan, 1996, pp. 141–157. MR1360500

[CLM2] Ciro Ciliberto, Angelo Felice Lopez, and Rick Miranda, *On the Wahl map of plane nodal curves*, Complex analysis and algebraic geometry, de Gruyter, Berlin, 2000, pp. 155–163. MR1760876

[Co] Marc Coppens, *Very ample linear systems on blowings-up at general points of smooth projective varieties*, Pacific J. Math. **202** (2002), no. 2, 313–327. MR1887767

[DM] Jeanne Duflot and Rick Miranda, *The Gaussian map for rational ruled surfaces*, Trans. Amer. Math. Soc. **330** (1992), no. 1, 447–459. MR1061775

[EV] Hélène Esnault and Eckart Viehweg, *Lectures on vanishing theorems*, DMV Seminar, vol. 20, Birkhäuser Verlag, Basel, 1992. MR1193913

[FKP] Flaminio Flamini, Andreas Leopold Knutsen, and Gianluca Pacienza, *Singular curves on a K3 surface and linear series on their normalizations*, Internat. J. Math. **18** (2007), no. 6, 671–693. MR2337399

[FKPS] Flaminio Flamini, Andreas Leopold Knutsen, Gianluca Pacienza, and Edoardo Sernesi, *Nodal curves with general moduli on K3 surfaces*, Comm. Algebra **36** (2008), no. 11, 3955–3971. MR2460395

[Ga] Concettina Galati, *On the existence of curves with a triple point on a K3 surface*, Atti Accad. Naz. Lincei Rend. Lincei Mat. Appl. **23** (2012), no. 3, 295–317. MR2960839

[GK] Concettina Galati and Andreas Leopold Knutsen, *On the existence of curves with A_k-singularities on K3 surfaces*, Math. Res. Lett. **21** (2014), no. 5, 1069–1109. MR3294563

[Go] Tomás L. Gómez, *Brill-Noether theory on singular curves and torsion-free sheaves on surfaces*, Comm. Anal. Geom. **9** (2001), no. 4, 725–756. MR1868919

[Hal1] Mihai Halic, *Modular properties of nodal curves on $K3$ surfaces*, Math. Z. **270** (2012), no. 3-4, 871–887. MR2892928

[Hal2] Mihai Halic, *Erratum to: Modular properties of nodal curves on $K3$ surfaces [MR2892928]*, Math. Z. **280** (2015), no. 3-4, 1203–1211. MR3369374

[Har] Robin Hartshorne, *Algebraic geometry*, Springer-Verlag, New York-Heidelberg, 1977. Graduate Texts in Mathematics, No. 52. MR0463157

[Ke] Michael Kemeny, *The moduli of singular curves on K3 surfaces* (English, with English and French summaries), J. Math. Pures Appl. (9) **104** (2015), no. 5, 882–920. MR3405539

[Kn] Andreas Leopold Knutsen, *Smooth curves on projective $K3$ surfaces*, Math. Scand. **90** (2002), no. 2, 215–231. MR1895612

[L] Robert Lazarsfeld, *Brill-Noether-Petri without degenerations*, J. Differential Geom. **23** (1986), no. 3, 299–307. MR852158

[MM] Shigefumi Mori and Shigeru Mukai, *The uniruledness of the moduli space of curves of genus* 11, Algebraic geometry (Tokyo/Kyoto, 1982), Lecture Notes in Math., vol. 1016, Springer, Berlin, 1983, pp. 334–353. MR726433

[W1] Jonathan M. Wahl, *The Jacobian algebra of a graded Gorenstein singularity*, Duke Math. J. **55** (1987), no. 4, 843–871. MR916123

[W2] Jonathan Wahl, *Gaussian maps on algebraic curves*, J. Differential Geom. **32** (1990), no. 1, 77–98. MR1064866

Dipartimento di Matematica e Fisica, Università Roma Tre, L.go S.L. Murialdo 1, 00146 Roma

Email address: sernesi@gmail.com

Contemporary Mathematics
Volume **712**, 2018
http://dx.doi.org/10.1090/conm/712/14353

Skoda's ideal generation from vanishing theorem for semipositive Nakano curvature and Cauchy-Schwarz inequality for tensors

Yum-Tong Siu

Dedicated to Lawrence Ein

ABSTRACT. The Bochner-Kodaira technique of completion of squares yields vanishing theorems and L^2 estimates of $\bar{\partial}$. Skoda's ideal generation, which is a crucial ingredient in the analytic approach to the finite generation of the canonical ring and the abundance conjecture, requires specially tailored analytic techniques for its proof. We introduce a new method of deriving Skoda's ideal generation which makes it and its formulation a natural consequence of the standard techniques of vanishing theorems and L^2 estimates of $\bar{\partial}$. Our method of derivation readily gives also other similar results on ideal generation. An essential rôle is played by one particular Cauchy-Schwarz inequality for tensors with a special factor which accounts for the exponent of the denominator in the formulation of the integral condition for Skoda's ideal generation.

Introduction

Since the nineteen sixties the Bochner-Kodaira technique of completion of squares, for manifolds either compact or with pseudoconvex boundary condition, has been a most important tool in algebraic and complex geometry. Sometimes ingenious *ad hoc* adaptations are required in its use. A remarkable example is the 1972 result of Skoda on ideal generation (Théorème 1 on pp.555–556 of [**3**]), which is a crucial ingredient in the analytic approach to the finite generation of the canonical ring and the abundance conjecture. Up to this point, special analytic techniques developed by Skoda, other than applications of the usual vanishing theorems and L^2 estimates for the $\bar{\partial}$-equation, are required for its proof.

In recent years analytic results from vanishing theorems and solvability of $\bar{\partial}$-equation have contributed to the solution of a number of longstanding open problems in algebraic geometry. The interaction between algebraic geometry and analytic methods in several complex variables and partial differential equations has been a very active and productive area of investigation. For such an interaction it is advantageous to minimize the use of *ad hoc* analytic methods in favor of approaches which are more amenable to adaptation to formulations in algebraic geometry.

In this note we give a simpler, more straightforward proof of Skoda's result on ideal generation which makes it and its formulation a natural consequence of

2010 *Mathematics Subject Classification.* Primary 32F32.

the standard techniques in vanishing theorems and solving $\bar\partial$-equation with L^2 estimates. Our proof readily gives other similar results on ideal generation. In §6 we present a number of such similar results. Our proof uses the following three ingredients.

(I) One particular Cauchy-Schwarz inequality for tensors

$$\left|\langle \mathbf{S}, \mathbf{T}\rangle_{\mathbb{C}^r\otimes\mathbb{C}^n}\right|^2 \leq \min(r,n)\,\|\langle \mathbf{S}, \mathbf{T}\rangle_{\mathbb{C}^n}\|^2_{\mathbb{C}^r\otimes\mathbb{C}^r}$$

for elements $\mathbf{S}$ and $\mathbf{T}$ of $\mathbb{C}^r\otimes\mathbb{C}^n$ with the special factor $\min(r,n)$. Detailed statement and proof will be given in §2. The usual Cauchy-Schwarz inequality is the special case of $n=1$. Our particular Cauchy-Schwarz inequality for tensors *with the factor* $\min(r,n)$ *on the right-hand side* is needed in order to handle the curvature operator condition for nonnegative Nakano curvature. Just like the usual Cauchy-Schwarz inequality, it is derived by a straightforward simple reduction to certain special orthonormal situation. The significance of our Cauchy-Schwarz inequality is the factor $\min(r,n)$ which plays an essential rôle the exponent of the denominator in the formulation of the integral condition for Skoda's ideal generation. Though $\mathbb{C}^r\otimes\mathbb{C}^n$ can be identified with $\mathbb{C}$-linear operators between the vector spaces $\mathbb{C}^r$ and $\mathbb{C}^n$ and there are generalizations of the usual Cauchy-Schwarz inequality to bounded $\mathbb{C}$-linear operators between Hilbert spaces, yet the nature of existent generalizations is different and the essential factor $\min(r,n)$ does not occur readily in them without repeating the kind of arguments used in deriving our Cauchy-Schwarz inequality (see Remark 2.3). Our Cauchy-Schwarz inequality for tensors should be a simple case of more general inequalities of Cauchy-Schwarz type for general tensors with special factors (obtained by reduction to orthonormal situations in Young tableau). In §7 where the manipulations in computation between our proof and Skoda's original proof are compared, one such inequality of Cauchy-Schwarz type for more general tensors is given.

(II) Nonnegativity of the Nakano curvature of the metric for the kernel bundle of $(g_1,\cdots,g_p):(\mathcal{O}_\Omega)^{\oplus p}\to\mathcal{O}_\Omega$ which is induced from the standard flat metric of $(\mathcal{O}_\Omega)^{\oplus p}$ and twisted by the weight function $\left(\sum_{j=1}^q |g_j|^2\right)^{-q}$, where Ω is a domain in $\mathbb{C}^n$ and $q=\min(n,p-1)$. This is verified by a simple straightforward computation in normal fiber coordinates at the point under consideration and the use of the above Cauchy-Schwarz inequality for tensors. Detailed statement and proof will be given in §3.

(III) Vanishing theorem (from solving a $\bar\partial$-equation with L^2 estimate) for a holomorphic vector bundle V with nonnegative Nakano curvature Θ on a strictly pseudoconvex domain Ω in an ambient complex manifold of complex dimension n for smooth $\bar\partial$-closed V-valued $(n,1)$-form F with finite $\left(\Theta^{-1}F,\,F\right)_{L^2(\Omega)}$, where Θ^{-1} is defined as the limit, as the positive number ε approaches 0, of the inverse of $\Theta+\varepsilon I$ with I being the identity operator. Actually what is needed is only the special case where X is a Stein domain spread over $\mathbb{C}^n$ where the canonical line bundle K_X is trivial and F can be replaced by a smooth $\bar\partial$-closed V-valued $(0,1)$-form. Detailed statement and proof will be given in §4. Usually an application of the technique of vanishing theorems to produce holomorphic sections indispensably requires strict positivity of curvature at least at some point. The special feature of the application here of the vanishing theorem from solving a $\bar\partial$-equation with L^2

estimate requires only semipositive Nakano curvature and *no strict positivity at any point*. The difficulty of producing holomorphic sections by solving $\bar{\partial}$-equation with L^2 estimate without positivity of curvature at any point is that it is not possible to get a usable right-hand side for the $\bar{\partial}$-equation by applying $\bar{\partial}$ to the product of a cut-off function and a local holomorphic section. For the problem of ideal generation, the nonnegative curvature condition is sufficient for the application, because there is a natural usable choice for the right-hand side of the $\bar{\partial}$-equation. Another known result requiring only nonnegative curvature condition is the extension result of Ohsawa-Takegoshi. Applications to produce holomorphic sections from only nonnegative curvature condition remain mostly an area yet to be explored.

Notations. The set of all positive integers is denoted by $\mathbb{N}$. We will use the notation $\mathcal{O}_X$ to denote the structure sheaf of X. To minimize the use of notations we will loosely refer to the locally free sheaf associated to a holomorphic vector bundle simply as a holomorphic vector bundle and vice versa. For example, we use the notation $(\mathcal{O}_{\mathbb{C}^n})^{\oplus p}$ to denote both the globally trivial vector bundle of rank p over $\mathbb{C}^n$ and the direct sum of p copies of the structure sheaf $\mathcal{O}_{\mathbb{C}^n}$ of $\mathbb{C}^n$. We use $\Gamma(Y, W)$ (or $\Gamma(Y, \mathcal{F})$) to denote the space of holomorphic sections of a bundle W (or a coherent sheaf $\mathcal{F}$) over Y.

For a holomorphic vector bundle V of rank r with a metric $H_{j\bar{k}}$ over a complex manifold X of complex dimension n with a Kähler metric $g_{\lambda\bar{\nu}}$, we will use the Latin letters j, k, i, ℓ etc. for the indices of the fiber coordinates of V and use the Greek letters λ, ν, μ, ρ etc. for the local coordinates $z^1, \cdots, z^n$ of X. We use $g^{\bar{\nu}\lambda}$ to mean the inverse of $g_{\lambda\bar{\nu}}$ and use $H^{\bar{k}j}$ to mean the inverse of $H_{j\bar{k}}$. Unless there is a possible confusion, without explicit mention we will use $H^{\bar{k}j}$ and $g^{\bar{\nu}\lambda}$ to raise the subscripts of tensors and use $H_{j\bar{k}}$ and $g_{\lambda\bar{\nu}}$ to lower the superscripts of tensors. The notation T_X denotes the (holomorphic) tangent bundle of X. The notation K_X denotes the canonical line bundle of X.

For a holomorphic subbundle W of V, when we use the metric of W which is induced from the metric $H_{j\bar{k}}$ of V, we simply say the metric $H_{j\bar{k}}$ of W instead of introducing a new notation.

When we refer to a positive $(1,1)$-form $\sqrt{-1}\sum_{1\leq\lambda,\nu\leq n}\eta_{\lambda\bar{\nu}}dz^\lambda \wedge \overline{dz^\nu}$, we will drop the factor $\sqrt{-1}$ and simply say that the $(1,1)$-form $\eta_{\lambda\bar{\nu}}$ is positive.

The notation ∂_λ means $\frac{\partial}{\partial z^\lambda}$ and the notation $\partial_{\bar{\lambda}}$ means $\frac{\partial}{\partial \overline{z^\lambda}}$. The notation $(\cdot,\cdot)$ or $\langle\cdot,\cdot\rangle$ means the inner product and the notation $\|\cdot\|$ means the norm, sometimes with a subscript to indicate in which space the inner product or the norm is taken.

The study of vanishing theorems and L^2 estimates for the $\bar{\partial}$-equation has a very long history and a very extensive literature. To avoid a long bibliography which nowadays can easily be fetched from readily available online searchable databases, we keep to a minimum the listing of references at the end of this note.

1. Algebraic Formulation and Application of Skoda's Ideal Generation

Though this is not part of our proof of Skoda's result on ideal generation, to draw attention to its use in algebraic geometric problems, we give here the algebraic geometric formulation of Skoda's result and how it is used in the analytic approach to the finite generation of the canonical ring before we present our proof.

We state in the following way a trivially more general form of Skoda's result than is given in Théorème 1 on pp.555–556 of [**3**] by using a Stein domain spread over $\mathbb{C}^n$ instead of a Stein domain in $\mathbb{C}^n$, in order to be able to formulate it in an algebraic geometric setting.

THEOREM 1.1 (Skoda's Theorem on Ideal Generation). *Let Ω be a domain spread over $\mathbb{C}^n$ which is Stein. Let ψ be a plurisubharmonic function on Ω, $g_1, \ldots, g_p$ be holomorphic functions on Ω (not all identically zero), $\alpha > 1$, $q = \min(n, p-1)$, and $|g|^2 = \sum_{j=1}^p |g_j|^2$. Then for any holomorphic function f on Ω with*

$$\int_\Omega \frac{|f|^2 e^{-\psi}}{|g|^{2(\alpha q+1)}} < \infty,$$

there exist holomorphic functions $h_1, \ldots, h_p$ on Ω with $f = \sum_{j=1}^p h_j g_j$ on Ω such that

$$\int_\Omega \frac{\sum_{k=1}^p |h_k|^2 e^{-\psi}}{|g|^{2\alpha q}} \leq \frac{\alpha}{\alpha - 1} \int_\Omega \frac{|f|^2 e^{-\psi}}{|g|^{2(\alpha q+1)}}$$

holds.

The following algebraic geometric formulation of Skoda's result is easily obtained by representing the compact complex algebraic manifold X of complex dimension n minus some complex hypersurface as a Stein domain over $\mathbb{C}$ and considering a holomorphic line bundle over X minus some complex hypersurface as globally trivial by division by a meromorphic section of it.

THEOREM 1.2 (Algebraic Geometric Formulation of Ideal Generation). *Let X be a compact complex algebraic manifold of complex dimension n and L and E be respectively holomorphic line bundles on X with (possibly singular) Hermitian metrics $e^{-\varphi_L}$ and $e^{-\varphi_E}$ such that φ_L and φ_E are plurisubharmonic. Let $s_1, \cdots, s_p \in \Gamma(X, L)$ and $0 < \gamma \leq 1$ and $q = \max(n, p-1)$. Then for any $s \in \Gamma(X, (q+2)L + E + K_X)$ with*

$$C_s := \int_X \frac{|s|^2 e^{-(1-\gamma)\varphi_L - \varphi_E}}{\left(\sum_{j=1}^p |s_j|^2\right)^{q+1+\gamma}} < \infty,$$

there exist $h_1, \cdots, h_p \in \Gamma(X, ((q+1)L + E + K_X)$ such that $s = \sum_{j=1}^p h_j s_j$ with

$$\int_X \frac{\sum_{k=1}^p |h_k|^2 e^{-(1-\gamma)\varphi_L - \varphi_E}}{\left(\sum_{j=1}^p |s_j|^2\right)^{q+\gamma}} \leq \left(1 + \frac{q}{\gamma}\right) C_s.$$

THEOREM 1.3 (Finite Generation of Section Module). *Let L be a holomorphic line bundle over a compact complex algebraic manifold X of complex dimension n. Let $s_1^{(m)}, \cdots, s_{q_m}^{(m)}$ be a $\mathbb{C}$-basis of $\Gamma(X, mL)$. Let N_0 and m_0 be positive integers such that $N_0 \geq m_0(n+2)$ and $\Gamma(X, m_0 L) \neq 0$. Let ε_m (for $m \in \mathbb{N}$) be a sequence of positive numbers which decrease rapidly enough to yield the convergence of the infinite series*

$$\Phi := \sum_{m=1}^\infty \varepsilon_m \sum_{j=1}^{q_m} \left| s_j^{(m)} \right|^{\frac{2}{m}}$$

on X. Suppose there exist elements $s_1^{(m_0)}, \cdots, s_{q_{m_0}}^{(m_0)}$ of $\Gamma(X, m_0 L)$ such that for any $m \geq N_0$ and for any $s^{(m)} \in \Gamma(X, mL + K_X)$ the (locally defined) function

$$\frac{\left|s^{(m)}\right|^2}{\Phi^{m-(n+2)m_0}\left(\sum_{j=1}^{q_{m_0}}\left|s_j^{(m_0)}\right|^2\right)^{n+2}} \tag{1.1}$$

is locally bounded on X. Then for $m \geq N_0$,

$$\Gamma(X, mL+K_X) = \sum_{\nu_1+\cdots+\nu_{q_{m_0}}=\ell_m} \left(s_1^{(m_0)}\right)^{\nu_1}\cdots\left(s_{q_{m_0}}^{(m_0)}\right)^{\nu_{q_{m_0}}}\Gamma\left((m-m_0\ell_m)L+K_X\right),$$

where ℓ_m the largest integer with $m - \ell_m m_0 \geq N_0$.

In particular, if one denotes by $R(X, F)$ the ring $\bigoplus_{k=1}^{\infty}\Gamma(X, kF)$ of all global holomorphic sections of positive tensor powers of a line bundle F over X, then the module $\bigoplus_{m=1}^{\infty}\Gamma(X, mL+K_X)$ is generated by a finite number of elements over the ring $R(X, m_0 L)$ and the finite number of elements can be taken to be a $\mathbb{C}$-basis of the finite-dimensional $\mathbb{C}$-vector space $\bigoplus_{m=1}^{N_0}\Gamma(X, mL+K_X)$. In other words,

$$\bigoplus_{m=1}^{\infty}\Gamma(X, mL+K_X) = R(X, m_0 L)\left(\bigoplus_{m=1}^{N_0}\Gamma(X, mL+K_X)\right).$$

In the special case of $L = K_X$, under the condition of the local boundedness of (1.1) the canonical ring $R(X, K_X)$ is finitely generated.

Proof. Let $e^{-\varphi_L} = \frac{1}{\Phi}$. For $m \geq N_0$ and any $s \in \Gamma(X, mL+K_X)$, by Theorem 1.2 from

$$\int_X \frac{|s|^2 e^{-(m-(n+2)m_0)\varphi_L}}{\left(\sum_{j=1}^{q_{m_0}}\left|s_j^{(m_0)}\right|^2\right)^{n+2}} < \infty,$$

it follows that there exist

$$h_1, \cdots, h_{q_{m_0}} \in \Gamma\left(X, (m-m_0)L\right)$$

such that $s = \sum_{j=1}^{q_{m_0}} h_j s_j^{(m_0)}$. If $m - (n+2)m_0$ is still not less than N_0, we can apply the argument to each h_j instead of s until by induction on ν we get

$$h_1^{(j_1,\cdots,j_\nu)}, \cdots, h_{q_{m_0}}^{(j_1,\cdots,j_\nu)} \in \Gamma\left(X, (m - m_0(\nu+1))L + K_X\right)$$

for $1 \leq j_1, \cdots, j_\nu \leq q_{m_0}$ with $1 \leq \nu \leq \ell_m$ such that

$$s = \sum_{1\leq j_1,\cdots,j_\nu\leq q_{m_0}} h_{j_\nu}^{(j_1,\cdots,j_{\nu-1})}\prod_{\lambda=1}^{\nu} s_{j_\lambda}^{(m_0)}$$

with $h_j^{(0)} = h_j$ for $1 \leq j \leq q_{m_0}$. □

We now introduce the three ingredients for our proof of Skoda's ideal generation.

2. Cauchy-Schwarz Inequality for Tensors

The following Cauchy-Schwarz inequality for tensors is introduced to handle inequalities for curvature operator conditions. Its proof will be done by applying the usual Cauchy-Schwarz inequality for vectors and a reduction to the special case of an orthonormal set of vectors by linear transformations.

PROPOSITION 2.1 (Cauchy-Schwarz Inequality for Tensors). *Let* $\mathbf{S}$ *and* $\mathbf{T}$ *be elements of* $\mathbb{C}^r \otimes \mathbb{C}^n$. *Then*

$$\left|\langle \mathbf{S},\, \mathbf{T}\rangle_{\mathbb{C}^r\otimes\mathbb{C}^n}\right|^2 \leq \min(r,n)\, \|\langle \mathbf{S},\, \mathbf{T}\rangle_{\mathbb{C}^n}\|^2_{\mathbb{C}^r\otimes\mathbb{C}^r}\,.$$

REMARK 2.2. When the tensors $\mathbf{S}$ and $\mathbf{T}$ are represented by $r\times n$ matrices

$$\mathbf{S} = \left(s_{\ell,k}\right)_{1\leq\ell\leq r,1\leq k\leq n} \quad \text{and} \quad \mathbf{T} = \left(t_{\ell,k}\right)_{1\leq\ell\leq r,1\leq k\leq n},$$

the inequality in Proposition 2.1 becomes

$$\left|\sum_{1\leq\ell\leq r,1\leq k\leq n} s_{\ell,k}t_{\ell,k}\right|^2 \leq \min(r,n) \sum_{1\leq m,\ell\leq r}\left|\sum_{k=1}^n s_{m,k}t_{\ell,k}\right|^2.$$

The special case of $n=1$ is simply the following usual Cauchy-Schwarz inequality for vectors

$$\left|\sum_{1\leq\ell\leq r} s_\ell t_\ell\right|^2 \leq \left(\sum_{1\leq\ell\leq r}|s_\ell|^2\right)\left(\sum_{1\leq\ell\leq r}|t_\ell|^2\right)$$

where $s_\ell = s_{\ell,1}$ and $t_\ell = t_{\ell,1}$.

PROOF OF PROPOSITION 2.1. Let

$$\mathbf{S} = \left(s_{\ell,k}\right)_{1\leq\ell\leq r,\,1\leq k\leq n} \quad \text{and} \quad \mathbf{T} = \left(t_{\ell,k}\right)_{1\leq\ell\leq r,\,1\leq k\leq n}.$$

The inequality

$$\left|\langle \mathbf{S},\, \mathbf{T}\rangle_{\mathbb{C}^r\otimes\mathbb{C}^n}\right|^2 \leq r\, \|\langle \mathbf{S},\, \mathbf{T}\rangle_{\mathbb{C}^n}\|_{\mathbb{C}^r\otimes\mathbb{C}^r}$$

is the same as

$$\left|\sum_{1\leq\ell\leq r,1\leq k\leq n} s_{\ell,k}t_{\ell,k}\right|^2 \leq r \sum_{1\leq m,\ell\leq r}\left|\sum_{k=1}^n s_{m,k}t_{\ell,k}\right|^2,$$

which simply follows from the following application of the Cauchy-Schwarz inequality with focus on the summation over the double index (ℓ, m) (where $\delta_{\ell m}$

for $1 \leq \ell, m \leq r$ is the Kronecker delta)

$$
\begin{aligned}
&\left| \sum_{1\leq \ell \leq r, 1\leq k\leq n} s_{\ell,k} t_{\ell,k} \right|^2 \\
&= \left| \sum_{1\leq \ell, m \leq r, 1\leq k\leq n} \delta_{\ell m} s_{m,k} t_{\ell,k} \right|^2 \\
&= \left| \sum_{1\leq \ell, m\leq r} \delta_{\ell m} \left(\sum_{k=1}^n s_{m,k} t_{\ell,k} \right) \right|^2 \\
&\leq \left(\sum_{1\leq m,\ell\leq r} (\delta_{m\ell})^2 \right) \left(\sum_{1\leq m,\ell\leq r} \left| \sum_{k=1}^n s_{m,k} t_{\ell,k} \right|^2 \right) \\
&= r \sum_{1\leq m,\ell\leq r} \left| \sum_{k=1}^n s_{m,k} t_{\ell,k} \right|^2 .
\end{aligned}
$$

The inequality

$$\left|\langle \mathbf{S}, \mathbf{T}\rangle_{\mathbb{C}^r\otimes\mathbb{C}^n}\right|^2 \leq n \left\| \langle \mathbf{S}, \mathbf{T}\rangle_{\mathbb{C}^n} \right\|_{\mathbb{C}^r\otimes\mathbb{C}^r}$$

is the same as

$$\left| \sum_{1\leq \ell\leq r, 1\leq k\leq n} s_{\ell,k} t_{\ell,k} \right|^2 \leq n \sum_{1\leq m,\ell\leq r} \left| \sum_{k=1}^n s_{m,k} t_{\ell,k} \right|^2 .$$

Instead of proving this inequality we prove the following equivalent inequality which is obtained by replacing $t_{\ell,k}$ by its complex conjugate $\overline{t_{\ell,k}}$.

$$\left| \sum_{1\leq \ell\leq r, 1\leq k\leq n} s_{\ell,k} \overline{t_{\ell,k}} \right|^2 \leq n \sum_{1\leq m,\ell\leq r} \left| \sum_{k=1}^n s_{m,k} \overline{t_{\ell,k}} \right|^2 . \tag{2.1}$$

Denote the n column r-vectors of the $r \times n$ matrix

$$\mathbf{S} = (s_{\ell,k})_{1\leq \ell\leq r, 1\leq k\leq n}$$

by $\vec{s}_1, \cdots, \vec{s}_n$. Likewise we denote the n column r-vectors of the $r \times n$ matrix

$$\mathbf{T} = (t_{\ell,k})_{1\leq \ell\leq r, 1\leq k\leq n}$$

by $\vec{t}_1, \cdots, \vec{t}_n$.

To prove (2.1) it suffices to assume that $n \leq r$ and by slightly perturbing $\vec{s}_1, \cdots, \vec{s}_n$ and then taking the limit to go back to the original set of $\vec{s}_1, \cdots, \vec{s}_n$, we can assume without loss of generality that the vectors $\vec{s}_1, \cdots, \vec{s}_n$ of $\mathbb{C}^r$ are $\mathbb{C}$-linearly independent. We apply a $\mathbb{C}$-linear transformation given by an $n \times n$ matrix R to the n elements $\vec{t}_1, \cdots, \vec{t}_n$ of $\mathbb{C}^r$ and at the same time the $\mathbb{C}$-linear transformation given by the $n \times n$ matrix $\overline{(R^t)^{-1}}$, which is the complex-conjugate transpose inverse of R, to the n elements $\vec{t}_1, \cdots, \vec{t}_n$ of $\mathbb{C}^r$ in order to preserve the inner product

$$\sum_{k=1}^n s_{j,k} \overline{t_{\ell,k}}$$

for any fixed $1 \leq j, \ell \leq r$. So without loss of generality we can assume that the set of n elements $\vec{s}_1, \cdots, \vec{s}_n$ of $\mathbb{C}^r$ are orthonormal with respect to $\langle \cdot, \cdot \rangle_{\mathbb{C}^r}$ and explicitly given by $s_{j,k}$ being equal to the Kronecker delta δ_{jk} for $1 \leq j, k \leq n$ and equal to 0 for $n < j \leq r$. In particular, the expression

$$\sum_{1 \leq m, \ell \leq r} \left| \sum_{k=1}^{n} s_{m,k} \overline{t_{\ell,k}} \right|^2$$

is equal to

$$\sum_{\ell=1}^{r} \sum_{m=1}^{n} \left| \sum_{k=1}^{n} \delta_{mk} \overline{t_{\ell,k}} \right|^2 = \sum_{\ell=1}^{r} \sum_{m=1}^{n} \left| \overline{t_{\ell,m}} \right|^2 = \sum_{k=1}^{n} \| \vec{t}_k \|^2_{\mathbb{C}^r} .$$

The left-hand of the inequality (2.1) can be rewritten as

$$\left| \sum_{1 \leq \ell \leq r, 1 \leq k \leq n} s_{\ell,k} \, \overline{t_{\ell,k}} \right|^2 = \left| \sum_{k=1}^{n} \langle \vec{s}_k, \vec{t}_k \rangle_{\mathbb{C}^r} \right|^2,$$

which by the Cauchy-Schwarz inequality is

$$\begin{aligned} &\leq n \sum_{k=1}^{n} \left| \langle \vec{s}_k, \vec{t}_k \rangle_{\mathbb{C}^r} \right|^2 \leq n \sum_{k=1}^{n} \| \vec{t}_k \|^2_{\mathbb{C}^r} \\ &= n \sum_{1 \leq m, \ell \leq r} \left| \sum_{k=1}^{n} s_{m,k} \overline{t_{\ell,k}} \right|^2 . \end{aligned}$$

□

REMARK 2.3. Elements of $\mathbb{C}^r \otimes \mathbb{C}^n$ can naturally be identified with $\mathbb{C}$-linear operators between the vector spaces $\mathbb{C}^r$ and $\mathbb{C}^n$. In the literature there are generalizations of the usual Cauchy-Schwarz inequality to bounded $\mathbb{C}$-linear operators between Hilbert spaces. For example, Lemma 2.4 on p.193 of [**1**] gives the following generalization of the usual Cauchy-Schwarz inequality

$$\left\| \sum_{j=1}^{n} a_j \otimes \overline{b_j} \right\|_{H \otimes \overline{K}} \leq \left\| \sum_{j=1}^{n} a_j \otimes \overline{a_j} \right\|_{H \otimes \overline{H}} \left\| \sum_{j=1}^{n} b_j \otimes \overline{b_j} \right\|_{K \otimes \overline{K}},$$

where $a_1, \cdots, a_n$ (respectively $b_1, \cdots, b_n$) are bounded $\mathbb{C}$-linear operators on the Hilbert space H (respectively K) over $\mathbb{C}$ (with appropriate interpretation for taking complex-conjugates). The norm

$$\left\| \sum_{j=1}^{n} a_j \otimes \overline{b_j} \right\|_{H \otimes \overline{K}}$$

means the norm of the $\mathbb{C}$-linear operator $\sum_{j=1}^{n} a_j \otimes \overline{b_j}$ on the Hilbert space $H \otimes \overline{K}$, which is the supremum of the norm of the image of any element of unit norm in $H \otimes \overline{K}$. However, the nature of such generalizations is different and the essential factor $\min(r, n)$ in the inequality in Proposition 2.1 is rather delicate and does not occur readily in such generalizations without repeating the kind of arguments used in the proof of Proposition 2.1.

3. Nakano Curvature of Bundle and Twisting by Metric of Trivial Line Bundle

We now straightforwardly compute the Nakano curvature in the case of a special subbundle of rank $p-1$ in the trivial vector bundle of rank p whose induced metric is twisted by a special weight function (*i.e.*, a metric of the trivial line bundle) to conclude the nonnegativity of the Nakano curvature.

3.1. *Nakano Curvature of Vector Bundles and Subbundles.* Let U be a domain in $\mathbb{C}^n$ with coordinates $z^1, \cdots, z^n$. Let V be a holomorphic vector bundle of rank r on U with smooth Hermitian metric H whose components are $H_{j\bar{k}}$ (for $1 \leq j, k \leq r$). We use the notation $\Theta^V_{j\bar{k}\lambda\bar{\nu}}$ (for $1 \leq j, k \leq r$ and $1 \leq \lambda, \nu \leq n$) to denote the components of the curvature tensor Θ^V of V which is defined as

$$\Theta^V_{j\bar{k}\lambda\bar{\nu}} = -\partial_\lambda \partial_{\bar{\nu}} H_{j\bar{k}} + \sum_{1 \leq i, \ell \leq r} H^{\bar{\ell} i} \left(\partial_\lambda H_{k\bar{\ell}}\right) \overline{\left(\partial_\nu H_{i\bar{j}}\right)}.$$

When there is no possible confusion, we will simply use $\Theta_{j\bar{k}\lambda\bar{\nu}}$ to denote $\Theta^V_{j\bar{k}\lambda\bar{\nu}}$ and use Θ to denote Θ^V.

The sign convention is chosen here so that when $r = 1$, the nonnegativity of the curvature of the metric $e^{-\varphi}$ means the plurisubharmonicity of the function φ.

When we have a holomorphic subbundle W of V with rank $1 \leq s < r$ and local frame $e_1, \cdots, e_s$ so that $e_1, \cdots, e_r$ is orthonormal at the point P under consideration (after the application of an P-dependent element of $GL(r, \mathbb{C})$), the curvature $\Theta^W_{j\bar{k}\lambda\bar{\nu}}$ for W (for $1 \leq j, k \leq s$ and $1 \leq \lambda, \nu \leq n$) is given by

$$\begin{aligned} \Theta^W_{j\bar{k}\lambda\bar{\nu}} &= -\partial_\lambda \partial_{\bar{\nu}} H_{j\bar{k}} + \sum_{\ell=1}^{s} \left(\partial_\lambda H_{j\bar{\ell}}\right) \overline{\left(\partial_\nu H_{\ell\bar{k}}\right)} \\ &= \Theta^V_{j\bar{k}\lambda\bar{\nu}} - \sum_{\ell=s+1}^{r} \left(\partial_\lambda H_{j\bar{\ell}}\right) \overline{\left(\partial_\nu H_{\ell\bar{k}}\right)} \end{aligned} \tag{3.1}$$

when W is given the metric induced from the metric of V.

By the Nakano curvature of V we mean the Hermitian form on $V \otimes T_U$ defined by

$$\left(v^{j\lambda}\right)_{1 \leq j \leq r, 1 \leq \lambda \leq n} \mapsto \sum_{1 \leq j, k \leq r, 1 \leq \lambda, \nu \leq n} \Theta^V_{j\bar{k}\lambda\bar{\nu}} v^{j\lambda} \overline{v^{k\nu}}$$

for $\left(v^{j\lambda}\right)_{1 \leq j \leq r, 1 \leq \lambda \leq n}$ in $V \otimes T_U$. This Hermitian form was introduced by S. Nakano in (2.10) on p.8 of [**2**]. When there is no possible confusion, we will simply use Θ^V or $\Theta^V_{j\bar{k}\lambda\bar{\nu}}$ to mean this Hermitian form on $V \otimes T_U$. We also use Θ^V to mean the self-adjoint operator on the vector bundle $V \otimes T_U$ and use $\left(\Theta^V\right)^{-1}$ to mean the self-adjoint operator on $V \otimes T_U$ which is the inverse of the self-adjoint operator Θ^V when Θ^V is strictly positive. When the self-adjoint operator Θ^V is only semipositive, we use the notation $\left(\Theta^V\right)^{-1}$ to mean the limit of $\left(\Theta^V + \varepsilon I\right)^{-1}$ as the positive number ε approaches 0, where I is the identity operator of $V \otimes T_U$. We will use this limit $\left(\Theta^V\right)^{-1}$ in the context of the inner product $\left(\left(\Theta^V\right)^{-1} F, F\right)$ for an element F of $V \otimes T_U$ which, defined as $\lim_{\varepsilon \to 0+} \left(\left(\Theta^V + \varepsilon I\right)^{-1} F, F\right)$, is allowed to assume the value $+\infty$.

For another element F' of $V\otimes T_U$ (over the same point of U), by letting $\varepsilon\to 0+$ in the Cauchy-Schwarz inequality

$$
\begin{aligned}
|(F,F')|^2 &= \left|\left(\left(\Theta^V+\varepsilon I\right)^{\frac{1}{2}}\left(\Theta^V+\varepsilon I\right)^{-\frac{1}{2}}F,\,F'\right)\right|^2\\
&= \left|\left(\left(\Theta^V+\varepsilon I\right)^{-\frac{1}{2}}F,\,\left(\Theta^V+\varepsilon I\right)^{\frac{1}{2}}F'\right)\right|^2\\
&\le \left\|\left(\Theta^V+\varepsilon I\right)^{-\frac{1}{2}}F\right\|^2\left\|\left(\Theta^V+\varepsilon I\right)^{\frac{1}{2}}F'\right\|^2\\
&= \left(\left(\Theta^V+\varepsilon I\right)^{-1}F,\,F\right)\left(\left(\Theta^V+\varepsilon I\right)F',\,F'\right),
\end{aligned}
$$

we conclude that

$$
|(F,F')|^2 \le \left(\left(\Theta^V\right)^{-1}F,\,F\right)\left(\left(\Theta^V\right)F',\,F'\right). \tag{3.2}
$$

This inequality (3.2) of Cauchy-Schwarz type also clearly holds for any semipositive tensors of the same tensor type as a Nakano curvature tensor.

Note that for a V-valued $(0,1)$-form F at a point of U, in the computation of $\left(\Theta^V\right)^{-1}F$ and $\left(\Theta^V\right)F$, the V-valued $(0,1)$-form F is naturally identified with an element of $V\otimes T_U$ at that point with the raising of the barred subscript of F to an unbarred superscript. The same kind of manipulation involving Nakano curvature and inner product is used when F is a V-valued $(n,1)$-form instead of a V-valued $(0,1)$-form.

When there is another tensor Ξ with components $\Xi_{j\bar{k}\lambda\bar{\nu}}$ which defines a self-adjoint operator on $V\otimes T_U$, we say that Ξ is is dominated by Θ^V or $\Xi\le\Theta^V$ if the Hermitian form on $V\otimes T_U$ defined by Ξ is $\le$ the Hermitian form on $V\otimes T_U$ defined by Θ^V.

For a $(1,1)$-form $\omega=\sum_{\lambda,\nu=1}^n\omega_{\lambda\bar{\nu}}dz^\lambda\wedge\overline{dz^\nu}$ we use the notation $H\otimes\omega$ to denote the tensor whose components are $H_{j\bar{k}}\omega_{\lambda\bar{\nu}}$.

PROPOSITION 3.1 (Nakano Curvature of Kernel Subbundle in Normal Fiber Coordinate). *Let $g_1,\cdots,g_p$ be holomorphic functions on a domain Ω in $\mathbb{C}^n$ and let $\mathcal{K}$ be the kernel of the bundle-homomorphism $(g_1,\cdots,g_p):(\mathcal{O}_\Omega)^{\oplus p}\to\mathcal{O}_\Omega$ which is given the metric induced from the standard flat metric of the trivial $\mathbb{C}$-vector bundle $(\mathcal{O}_\Omega)^{\oplus p}$ of rank p. Let P be a point of Ω with $g_1(P)=\cdots=g_{p-1}(P)=0$ and $g_p(P)=1$. Then at the point P the Nakano curvature $\Theta_{j\bar{k}\lambda\bar{\nu}}$ of the kernel vector subbundle $\mathcal{K}$ of rank $p-1$ is given by*

$$
\begin{aligned}
\left(v^{j\lambda}\right)_{1\le j\le p-1,1\le\lambda\le n}\mapsto &\sum_{1\le j,k\le p-1,1\le\lambda,\nu\le n}\Theta_{j\bar{k}\lambda\bar{\nu}}v^{j\lambda}\overline{v^{k\nu}}\\
&=-\left|\sum_{1\le j\le p-1,1\le\lambda\le n}\left(\partial_\lambda g_j\right)v^{j\lambda}\right|^2
\end{aligned}
$$

for $\left(v^{j\lambda}\right)_{1\le j\le p-1,1\le\lambda\le n}$.

PROOF. At points of Ω where $g_1,\cdots,g_p$ are not all zero, the kernel subbundle $\mathcal{K}$ is spanned by the local basis of $p-1$ local holomorphic sections

$$
e_j=\left(0,\cdots,0,\underbrace{g_p}_{j\text{ th}},0,\cdots,0,-g_j\right)
$$

for $1 \leq j \leq p-1$ so that the metric induced on $\mathcal{K}$ by the trivial $\mathbb{C}$-vector bundle $(\mathcal{O}_\Omega)^{\oplus p}$ is given by

$$H_{j\bar{k}} = \langle e_j, e_k \rangle = \delta_{jk} |g_p|^2 + g_j \overline{g_k} \tag{3.3}$$

for $1 \leq j, k \leq p-1$. Let $e_p = (g_1, \cdots, g_p)$. Since $g_1(P) = \cdots = g_{p-1}(P) = 0$ and $g_p(P) = 1$, it follows from (3.3) that the frame $e_1, \cdots, e_p$ of $(\mathcal{O}_\Omega)^{\oplus p}$ is orthonormal at P. We also denote the inner product $\langle e_j, e_k \rangle$ by $H_{j\bar{k}}$ when either j or k is p.

We now compute the curvature $\Theta_{j\bar{k}\lambda\bar{\nu}}$ of $H_{j\bar{k}}$ at the point P by using the formula (3.1) with $V = (\mathcal{O}_\Omega)^{\oplus p}$ and $W = \mathcal{K}$ as follows. Since V is the trivial $\mathbb{C}$-vector bundle of rank p, we have $\Theta^V_{j\bar{k}\alpha\bar{\beta}} \equiv 0$ and $r = p$ and $s = p-1$ so that

$$\Theta^W_{j\bar{k}\alpha\bar{\beta}} = -\left(\partial_\alpha H_{j\bar{p}}\right)\overline{\left(\partial_\beta H_{k\bar{p}}\right)}.$$

Since

$$H_{j\bar{p}} = \langle e_j, e_p \rangle = g_j \overline{g_p}$$

for $1 \leq j, k \leq p-1$ by (3.3), it follows that at P,

$$\Theta^W_{j\bar{k}\lambda\bar{\nu}} = -\left(\partial_\lambda H_{j\bar{p}}\right)\overline{\left(\partial_\nu H_{k\bar{p}}\right)} = -\left(\partial_\lambda g_j\right)\overline{\left(\partial_\nu g_k\right)}$$

for $1 \leq j, k \leq p-1$ and for

$$\left(v^{j\lambda}\right)_{1 \leq j \leq p-1, 1 \leq \lambda \leq n}$$

at P we have

$$\begin{aligned}\sum_{1 \leq j,k \leq p-1, 1 \leq \lambda,\nu \leq n} \Theta^W_{j\bar{k}\lambda\bar{\nu}} v^{j\lambda}\overline{v^{k\nu}} &= -\sum_{1 \leq j,k \leq p-1, 1 \leq \lambda,\nu \leq n} \left(\partial_\lambda g_j\right)\overline{\left(\partial_\nu g_k\right)} v^{j\lambda}\overline{v^{k\nu}} \\ &= -\left|\sum_{1 \leq j \leq p-1, 1 \leq \lambda \leq n} \left(\partial_\lambda g_j\right) v^{j\lambda}\right|^2.\end{aligned}$$

□

3.2. *Curvature Contribution from Twisting by Line Bundle.* We twist the induced metric H of the kernel vector bundle $\mathcal{K}$ by $\frac{1}{|g|^{2\gamma}}$ for some $\gamma > 0$, where $|g|^2 = \sum_{j=1}^p |g_j|^2$. At the point P of Ω where $g_1(P) = \cdots = g_{p-1}(P) = 0$, the curvature contribution from the metric $\frac{1}{|g|^2}$ of the trivial line bundle is

$$\partial\bar{\partial} \log |g|^2 = \frac{1}{|g|^4} \sum_{1 \leq j < k \leq p} \left((\partial g_j) g_k - (\partial g_k) g_j\right) \wedge \overline{\left((\partial g_j) g_k - (\partial g_k) g_j\right)}$$

which at P is reduced to

$$\partial\partial \log |g|^2 = \frac{1}{|g_p|^2} \sum_{1 \leq j \leq p-1} (\partial g_j) \wedge \overline{(\partial g_j)} \tag{3.4}$$

(because $(\partial g_j) g_k - (\partial g_k) g_j$ can only be nonzero when $k = p$ and $j \neq k$ on account of $j < k$). In addition, when $g_p(P)$ is assumed to be 1,

$$\partial\bar{\partial} \log |g|^2 = \sum_{1 \leq j \leq p-1} (\partial g_j) \wedge \overline{(\partial g_j)} \tag{3.5}$$

at the point P. Thus, combined with the computation in Proposition 3.1, under the assumption of $g_1(P) = \cdots = g_p(P) = 0$ and $g_p(P) = 1$ the value at

$\left(v^{j\lambda}\right)_{1\leq j\leq p-1,1\leq\lambda\leq n}$ of the Hermitian form at P defined by the Nakano curvature of the metric $\frac{H_{j\bar{k}}}{|g|^{2\gamma}}$ for $\gamma>0$ is

$$\gamma\sum_{1\leq j,\ell\leq p-1}\left|\sum_{1\leq\lambda\leq n}(\partial_\lambda g_j)v^{\ell\lambda}\right|^2-\left|\sum_{1\leq j\leq p-1,1\leq\lambda\leq n}(\partial_\lambda g_j)\,v^{j\lambda}\right|^2.$$

By applying the Cauchy-Schwarz inequality for tensors as formulated in Remark 2.2

$$\left|\sum_{1\leq\ell\leq r,1\leq k\leq n}s_{\ell,k}t_{\ell,k}\right|^2\leq\min(r,n)\sum_{1\leq m,\ell\leq r}\left|\sum_{k=1}^n s_{m,k}t_{\ell,k}\right|^2$$

with $r=p-1$ and $s_{\ell,\lambda}=\partial_\lambda g_\ell$ and $t_{\ell,\lambda}=v^{\ell\lambda}$, we conclude that

$$\gamma\sum_{1\leq j,\ell\leq p-1}\left|\sum_{1\leq\lambda\leq n}(\partial_\lambda g_j)v^{\ell\lambda}\right|^2\geq\left|\sum_{1\leq j\leq p-1,1\leq\lambda\leq n}(\partial_\lambda g_j)\,v^{j\lambda}\right|^2$$

for $\gamma\geq\min(n,p-1)$ at a point P with $g_1(P)=\cdots=g_p(P)=0$ and $g_p(P)=1$.

We now drop the assumption of $g_1(P)=\cdots=g_p(P)=0$ and $g_p(P)=1$ so that when an element v of $\mathcal{K}$ is regarded as an element v of $(\mathcal{O}_\Omega)^{\oplus p}$ with components $(v_1,\cdots,v_p)$, we can apply the above computation to the general case with v_p not necessarily 0. The following conclusion now holds.

PROPOSITION 3.2. *The Hermitian form*

$$\left(v^{j\lambda}\right)_{1\leq j\leq p,1\leq\lambda\leq n}\mapsto\sum_{1\leq j,k\leq p,1\leq\lambda,\nu\leq n}\Theta(\gamma)_{j\bar{k}\lambda\bar{\nu}}v^{j\lambda}\overline{v^{k\nu}}$$

on $V\otimes T_\Omega$ *defined by the curvature* $\Theta(\gamma)_{j\bar{k}\lambda\bar{\nu}}$ *of the metric*

$$H(\gamma):=\frac{H_{j\bar{k}}}{|g|^{2\gamma}}$$

for $\gamma\geq\min(n,p-1)$ *dominates* $\gamma-\min(n,p-1)$ *times the Hermitian form*

$$\left(v^{j\lambda}\right)_{1\leq j\leq p,1\leq\lambda\leq n}\mapsto\left(\sum_{1\leq\lambda,\nu\leq n}\partial_\lambda\partial_{\bar{\nu}}\log|g|^2\right)\left(\sum_{j=1}^p v^{j\lambda}\overline{v^{j\nu}}\right)$$

defined by (the tensor product of the identity operator of V *and)* $\partial\bar{\partial}\log|g|^2$ *on* $V\otimes T_\Omega$.

4. Vanishing Theorem and Solution of $\bar{\partial}$ Equation for Nonnegative Nakano Curvature

We now present the vanishing theorem and solution of the $\bar{\partial}$-equation for a holomorphic vector bundle of nonnegative Nakano curvature on a strictly pseudoconvex manifold, with emphasis on the lack of strict positivity of the curvature at any point. Even for application to algebraic geometry, the strictly pseudoconvex situation is needed to enable the removal of an ample complex hypersurface to handle the singularity of the metrics of the vector bundle which is inherent to our problem at hand.

Though the setting of Skoda's ideal generation is for a Stein domain spread over $\mathbb{C}^n$ whose canonical line bundle is trivial, in this section we formulate the vanishing theorem in the setting of a general complex manifold X of complex dimension n whose canonical line bundle K_X may not be trivial, where the solution of the $\bar\partial$-equation has to be formulated with the right-hand side being a vector-bundle-valued $(n,1)$-form and the unknown being a vector-bundle-valued $(n,0)$-form. Later in the application of the vanishing theorem to Skoda's ideal generation, the vector-bundle-valued $(n,1)$-form on the right-hand side and the vector-bundle-valued $(n,0)$-form as the unknown will be changed respectively to the vector-bundle-valued $(0,1)$-form on the right-hand side and the vector-bundle-valued function as the unknown.

THEOREM 4.1 (Vanishing Theorem of Kodaira-Nakano with L^2 Estimate for Pseudoconvex Manifolds). *Let Ω be a relatively compact domain in an n-dimensional complex manifold X with Kähler metric $g_{\lambda\bar\nu}$ (for $1 \le \lambda, \nu \le n$) such that the boundary of Ω is smooth and strictly psuedoconvex. Let V be a holomorphic vector bundle of rank r on X with smooth metric $H_{j\bar k}$ (for $1 \le j, k \le r$) whose Nakano curvature $\Theta_{j\bar k\lambda\bar\nu}$ is semipositive on X. Let F be a smooth V-valued $\bar\partial$-closed $(n,1)$-form on Ω such that $\left(\Theta^{-1}F,\, F\right)_{L^2(\Omega)}$ is finite (in the sense of* **3.1***). Then the equation $\bar\partial u = F$ can be solved for a smooth V-valued $(n,0)$-form over Ω with $\|u\|^2_{L^2(\Omega)} \le \left(\Theta^{-1}F,\, F\right)_{L^2(\Omega)}$.*

4.1. Though this statement may not be found in the literature as stated, it can be routinely derived from known available techniques. Instead of giving a detailed proof of it here, we will only comment on the noteworthy features of the statement and explain the main lines of arguments for its proof. The most important feature of the statement is that there may be no point in X where the Nakano curvature Θ of the holomorphic vector bundle V is strictly positive. At a point P of X the Hermitian form of the Nakano curvature Θ may be positive when evaluated at some elements $V \otimes T_X$ at P and may be zero at some other elements of $V \otimes T_X$ at P. This may be the situation at every point P of X. The assumption of the finiteness of $\left(\Theta^{-1}F,\, F\right)_{L^2(\Omega)}$ means the vanishing of components of F which correspond to elements of $(V \otimes T_X)|_\Omega$ where the Hermitian form of the Nakano curvature Θ vanishes.

The key argument of solving the equation $\bar\partial u = F$ on Ω for a smooth V-valued $(n,0)$-form u on Ω is to use the following basic estimate obtained by completion of squares by integration by parts

$$\left\|\bar\partial v\right\|^2_{L^2(\Omega)} + \left\|\bar\partial^* v\right\|^2_{L^2(\Omega)} = \left\|\bar\nabla v\right\|^2_{L^2(\Omega)} + \left(\mathrm{Levi}_{\partial\Omega},\, v \wedge \bar v\right)_{L^2(\partial\Omega)} + \left(\Theta v, v\right)_{L^2(\Omega)}$$

for a test V-valued $(n,1)$-form v on Ω which belongs to the domain of $\bar\partial$ and $\bar\partial^*$ on Ω, where $\mathrm{Levi}_{\partial\Omega}$ is the Levi form $\partial\bar\partial\rho$ of $\partial\Omega$ (with ρ being a smooth function on $\partial\Omega$ defining Ω as $\Omega = \{r < 0\}$ and $d\rho$ identically 1 on $\partial\Omega$) and $\bar\nabla$ means the covariant differential operator along tangent vectors of type $(0,1)$. If one has the estimate

$$\left|(v,\, F)_{L^2(\Omega)}\right| \le \hat C \left\|\bar\partial^* v\right\|_{L^2(\Omega)} \tag{4.1}$$

for all test V-valued $(n,1)$-form v on Ω in the domains of the two operators $\bar\partial$ and $\bar\partial^*$, one can apply Riesz's representation theorem to the $\mathbb{C}$-linear functional

$$\bar\partial^* v \mapsto (v, F)_{L^2(\Omega)} \tag{4.2}$$

with bound $\hat{C}$ to write the $\mathbb{C}$-linear functional (4.2) in the form $w \mapsto (w, u)_{L^2(\Omega)}$ for some u with $\|u\|_{L^2(\Omega)} \leq \hat{C}$ so that from

$$(v,\, F)_{L^2(\Omega)} = (\bar{\partial}^* v,\, u)_{L^2(\Omega)} = (v, \bar{\partial} u)_{L^2(\Omega)}$$

for all test V-valued $(n,1)$-form v on Ω in the domains of $\bar{\partial}$ and $\bar{\partial}^*$ it follows that $\bar{\partial} u = F$ with $\|u\|_{L^2(\Omega)} \leq \hat{C}$. To obtain the estimate (4.1), since F is $\bar{\partial}$-closed, it suffices to get the estimate for test V-valued $(n,1)$-forms v which are $\bar{\partial}$-closed. By the inequality (3.2) of Cauchy-Schwarz type we have

$$\begin{aligned}
\left|(v,\, F)_{L^2(\Omega)}\right|^2 &\leq \left|\left(\Theta^{-1} F,\, F\right)_{L^2(\Omega)}\right| \cdot \left|(\Theta v,\, v)_{L^2(\Omega)}\right| \\
&\leq \left|\left(\Theta^{-1} F,\, F\right)_{L^2(\Omega)}\right| \left(\left\|\bar{\partial} v\right\|^2_{L^2(\Omega)} + \left\|\bar{\partial}^* v\right\|^2_{L^2(\Omega)}\right) \\
&= \left|\left(\Theta^{-1} F,\, F\right)_{L^2(\Omega)}\right| \cdot \left\|\bar{\partial}^* v\right\|^2_{L^2(\Omega)}
\end{aligned}$$

so that the estimate (4.1) holds with $\hat{C} = \left|\left(\Theta^{-1} F,\, F\right)_{L^2(\Omega)}\right|^{\frac{1}{2}}$, which according to the assumption is finite. This finishes the argument, because $\hat{C}$ is equal to the square root of $C = \left|\left(\Theta^{-1} F,\, F\right)_{L^2(\Omega)}\right|$ and $\|u\|^2_{L^2(\Omega)} \leq C$.

Here we have left out the routine details about the use of convolution to handle the problem of proving that smooth V-valued $(n,1)$-forms on Ω up to the boundary in domain of $\bar{\partial}^*$ are dense, with respect to the graph norm, in the space of all L^2 V-valued $(n,1)$-forms on Ω in the domains of $\bar{\partial}$ and $\bar{\partial}^*$.

More general statements than that given in Theorem 4.1 hold by the same arguments. For example, the V-valued $(n,1)$-form F can be replaced by a V-valued (n,q)-form with $q \geq 1$ and the condition of $\partial\Omega$ being smooth and strictly pseudoconvex can be weakened. Here we only give the statement which suffices for our purpose.

COROLLARY 4.2 (Special Case of Semipositive $(1,1)$-Form as Lower Bound). *In Theorem* 4.1 *suppose there are a smooth semipositive* $(1,1)$*-form* ω *on* X *and a positive* (n,n)*-form* Φ *on* X *such that the Nakano curvature* Θ *of* V *dominates* $H \otimes \omega$ *on* X *and* $\Phi\omega$ *dominates (the nonnegative quadratic form defined by the coefficient of) the fiber trace*

$$\mathrm{FbTr}(F) = \sum_{\lambda,\nu=1}^{n} \left(\sum_{j,k=1}^{r} H_{j\bar{k}} F^j_{\bar{\nu}} \overline{F^k_{\bar{\lambda}}}\right) \left(dz^\lambda \wedge \overline{dz^\nu}\right) \left|dz^1 \wedge \cdots \wedge dz^n\right|^2$$

of the $\bar{\partial}$*-closed* V*-valued* $(n,1)$*-form* F *on* Ω*, where the components of* F *with respect to the local fiber coordinates of* V *are* $F^1, \cdots, F^r$ *and*

$$F^j = \sum_{\lambda=1}^{n} F^j_{\bar{\lambda}} (dz^1 \wedge \cdots \wedge dz^n) \wedge \overline{dz^\lambda}.$$

Let n_ω *be a positive integer* $\leq n$ *such that the number of nonzero eigenvalues of* ω *is* $\leq n_\omega$ *at every point of* Ω*. If the integral* $\int_\Omega \Phi$ *is finite, then the* $\bar{\partial}$*-equation* $\bar{\partial} u = F$ *on* Ω *can be solved for the unknown smooth* V*-valued* $(n,0)$*-form* u *over* Ω *such that the square of the* L^2 *norm of* u *on* Ω *with respect to* H *is* $\leq n_\omega \int_\Omega \Phi$.

PROOF. At any point P of Ω we can choose local coordinates $(z^1, \cdots, z^n)$ centered at P such that both ω and H are diagonalized at P and

$$\Phi\,\omega = \left(\sum_{\lambda=1}^{n_\omega} a_\lambda dz^\lambda \wedge \overline{dz^\lambda}\right) \left|dz^1 \wedge \cdots \wedge dz^n\right|^2$$

and

$$\operatorname{FrTr}(F) = \left(\sum_{\lambda=1}^{n_\omega} b_\lambda dz^\lambda \wedge \overline{dz^\lambda}\right) \left|dz^1 \wedge \cdots \wedge dz^n\right|^2$$

at P with $0 \leq b_\lambda \leq a_\lambda$ for $1 \leq \lambda \leq n_\omega$. Since the pointwise inner product $\left\langle (H \otimes \omega)^{-1} F,\, F \right\rangle_P$ at the point P is equal to

$$\sum_{\lambda=1}^{n_\omega} \frac{\Phi}{a_\lambda} \sum_{j,k=1}^{r} H^{\bar{k}j} F_{\bar{k}\bar{\lambda}} \overline{F_{\bar{j}\bar{\lambda}}} = \sum_{\lambda=1}^{n_\omega} \frac{b_\lambda \Phi}{a_\lambda} \leq n_\omega \Phi,$$

it follows that

$$\begin{aligned}\left(\Theta^{-1} F,\, F\right)_{L^2(\Omega)} &\leq \left((H \otimes \omega)^{-1} F,\, F\right)_{L^2(\Omega)} \\ &= \int_{P \in \Omega} \left\langle (H \otimes \omega)^{-1} F,\, F \right\rangle_P \leq n_\omega \int_\Omega \Phi\end{aligned}$$

and the conclusion follows from Theorem 4.1. □

COROLLARY 4.3 (Solution Estimate from Base Trace of Fiber Trace of Kernel Bundle Valued $(n,1)$-Form). *In Theorem 4.1 suppose there are a smooth semipositive $(1,1)$-form ω on X and a positive smooth (n,n)-form Φ on X such that the Nakano curvature Θ of V dominates $H \otimes \omega$ on X and at every point of Ω the null space of ω is contained in the null space of the fiber trace* $\operatorname{FbTr}(F)$ *of the $\bar\partial$-closed V-valued $(n,1)$-form F on Ω. Let* $\operatorname{Tr}_\omega(\operatorname{FbTr}(F))$ *be the trace of* $\operatorname{FbTr}(F)$ *with respect to ω (which means the limit, as $\varepsilon \to 0+$, of the trace of* $\operatorname{FbTr}(F)$ *with respect to the sum of ω and ε times the Kähler form of X). If the integral* $\int_\Omega \operatorname{Tr}_\omega(\operatorname{FbTr}(F))$ *is finite, then the $\bar\partial$-equation $\bar\partial u = F$ on Ω can be solved for the unknown smooth $(n,0)$-form u of V over Ω such that the square of the L^2 norm of u on Ω with respect to H is* $\leq \int_\Omega \operatorname{Tr}_\omega(\operatorname{FbTr}(F))$.

PROOF. It follows from Theorem 4.1 and

$$\left(\Theta^{-1} F,\, F\right)_{L^2(\Omega)} \leq \left((H \otimes \omega)^{-1} F,\, F\right)_{L^2(\Omega)} = \int_\Omega \operatorname{Tr}_\omega(\operatorname{FbTr}(F)).$$

□

5. Proof of Skoda's Result on Ideal Generation

We now present our proof of Skoda's result on ideal generation by using the above three ingredients.

5.1. *Setup and Solution of $\bar\partial$ Equation.* Let Ω be a Stein domain spread over $\mathbb{C}^n$ and let $g_1, \cdots, g_p$ be holomorphic functions on Ω and $\mathcal{K}$ be the kernel subbundle of the bundle-homomorphism $(g_1, \cdots, g_p) : (\mathcal{O}_\Omega)^{\oplus p} \to \mathcal{O}_\Omega$. Let ψ be a plurisubharmonic function on Ω. Let f be a holomorphic function on Ω. Let Z be a complex hypersurface in Ω which contains the common zero-set of $g_1, \cdots, g_p$ such that $\Omega - Z$ is Stein. On $\Omega - Z$ we can write

$$f = \sum_{j=1}^{n} \frac{\overline{g_j}}{|g|^2} f g_j \tag{5.1}$$

and let

$$F = (F_1, \cdots, F_p) = \left(\bar{\partial}\left(\frac{\overline{g_1}}{|g|^2}\right), \cdots, \bar{\partial}\left(\frac{\overline{g_p}}{|g|^2}\right)\right) f. \tag{5.2}$$

Then F is a $\mathcal{K}$-valued $(0,1)$-form on $\Omega - Z$, because $\sum_{j=1}^{p} F_j g_j \equiv 0$ follows by applying $\bar{\partial}$ to (5.1). We consider solving $\bar{\partial} u = F$ for u in the kernel bundle $\mathcal{K}$ with L^2 estimates on the Stein domain $\Omega - Z$ spread over $\mathbb{C}^n$. As remarked earlier, when we apply the vanishing theorem in §4, we will simply regard the $\mathcal{K}$-valued $(0,1)$-form F on Ω naturally as a $\mathcal{K}$-valued $(n,1)$-form on Ω and regard the $\mathcal{K}$-valued function u on Ω naturally as a $\mathcal{K}$-valued $(n,0)$-form on Ω.

We find relatively compact subdomains Ω_m of Ω with smooth strictly pseudoconvex boundary (for $m \in \mathbb{N}$) such that (i) Ω_m is relatively compact in Ω_{m+1} for $m \in \mathbb{N}$ and (ii) the union of all Ω_m for $m \in \mathbb{N}$ is equal to $\Omega - Z$. By using smoothing by convolution we can find a smooth plurisubharmonic function ψ_m on an open neighborhood of the closure of Ω_m in Ω_{m+1} such that $\psi \le \psi_{m+1} \le \psi_m$ on Ω_m and ψ_m approaches ψ on $\Omega - Z$ as $m \to \infty$.

Let $q = \min(n, p-1)$ and $\gamma > 0$. Assume that

$$\hat{C} = \int_{\Omega} \frac{|f|^2 e^{-\psi}}{|g|^{2(q+1+\gamma)}} < \infty. \tag{5.3}$$

We are going to apply Corollary 4.2 to the kernel vector subbundle $\mathcal{K}$ on Ω_m with the metric

$$H(\gamma, m)_{j\bar{k}} := \frac{H_{j\bar{k}} e^{-\psi_m}}{|g|^{2(q+\gamma)}},$$

where $H_{j\bar{k}}$ is the metric for $\mathcal{K}$ induced from the standard flat metric of $(\mathcal{O}_\Omega)^{\oplus p}$. Let $\omega(\gamma) = \gamma \partial\bar{\partial} \log |g|^2$. By (3.5) we have the bound $n_{\omega(\gamma)} \le p-1$ for the number $n_{\omega(\gamma)}$ defined in Corollary 4.2 when we do the computation of the number of nonzero eigenvalues of $\omega(\gamma)$ at a point P by using the normal fiber coordinates with $g_1(P) = \cdots = g_{p-1}(P) = 0$ and $g_p(P) = 1$ as in **3.2**. Since clearly $n_{\omega(\gamma)}$ cannot greater than n, it follows that $n_{\omega(\gamma)} \le q$. By (**??**) we have

$$\Phi_m \omega(\gamma) = \mathrm{FbTr}(F) \quad \text{with} \quad \Phi_m = \frac{|f|^2 e^{-\psi_m}}{|g|^{2(q+\gamma)}} \frac{1}{\gamma |g|^2}, \tag{5.4}$$

where the factor $\frac{|f|^2 e^{-\psi_m}}{|g|^{2(q+\gamma)}}$ comes from the metric $\frac{\delta_{jk} e^{-\psi_m}}{|g|^{2(q+\gamma)}}$ of $\mathcal{O}_\Omega^{\oplus p}$ which induces the metric $H(\gamma, m)$ on the subbundle $\mathcal{K}$ of $\mathcal{O}_\Omega^{\oplus p}$. Here, as remarked earlier, because of the triviality of the canonical line bundle of the Stein domain Ω spread over $\mathbb{C}^n$, we simply represent naturally the (n,n)-form Φ_m as a function. By Corollary 4.2, (5.3), and (5.4), from $\psi \le \psi_m$ on Ω_m it follows that there exists some smooth section $u^{(m)}$ of $\mathcal{K}$ over Ω_m such that $\bar\partial u_m = F$ on Ω_m with

$$\int_{\Omega_m} \frac{\sum_{j=1}^p \left|u_j^{(m)}\right|^2 e^{-\psi_m}}{|g|^{2(q+\gamma)}} \leq n_{\omega(\gamma)} \int_\Omega \Phi_m$$
$$\leq \frac{n_{\omega(\gamma)}}{\gamma} \int_{\Omega_m} \frac{|f|^2 e^{-\psi_m}}{|g|^{2(q+1+\gamma)}} = \frac{n_{\omega(\gamma)}}{\gamma}\hat{C} \leq \frac{q}{\gamma}\hat{C},$$

where $u_j^{(m)}$ is the j-component of $u^{(m)}$ when it is regarded as an element of $\mathcal{O}_\Omega^{\oplus p}$ over Ω_m. By the standard process of using a subsequence $\{m_\nu\}$ of $\{m\}$ to pass to the limit of of $u^{(m_\nu)}$ when $\nu \to \infty$, we conclude that there exists a smooth section u of $\mathcal{K}$ over $\Omega - Z$ such that $\bar\partial u = F$ on $\Omega - Z$ and

$$\int_{\Omega - Z} \frac{\sum_{j=1}^p |u_j|^2 e^{-\psi}}{|g|^{2(q+\gamma)}} \leq \frac{q}{\gamma}\hat{C},$$

where u_j is the j-component of u when it is regarded as an element of $\mathcal{O}_\Omega^{\oplus p}$ over $\Omega - Z$. Let $h_j = \frac{\bar g_j}{|g|^2} f - u_j$. Then h_j is a holomorphic section of $\mathcal{K}$ over $\Omega - Z$ with $\sum_{j=1}^p h_j g_j = f$ on $\Omega - Z$ and

$$\int_{\Omega - Z} \frac{\sum_{j=1}^p |h_j|^2 e^{-\psi}}{|g|^{2(q+\gamma)}}$$
$$= \int_{\Omega - Z} \frac{\sum_{j=1}^p \left|\frac{\bar g_j}{|g|^2} f - u_j\right|^2 e^{-\psi}}{|g|^{2(q+\gamma)}}$$
$$\leq 2 \int_{\Omega - Z} \frac{\sum_{j=1}^p \left|\frac{\bar g_j}{|g|^2} f\right|^2 e^{-\psi} + \sum_{j=1}^p |u_j|^2 e^{-\psi}}{|g|^{2(q+\gamma)}}$$
$$= 2 \int_{\Omega - Z} \frac{|f|^2 e^{-\psi}}{|g|^{2(q+1+\gamma)}} + \int_\Omega \sum_{j=1}^p \frac{|u_j|^2 e^{-\psi}}{|g|^{2(q+\gamma)}}$$
$$\leq 2\left(1 + \frac{q}{\gamma}\right)\hat{C}.$$

We can now extend h_j to be a holomorphic section of $\mathcal{K}$ over Ω from the L^2 estimate. This finishes our proof of Skoda's result on ideal generation, except for a factor of 2 in the L^2 estimate for the solution $h_1, \cdots, h_p$, which we now discuss.

5.2. Though it is an insignificant point for our purpose, there is a factor of 2 which should not be there on the right-hand side of the estimate

$$\int_\Omega \frac{\sum_{j=1}^p |h_j|^2 e^{-\psi}}{|g|^{2(q+\gamma)}} \leq 2\left(1 + \frac{q}{\gamma}\right)\hat{C}$$

in **5.1**. The constant γ is related to the constant α in Theorem 1.1 by $q + 1 + \gamma = \alpha q + 1$ which means $\alpha = 1 + \frac{\gamma}{q}$ so that $\frac{\alpha}{\alpha - 1} = 1 + \frac{q}{\gamma}$. In its computation in **5.1** the estimate of

$$\left|2\mathrm{Re}\left(\frac{\bar g_j}{|g|^2} f\, u_j\right)\right| \leq \left|\frac{\bar g_j}{|g|^2} f\right|^2 + |u_j|^2$$

can be replaced by the better inequality of

$$\left|2\mathrm{Re}\left(\frac{\bar{g}_j}{|g|^2} f\, \overline{u_j}\right)\right| \leq \beta \left|\frac{\bar{g}_j}{|g|^2} f\right|^2 + \frac{1}{\beta}\, |u_j|^2$$

with $\beta = \sqrt{\frac{q}{\gamma}}$ to give the smaller bound of $\left(1 + \sqrt{\frac{q}{\gamma}}\right)^2$ which is still greater than $1 + \frac{q}{\gamma}$.

One possible way to optimize better the bound is to use the fact that since each solution section $u^{(m)} = \left(u_1^{(m)}, \cdots, u_p^{(m)}\right)$ of the bundle $\mathcal{K}$ over Ω_m is obtained from the Riesz representation theorem, the section $u^{(m)}$ is orthogonal to the kernel of the $\bar{\partial}$ operator on Ω_m with respect to the metric $H(\gamma, m)_{j\bar{k}}$ on $\Omega^{(m)}$. In this note we will not go further into the question of optimization of the constant in the integral bound of the solution of ideal generation.

6. Variants of Skoda's Ideal Generation

From our proof variants of Skoda's ideal generation result can readily be obtained by different choices of the weight function (which is the twisting by the metric of the trivial line bundle). Here we give three examples in the following theorem. The example in Theorem 6.1(b) yields Théorème 2 on p.571 in Skoda's paper [**3**] with the careful choice of φ given there.

THEOREM 6.1 (Variants of Skoda's Ideal Generation). *Let Ω be a domain spread over $\mathbb{C}^n$ which is Stein. Let ψ be a plurisubharmonic function on Ω, $g_1, \ldots, g_p$ be holomorphic functions on Ω (not all identically zero), $\gamma > 0$, $q = \min(n, p-1)$, $|g|^2 = \sum_{j=1}^p |g_j|^2$, and f be a holomorphic function on Ω.*

(a) *If the integral*

$$C_1 := \int_\Omega \frac{|f|^2 \left(|g|^2 + q(1+|g|^2)\right) e^{-\psi}}{|g|^{2(q+2)}(1+|g|^2)}$$

is finite, then there exist holomorphic functions $h_1, \ldots, h_p$ on Ω with $f = \sum_{j=1}^p h_j g_j$ on Ω such that

$$\int_\Omega \frac{\sum_{j=1}^p |h_j|^2 e^{-\psi}}{|g|^{2q}(1+|g|^2)} \leq 2C_1.$$

(b) *Let φ be a smooth plurisubharmonic function on Ω such that $\omega := \partial\bar{\partial}\varphi$ is not identically zero on Ω. If the integral*

$$C_2 := \int_\Omega \frac{|f|^2 (1 + \Delta_\omega \log |g|^2) e^{-(\varphi+\psi)}}{|g|^{2(q+1)}}$$

is finite, then there exist holomorphic functions $h_1, \ldots, h_p$ on Ω with $f = \sum_{j=1}^p h_j g_j$ on Ω such that

$$\int_\Omega \frac{\sum_{j=1}^p |h_j|^2 e^{-(\varphi+\psi)}}{|g|^{2q}} \leq 2C_2,$$

where $\Delta_\omega \log |g|^2$ means the Laplacian of the function $\log |g|^2$ with respect to ω (which means the limit as $\varepsilon \to 0+$ of the Laplacian of the function $\log |g|^2$ with respect to the sum of ω and ε times the standard Euclidean Kähler form of $\mathbb{C}^n$).

(c) *If* $|g_j| < 1$ *on* Ω *for* $1 \leq j \leq p$ *and*

$$C_3 := \int_\Omega \frac{|f|^2 \left(\left(\log\left(\frac{1}{|g|^2}\right)\right) + q\left(\log\left(\frac{1}{|g|^2}\right)\right)^2\right) e^{-\psi}}{|g|^{2(q+1)}}$$

is finite, then there exist holomorphic functions $h_1, \ldots, h_p$ *on* Ω *with* $f = \sum_{j=1}^p h_j g_j$ *on* Ω *such that*

$$\int_\Omega \frac{\sum_{j=1}^p |h_j|^2 \left(\log\left(\frac{1}{|g|^2}\right)\right) e^{-\psi}}{|g|^{2q}} \leq 2C_3.$$

PROOF. For the proof of Part (a), since

$$\begin{aligned}\partial\bar\partial \log(1+|g|^2) &= \frac{|g|^2 \partial\bar\partial \log |g|^2 + \sum_{k=1}^p \partial g_k \wedge \overline{\partial g_k}}{(1+|g|^2)^2} \\ &\geq \frac{|g|^2 \partial\bar\partial \log |g|^2}{1+|g|^2},\end{aligned}$$

it follows that Corollary 4.2 can be applied to the subbundle $\mathcal{K}$ of $(\mathcal{O}_\Omega)^{\oplus p}$ with the metric induced by the standard flat metric of $(\mathcal{O}_\Omega)^{\oplus p}$ multiplied by the weight function

$$\frac{|f|^2 e^{-\psi}}{|g|^{2q}(1+|g|^2)}$$

so that the function Φ can be chosen to be

$$\frac{|f|^2 e^{-\psi}}{|g|^{2q+4}}$$

to solve the equation $\bar\partial u = F$ for a smooth section of $\mathcal{K}$ on Ω with F given by (5.2) as described in **5.1**.

The proof of Part (b) follows immediately from Corollary 4.3 with the same meaning for ω as in Corollary 4.3 and with the fiber trace $\mathrm{FbTr}(F)$ of F of (5.2) being equal to $\frac{1}{|g|^2}\Delta_\omega \log |g|^2$.

For the proof of Part (c), since

$$\begin{aligned}-\partial\bar\partial \log\log\left(\frac{1}{|g|^2}\right) &= \frac{\partial\bar\partial \log |g|^2}{\log\left(\frac{1}{|g|^2}\right)} + \frac{(\partial \log |g|^2) \wedge (\bar\partial \log |g|^2)}{\left(\log\left(\frac{1}{|g|^2}\right)\right)^2} \\ &\geq \frac{\partial\bar\partial \log |g|^2}{\log\left(\frac{1}{|g|^2}\right)},\end{aligned}$$

it follows that Corollary 4.2 can be applied to the subbundle $\mathcal{K}$ of $(\mathcal{O}_\Omega)^{\oplus p}$ with the metric induced by the standard flat metric of $(\mathcal{O}_\Omega)^{\oplus p}$ multiplied by the weight function

$$\frac{\left(\log\left(\frac{1}{|g|^2}\right)\right) e^{-\psi}}{|g|^{2q}}$$

so that the function Φ can be chosen to be

$$\frac{1}{|g|^2} \frac{\left(\log\left(\frac{1}{|g|^2}\right)^2\right) e^{-\psi}}{|g|^{2q}} |f|^2$$

to solve the equation $\bar{\partial} u = F$ for a smooth section of $\mathcal{K}$ on Ω with F given by (5.2) as described in **5.1**. $\square$

REMARK 6.2. Theorem 6.1(b) can be applied to any $\varphi = \frac{1}{(1+|z|^2)^m}$ (for $m \in \mathbb{N}$) or $\varphi = e^{-|z|^2}$ or any other global smooth strictly plurisubharmonic function on $\mathbb{C}^n$, but the use of any weight function $e^{-\varphi}$ with φ strictly plurisubharmonic defeats the original purpose of applying vanishing theorems and L^2 estimates of $\bar{\partial}$ without the Nakano curvature being strictly positive at any point. Unfortunately, unlike Theorem 1.1 (with Theorem 1.2 as its algebraic geometric formulation), none of the three parts of Theorem 6.1 can be formulated and used in an algebraic geometric setting.

7. Comparison with Original Proof of Skoda's Ideal Generation

The proof given here is simpler and more straightforward than Skoda's original proof, because (i) we can quote directly well-developed known techniques and geometric notions and (ii) we can use normal coordinates to simplify the computations on account of the coordinate-independence of the geometric entities involved. However, behind the facade of geometric and analytic arguments, the manipulations in computation between our proof and Skoda's original proof can be put in parallel correspondence, with expressions in our proof neater and simpler than in Skoda's proof due to the use of normal coordinates. For the discussion on the comparison, we start out with the following Cauchy-Schwarz inequality (with a special factor) for tensors more general than those in Proposition 2.1.

PROPOSITION 7.1 (Cauchy-Schwarz Inequality for More General Tensors). *Let $\vec{a} \in \mathbb{C}^p$. Let $\mathbf{b}$ and $\mathbf{c}$ be elements of $\mathbb{C}^p \otimes \mathbb{C}^n$. Let $q = \min(n, p-1)$. Then*

$$\left| \langle \vec{a} \wedge \mathbf{b},\, \vec{a} \wedge \mathbf{c} \rangle_{(\bigwedge^2 \mathbb{C}^p) \times \mathbb{C}^n} \right|^2 \leq q\, \|\vec{a}\|^2_{\mathbb{C}^p} \left\| \langle \vec{a} \wedge \mathbf{b},\, \mathbf{c} \rangle_{\mathbb{C}^n} \right\|^2_{(\bigwedge^2 \mathbb{C}^p) \times \mathbb{C}^p}.$$

Here (i) $\vec{a} \wedge \mathbf{b}$ is the wedge product of $\vec{a}$ separately with each of the n elements of $\mathbb{C}^p$ defined by $\mathbf{b}$ to end up with an element of $\left(\bigwedge^2 \mathbb{C}^p\right) \times \mathbb{C}^n$ and (ii) the inner product $\langle \vec{a} \wedge \mathbf{b},\, \mathbf{c} \rangle_{\mathbb{C}^n}$ is taken when the element $\vec{a} \wedge \mathbf{b}$ of $\left(\bigwedge^2 \mathbb{C}^p\right) \times \mathbb{C}^n$ is regarded as an $\left(\bigwedge^2 \mathbb{C}^p\right)$-valued element of $\mathbb{C}^n$ and the element $\mathbf{c}$ is regarded as a $\mathbb{C}^p$-valued element of $\mathbb{C}^n$ so that $\langle \vec{a} \wedge \mathbf{b},\, \mathbf{c} \rangle_{\mathbb{C}^n}$ becomes an element of $\left(\bigwedge^2 \mathbb{C}^p\right) \times \mathbb{C}^p$.

7.1. *Equivalence of Proposition* 7.1 *and Proposition* 2.1. Proposition 7.1 is derived from Proposition 2.1 simply by choosing an orthonormal basis in $\mathbb{C}^p$ with respect to which $\vec{a}$ becomes the element $(0, \cdots, 0, 1)$ of $\mathbb{C}^p$. On the other hand, Proposition 2.1 is derived from Proposition 7.1 simply by choosing $r = p - 1$ and $\vec{a} = (0, \cdots, 0, 1) \in \mathbb{C}^p$ and $\mathbf{S} = \vec{a} \wedge \mathbf{b}$ and $\mathbf{T} = \vec{a} \wedge \mathbf{c}$.

REMARK 7.2. In terms of the formulation of tensors with indices, when the tensors $\mathbf{b}$ and $\mathbf{c}$ are represented by $p \times n$ matrices

$$\mathbf{b} = \left(b_{\ell,k}\right)_{1 \leq \ell \leq p,\, 1 \leq k \leq n} \quad \text{and} \quad \mathbf{c} = \left(c_{\ell,k}\right)_{1 \leq \ell \leq p,\, 1 \leq k \leq n},$$

the inequality in Proposition 7.1 becomes

$$\left|\sum_{1\le j<\ell\le p}\sum_{k=1}^{n}(a_j\,b_{\ell,k}-a_\ell\,b_{j,k})\,\overline{(a_j\,c_{\ell,k}-a_\ell\,c_{j,k})}\right|^2$$

$$\le q\left(\sum_{j=1}^{p}|a_j|^2\right)\sum_{\ell=1}^{p}\sum_{1\le m<j\le p}\left|\sum_{k=1}^{n}(a_m\,b_{j,k}-a_j\,b_{m,k})\,\overline{c_{\ell,k}}\right|^2,$$

because

$$\vec{a}\wedge\mathbf{b}=(a_j\,b_{\ell,k}-a_\ell\,b_{j,k})_{1\le j<\ell\le p,\,1\le k\le n}\,.$$

Since

$$\sum_{1\le j<\ell\le p}\sum_{k=1}^{n}(a_j\,b_{\ell,k}-a_\ell\,b_{j,k})\,\overline{(a_j\,c_{\ell,k}-a_\ell\,c_{j,k})}$$

$$=\sum_{1\le j,\ell\le p}\sum_{k=1}^{n}(a_j\,b_{\ell,k}-a_\ell\,b_{j,k})\,\overline{a_j}\,\overline{c_{\ell,k}}$$

(from the fact that the inner product of a 2-tensor C with a skew-symmetric 2-tensor B is unchanged when C is replaced by its skew-symmetrization), by replacing $c_{\ell,k}$ by its complex-conjugate we conclude that the inequality in Proposition 7.1 is equivalent to

$$\left|\sum_{j,\ell=1}^{p}\sum_{k=1}^{n}\overline{a_j}\,(a_j\,b_{\ell,k}-a_\ell\,b_{j,k})\,c_{\ell,k}\right|^2$$

$$\le q\left(\sum_{j=1}^{p}|a_j|^2\right)\sum_{\ell=1}^{p}\sum_{1\le m<j\le p}\left|\sum_{k=1}^{n}(a_m\,b_{j,k}-a_j\,b_{m,k})\,c_{\ell,k}\right|^2,$$

which is precisely the inequality in Lemme 1 on p.552 of [**3**].

We give in Proposition 3.1 the formula for the Nakano curvature of the kernel subbundle $\mathcal{K}$ in normal fiber coordinates. Now for comparison between our proof and Skoda's original proof, we give the formula for it in general fiber coordinates as follows.

PROPOSITION 7.3 (Nakano Curvature of Kernel Subbundle). *Let $g_1,\cdots,g_p$ be holomorphic functions on a domain Ω in $\mathbb{C}^n$ and let $\mathcal{K}$ be the kernel of the bundle-homomorphism $(g_1,\cdots,g_p):(\mathcal{O}_\Omega)^{\oplus p}\to\mathcal{O}_\Omega$ which is given the metric induced from the standard flat metric of the trivial $\mathbb{C}$-vector bundle $(\mathcal{O}_\Omega)^{\oplus p}$ of rank p. The Nakano curvature of $\mathcal{K}$ is given by*

$$\left(v^{j\lambda}\right)_{1\le j\le p,1\le\lambda\le n}\mapsto\sum_{1\le j,k\le p,1\le\lambda,\nu\le n}\Theta_{j\bar{k}\lambda\bar{\nu}}v^{j\lambda}\overline{v^{k\nu}}$$

$$=-\frac{1}{\left(\sum_{j=1}^{p}|g_j|^2\right)^3}\left|\sum_{j,\ell=1}^{p}\sum_{\lambda=1}^{n}\overline{g_\ell}\,(g_\ell(\partial_\lambda g_j)-g_j(\partial_\lambda g_\ell))\,v^{j\lambda}\right|^2$$

for $\left(v^{j\lambda}\right)_{1\le j\le p,1\le\lambda\le n}$ satisfying $\sum_{j=1}^{p}g_jv^{j\lambda}=0$ for $1\le\lambda\le n$.

PROOF. Proposition 3.1 is the special case where $g_1 = \cdots = g_{p-1} = 0$ and $g_p = 1$ at the point under consideration. To prove Proposition 7.3, we can argue as follows by applying a (constant) unitary transformation U of order p to the p-tuple of holomorphic functions $(g_1, \cdots, g_p)$.

Let $(\tilde{g}_1, \cdots, \tilde{g}_p)$ be the image of $(g_1, \cdots, g_p)$ under the (constant) unitary transformation U of order p. For fixed $1 \leq \lambda \leq n$ let $\left(\tilde{v}^{1\lambda}, \cdots, \tilde{v}^{p\lambda}\right) \in \mathbb{C}^p$ be the image of $\left(v^{1\lambda}, \cdots, v^{p\lambda}\right) \in \mathbb{C}^n$ under the complex conjugate $\bar{U}$ of U so that $\sum_{j=1}^p \tilde{w}_j \tilde{v}^{j\lambda} = \sum_{j=1}^p w_j v^{j\lambda}$ for any $(w_1, \cdots, w_p) \in \mathbb{C}$ whose image under U is $(\tilde{w}_1, \cdots, \tilde{w}_p)$, because the image of $\left(\overline{v^{1\lambda}}, \cdots, \overline{v^{p\lambda}}\right)$ under U is $\left(\overline{\tilde{v}^{1\lambda}}, \cdots, \overline{\tilde{v}^{p\lambda}}\right)$. Hence for fixed $1 \leq \lambda \leq n$ we have

$$\sum_{j,\ell=1}^{p} \overline{g_\ell}\left(g_\ell(\partial_\lambda g_j) - g_j(\partial_\lambda g_\ell)\right) v^{j\lambda} = \sum_{j,\ell=1}^{p} \overline{\tilde{g}_\ell}\left(\tilde{g}_\ell(\partial_\lambda \tilde{g}_j) - \tilde{g}_j(\partial_\lambda \tilde{g}_\ell)\right) \tilde{v}^{j\lambda}$$

and summation over $1 \leq \lambda \leq n$ yields

$$\sum_{j,\ell=1}^{p} \sum_{\lambda=1}^{n} \overline{g_\ell}\left(g_\ell(\partial_\lambda g_j) - g_j(\partial_\lambda g_\ell)\right) v^{j\lambda} = \sum_{j,\ell=1}^{p} \sum_{\lambda=1}^{n} \overline{\tilde{g}_\ell}\left(\tilde{g}_\ell(\partial_\lambda \tilde{g}_j) - \tilde{g}_j(\partial_\lambda \tilde{g}_\ell)\right) \tilde{v}^{j\lambda}$$

and

$$\begin{aligned} &\frac{1}{\left(\sum_{j=1}^p |g_j|^2\right)^3} \left| \sum_{j,\ell=1}^{p} \sum_{\lambda=1}^{n} \overline{g_\ell}\left(g_\ell(\partial_\lambda g_j) - g_j(\partial_\lambda g_\ell)\right) v^{j\lambda} \right|^2 \\ &= \frac{1}{\left(\sum_{j=1}^p |\tilde{g}_j|^2\right)^3} \left| \sum_{j,\ell=1}^{p} \sum_{\lambda=1}^{n} \overline{\tilde{g}_\ell}\left(\tilde{g}_\ell(\partial_\lambda \tilde{g}_j) - \tilde{g}_j(\partial_\lambda \tilde{g}_\ell)\right) \tilde{v}^{j\lambda} \right|^2. \end{aligned}$$

Let $\tilde{\Theta}_{j\bar{k}\lambda\bar{\nu}}$ be the curvature tensor of the kernel vector subbundle when $(g_1, \cdots, g_p)$ is replaced by its image $(\tilde{g}_1, \cdots, \tilde{g}_p)$ under U. Then

$$\sum_{1 \leq j,k \leq p, 1 \leq \lambda,\nu \leq n} \Theta_{j\bar{k}\lambda\bar{\nu}} v^{j\lambda} \overline{v^{k\nu}} = \sum_{1 \leq j,k \leq p, 1 \leq \lambda,\nu \leq n} \tilde{\Theta}_{j\bar{k}\lambda\bar{\nu}} \tilde{v}_{j\lambda} \overline{\tilde{v}^{k\nu}}$$

for $\left(v^{k\lambda}\right)_{1 \leq j \leq p,\, 1 \leq \lambda \leq n}$ satisfying $\sum_{j=1}^p g_j v^{j\lambda} = 0$ for $1 \leq \lambda \leq n$. Hence at any prescribed point P (where $g_1(P), \cdots, g_p(P)$ are not all zero) we can choose a (constant) unitary transformation U of order p so that the image $(\tilde{g}_1(P), \cdots, \tilde{g}_p(P))$ of $(g_1(P), \cdots, g_p(P))$ satisfies the condition that $\tilde{g}_j(P) = 0$ for $1 \leq j \leq p-1$ and we can obtain the general case from the special case of $g_1(P) = \cdots = g_{p-1}(P) = 0$ and $g_p(P) = 1$. □

REMARK 7.4. The arguments with inequalities centered around p.552 of [**3**] correspond to Proposition 3.2 when the formula in Proposition 7.3 is used.

References

[1] Uffe Haagerup, *Injectivity and decomposition of completely bounded maps*, Operator algebras and their connections with topology and ergodic theory (Buşteni, 1983), Lecture Notes in Math., vol. 1132, Springer, Berlin, 1985, pp. 170–222, DOI 10.1007/BFb0074885. MR799569

[2] Shigeo Nakano, *On complex analytic vector bundles*, J. Math. Soc. Japan **7** (1955), 1–12, DOI 10.2969/jmsj/00710001. MR0073263

[3] Henri Skoda, *Application des techniques L^2 à la théorie des idéaux d'une algèbre de fonctions holomorphes avec poids* (French), Ann. Sci. École Norm. Sup. (4) **5** (1972), 545–579. MR0333246

Department of Mathematics, Harvard University, 1 Oxford Street, Cambridge, Massachusetts 02138

Email address: siu@math.harvard.edu

Contemporary Mathematics
Volume **712**, 2018
http://dx.doi.org/10.1090/conm/712/14354

Hyper-Kähler compactification of the intermediate Jacobian fibration of a cubic fourfold: The twisted case

Claire Voisin

Pour Lawrence, avec beaucoup d'estime et d'amitié

ABSTRACT. The starting point of this note is our recent paper with Laza and Saccà constructing deformations of O'Grady's 10-dimensional hyper-Kähler manifolds as compactifications of intermediate Jacobian fibrations associated to cubic fourfolds. The note provides a complement to that paper consisting in the analogous construction in the twisted case, leading to isogenous but presumably not isomorphic or birational hyper-Kähler manifolds.

1. Introduction

Hyper-Kähler geometry is a geometry of a very restricted type which is part of the more general setting of K-trivial compact Kähler geometry. The existence of hyper-Kähler manifolds rests on Yau's theorem [**28**]. Hyper-Kähler manifolds are compact simply connected complex manifolds of even complex dimension $2n$ with a Ricci-flat Kähler metric and parallel everywhere nondegenerate holomorphic 2-form, unique up to a coefficient. Forgetting about the metric, the complex manifolds one obtains can always be deformed to projective complex manifolds. Hodge theory plays a major role in the deformation theory of these complex manifolds. In fact they are not only locally but also globally determined by their period point (see [**3**], [**25**]), namely the de Rham cohomology class of the closed holomorphic 2-form. It is also remarkable that studying the period map for these manifolds led Beauville and Bogomolov to the discovery of the so-called Beauville-Bogomolov quadratic form, whose existence is their most striking topological property. The situation concerning the construction and classification of deformation types of hyper-Kähler manifolds is very strange: Two infinite series are known (see [**3**]), each one having one type for each even dimension, and furthermore two sporadic (families of) examples in dimension 6 and 10 were constructed by O'Grady ([**22**], [**23**]). Another strange feature of the theory is the following: the simplest kyper-Kähler manifolds are $K3$ surfaces, with particular examples constructed as Kummer surfaces, hence associated with abelian surfaces or 2-dimensional complex tori. There are many different ways of associating to $K3$ surfaces or abelian surfaces higher dimensional hyper-Kähler manifolds with $b_2 = 23$ (and also 24 for the O'Grady examples) built as (desingularizations of) moduli spaces of simple sheaves on them.

2010 *Mathematics Subject Classification.* Primary 14J99, 14D06; Secondary 14K99.
Key words and phrases. Hyper-Kähler manifolds, cubic fourfolds, Lagrangian fibrations.

Unfortunately algebraic $K3$ surfaces have only 19 parameters, while these algebraic hyper-Kähler manifolds have their deformation spaces (as polarized manifolds) of dimension $b_2 - 3 > 19$, so the general one does not come from an (algebraic) $K3$ surface. Very curiously, cubic hypersurfaces in $\mathbb{P}^5$, which have 20 parameters, have also been very much used for the construction of various 20 parameters families of algebraic hyper-Kähler manifolds. This is well-understood and even expected Hodge-theoretically, but rather unexpected geometrically. In fact the variation of Hodge structure on the cohomology of degree 4 of a cubic fourfold exactly looks like the variation of Hodge structure on the degree 2 cohomology of a polarized hyper-Kähler manifold with $b_2 = 23$. We will describe several instances of these constructions in Section 2.2. The most recent such construction has been provided in [**16**] and we will achieve in Sections 3 and 4 a twisted variant of that construction. Let $X \subset \mathbb{P}^5$ be a smooth cubic fourfold. Let $U \subset B := (\mathbb{P}^5)^*$ be the open set parametrizing smooth hyperplane sections $Y \subset X$. The family of intermediate Jacobians $J(Y_t)_{t\in U}$ is a smooth projective fibration $\pi_U : \mathcal{J}_U \to U$ which according to [**9**] has a nondegenerate closed holomorphic 2-form making the fibration Lagrangian. The following is the main result of [**16**]:

THEOREM 1.1. *There exists a flat projective fibration $\pi : \overline{\mathcal{J}} \to B$ extending π_U, such that the total space $\overline{\mathcal{J}}$ is smooth and hyper-Kähler. Furthermore, $\overline{\mathcal{J}}$ is a deformation of a 10-dimensional O'Grady hyper-Kähler manifold.*

We can be slightly more precise, introducing the open set $U_1 \subset B$ parametrizing at worst 1-nodal hyperplane sections of X. The motivation for introducing U_1 is the fact that $\operatorname{codim}(B \setminus U_1 \subset B) \geq 2$, and this is a key point in the strategy of [**16**]. The family of intermediate Jacobians has a standard extension $\mathcal{J}_{U_1} \to U_1$ over U_1, first as a family of quasiabelian schemes $\mathcal{J}^{\circ}_{U_1} \to U_1$, and then by applying the Mumford compactification: the fibers of $\pi_{U_1} : \mathcal{J}^{\circ}_{U_1} \to U_1$ are $\mathbb{C}^*$-bundles over the four-dimensional intermediate Jacobians $J(\widetilde{Y}_t)$ for $t \in U_1 \setminus U$, and the Mumford compactification is obtained by compactifying the $\mathbb{C}^*$-bundle to a $\mathbb{P}^1$-bundle with the 0- and ∞-sections glued via a translation. The compactified hyper-Kähler manifold $\overline{\mathcal{J}}$ is in fact a compactification of $\mathcal{J}_{U_1}$.

The intermediate Jacobian fibration $\mathcal{J}_U$ has a twisted version $\mathcal{J}^T_U$ (which appears in [**27**] and plays an important role there, although it is not defined very carefully). There are several ways of understanding it (see Section 3). The set of points in the fiber of $\mathcal{J}^T_U$ over a point $t \in U$ identifies with the set of 1-cycles of degree 1 in the fiber Y_t modulo rational equivalence (see Section 3). We will construct in Section 3 $\mathcal{J}^T_U$ as an algebraic variety (a torsor over $\mathcal{J}_U$) and a natural extension $\mathcal{J}^T_{U_1}$ of $\mathcal{J}^T_U$ over U_1 which is étale (or analytically) locally isomorphic to $\mathcal{J}_{U_1}$ over U_1, thus getting a twisted version of $\mathcal{J}_{U_1}$. Note that $\mathcal{J}^T_{U_1}$ carries a nondegenerate closed holomorphic 2-form, for exactly the same reasons $\mathcal{J}_{U_1}$ does. The goal of this note is to prove the following twisted analogue of Theorem 1.1:

THEOREM 1.2. *Let X be general cubic fourfold. There exists a flat projective fibration $\pi^T : \overline{\mathcal{J}}^T \to B$ extending $\pi^T_{U_1} : \mathcal{J}^T_{U_1} \to U_1$, such that the total space $\overline{\mathcal{J}}^T$ is smooth and hyper-Kähler.*

The conclusion above holds for general X, that is, all cubic fourfolds parametrized by a certain Zariski open set of the space of all smooth cubic fourfolds. When the cubic carries a degree 4 integral Hodge class which restricts to the generator of $H^4(Y_t,\mathbb{Z})$ on its hyperplane sections, but is still general in the sense that it belongs to this Zariski open set, which happens along a countable dense union of hypersurfaces in the moduli space, the two fibrations $\mathcal{J}_U$ and $\mathcal{J}_U^T$ are isomorphic over U and the two varieties $\overline{\mathcal{J}}^T$ and $\overline{\mathcal{J}}$ are thus birational. It is not clear that they are then isomorphic, but Huybrechts [**13**] shows that they are deformation equivalent. In particular the varieties $\overline{\mathcal{J}}^T$ are deformation equivalent to OG10 manifolds. One may wonder if the varieties $\overline{\mathcal{J}}^T$ and $\overline{\mathcal{J}}$ are birational for very general, nonspecial X. It is clear that they are not birational as Lagrangian fibrations, since one has a rational section, while the other does not have a rational section. It is likely that the varieties we are considering have a unique Lagrangian fibration so in fact are not isomorphic or even birational, but we have not pursued this.

Thanks. *I am happy to thank Radu Laza and Giulia Saccà with whom I had many interesting discussions at the origin of this note.*

2. Deforming and constructing hyper-Kähler manifolds

2.1. Deformation theory and the period map. The Bogomolov-Tian-Todorov theorem says that a compact Kähler manifold X with trivial canonical bundle has unobstructed deformations. This means that a first order deformation of the complex structure of X, which is given by an element of $H^1(X,T_X)$ called the Kodaira-Spencer class, see [**26**, 9.1.2], extends to a deformation of arbitrarily large order, and in fact there exists in this case a universal family $\mathcal{X}\to B$ where B is a ball in $H^1(X,T_X)\cong\mathbb{C}^N$, $\mathcal{X}$ is a complex manifold and ϕ is smooth proper holomorphic with central fiber $\mathcal{X}_0\cong X$ such that the Kodaira-Spencer map $T_{B,0}\to H^1(X,T_X)$ (the classifying map for first order deformations) is an isomorphism.

Assume now that X is hyper-Kähler. It is a general fact that the fibers $\mathcal{X}_t$ for t small are still hyper-Kähler: indeed, the close fibers are still Kähler as small deformations of a compact Kähler manifold, and the holomorphic 2-form still exists on $\mathcal{X}_t$ for small t because the Hodge numbers $h^{p,q}(X):=\dim H^{p,q}(X)$, $H^{p,q}(X)=H^q(X,\Omega_X^p)$, are constant under a deformation of compact Kähler manifolds. Ehresmann's fibration theorem tells us that the family $\phi:\mathcal{X}\to B$ is $\mathcal{C}^\infty$ trivial and in particular topologically trivial: $\mathcal{X}\overset{homeo}{\cong}\mathcal{X}_0\times B$. In particular, we have canonical identifications

$$H^2(\mathcal{X}_t,\mathbb{C})\cong H^2(\mathcal{X}_0,\mathbb{C})=H^2(X,\mathbb{C}). \tag{2.1}$$

The period map $\mathcal{P}$ associates to $t\in B$ the class $[\sigma_t]$ of the closed holomorphic 2-form σ_t on $\mathcal{X}_t$, seen as an element of $H^2(X,\mathbb{C})$ via the isomorphism (2.1). Note that σ_t is defined up to a multiplicative coefficient, hence $[\sigma_t]$ is well-defined only in $\mathbb{P}(H^2(X,\mathbb{C}))$. Hence the period map takes value in $\mathbb{P}(H^2(X,\mathbb{C}))$. It is a general result due to Griffiths that the period map is holomorphic. Furthermore the computation of its differential shows in our case that the period map is an immersion. Note that $\dim H^1(X,T_X)=\dim H^1(X,\Omega_X)=b_2(X)-2$, as Hodge theory provides the Hodge decomposition $H^2(X,\mathbb{C})=H^{2,0}(X)\oplus H^{1,1}(X)\oplus H^{0,2}(X)$ with $H^{1,1}(X)\cong H^1(X,\Omega_X)$, the two other spaces being 1-dimensional. It follows that $\mathcal{P}(B)\subset\mathbb{P}(H^2(X,\mathbb{C}))$ is a germ $\mathcal{D}_{loc}$ of analytic hypersurface. One can use it (see [**3**], [**12**]) to construct the Beauville-Bogomolov quadratic form q on $H^2(X,\mathbb{Q})$.

THEOREM 2.1. *(see [**3**]) Let X be a hyper-Kähler manifold of dimension $2n$. There exists a nondegenerate quadratic form q of signature $(3, b_2(X) - 3)$ on $H^2(X, \mathbb{Q})$ which is defined up to a multiplicative coefficient by the following condition: for some positive rational number λ, one has*

$$\int_X \alpha^{2n} = \lambda q(\alpha)^n \tag{2.2}$$

for any $\alpha \in H^2(X, \mathbb{Q})$.

This form is usually normalized in such a way that it takes integral values on $H^2(X, \mathbb{Z})$ and is not divisible as an intersection form on $H^2(X, \mathbb{Z})$; it is then uniquely determined. Although Beauville gives an explicit formula for q, the existence of q follows directly from the study of the period map. Indeed, we observe that classes $[\sigma_t] \in \mathcal{D}_{loc}$ satisfy $[\sigma_t]^{n+1} = 0$ in $H^2(X, \mathbb{C})$. Indeed, they are classes of holomorphic 2-forms σ_t on (some deformation X_t of) X, and clearly $\sigma_t^{n+1} = 0$ as a form on X. It follows that the hypersurface $\mathcal{H}_{2n}$ of degree $2n$ in $\mathbb{P}(H^2(X, \mathbb{C}))$ which is defined by the degree $2n$ homogeneous form f given by the formula $f(\alpha) = \int_X \alpha^{2n}$ contains $\mathcal{D}_{loc}$ and has multiplicity at least n along it, hence has a component of multiplicity $\geq n$, which is not a hyperplane. One then concludes that this component is a quadric hypersurface Q and that $\mathcal{D}_{loc}$ is open in Q. Thus the equation q of the quadric Q and f are related by $f = \lambda q^n$ for some coefficient λ. Finally, using the fact that f is rational, one sees that both λ and q can be taken to be rational. The statement concerning the signature of q follows from the Hodge index theorem and further identities derived from (2.2).

The Verbitsky Torelli theorem for marked hyper-Kähler manifolds involves the integral structure on cohomology. It says the following:

THEOREM 2.2. *([**25**], [**14**]) Let X, X' be two hyper-Kähler manifolds which are deformation equivalent. Assume there is an isomorphism $\phi : H^2(X, \mathbb{Z}) \cong H^2(X', \mathbb{Z})$ which is obtained by transporting cohomology along a path of deformations from X to X' and such that $\phi([\sigma_X]) = [\sigma_{X'}]$. Then X and X' are birationally equivalent. If furthermore ϕ sends one Kähler class on X to a Kähler class on X', X and X' are isomorphic.*

2.2. Constructing hyper-Kähler manifolds. One particularity of hyper-Kähler geometry is the fact that although their deformation theory has an analytic and transcendental character, all known examples have been constructed by means of algebraic geometry. Note again that algebraic geometry provides constructions for the underlying complex manifolds, but of course not for the hyper-Kähler metrics. In dimension 2, hyper-Kähler manifolds are $K3$ surfaces (see [**29**]). These surfaces are all obtained by deforming smooth quartic surfaces in $\mathbb{P}^3$, defined by one degree 4 homogeneous equation $f(X_0, \ldots, X_3)$. The projective dimension of the space of such polynomials is 34, while $\mathrm{Aut}\,\mathbb{P}^3 = PGl(4)$ has dimension 15, so that the family of isomorphism classes of quartic $K3$ surfaces has dimension 19. This is a general fact of hyper-Kähler geometry. The number $h^{1,1}$ in this case is equal to 20 and this is the dimension of the space of all deformations of the $K3$ surface S. The deformations of S as a quartic surface are restricted since these $K3$ surfaces carry a holomorphic line bundle L such that $c_1(L)^2 = 4$. The class $c_1(L)$ is a Hodge class, that is integral of type $(1,1)$, and equivalently it is integral and satisfies $q(c_1(L), \sigma_S) = 0$. Note that the Beauville-Bogomolov form q is in this case the intersection pairing on middle cohomology. The general situation will be similar: given a hyper-Kähler manifold X equipped with an ample line bundle L, the polarized deformations of X, that is the deformations of the pair (X, L) consisting of a complex manifold and a holomorphic line bundle on it, have $h^{1,1}(X) - 1$ parameters, and are parameterized by the period domain $\mathcal{D}_l \subset \mathcal{D}$ determined by the condition $q(\sigma, l) = 0$ where $l = c_1(L)$.

In dimension 4, there are two known topological types of hyper-Kähler manifolds, namely $S^{[2]}$ and $K_2(A)$, where S is a $K3$ surface and A is a 2-dimensional abelian surface (or complex torus). The second punctual Hilbert scheme $X^{[2]}$ is defined for any algebraic variety or complex manifold X as the set of length 2 subschemes of X, which consist either of two distinct points of X or a point with a tangent vector. If X is smooth, this is the desingularization of the symmetric product $X^{(2)}$ obtained by blowing-up the diagonal. The generalized Kummer variety $K_2(A)$ is a fourfold obtained as follows: using the group structure of A, the smooth variety (or complex manifold) $A^{[3]}$ parametrizing length 3 subschemes of A admits the sum morphism $A^{[3]} \to A$, and $K_2(A)$ is any fiber of this morphism. These two examples appear in [**3**], as their next generalizations: It is a general theorem due to Fogarty [**10**] that the n-th Hilbert scheme $S^{[n]}$ is smooth for any n and any smooth surface S. For $n \leq 3$, the result is true without any assumptions on the dimension because length 3 subschemes are supported on smooth surfaces. Beauville shows that $S^{[n]}$ and $K_n(A) \subset A^{[n+1]}$ are hyper-Kähler manifolds of dimension $2n$ for S a $K3$ surface, A a 2-dimensional complex torus. The holomorphic 2-form on $S^{[n]}$ comes from the holomorphic 2-form σ_S on S by descending the $(2,0)$-form $\sum_i pr_i^* \sigma_S$ on S^n, which is invariant under the action of the symmetric group $\mathfrak{S}_n$, to the smooth part of the quotient $S^{(n)} = S^n/\mathfrak{S}_n$ and then showing that it extends to a $(2,0)$-form on the desingularization $S^{[n]}$.

Apart from these two series of examples, only two other deformation types are known, of respective dimensions 6 and 10, and they have been constructed by O'Grady [**22**], [**23**]. The 10-dimensional example is constructed as follows: Consider the moduli space of simple coherent sheaves $\mathcal{E}$ of rank 2 on a $K3$ surface S, satisfying $\det \mathcal{E} = \mathcal{O}_S$ (equivalently $c_1(\mathcal{E}) = 0$ in $H^2(S, \mathbb{Z})$), and $\deg c_2(\mathcal{E}) = 4$. By results of Mukai [**19**], this is a smooth algebraic variety with an everywhere nondegenerate closed holomorphic (in fact algebraic) 2-form. The variety is quasiprojective and admits a natural projective completion which is a moduli space of semistable sheaves (with respect to a given polarization). The later is singular along the locus of nonstable objects (for example $\mathcal{I}_z \oplus \mathcal{I}_{z'}$, where z and z' are two subschemes of length 2 in S). O'Grady constructs a hyper-Kähler desingularization of this singular moduli space. This variety has $b_2 = 24$, hence its algebraic deformations have 21 parameters. Those constructed starting from an algebraic $K3$ surface have 19 parameters.

Cubic fourfolds, that is smooth cubic hypersurfaces in $\mathbb{P}^5$, play an unexpected role in this study. Although they are Fano varieties, hence have a priori nothing to do with hyper-Kähler varieties, the Hodge decomposition on their degree 4 cohomology takes the form

$$H^4(X, \mathbb{C}) = H^{3,1}(X) \oplus H^{2,2}(X) \oplus H^{1,3}(X),$$

with $\dim H^{3,1}(X) = \dim H^{1,3}(X) = 1$. Furthermore, the space $H^{2,2}$ has dimension 21, and it contains one Hodge (in fact algebraic) class, namely h^2, $h = c_1(\mathcal{O}_X(1))$.

This Hodge structure is polarized by the intersection form $\langle\,,\,\rangle_X$ on the middle degree cohomology $H^4(X,\mathbb{Z})$ (which will thus play the role of the Beauville-Bogomolov form). This notion of polarization of the Hodge structure says that the line $H^{3,1}(X)\subset H^4(X,\mathbb{C})$ is isotropic for $\langle\,,\,\rangle_X$, and futhermore there is a (open) sign condition saying in our case that $\langle\alpha,\overline{\alpha}\rangle_X<0$ for $0\neq\alpha\in H^{3,1}(X)$. Thus up to shift of bigrading by $(1,1)$, (and change of sign for the quadratic form), what we get is a polarized Hodge structure of hyper-Kähler type. Furthermore, the cubic has 20 parameters (this is the dimension of the projective space of cubic homogeneous polynomials in 6 variables minus the dimension of $PGl(6)$) and in fact the period map $t\mapsto H^{3,1}(X_t)\subset H^4(X_t,\mathbb{C})=H^4(X,\mathbb{C})$ is a local isomorphism to the polarized period domain $\mathcal{D}_{h^2}$ which as before is an open set in the quadric defined by the Poincaré intersection pairing on $H^4(X,\mathbb{Z})^{\perp h^2}$. The cubic is there only to give us a polarized variation of Hodge structure of hyper-Kähler type with 20 parameters. Surprisingly enough, there are many hyper-Kähler manifolds associated with a cubic fourfold, which have their variation of rational Hodge structure isomorphic to the one described above on the degree 4 cohomology of the cubic (with a shift of degree). The known examples are:

(1) The variety of lines of a cubic fourfold X [**4**]. This is a hyper-Kähler fourfold $F(X)$ and the incidence correspondence $P\subset F(X)\times X$ (which is the universal $\mathbb{P}^1$-bundle over $F(X)$) induces an isomorphism $P^*:H^4(X,\mathbb{Z})\cong H^2(F(X),\mathbb{Z})$ of Hodge structures. Beauville and Donagi show that for some special cubic fourfolds X (more precisely "Pfaffian" cubic fourfolds), $F(X)$ becomes isomorphic to $S^{[2]}$ for some $K3$ surface S.

(2) The variety of rational curves of degree 3 in X [**17**]. This is a $\mathbb{P}^2$-bundle on a variety which is the blow-up of a hyper-Kähler 8-fold $Z(X)$ along a Lagrangian embedding of X in $Z(X)$. Again, the variations of Hodge structures on $H^4(X)$ and $H^2(Z(X))$ coincide, using the fact that up to a unininteresting summand, the H^2 of $Z(X)$ and $F_3(X)$ coincide, and then using the universal correspondence between $F_3(X)$ and X. This variety $F(X)$ has been shown in [**1**] to be of the same deformation type as $S^{[4]}$, where S is a $K3$ surface. Another proof of this last statement has been also given in [**15**].

(3) The intermediate Jacobian fibration of a cubic fourfold has been already mentioned in the introduction. This is a quasiprojective holomorphically symplectic 10-fold. It is proved in [**16**] that it admits a hyper-Kähler compactification $\overline{\mathcal{J}}$, and furthermore that $\overline{\mathcal{J}}$ is deformation equivalent to O'Grady's 10-dimensional varieties. One remark is that contrarily to the previous cases where we exhibited a complete family of projective hyper-Kähler manifolds, (hence a family with b_2-3 parameters), the expected dimension for the O'Grady examples should be 21 while our family has 20 parameters. This is due to the fact that the varieties that we construct are by definition Lagrangian fibered over $\mathbb{P}^5$. Hence their Picard number is at least 2, containing one class pulled-back from $\mathbb{P}^5$ and one ample class, and in fact it has to be generally equal to 2, since the family we construct has 20 parameters and the manifolds have $b_2=24$.

The rest of this paper is devoted to the description of the twisted version of (3), giving rise to a second 20 parameters family of deformations of O'Grady 10-dimensional varieties.

REMARK 2.3. The cubic fourfold is not the only Fano variety X which has associated hyper-Kähler manifolds whose variation of Hodge structure on H^2 is isomorphic (up to a shift of degree) to the variation of Hodge structure on some cohomology group of X. Another example can be found in [**8**], where X is a Plücker hyperplane section of the Grassmannian $G(3, V_{10})$ of 3-dimensional vector subspaces of a given 10-dimensional vector space V_{10}, so that X is defined by a general element s of $\bigwedge^3 V_{10}^*$. It is proved in *loc. cit.* that the fourfold $K_s \subset G(6, V_{10})$ consisting of 6-dimensional vector subspaces of V_{10} on which s vanishes identically is a hyper-Kähler fourfold, with VHS on H^2_{prim} isomorphic to the VHS on $H^{20}(X)_{prim}$.

3. The twisted intermediate Jacobian fibration: smooth and 1-nodal case

Let $X \subset \mathbb{P}^5$ be a smooth cubic fourfold, $B = (\mathbb{P}^5)^\vee$ be the set of hyperplane sections of X, and let $\mathcal{Y} \subset B \times X$ be the universal hypersurface. Let $U \subset U_1 \subset B$ be the Zariski open sets of B parametrizing smooth, resp. 1-nodal, hyperplane sections of X. The reason for introducing U_1 is, as in [**16**] which we follow closely, the fact that $B \setminus U_1$ has codimension 2 in B, so that the flat fibration with smooth, hence normal, total space $\overline{\mathcal{J}}^T$ will be determined (see Section 4) by its restriction $\mathcal{J}^T_{U_1}$ over U_1.

As we mentioned in the introduction, the twisted intermediate Jacobian fibration $\mathcal{J}^T_U$ over U can be interpreted as parametrizing 1-cycles of degree 1 in the fibers of the universal (smooth) hypersurface $u : \mathcal{Y}_U \to U$ over U. If we want to describe it as an abstract torsor over $\mathcal{J}_U$ or even better as a complex manifold, it can be defined by considering Deligne-Beilinson cohomology along the fibers of $u : \mathcal{Y}_U \to U$. This gives an exact sequence of sheaves of groups on U

$$0 \to \mathcal{J}_U \to \mathcal{H}^4_{\mathcal{D}} \stackrel{c}{\to} R^4 u_* \mathbb{Z} \to 0,$$

where the sheaf $R^4 u_* \mathbb{Z}$ is canonically isomorphic to $\mathbb{Z}$. We can then define $\mathcal{J}^T_U$ as $c^{-1}(1)$ (note that we identify here the analytic group fibration and its sheaf of holomorphic sections). The definition above is not good to understand the extension $\mathcal{J}^T_{U_1}$ and does not describe $\mathcal{J}^T_U$ as an algebraic object. We will thus give here an alternative description of $\mathcal{J}^T_{U_1}$ actually as a twist of $\mathcal{J}_{U_1}$. Our goal in this section is to establish the following result:

PROPOSITION 3.1. *There exists a quasiprojective variety $\mathcal{J}^T_{U_1}$ with a projective morphism $\pi^T_{U_1} : \mathcal{J}^T_{U_1} \to U_1$ which has the following properties.*

(1) The family $\mathcal{J}^T_{U_1}$ is étale locally isomorphic to $\mathcal{J}_{U_1}$ over U_1.

(2) Let $f : U' \to U$ be a base change, with U' smooth, and assume that there is a codimension 2 cycle $\mathcal{Z} \in \mathrm{CH}^2(\mathcal{Y}_{U'})$ that has degree 1 in the fibers of $\mathcal{Y}_{U'} \to U'$. Then there is a canonical section $U' \to \mathcal{J}^T_{U'}$ of the fibration $\mathcal{J}^T_{U'} := f^* \mathcal{J}^T_U \to U'$. Equivalently, there is a morphism $\Phi_{\mathcal{Z}} : U' \to \mathcal{J}^T_U$ over U.

More precisely, the morphism $\Phi_{\mathcal{Z}}$ is in fact compatible with the Abel-Jacobi map in the sense that for two points u', $u'' \in U'$ with $f(u') = f(u'') = u$, $\Phi_{\mathcal{Z}}(u') = \Phi_{\mathcal{Z}}(u'') + \Phi_{\mathcal{Y}_u}(\mathcal{Z}_{u'} - \mathcal{Z}_{u''})$, where $\Phi_{\mathcal{Y}_u}$ is the Abel-Jacobi map of $\mathcal{Y}_u$. This last fact is automatic, due to the universal property of the Abel-Jacobi map (see [**21**]). The proof of the proposition will be done through a few lemmas.

To start with, we observe that $\mathcal{J}_{U_1}$ contains a Zariski open set $\mathcal{J}^\circ_{U_1}$, which is a group scheme over U_1 and differs from $\mathcal{J}_{U_1}$ only over $U_1 \setminus U$. Over $U_1 \setminus U$, the fibers of $\mathcal{J}^\circ_{U_1}$ are quasiabelian varieties, and more precisely, the fiber $\mathcal{J}^\circ_{U_1,t}$ over $t \in U_1 \setminus U$ is a $\mathbb{C}^*$-bundle over the intermediate Jacobian $J(\widetilde{\mathcal{Y}_t})$, where $\widetilde{\mathcal{Y}_t}$ is the desingularization of $\mathcal{Y}_t$ obtained by blowing-up the node. Denoting by o_t the singular point of $\mathcal{Y}_t$, the fiber $\mathcal{J}^\circ_{U_1,t}$ can be understood in terms of algebraic cycles as the group of 1-cycles homologous to 0 in $\mathcal{Y}_t \setminus \{o_t\}$ modulo rational equivalence relative to o_t, that is modulo the cycles of the form div ϕ, where ϕ is a rational function on $W \subset \mathcal{Y}_t$ which is well defined and invertible near o_t if $o_t \in W$ (see [**18**], [**5**, Lecture 3]). The fibers $\mathcal{J}_{U_1,t}$ of the compactified Jacobian fibration $\mathcal{J}_{U_1}$ are obtained as the Mumford compactification [**20**], obtained by replacing the $\mathbb{C}^*$-bundle mentioned above by the corresponding $\mathbb{P}^1$-bundle with the sections 0 and ∞ glued via a translation of the base $J(\widetilde{\mathcal{Y}_t})$. The translation is given by an element $\eta \in J(\widetilde{\mathcal{Y}_t})$ and the $\mathbb{C}^*$-bundle is given by its class $\eta' \in \mathrm{Pic}^0 J(\widetilde{\mathcal{Y}_t})$; the two classes η and η' must be identified via the (principal) Theta divisor of $J(\widetilde{\mathcal{Y}_t})$. This is indeed the condition that this compactification admits an ample divisor which is the limit of the Theta divisors on the smooth fibers $J(\mathcal{Y}_s)$. (The reader will find in [**2**], [**7**] the geometric interpretation of this class in the case of the family of intermediate Jacobians associated to a Lefschetz degeneration of threefolds.) Notice that the (sheaf of holomorphic sections of) the analytic group scheme $\mathcal{J}^\circ_{U_1}$ is given by the formula

$$\mathcal{J}^\circ_{U_1} = \mathcal{H}^{1,2}/R^3u_{1*}\mathbb{Z}, \tag{3.1}$$

where $u_1 : \mathcal{Y}_{U_1} \to U_1$ is the universal hypersurface over U_1, (thus, by Picard-Lefschetz theory, $R^3u_{1*}\mathbb{Z} = j_{1*}H^3_{\mathbb{Z}}$ is the natural extension of the local system $H^3_{\mathbb{Z}} = R^3u_*\mathbb{Z}$ existing on U,) and $\mathcal{H}^{1,2}$ is the Deligne extension of the Hodge bundle

$$\mathcal{H}^{1,2}_U = R^1u_*\Omega^2_{\mathcal{Y}_U/U}$$

existing on U. Let us prove the following lemma:

LEMMA 3.2. *The group scheme $\mathcal{J}^\circ_{U_1}$ acts (over U_1) on the compactification $\mathcal{J}_{U_1}$.*

PROOF. One easy way to prove this is to observe that we have a family $\pi_U : \mathcal{J}_U \to U$ of principally polarized varieties with a canonical Theta divisor $\Theta_U \subset \mathcal{J}_U$ inducing a relative isomorphism

$$\mathcal{J}_U \cong \mathrm{Pic}^0(\mathcal{J}_U/U).$$

Then $\mathcal{J}^\circ_{U_1}$ identifies to $\mathrm{Pic}^0(\mathcal{J}_{U_1/U_1})$ and by uniqueness, the compactification $\mathcal{J}_{U_1}$ identifies with the Mumford compactification of $\mathrm{Pic}^0(\mathcal{J}_{U_1/U_1})$ parameterizing rank 1 torsion free sheaves on fibers of $\pi_{U_1} : \mathcal{J}_{U_1} \to U_1$. The action over U_1 of $\mathcal{J}^\circ_{U_1}$ on $\mathcal{J}_{U_1}$ is then the action of Pic^0 on sheaves by tensor product. □

Formula (3.1) shows that $R^3u_{1*}(\mathbb{Z}/3\mathbb{Z}) \subset \mathcal{J}^\circ_{U_1}$ identifies with the sheaf ${}_3\mathcal{J}^\circ_{U_1}$ of 3-torsion points and Lemma 3.2 shows that $\mathcal{J}^\circ_{U_1} \subset \mathcal{A}ut\,(\mathcal{J}_{U_1}/U_1)$. Let us now exhibit the twisting class in $H^1_{et}(U_1, \mathcal{A}ut\,(\mathcal{J}_{U_1}/U_1))$ needed to construct $\mathcal{J}^T_{U_1}$. The twisting class will be of 3-torsion and more precisely will come from a class in $H^1_{et}(U_1, {}_3\mathcal{J}^\circ_{U_1}) = H^1(U_1, {}_3\mathcal{J}^\circ_{U_1})$, where in the right hand side we consider cohomology with respect to the usual topology. Consider the morphism $u_1 : \mathcal{Y}_{u_1} \to U_1$. As they admit at worst one node, the fibers of u_1 satisfy $H^4(\mathcal{Y}_t, \mathbb{Z}) = \mathbb{Z}$ with generator at the point t the class of a line (not passing through the node of $\mathcal{Y}_t$ if $\mathcal{Y}_t$ is singular). Let $h = c_1(\mathcal{O}_X(1))$ and $h_{\mathcal{Y}} \in H^2(\mathcal{Y}, \mathbb{Z})$ be its pull-back to $\mathcal{Y}$. The image $\widetilde{h^2_{\mathcal{Y}}}$ of the class $h^2_{\mathcal{Y}}$ in $H^0(U_1, R^4u_{1*}\mathbb{Z})$ is equal to 3 times the above generator of $R^4u_{1*}\mathbb{Z} = \mathbb{Z}$. The twisting class comes from the lack of degeneracy at E_2 of the Leray spectral sequence of u_1 with integral coefficients.

LEMMA 3.3. *Let $\sigma \in H^0(U_1, R^4u_{1*}\mathbb{Z})$ be the natural generator. Then the image $d_2\sigma \in H^2(U_1, R^3u_{1*}\mathbb{Z})$ is of 3-torsion, and comes from a canonically defined element*

$$t \in H^1(U_1, R^3u_{1*}(\mathbb{Z}/3\mathbb{Z})).$$

PROOF. The class $d_2\sigma \in H^2(U_1, R^3u_{1*}\mathbb{Z})$ is of 3-torsion because $3\sigma = \widetilde{h^2_{\mathcal{Y}}}$ is the image of $h^2_{\mathcal{Y}} \in H^4(\mathcal{Y}_{U_1}, \mathbb{Z})$ in $H^0(U_1, R^4u_{1*}\mathbb{Z})$. The exact sequence

$$0 \to \mathbb{Z} \stackrel{3}{\to} \mathbb{Z} \to \mathbb{Z}/3\mathbb{Z} \to 0$$

of constant sheaves on $\mathcal{Y}_{U_1}$ induces the exact sequence

$$0 \to R^3u_{1*}\mathbb{Z} \stackrel{3}{\to} R^3u_{1*}\mathbb{Z} \to R^3u_{1*}(\mathbb{Z}/3\mathbb{Z}) \to 0 \tag{3.2}$$

because $R^3u_{1*}\mathbb{Z}$ and $R^4u_{1*}\mathbb{Z}$ have no torsion. The long exact sequence associated to (3.2) thus shows that $d_2\sigma$ lifts to an element of $H^1(U_1, R^3u_{1*}(\mathbb{Z}/3\mathbb{Z})) = H^1(U_1, {}_3\mathcal{J}^\circ_{U_1})$. This does not prove however that this lift is canonical since in the considered long exact sequence of Betti cohomology, there is a nontrivial cohomology group $H^1(U_1, R^3u_{1*}\mathbb{Z})$ (this is due to the non-triviality of $H^4(X, \mathbb{Z})$). We can however go around this difficulty by considering the universal situation where instead of considering the universal family $\mathcal{Y}_{U_1}$ of smooth or 1-nodal hyperplane sections of the given cubic fourfold X, we consider the universal family $v_1 : \mathcal{Y}^{univ}_{W_1} \to W_1$ of smooth or 1-nodal cubic threefolds contained in some hyperplane $\mathbb{P}^4 \subset \mathbb{P}^5$. Here the whole parameter space B_{univ} is the projective bundle over $(\mathbb{P}^5)^\vee$ with fiber over $[H]$ the projective space $\mathbb{P}(H^0(\mathcal{O}_H(3)))$ and $W_1 \subset B_{univ}$ is a Zariski open set in it. In this case, we have

SUBLEMMA 3.4. *One has $H^1(W_1, R^3v_{1*}\mathbb{Z}) = 0$.*

PROOF. One proves the result first of all with $\mathbb{Q}$-coefficients, using the fact that the Leray spectral sequence degenerates in E_2 and that the degree 4 cohomology of $\mathcal{Y}^{univ}_{W_1}$ is algebraic and very simple. Then one concludes using the exact sequence on W_1

$$0 \to R^3v_{1*}\mathbb{Z} \to R^3v_{1*}\mathbb{Q} \to R^3v_{1*}(\mathbb{Q}/\mathbb{Z}) \to 0 \tag{3.3}$$

and the fact that $H^0(W_1, R^3v_{1*}(\mathbb{Q}/\mathbb{Z})) = 0$. □

This fact gives us a canonical lift of $d_2\sigma^{univ}$ to an element

$$t^{univ} \in H^1(W_1, R^3v_{1*}(\mathbb{Z}/3\mathbb{Z})),$$

hence by pull-back to U_1 (using the natural map $g : U_1 \to W_1$), the desired canonical lift t of $d_2\sigma$ to an element of $H^1(U_1, R^3u_{1*}(\mathbb{Z}/3\mathbb{Z}))$. □

Lemma 3.3 gives us a class in $H^1(U_1, {}_3\mathcal{J}_{U_1}) = H^1_{et}(U_1, {}_3\mathcal{J}_{U_1})$ which by Lemma 3.2 allows us to construct a twisted family $\pi_{U_1} : \mathcal{J}^T_{U_1} \to U_1$, which is above U a torsor over the group scheme $\mathcal{J}_U$. Note that the proof of Lemma 3.3 also shows that $\mathcal{J}^T_{U_1}$ is the pull-back under $g_1 : U_1 \to W_1$ of the corresponding object $\mathcal{J}^{T,univ}_{W_1}$ over W_1. We now prove that the torsor so constructed is actually the object we want, namely, the target of the Abel-Jacobi map for 1-cycles of degree 1 along the smooth fibers of u. This will follow from the following result: Let as before $W_1 \subset B_{univ}$ be the Zariski open set parameterizing cubic threefolds in $\mathbb{P}^5$ with at most one ordinary double point. Let $o_1 : \mathcal{F}^{univ}_1 \to W_1$ be the family of lines in the fibers of $v_1 : \mathcal{Y}^{univ}_{W_1} \to W_1$ and $\mathcal{F}^{0,univ}_1 \subset \mathcal{F}^{univ}_1$ be the Zariski open set consisting of lines not passing through the node when the corresponding cubic threefold is singular. Let $o_{1,0}$ be the restriction of o_1 to $\mathcal{F}_1^0$.

LEMMA 3.5. *There exists an isomorphism*

$$o_{1,0}^*\mathcal{J}_{W_1}^{univ} \cong o_{1,0}^*\mathcal{J}_{W_1}^{T,univ} \tag{3.4}$$

of quasi-projective varieties over $\mathcal{F}_1^{0,univ}$.

PROOF. It suffices to show that the pull-back $o_{1,0}^*(t)$ vanishes in the cohomology group $H^1(\mathcal{F}_1^{0,univ}, o_{1,0}^*({}_3\mathcal{J}_{W_1}^{univ}))$. Note that the morphism $o_{1,0}$ is smooth, because the family of lines in the smooth locus of a cubic hypersurface is smooth. It follows that the total space of the universal family

$$\mathcal{Y}_{\mathcal{F}_1^{0,univ}}^{univ} := \mathcal{Y}_{W_1}^{univ} \times_{W_1} \mathcal{F}_1^{0,univ}$$

is smooth. Let $v_1' : \mathcal{Y}_{\mathcal{F}_1^{0,univ}}^{univ} \to \mathcal{F}_1^{0,univ}$ be the natural morphism. Let $H_{\mathbb{Z}}^4$, $H_{\mathbb{Z}}^3$, $H_{\mathbb{Z}/3\mathbb{Z}}^3$ be respectively the sheaves

$$R^4v'_{1*}\mathbb{Z}=o_{1,0}^*(R^4v_{1*}\mathbb{Z}),\ R^3v'_{1*}\mathbb{Z}=o_{1,0}^*(R^3v_{1*}\mathbb{Z}),\ R^3v'_{1*}(\mathbb{Z}/3\mathbb{Z})=o_{1,0}^*(R^3v_{1*}(\mathbb{Z}/3\mathbb{Z}))$$

on $\mathcal{F}_1^{0,univ}$. The class

$$o_{1,0}^*\sigma \in H^0(\mathcal{F}_1^{0,univ}, o_{1,0}^*(R^4v_{1*}\mathbb{Z})) = H^0(\mathcal{F}_1^{0,univ}, H_{\mathbb{Z}}^4)$$

comes from a class in $H^4(\mathcal{Y}_{\mathcal{F}_1^{0,univ}}^{univ}, \mathbb{Z})$, namely the class $[\Delta^{univ}]$ of the universal line $\Delta^{univ} \subset \mathcal{Y}_{\mathcal{F}_1^{0,univ}}^{univ}$. We thus get that $d_2(o_{1,0}^*\sigma) = 0$ in $H^2(\mathcal{F}_1^{0,univ}, H_{\mathbb{Z}}^3)$. This means that the class

$$o_{1,0}^*(t) \in H^1(\mathcal{F}_1^{0,univ}, H_{\mathbb{Z}/3\mathbb{Z}}^3) = H^1(\mathcal{F}_1^{0,univ},\ {}_3\mathcal{J}_{\mathcal{F}_1^0}^{univ})$$

vanishes in $H^2(\mathcal{F}_1^{0,univ}, H_{\mathbb{Z}}^3)$ (using as before the long exact sequence associated to the short exact sequence

$$0 \to H_{\mathbb{Z}}^3 \to H_{\mathbb{Z}}^3 \to H_{\mathbb{Z}/3\mathbb{Z}}^3 \to 0.$$

In order to prove that $o_{1,0}^*(t)$ vanishes in $H^1(\mathcal{F}_1^{0,univ},\ {}_3\mathcal{J}_{\mathcal{F}_1^0}^{univ})$, it thus suffices to show that $H^1(\mathcal{F}_1^{0,univ}, o_{1,0}^*H_{\mathbb{Z}}^3) = 0$, which is done exactly as before. ☐

REMARK 3.6. The isomorphism between the pullbacks to $\mathcal{F}_1^{0,univ}$ of the twisted and the untwisted families given in Lemma 3.5 is not canonical since over $\mathcal{F}_W^{univ}$ the intermediate Jacobian fibration has a nonzero section. Indeed, for each cubic smooth cubic threefold Y with a line $\Delta \subset Y$, the 1-cycle $3\Delta - h^2$ is cohomologous to zero on Y, hence has a nontrivial Abel-Jacobi invariant. This provides a nontrivial section which acts by translation on $o^*\mathcal{J}_W^{univ}$ over $\mathcal{F}_W$. One can show that the isomorphism is unique up to the action of the group generated by this translation.

PROOF OF PROPOSITION 3.1. We already constructed $\mathcal{J}_{U_1}^T$ as the pull-back via $g_1 : U_1 \to W_1$ of the twisted family $\mathcal{J}_{W_1}^T$. That it is étale locally isomorphic to $\mathcal{J}_{U_1}$ over U_1 is by construction. The only thing to prove is point (2). However, Lemma 3.5 shows that for any base change morphism $f : U' \to U$, assuming there is a family of lines $\Delta_{U'} \subset \mathcal{Y}_{U'}$, there is a canonical morphism $\Phi_\Delta : U' \to \mathcal{J}_U^T$ over U. Indeed, the data above give a morphism $c : U' \to \mathcal{F}_W^{univ}$ so we can use Φ_Δ^{univ} of the previous lemma, and define $\Phi_\Delta := \Phi_{\Delta^{univ}} \circ c$. To see that a family of 1-cycles of degree 1 in the fibers of $\mathcal{Y}_{U'} \to U'$ in fact suffices to trivialize the twisted Jacobian fibration, we can use the universal generation result of [**24**] which says that any 1-cycle on a smooth cubic hypersurface Y over a field K comes from a 0-cycle on the surface of lines $F(Y)$, also defined over K. In our case, K is the function field of U'. Using this result, we get a correspondence over U' between U' and $\mathcal{F}_{U'}$, which has to be of degree 1. Using this correspondence and the previous step, we construct the section $U' \to \mathcal{J}^T_{U'}$. □

4. Descent

Our goal in this section is to explain how to mimic the arguments from [**16**] in order to construct for a general cubic fourfold X a hyper-Kähler compactification of the twisted intermediate Jacobian fibration $\mathcal{J}^T_{U_1}$ constructed in the previous section. Recall the notion of very good line in a cubic threefold Y, introduced and used in [**16**] (see also [**6**] for a slightly weaker notion): A line $\Delta \subset Y$ is very good if Δ does not pass through the singular point of Y (equivalently the surface of lines $F(Y)$ is smooth at $[\Delta]$), the curve $\widetilde{C}_\Delta$ of lines in Y meeting Δ is irreducible and the natural involution acting on $\widetilde{C}_\Delta$ has no fixed points. One of the results proved in [**16**] (improving previous results of [**6**]) is:

PROPOSITION 4.1. *If X is a general cubic 4-fold, and $Y \subset X$ is any hyperplane section, Y has a very good line.*

Let $o_{vg} : \mathcal{F}^{vg} \to B$ be the family of very good lines in the fibers of $u : \mathcal{Y} \to B$. Proposition 4.1 says that o_{vg} is surjective and by definition it is smooth. Restricting over U_1, we get a Zariski open set

$$\mathcal{F}_1^{vg} \subset \mathcal{F}_1^0,\ o_{vg,1} : \mathcal{F}_1^{vg} \to U_1$$

with its pulled-back intermediate Jacobian fibration $o^*_{vg,1}\mathcal{J}_{U_1} := \mathcal{F}_1^{vg} \times_{U_1} \mathcal{J}_{U_1}$, which is also isomorphic to $o^*_{vg,1}\mathcal{J}^T_{U_1}$ by Lemma 3.5. In [**16**], a smooth quasiprojective compactification $\pi_{\mathcal{F}^{vg}} : \overline{\mathcal{P}}_{\mathcal{F}^{vg}} \to \mathcal{F}^{vg}$ of $o^*_{vg,1}\mathcal{J}_{U_1} \to \mathcal{F}_1^{vg}$ is constructed, with a morphism $\pi_{\mathcal{F}^{vg}}$ which is flat and projective. The compactified intermediate Jacobian fibration $\overline{\mathcal{J}}$ is then obtained by descending the fibration $\pi_{\mathcal{F}^{vg}}$. The statement which makes it possible is:

LEMMA 4.2. *There exists a line bundle $\mathcal{O}(\Theta_1)$ on $\mathcal{J}_{U_1}$ whose pull-back $o^*_{vg,1}\mathcal{O}(\Theta_1)$ to $o^*_{vg,1}\mathcal{J}_{U_1}$ extends uniquely to a relatively ample line bundle $\mathcal{O}(\Theta^{vg})$ on $\overline{\mathcal{P}}_{\mathcal{F}^{vg}}$.*

Note that the extension of the pull-back $o^*_{vg,1}\mathcal{O}(\Theta_1)$ to a line bundle on $\overline{\mathcal{P}}_{\mathcal{F}^{vg}}$ exists and is unique because $\overline{\mathcal{P}}_{\mathcal{F}^{vg}}$ is smooth and $o^*_{vg,1}\mathcal{J}_{U_1} \subset \overline{\mathcal{P}}_{\mathcal{F}^{vg}}$ is Zariski open with complement of codimension at least 2. This last point follows from the flatness of $\pi_{\mathcal{F}^{vg}}$ because $\mathcal{F}_1^{vg} \subset \mathcal{F}^{vg}$ is Zariski open with complement of codimension ≥ 2. The important point in Lemma 4.2 is thus relative ampleness. Once we have the line bundle $\mathcal{O}(\Theta_1)$ on $\mathcal{J}_{U_1}$ as in the lemma, we get the formula defining $\overline{\mathcal{J}}$ as a Proj over B, namely, letting $j_1 : U_1 \to B$ be the natural inclusion, we set

$$\overline{\mathcal{J}} = Proj\,(\oplus_k j_{1*}(R^0\pi_{U_1*}\mathcal{O}(k\Theta_1))). \tag{4.1}$$

Using the fact that $B \setminus U_1$ has codimension 2 in B, flatness of $\pi_{\mathcal{F}^{vg}}$ and relative ampleness of (the extension of) the pull-back of Θ_1, we see that the sheaf of algebras $\oplus_k j_{1*}(R^0\pi_{U_1*}\mathcal{O}(k\Theta_1))$ is a sheaf of finitely generated algebras over $\mathcal{O}_B$ whose graded pieces are locally free and that the associated Proj is smooth, because all these properties become true after pull-back to $\mathcal{F}^{vg}$, thanks to the existence of the flat compactification $\pi_{\mathcal{F}^{vg}}$. Indeed, the sheaf of algebras $\oplus_k j_{1*}(R^0\pi_{U_1*}\mathcal{O}(k\Theta_1))$ pulls-back to $\oplus_k R^0\pi_{\mathcal{F}^{vg}*}\mathcal{O}(k\Theta^{vg})$ on $\mathcal{F}^{vg}$. (To be completely rigorous here, we should in fact replace the line bundle Θ_1 by a multiple.)

The way this descent construction works also makes clear what ingredient we need in order to treat the twisted case. Indeed, what we are going to do is to descend the same fibration $\pi_{\mathcal{F}^{vg}} : \overline{\mathcal{P}}_{\mathcal{F}^{vg}} \to \mathcal{F}^{vg}$ to a flat fibration $\pi^T : \overline{\mathcal{J}}^T \to B$ with a different descent data, given by a different relatively ample divisor. More precisely, consider $\pi_{\mathcal{F}^{vg}} : \overline{\mathcal{P}}_{\mathcal{F}^{vg}} \to \mathcal{F}^{vg}$ as a relative flat projective compactification of $o^*_{vg,1}\mathcal{J}^T_{U_1} \to \mathcal{F}^{vg}_1$ using Lemma 3.5. The only ingredient needed is the following:

Proposition 4.3. *There exists a line bundle $\mathcal{O}(\Theta^{vg,T})$ on $\overline{\mathcal{P}}_{\mathcal{F}^{vg}}$, which is relatively ample over $\mathcal{F}^{vg}$, and whose restriction to $\mathcal{F}^{vg}_1$ is the pull-back (using the isomorphism (3.4)) of a line bundle $\mathcal{O}(\Theta^T_1)$ on $\mathcal{J}^T_{U_1}$.*

Indeed, once one has the line bundles $\mathcal{O}(\Theta^{vg,T})$ and $\mathcal{O}(\Theta^T_1)$ as above, one defines $\overline{\mathcal{J}}^T$ as

$$\overline{\mathcal{J}}^T = Proj\,(\oplus_k j_{1*}(R^0\pi_{U_1*}\mathcal{O}(k\Theta^T_1))) \tag{4.2}$$

and the same arguments as above show that this is a smooth projective variety, flat over B and extending $\mathcal{J}^T_{U_1}$.

Proof of Proposition 4.3. It is proved in [**16**, Section 4] that, X being general, the fibers of $\pi_{\mathcal{F}_{vg}} : \overline{\mathcal{P}}_{\mathcal{F}^{vg}} \to \mathcal{F}^{vg}$ are irreducible. We now use the following lemma:

Lemma 4.4. *Let M be a smooth irreducible quasiprojective variety, $f : M \to N$ be a flat projective morphism with irreducible fibers M_t, $\forall t \in N$, and let $\mathcal{L} \in \operatorname{Pic} M$. If $\mathcal{L}_{|M_t}$ is topologically trivial for the general point $t \in N$, $\mathcal{L}_{|M_t}$ is numerically trivial for all $t \in N$.*

Proof. The statement is local on N. Let $t_0 \in N$ and let $C \subset M_{t_0}$ be a curve. Choosing a sufficiently relatively ample line bundle H on M and using the fact that M_{t_0} is irreducible, we can assume that C is contained in an irreducible surface $S_{t_0} \subset M_{t_0}$ which is a complete intersection of members of $|H_{|M_{t_0}}|$. Up to replacing H by a multiple and shrinking N if necessary, we can construct a flat family $f_S : S \to N$, $S \subset M$ of complete intersection surfaces with fiber S_{t_0} over the point t_0. As the line bundle $\mathcal{L}_{|M_t}$ is topologically trivial for general t, we conclude that

$$c_1(\mathcal{L}) \cdot c_1(H) \cdot S_t = 0,\ c_1(\mathcal{L})^2 \cdot S_t = 0$$

for general t, and by flatness, this is also true for S_{t_0}. Let $\tau : \widetilde{S_{t_0}} \to S_{t_0}$ be a desingularization. The line bundle $\mathcal{L}'_0 := \tau^*\mathcal{L}_{|S_{t_0}}$ on the smooth connected surface $\widetilde{S_{t_0}}$ satisfies

$$c_1(\mathcal{L}'_0)^2 = 0,\ c_1(\mathcal{L}'_0) \cdot c_1(\tau^* H) = 0.$$

As $c_1(\tau^*H)^2 > 0$, it follows from the Hodge index theorem that $\mathcal{L}'_0$ is topologically trivial modulo torsion on $\widetilde{S_{t_0}}$. In particular, if $C' \subset \widetilde{S_{t_0}}$ is a curve mapping onto $C \subset S_{t_0}$, we have $\deg \mathcal{L}'_{0|C'} = 0 = D \deg \mathcal{L}_{|C}$, where D is the degree of C' over C. Thus $\deg \mathcal{L}_{|C} = 0$. ∎

Let now Θ_1^T be any relatively ample line bundle on $\mathcal{J}_{U_1}^T \to U_1$. Its pull-back to $\mathcal{J}_{\mathcal{F}_1^{vg}}^T \cong \mathcal{J}_{\mathcal{F}_1^{vg}}$ extends to a line bundle $\mathcal{O}(\Theta^{vg,T})$ on the compactified family of Prym varieties $\pi_{\mathcal{F}^{vg}} : \overline{\mathcal{P}}_{\mathcal{F}^{vg}} \to \mathcal{F}^{vg}$. We also have on $\mathcal{J}_{\mathcal{F}_1^{vg}}$ the pull-back of the relatively ample line bundle Θ_1 on $\mathcal{J}_{U_1} \to U_1$. The later extends (in fact uniquely) to a relatively ample line bundle $\mathcal{O}(\Theta^{vg})$ on $\overline{\mathcal{P}}_{\mathcal{F}^{vg}} \to \mathcal{F}^{vg}$. Next, an easy monodromy argument shows (see [**16**, Section 5]) that the Néron-Severi group of the intermediate Jacobian of a very general cubic 3-fold is isomorphic to $\mathbb{Z}$. It follows that for adequate positive integers a, b, the line bundle $\mathcal{O}(a\Theta^{vg,T} - b\Theta^{vg})$ is topologically trivial on the general fibers of $\pi_{\mathcal{F}^{vg}}$. We then conclude by Lemma 4.4 that $\mathcal{O}(a\Theta^{vg,T} - b\Theta^{vg})$ is numerically trivial on all the fibers of $\pi_{\mathcal{F}^{vg}}$. Thus $\mathcal{O}(a\Theta^{vg,T})$ is the sum of an ample line bundle and a numerically trivial line bundle on any fiber of $\pi_{\mathcal{F}^{vg}}$, hence it is relatively ample. □

This concludes the construction of the smooth projective variety $\overline{\mathcal{J}}^T$. The fact that this is a hyper-Kähler manifold follows easily, as in [**16**]: The existence of the holomorphic 2-form works as in [**16**]. Indeed, according to Proposition 3.1, $\mathcal{J}_U^T$ is naturally the set of 1-cycles of degree 1 modulo rational equivalence in the fibers of the universal family $\mathcal{Y}_U \to U$, in the sense that for any smooth morphism $\phi : \mathcal{M} \to U$, and codimension 2 cycle $\mathcal{Z} \in \mathrm{CH}^2(\mathcal{M} \times_U \times \mathcal{Y}_U)$ such that $\mathcal{Z}_m$ is of degree 1 on the fiber $\mathcal{Y}_t$, $t = \phi(m)$, there exists a morphism (twisted Abel-Jacobi map) $\mathcal{M} \to \mathcal{J}_U^T$. A universal cycle, for which $\mathcal{M}$ is $\mathcal{J}_U^T$ and the map is the identity, may not exist with integer coefficients, but it always exists with rational coefficients, so we have a cycle $\mathcal{Z} \in \mathrm{CH}^2(\mathcal{J}_U^T \times_U \times \mathcal{Y}_U)_{\mathbb{Q}}$ with rational coefficients such that $[\mathcal{Z}]^*$ is the natural isomorphism $R^3u_*\mathbb{Q} \cong R^1\pi_{U*}\mathbb{Q}$. Using the natural proper map (inclusion) $\Phi : \mathcal{J}_U^T \times_U \times \mathcal{Y}_U \to \mathcal{J}_U^T \times X$, we get a codimension 3 cycle $\mathcal{Z}' = \Phi_*(\mathcal{Z}) \in \mathrm{CH}^3(\mathcal{J}_U^T \times X)_{\mathbb{Q}}$. We construct the holomorphic 2-form on $\mathcal{J}_U^T$ as

$$\sigma_U = [\mathcal{Z}']^*\alpha_X \in H^0(\mathcal{J}_U^T, \Omega^2_{\mathcal{J}_U^T}),$$

where α_X is a generator of $H^1(X, \Omega_X^3)$. For any algebraic extension W of $\mathcal{J}_U^T$, we can extend the cycle $\mathcal{Z}'$ to $W \times X$, showing that the $(2,0)$-form σ_U extends to W. In particular, we get the desired holomorphic 2-forms σ_{U_1} on $\mathcal{J}_{U_1}^T$ and σ on $\overline{\mathcal{J}}^T$. The fact that σ is nondegenerate on $\overline{\mathcal{J}}^T$ is a consequence of the fact that σ_{U_1} is nondegenerate on $\mathcal{J}_{U_1}^T$, since $B \setminus U_1$ has codimension ≥ 2 in B and $\pi : \overline{\mathcal{J}}^T \to B$ is flat. The fact that σ_{U_1} is nondegenerate on $\mathcal{J}_{U_1}^T$ follows from the similar statement for the untwisted family since they are étale locally isomorphic. Finally, the fact that the variety we construct is actually hyper-Kähler (i.e. simply connected with only one holomorphic 2-form up to a coefficient) follows from the fact that, by Proposition 3.1, the two varieties are birational (this will also imply *a posteriori* that they are deformation equivalent) when the cubic fourfold X acquires an integral Hodge class $\alpha \in H^4(X, \mathbb{Z})$ which has degree 1 on its hyperplane sections. Note that the above construction of $\overline{\mathcal{J}}^T$ and $\overline{\mathcal{J}}$ works a priori only for general X, but the set of special X's as above is Zariski dense (this is a countable union of hypersurfaces, dense for the usual topology in the moduli space of cubic fourfolds), hence there are points which correspond to special X's in Hassett's sense (see [**11**]), with a special class of degree 1 along hyperplane sections, and for which the constructions of $\overline{\mathcal{J}}^T$ and $\overline{\mathcal{J}}$ specialize well.

REMARK 4.5. The two varieties $\overline{\mathcal{J}}^T$ and $\overline{\mathcal{J}}$ are isogenous in the sense that there is a rational map (of degree 3^{10}) $\overline{\mathcal{J}}^T \dashrightarrow \overline{\mathcal{J}}$. This is obvious from the fact that the open part $\mathcal{J}_U^T$ is constructed as a torsor over the group scheme $\mathcal{J}_U^T$, with a twisting class of order 3. We believe but did not prove that the two varieties are not birational. We only note that by construction they are not birational over B.

References

[1] Nicolas Addington and Manfred Lehn, *On the symplectic eightfold associated to a Pfaffian cubic fourfold*, J. Reine Angew. Math. **731** (2017), 129–137. MR3709062

[2] Fabio Bardelli, *Polarized mixed Hodge structures: on irrationality of threefolds via degeneration* (English, with Italian summary), Ann. Mat. Pura Appl. (4) **137** (1984), 287–369. MR772264

[3] Arnaud Beauville, *Variétés Kähleriennes dont la première classe de Chern est nulle* (French), J. Differential Geom. **18** (1983), no. 4, 755–782 (1984). MR730926

[4] Arnaud Beauville and Ron Donagi, *La variété des droites d'une hypersurface cubique de dimension* 4 (French, with English summary), C. R. Acad. Sci. Paris Sér. I Math. **301** (1985), no. 14, 703–706. MR818549

[5] Spencer Bloch, *Lectures on algebraic cycles*, Duke University Mathematics Series, IV, Duke University, Mathematics Department, Durham, N.C., 1980. MR558224

[6] Sebastian Casalaina-Martin and Radu Laza, *The moduli space of cubic threefolds via degenerations of the intermediate Jacobian*, J. Reine Angew. Math. **633** (2009), 29–65. MR2561195

[7] Alberto Collino, *A cheap proof of the irrationality of most cubic threefolds* (English, with Italian summary), Boll. Un. Mat. Ital. B (5) **16** (1979), no. 2, 451–465. MR546468

[8] Olivier Debarre and Claire Voisin, *Hyper-Kähler fourfolds and Grassmann geometry*, J. Reine Angew. Math. **649** (2010), 63–87. MR2746467

[9] Ron Donagi and Eyal Markman, *Cubics, integrable systems, and Calabi-Yau threefolds*, Proceedings of the Hirzebruch 65 Conference on Algebraic Geometry (Ramat Gan, 1993), Israel Math. Conf. Proc., vol. 9, Bar-Ilan Univ., Ramat Gan, 1996, pp. 199–221. MR1360503

[10] John Fogarty, *Algebraic families on an algebraic surface*, Amer. J. Math **90** (1968), 511–521. MR0237496

[11] Brendan Hassett, *Special cubic fourfolds*, Compositio Math. **120** (2000), no. 1, 1–23. MR1738215

[12] D. Huybrechts. Compact hyperkähler manifolds. Habilitationsschrift Essen (1997).

[13] Daniel Huybrechts, *Compact hyper-Kähler manifolds: basic results*, Invent. Math. **135** (1999), no. 1, 63–113. MR1664696

[14] Daniel Huybrechts, *A global Torelli theorem for hyperkähler manifolds [after M. Verbitsky]*, Astérisque **348** (2012), Exp. No. 1040, x, 375–403. Séminaire Bourbaki: Vol. 2010/2011. Exposés 1027–1042. MR3051203

[15] J. Kollár, R. Laza, G. Saccà, C. Voisin. Remarks on degenerations of hyper-Kähler manifolds, arXiv:1704.02731.

[16] Radu Laza, Giulia Saccà, and Claire Voisin, *A hyper-Kähler compactification of the intermediate Jacobian fibration associated with a cubic 4-fold*, Acta Math. **218** (2017), no. 1, 55–135. MR3710794

[17] Christian Lehn, Manfred Lehn, Christoph Sorger, and Duco van Straten, *Twisted cubics on cubic fourfolds*, J. Reine Angew. Math. **731** (2017), 87–128. MR3709061

[18] Marc Levine and Chuck Weibel, *Zero cycles and complete intersections on singular varieties*, J. Reine Angew. Math. **359** (1985), 106–120. MR794801

[19] Shigeru Mukai, *Symplectic structure of the moduli space of sheaves on an abelian or K3 surface*, Invent. Math. **77** (1984), no. 1, 101–116. MR751133

[20] David Mumford, *An analytic construction of degenerating abelian varieties over complete rings*, Compositio Math. **24** (1972), 239–272. MR0352106

[21] J. P. Murre, *Applications of algebraic K-theory to the theory of algebraic cycles*, Algebraic geometry, Sitges (Barcelona), 1983, Lecture Notes in Math., vol. 1124, Springer, Berlin, 1985, pp. 216–261. MR805336

[22] Kieran G. O'Grady, *Desingularized moduli spaces of sheaves on a K3*, J. Reine Angew. Math. **512** (1999), 49–117. MR1703077

[23] Kieran G. O'Grady, *A new six-dimensional irreducible symplectic variety*, J. Algebraic Geom. **12** (2003), no. 3, 435–505. MR1966024

[24] M. Shen. Rationality, universal generation and the integral Hodge conjecture, arXiv:1602.07331.

[25] Misha Verbitsky, *Mapping class group and a global Torelli theorem for hyperkähler manifolds*, Duke Math. J. **162** (2013), no. 15, 2929–2986. Appendix A by Eyal Markman. MR3161308

[26] Claire Voisin, *Hodge theory and complex algebraic geometry. I*, Reprint of the 2002 English edition, Cambridge Studies in Advanced Mathematics, vol. 76, Cambridge University Press, Cambridge, 2007. Translated from the French by Leila Schneps. MR2451566

[27] Claire Voisin, *Abel-Jacobi map, integral Hodge classes and decomposition of the diagonal*, J. Algebraic Geom. **22** (2013), no. 1, 141–174. MR2993050

[28] Shing Tung Yau, *On the Ricci curvature of a compact Kähler manifold and the complex Monge-Ampère equation. I*, Comm. Pure Appl. Math. **31** (1978), no. 3, 339–411. MR480350

[29] *Géométrie des surfaces K3: modules et périodes*, Astérisque No. 126 (1985), Société Mathématique de France, Paris.

COLLÈGE DE FRANCE, 3 RUE D'ULM, 75005 PARIS, FRANCE